OXFORD SERIES ON NEUTRON SCATTERING IN CONDENSED MATTER

General Editors
S.W. Lovesey, E.W.J. Mitchell

OXFORD SERIES ON NEUTRON SCATTERING IN CONDENSED MATTER

---

1. W.G. Williams: *Polarized neutrons*
2. E. Balcar and S.W. Lovesey: *Theory of magnetic neutron and photon scattering*
3. V.F. Sears: *Neutron optics*
4. M.F. Collins: *Magnetic critical scattering*
5. V.K. Ignatovich: *The physics of ultracold neutrons*

THE PHYSICS OF

Ultracold Neutrons

V.K. Ignatovich

Joint Institute for Nuclear Research, Dubna

Translated by G.B. Pontecorvo

Joint Institute for Nuclear Research, Dubna

CLARENDON PRESS · OXFORD

This book has been printed digitally and produced in a standard specification in order to ensure its continuing availability

OXFORD
UNIVERSITY PRESS

Great Clarendon Street, Oxford OX2 6DP

Oxford University Press is a department of the University of Oxford.
It furthers the University's objective of excellence in research, scholarship,
and education by publishing worldwide in

Oxford New York

Auckland Cape Town Dar es Salaam Hong Kong Karachi
Kuala Lumpur Madrid Melbourne Mexico City Nairobi
New Delhi Shanghai Taipei Toronto
With offices in
Argentina Austria Brazil Chile Czech Republic France Greece
Guatemala Hungary Italy Japan South Korea Poland Portugal
Singapore Switzerland Thailand Turkey Ukraine Vietnam

Oxford is a registered trade mark of Oxford University Press
in the UK and in certain other countries

Published in the United States
by Oxford University Press Inc., New York

Reprinted 2004

ISBN 0-19-851015-2

Antony Rowe Ltd., Eastbourne

Preface

This book is about very slow neutrons, that is neutrons with velocities less than $10\,\mathrm{m\,s^{-1}}$. One remarkable property of such slow neutrons is that they can be kept in a sealed vessel. To obtain a free neutron gas is exciting enough in itself, but the question is what to do with it and how this investigation enriches our knowledge of nature. To answer this question, it is sufficient to point out that ultracold neutrons (UCN) represent a unique object with which one may literally achieve some contact with all modern physics. Many abstract concepts become quite illustrative when applied to UCN. It would be extremely tempting to take advantage of the properties of UCN and use them to illustrate all contemporary physics, for example phenomena such as superconductivity and neutron stars, gauge interactions and the model of a skyrmion. These illustrations are very useful because they may reveal new facets of known physical processes and thus facilitate the discovery of yet unknown phenomena. However, for the time being I shall digress on these interesting things only slightly, because this book is not about physics in general, but mainly about UCN: about how they are produced (Chapter 2), how they are detected and analysed (Chapter 3), and how they are transported from the reactor core to the experimental equipment (Chapter 4); about experiments that are carried out with them; and about the problems that arise in performing such experiments (Chapters 6–8). Some of these problems still remain unsolved, even though 30 years have passed since the first publication on UCN. Occasionally I have taken the liberty of digressing into fields adjacent to UCN (for example, paradoxical as it may seem, into relativistic astrophysics) to make the book interesting to a large readership. Most likely, Chapter 1 will appeal to many readers, because unexpected connections with other branches of physics are followed up in it. A wide circle of physicists may also be interested in Chapter 5, in which a novel approach to the solution of problems for a particle in a periodic potential is developed that provides a new insight into dynamical diffraction theory. An extremely interesting problem that has only been partially solved so far is the storage of a neutron in a magnetic trap (Chapter 7). A neutron in a magnetic trap represents an unstable quantum-mechanical system, the decay of which is governed by laws that are different from the ones governing the well-known α-decay. Specialists in molecular gas dynamics may also find sections in this book appealing (Chapter 4, for instance). On the whole, the problems touched upon are of most interest to neutron and solid-state physicists.

The material is presented as at a seminar, meaning that throughout the text only concepts, elementary mathematical computations, and easily comprehensible formulae are given. At the same time, readers desiring to verify the validity of the formulae are referred to mathematical appendices at the end of each chapter. Each appendix consists of two parts: formulae written out in succession, immediately followed by their description. This arrangement of the appendices seems to me to be the most reasonable way, since for a theoretician it will be sufficient to look through the formulae to understand how they are derived, without actually referring to their description.

For the specialists in UCN I have added Chapter 9, which contains specific information, but without detailed comments. The Bibliography contains a list of all the references on UCN known to me. Readers may find here three kinds of references. Some are supplied with brief summaries which, of course, cannot reveal the full content of a given paper, but are intended to help specialists find their way through the vast literature on UCN. Other references are only supplied with headlines, because in certain cases a headline reflects the content of a paper properly, or else the paper was not available to me. Finally, only the sources of some publications could be referred to. References that are not related to UCN are placed in Further Reading and are marked in the text by an asterisk.

A few words are in order about the units chosen. Energy is conveniently expressed in units of $\hbar^2/2m$, where m is the mass of the neutron; in this case the dimensions of all other quantities of interest are readily found. Thus, for example, dividing $\hbar\omega$, $k_B T$, and $\hbar/t$ by $\hbar^2/2m$, we obtain in the chosen set of units $\omega = 2m[\omega]/\hbar$, $T = 2mk_B[T]/\hbar$, $t = \hbar[t]/2m$, where the quantities in the square brackets have conventional dimensions.

This book might not have been completed without the help of numerous participants in investigations in the field of UCN. I would like to acknowledge the following with gratitude: Yu.G. Abov, I. Berceanu, B.G. Erozolimsky, A.I. Frank, I.M. Frank, V.V. Golikov, R. Golub, I.I. Gurevich, M. Hetzelt, M.V. Kazarnovsky, V.I. Luschikov, V.I. Morozov, Yu.V. Nikitenko, Yu.N. Pokotilovskii, G.D. Pozsev, L.M. Satarov, A.P. Serebrov, A.V. Stepanov, A. Steyerl, A.V. Strelkov, Yu.V. Taran, G.I. Terekhov.

I dedicate this book to the dear memory of my mother, father, and my teacher Fyodor L'vovich Shapiro. Professor F.L. Shapiro initiated the practical investigation of UCN and invited me to participate in this work. Above all, this dedication is a tribute to his exceptionally high moral standards.

06.04.1986 V.K.I.

Preface to the English edition

I am extremely grateful to Professor Sir Rudolf Peierls and to the Oxford University Press for noticing my book and for wishing to have it published in English. This would be flattering for any author, and, besides, it provided the possibility of improving the text. In the preparation of the English version amendments were made to all the chapters. Certain misreadings were corrected and new achievements were taken into account. Chapters 4, 5, and 6 underwent extensive alterations and the bibliography of UCN was extended.

I am sincerely grateful to G.B. Pontecorvo for his collaboration, in translating the book, that was both pleasant and very useful.

17.11.1989 V.K.I.

Contents

1

Introduction to the physics of ultracold neutrons

1.1 SOME HISTORY

Ultracold neutrons (UCN) are neutrons with such low energies (10^{-7} eV) that they can undergo total reflection from many materials at arbitrary angles of incidence, and so they can be stored in sealed vessels.

In 1920, long before the discovery of the neutron, Ernest Rutherford, in predicting its existence (*Rutherford 1920), pictured the neutron in the form of a compact hydrogen atom. 'Such an atom', he wrote, 'would have very novel properties. Its external field would be practically zero, except very close to the nucleus, and in consequence it should be able to move freely through matter. Its presence would probably be difficult to detect by the spectroscope, and it may be impossible to contain it in a sealed vessel.'

The neutron was discovered in 1932, and it was observed precisely because of its ability to penetrate relatively thick layers of matter. It therefore seemed impossible to store neutrons in vessels (*Abramov 1972).

In 1959 Ya.B. Zel'dovich was the first to make an opposite assertion (Zel'dovich 1959). Apparently, the notion that neutrons can actually be stored already existed, because the text of Zel'dovich (1959) begins with the words: 'The idea of storing neutrons has been expressed repeatedly; however, no experiments have yet been performed, and in the literature there do not exist even rough estimates pertinent to the subject.'

Zel'dovich took advantage of the fact that at small grazing angles neutrons undergo total reflection from the surface of most materials. This occurs when the component of their velocity normal to the surface does not exceed a certain limiting value characteristic of the given material, v_{lim} (about 5 m s^{-1} for graphite and 7 m s^{-1} for beryllium). Hence, Zel'dovich arrived at the conclusion that a neutron will be totally reflected from a material at any angle of incidence if its total velocity v is less than v_{lim}. Such neutrons as these can be kept in a sealed bottle.

The storage time (the time during which the neutron can be kept in the bottle) was estimated as follows. At each collision with the wall the neutron penetrates the material to a depth of the order of magnitude of the wavelength $\lambda \approx 10^{-6}$ cm; hence the probability of its being absorbed in the wall is approximately $N_0\sigma_a\lambda$, where σ_a is the cross-section for the neutron to be captured by the nucleus and N_0 is the atomic density of the material. The number of collisions which the neutron may undergo before actually being captured is $n = 1/N_0\sigma_a\lambda$. If the average dimension of the vessel is l, then the average time during which the neutron travels freely between two collisions with the walls is $t = l/v$, so the storage time is $\tau_a = nt = l/vN_0\sigma_a\lambda$. If the vessel is made of graphite, then $1/N_0\sigma_a v \approx 0.01$ s and for $l = 10$ cm the storage time $\tau_a = 0.01l/\lambda$ s $= 10^5$ s, which is significantly larger than the lifetime of the neutron itself, $\tau_\beta \approx 10^3$ s. Thus, we may expect the resulting storage time to be $\tau \approx 10^3$ s (actually, it turned out to be significantly smaller, and a great deal of effort was required to understand the reason for such a discrepancy).

Zel'dovich also considered the possibility of producing UCN with liquid helium. (At the beginning this method did not seem quite feasible, but today there are helium sources of UCN in France and Japan.) The estimates presented by Zel'dovich show that with the aid of helium the density of UCN stored in the trap can be brought to 50 cm^{-3} (such a density has now become a reality) and the stored neutrons can also be transported together with the trap (eventually this method of transporting UCN was actually realized.)

For detecting the neutrons present in the trap at a given time Zel'dovich proposed to insert an absorber into the vessel and to detect the γ-quanta from the (n, γ) reaction. This method of detection can be used for the investigation of UCN interaction with various materials. Zel'dovich also noted that the neutrons in the vessel can be heated to velocities higher than v_{lim} by moving the walls of the vessel mechanically. (Nowadays mechanical motion is used not for heating but for cooling faster neutrons to obtain UCN. See the description of the Steyerl and Kashoukeev turbines and Dombeck's moving crystal in the next chapter.)

In 1960 two papers on UCN were published: Drabkin and Zhitnikov (1960) and Vladimirskii (1960). Drabkin and Zhitnikov proposed to cool the neutrons with the aid of a magnetic field variable in space. The neutron possesses a magnetic moment μ, and therefore in a magnetic field $\boldsymbol{B}$ it acquires a potential energy equal to $-\boldsymbol{\mu B}$. When a neutron is transferred from the region of $\mu\uparrow\uparrow B$ to the region of $\mu\uparrow\downarrow B$, its potential energy increases by $2\mu B$. Its kinetic energy decreases by the same amount. If a radio-frequency field is now applied to change the direction of the magnetic moment to the opposite one ($\mu\downarrow\downarrow B$), then under transition to the region with the previous magnetic field ($\mu\downarrow\uparrow B$) the kinetic energy is once again decreased by $2\mu B$. In a magnetic field periodic in space and of amplitude B_0, the energy of a neutron crossing n periods can be decreased by $2n\cdot 2\mu B_0$. For $B_0 = 2$ T we have $2\mu B_0 \sim 10^{-7}$ eV. (Such cooling has not yet been realized in practice.) Drabkin and Zhitnikov (1960) also pointed out that neutrons from pulsed sources can be cooled with the aid of a travelling wave. Although this form of cooling has also not yet been realized in practice, some form of it is being considered seriously at presnet (*Rauch *et al.*, 1985; *Rauch, 1987; *Summhammer *et al.* 1986).

Vladimirskii (1960) considered many issues pertinent to UCN in his work. He proposed magnetic neutron guides and traps, for instance multipole, toroidal (subsequently constructed at Grenoble), the straight-line current field, and the horizontal magnetic mirror (traps of this kind were subsequently constructed in the USSR). Vladimirskii estimated the rate of UCN depolarization in the trap's magnetic field and neutron heating by low-frequency field oscillations. Furthermore, he also considered methods of UCN production. Liquid hydrogen sources are shown to be promising (in the Leningrad Institute of Nuclear Physics record fluxes of UCN were obtained by using liquid hydrogen). It is shown that, since the interaction of a neutron with the material can be described by the potential $u = (\hbar^2/2m)4\pi N_0 b$ (where b is the amplitude of coherent scattering on the nuclei of the material and m is the mass of the neutron), the energy of an outgoing neutron from the moderator is not less than u. Therefore, to produce neutrons of low energies it is necessary for the moderator to consist of a material with a low potential u, or the neutrons must be additionally moderated by the gravitational field (vertical and inclined channels for UCN have become widespread). The potential of the material must also be taken into account in constructing UCN detectors. The windows of the detectors must be chosen to be transparent for UCN, i.e. they have to be made of materials with a small or even negative scattering amplitude b.

Vladimirskii obtained a realistic estimate for the attainable density of neutrons stored in the bottle—10–100 neutrons per litre (such neutron densities are now surpassed by three orders of magnitude). We note also that the term 'ultracold neutrons' was coined by Vladimirskii. Zel'dovich called these neutrons 'slow' and 'cold', while Drabkin and Zhitnikov termed them cold and supercold.

In the following year, 1961, only a single paper concerning UCN was published (Gurevich and Nemirovsky 1961). In this paper it was demonstrated that if neutrons are strongly absorbed in a substance, then the amplitude b in the potential u has a large imaginary part; in this case neutrons of sufficiently low energies will be almost totally reflected from the substance even if the real part of b is zero. The stronger the reflection, the stronger is the absorption in the substance. In analogy with the optics of metals, this reflection was termed by Gurevich and Nemirovsky the metallic reflection (at present several experiments have been performed to study the reflection).

The next publication on UCN appeared in 1965. It was the monograph by I.I. Gurevich and L.V. Tarasov entitled *The physics of low-energy neutrons*, in which a whole chapter was devoted to UCN. In this book, all the four works listed above were discussed and it was noted that with the aid of UCN the precision of measurement of such quantities as the lifetime of the neutron, its electric dipole moment, and its charge can be enhanced. However, at the same time the authors pointed out that if it were possible to extract UCN from a reactor, 'their extremely low intensity would most likely make it impossible to utilize them in experiments'. (Luckily, this prediction turned out to be wrong, and applications of UCN in fundamental investigations are now in progress.)

Another theoretical work on UCN (Foldy 1966) was published in 1966 in the book *Preludes in theoretical physics*, dedicated to W. Pauli. Apparently, the author of this article was unfamiliar with the work of Soviet physicists and he arrived independently at the conclusion that neutrons can be kept in vessels. In Foldy's opinion, however, the walls of the vessel must necessarily not absorb neutrons and have a temperature lower than the temperature of UCN, which can be considered to be of the order of 10^{-3} K. The only substance that meets such requirements is helium. Therefore, Foldy concluded that the walls of a neutron bottle should be covered with a thin layer of superfluid helium. He estimated the maximum attainable density of UCN in such a helium trap. If the trap is considered to be filled with a degenerate neutron gas of maximum energy E_{lim}, characteristic of helium, then the neutron density can reach $2.2 \times 10^{14}\ \text{cm}^{-3}$. Such a neutron gas confined in a vessel would be equivalent to a radioactive source of $10\ \text{Ci}\,\text{cm}^{-3}$.

Meanwhile, experimental physics was getting closer to the apparently unattainable energy region of 10^{-7} eV. In 1962 a bent copper neutron guide was first used to extract 10^{-3} eV neutrons from a reactor core (*Mayer-Leibnitz and Springer 1963). The fast neutrons travelling along straight lines passed freely through the walls of the bent neutron guide, while the slower neutrons followed the neutron guide and underwent reflection at small grazing angles from its walls. With the aid of such a neutron guide it was possible to steer the slow neutrons away from the reactor to a point where the reactor background did not hinder the experiments. The main purpose of the experiments consisted of measuring the interaction cross-sections of slow neutrons with various nuclei.

In 1969 A. Steyerl (who subsequently became one of the most active investigators in the field of UCN), using a vertical neutron guide and the time-of-flight method, succeeded in identifying and measuring the scattering of neutrons with velocities down to $5\ \text{m}\,\text{s}^{-1}$ (Steyerl 1969). In fact, he observed (Steyerl 1977, p. 62) even slower neutrons, meaning that energies characteristic of UCN were achieved, but at the time no particular significance was attributed to UCN by this author.

1.2 THE FIRST EXPERIMENT WITH UCN

However, just prior to this, in 1968, F.L. Shapiro's group at Dubna had performed the first experiment with UCN in which the specific properties of these neutrons were precisely utilized (Luschikov *et al.* 1969). The experiments with UCN were initiated with the aim of preparing for a highly sensitive search of the electric dipole moment (EDM) of the neutron, since, as shown somewhat earlier (Shapiro 1968), this was the only way to improve the upper limit imposed on the neutron EDM.

The layout of the first experiment is shown in Fig. 1.1. The 6 kW pulsed reactor IBR served as the neutron source. The neutrons from the reactor were slowed down in paraffin into which was immersed the end of a long sharply bent copper neutron guide used for guiding the UCN away from the reactor core to the detectors in the experimental hall. A neutron guide must be evacuated, otherwise the UCN will undergo collisions with the molecules of air along the long path (10.5 m) from the reactor to the detectors, and as a result the neutrons will be heated to higher energies, escape through the walls of the neutron guide, and be lost. (It will be shown below that in order to keep UCN, a vacuum of the order of 10^{-3} Torr is sufficient.) For creating a vacuum both ends of the neutron guide must be sealed. The end immersed in the paraffin was made entirely of copper, so the UCN leaving the paraffin moderator could not enter the neutron guide because they were totally reflected from the copper. In order to have UCN inside the neutron guide, an additional moderator was placed in it, termed the 'converter'. (In the first experiment it was a polyethylene plate 1 mm thick.)

At the other end there were two detectors which could alternately be shielded by a copper shutter. The idea of the experiment was very simple: since the copper shutter reflects UCN, the neutrons would be counted only by that detector in front of which there was no shutter. Naturally, as there was always a background of fast neutrons present, the count of the detector closed by the shutter was not zero, but the difference between the counting rates of a single detector when it is open and when it is closed by the shutter (this quantity will henceforth be called the shutter difference) equals the number of UCN (strictly speaking, this is not quite so, but this issue will be dealt with later on). Actually, in the above experiment the counting rate of the open detector was three times as high as the counting rate of the closed one. It must

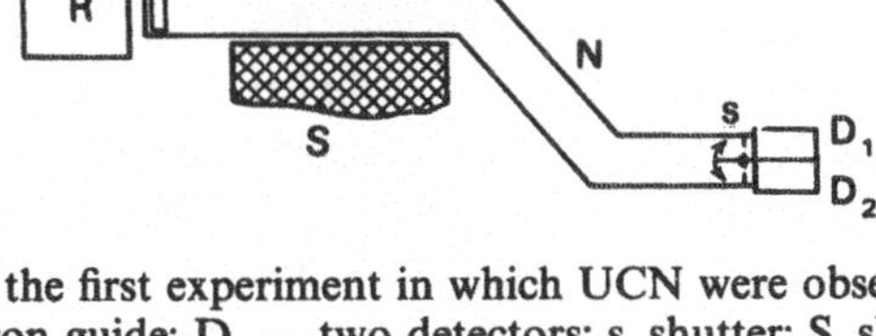

Fig. 1.1 Layout of the first experiment in which UCN were observed: R, reactor; C, converter; N, neutron guide; $D_{1,2}$, two detectors; s, shutter; S, shielding (Luschikov *et al.* 1969).

be noted that the reactor was pulsed with a frequency of 5 bursts per second, but the UCN were counted uniformly during the time intervals between the bursts. Hence the idea arose that UCN diffuse slowly along the neutron guide and are scattered out in all directions each time they collide with its walls. (This picture was subsequently altered significantly, since the probability for a neutron colliding with the wall to be totally reflected specularly may in some cases be quite high.)

Scintillation detectors were used in this experiment. Their surface was covered with a thin layer of $LiOH.H_2O$ with a limiting energy close to zero, so the UCN easily penetrate this layer and are captured by the ^6Li nuclei giving rise to scintillations. To verify that it was the UCN that were detected in the experiment, the $LiOH.H_2O$ was substituted by LiF with a higher limiting energy E_{lim}. The higher the value of E_{lim}, the better are UCN reflected from the substance and the less they penetrate it. Therefore, detectors with LiF should count less UCN than detectors with $LiOH.H_2O$. The observed change of counting rates of the detectors (the counting rate of the detectors with LiF was lower) was in good agreement with this view. If, on the other hand, the detectors are completely stripped of the surface layers, their counting rates should become zero. Such a check was also performed, and it was revealed that the detectors did indeed register neutrons, actually neutrons with energies of UCN.

As we already know, the neutron guide was evacuated. The pressure of the residual gases amounted to 5×10^{-3} Torr. If the pressure were increased, the counting rate of the UCN should fall. In Luschikov *et al.* (1969) the pressure was raised with the aid of helium. The counting rate of the UCN fell as the helium pressure increased, as expected. (Subsequently, the dependence of the counting rate on the helium pressure was measured in many experiments, and the corresponding curve was termed the 'helium curve'.)

Although the ideas of all measurements with UCN performed in the first experiment were extremely simple, the experiment itself was quite complicated. Most of the difficulties related to the extremely low intensity of the UCN. The counting rate of the detector 12 cm^2 in area was 1 neutron per 200 s. Nevertheless, even with such a low intensity experiments with UCN turned out to be quite performable, this is the main result of the first experiment.

The intensities of UCN were subsequently increased by several orders of magnitude. At present the record flux is $3 \times 10^6\ s^{-1}$. It was achieved at the Institut Laue–Langevin (ILL) in Grenoble. A somewhat lower flux of about $10^6\ s^{-1}$ was achieved at the Leningrad Institute of Nuclear Physics (Gatchina, USSR) under the leadership of V.M. Lobashov. There also, the high-sensitivity experiment was performed in which the EDM of the neutron was sought, which was actually the real purpose for starting the whole enterprise.

1.3 THE REFLECTION OF A NEUTRON FROM THE SURFACE OF A MATERIAL

The first thing one must do when proceeding to study the physics of UCN is to clarify how UCN are reflected from a wall (Fig. 1.2). We shall postpone till Chapter 5 the discussion of why the reflection takes place at all. Here we shall only postulate that the interaction of a neutron with matter is described by the potential

$$u = (\hbar^2/2m)4\pi N_0 b,$$

where $\hbar$ is the Planck constant, m is the neutron mass, N_0 is the number of atoms per unit volume, and b is the coherent scattering amplitude of a neutron on the nucleus of the material. When a neutron of momentum p and, respectively, of kinetic energy $E = p^2/2m$ penetrates a material, it acquires a potential energy u and, therefore, its kinetic energy becomes equal to $E - u$, i.e. $p'^2 = \hbar^2(p^2/\hbar^2 - u_0)$, where $u_0 = 4\pi N_0 b$. It is more convenient to pass from the momentum p to the wavenumber $k = p/\hbar$, in which case the previous relation assumes the form $k'^2 = k^2 - u_0$†. From the energy conservation law it follows that if the incident neutron energy is smaller than the potential u, $E < E_{\text{lim}} = u_0$, then the neutron cannot penetrate the material. Thus, neutrons with energies $E < E_{\text{lim}}$ will be totally reflected from the surface of the material. This conclusion follows directly from the laws of classical mechanics. The neutron, however, is a quantum-mechanical particle, and for describing its reflection it is necessary to apply the Schrödinger wave equation. Assuming the material to occupy the semi-space $z > 0$, we shall represent the Schrödinger equation in the form

$$[-\Delta + u_0\theta(z > 0) - k^2]\Psi = 0, \tag{1.1}$$

where $\theta\,(z > 0)$ is a step function equal to unity when the condition indicated in the argument is fulfilled and equal to zero otherwise.

Now let the neutron incident on the surface $z = 0$ be described by a plane wave $\exp(i\mathbf{kr})$ with the wavevector $\mathbf{k}$. Since the potential in eqn (1.1) depends

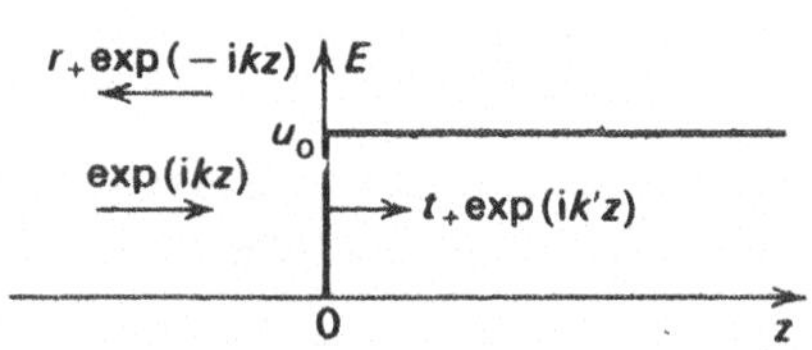

Fig. 1.2 Reflection of a neutron from a potential step.

† The quantities u and u_0 differ only in the factor $\hbar^2/2m$; henceforth we shall call them both, identically, potentials.

only on z, the components of the wavevector, $\mathbf{k}_{\parallel}$, that are parallel to the interface, must be conserved as the neutron penetrates the medium. The total wavefunction of the neutron, $\Psi(\mathbf{r})$, is equal to the combination of three waves: the incident, the reflected, and the refracted wave. Taking into account the conservation of $\mathbf{k}_{\parallel}$, we write $\Psi(\mathbf{r})$ in the form

$$\Psi(\mathbf{r}) = e^{i\mathbf{k}_{\parallel}\mathbf{r}_{\parallel}}[(e^{ik_{\perp}z} + r_{+}e^{-ik_{\perp}z})\theta(z < 0) + t_{+}e^{ik'_{\perp}z}\theta(z > 0)]. \quad (1.2)$$

Here $\mathbf{r}_{\parallel} = (x, y)$, $k'_{\perp}$ and $k_{\perp}$ are the normal components of the wavevector inside and outside the medium, respectively, $k'_{\perp} = (k_{\perp}^2 - u_0)^{1/2}$, and r_{+} and t_{+} are the amplitudes of the reflected and refracted waves. The index + indicates that the incident wave travels in the direction towards the half-space $z > 0$. The amplitudes r_{+} and t_{+} are determined from the matching condition for the wavefunction and its normal derivative at the interface. In Appendix 1.1 they are shown to be

$$r_{+} = (k_{\perp} - k'_{\perp})/(k_{\perp} + k'_{\perp}); \qquad t_{+} = 2k_{\perp}/(k_{\perp} + k'_{\perp}). \quad (1.3)$$

If $k_{\perp}^2 < u_0$, then $k'_{\perp} = ik''_{\perp} = i(u_0 - k_{\perp}^2)^{1/2}$ and for real u_0 it is a purely imaginary quantity. Therefore, $|r_{+}| = 1$. This case corresponds to total reflection. Thus, for total reflection only that part of the energy corresponding to the velocity component perpendicular to the interface is required to be less than the potential u. But if *all* the energy is less than u, then the reflection will be total at all angles of incidence. The potential u and, correspondingly, the limiting energy $E_{\text{lim}} = u$ differ for different materials, and therefore the energy E of the neutron may turn out to be less than the E_{lim} of one material, but higher than the E_{lim} of another material. In such cases the UCN will be referred to as sub-limiting relative to the first medium and over-limiting with respect to the second.

Now, a neutron is totally reflected when $k_{\perp}^2 < u_0$.† Naturally, for this condition to be satisfied at least for some value of $k_{\perp}$, the potential u_0 has to be positive. If the potential is complex, $u_0 = u'_0 - iu''_0$, $u''_0 > 0$, then even for $k_{\perp}^2 < u'_0$ the reflection is not total, since the reflection coefficient $|r_{+}|^2$ in this case is smaller than unity. The quantity $\mu = 1 - |r_{+}|^2$ is known as the loss coefficient. Its expression is also presented in Appendix 1.1.

1.4 UCN IN ASTROPHYSICS. THE REFLECTION OF RELATIVISTIC POINTLIKE PARTICLES FROM A POTENTIAL BARRIER

The potential u_0 is usually small and lies in the energy range of $u_0 \approx 10^{-7}$ eV. However, we can imagine a situation when it is large. An example corresponding to this case may be found in astrophysics. A neutron star, which is a

† We shall see below that in the case of crystals with a simple lattice described by a periodic potential with a period a, total reflection occurs also at those $k_{\perp}$ which satisfy the condition $k_n^2 < k_{\perp}^2 < k_n^2 + 2u_0$ for $u_0 > 0$, or $k_n^2 + 2u_0 < k_{\perp}^2 < k_n^2$ for $u_0 < 0$, where $k_n = \pi n/a$ and n is an arbitrary natural number different from zero. Such reflections are known as Bragg diffraction.

well-known astrophysical object, can be considered as a cluster of UCN trapped by its own gravitational field.

Neutrons are confined in a star not only by the gravitational field, but also by quantum forces similar to electrostrictive forces, which thus may be said to represent neutronstriction. The nature of these forces can be understood by using the example of the interaction of a neutron with matter, characterized by a negative scattering amplitude, $b < 0$. In this case, for the neutron the medium represents a potential well with a depth $4\pi N_0|b|$, in which discrete levels of bound states may exist. Here, the neutrons are not only confined in the medium, but may themselves bring about a contraction of the medium, since expansion leads to a decrease of N_0 and, consequently, to a decrease of the well depth. This in turn leads to a rise in the energy levels of bound states, which is not favoured from an energy point of view.

The neutron–neutron scattering amplitude is $b = -1.7 \times 10^{-12}$ cm. The density N_0 in a neutron star is higher than the density of ordinary matter by 14–15 orders of magnitude, and therefore the neutronstrictive forces in a star are exceptionally large. The potential well is so deep that if the neutrons are considered a degenerate gas, the Fermi level in it amounts to tens of MeV. Of course, at such high energies it is not correct to describe the potential energy by the expression $4\pi N_0 b$, because the amplitude b depends on the neutron energy. Nevertheless, the potential of a neutron star is quite deep.

In principle, we could also imagine a situation when, for describing the reflection of a particle from a potential barrier, relativistic equations must be applied. It is interesting to see whether in this case any peculiarities arise in the reflection compared with the non-relativistic case. Relativistic equations for example are required when neutrinos are considered (∗ Bel'aev 1978).

For this reason we shall consider the reflection of a relativistic pointlike particle from a barrier without going into details of the particle's nature. The particle may satisfy the Klein–Gordon equation or the Dirac equation, and it will sometimes be called a neutron; this means that the neutron will be understood to be a neutral pointlike particle without its complex structure, since taking into account the structure requires considerations beyond the scope of this book. We refer readers interested in such topics to ∗ Greiner *et al.* (1985) and ∗ Coutinho *et al.* (1988) and to references therein.

Before we apply a relativistic equation, it is necessary to clarify the nature of the potential u_0: whether it is really a scalar or whether it only represents, as in the case of an electromagnetic field, the fourth component of the four-dimensional vector potential. We shall consider both cases since they generally lead to significantly different results.

1.4.1 The Klein–Gordon equation

In handling the relativistic equations it is convenient to adopt new units of length and time so as to eliminate from them the velocity of light c. Since we shall not have to perform numerical calculations, we shall assume below that c is equal to unity and, to avoid confusion, we shall denote the interaction potential of a particle by φ, instead of u_0.

The Klein–Gordon equation can then be written in the form

$$[k_0^2 + \Delta - (m^2 + \varphi)]\Psi = 0, \tag{1.4}$$

where $k_0 = (k^2 + m^2)^{1/2}$ is the energy of the particle. The solution of this equation will be sought, as before, in the form (1.2). The amplitudes r_+ and t_+ will, then, have the same form (1.3), but for $k'_\perp$ we obtain a somewhat different expression, $k'_\perp = (k_\perp^2 - \varphi)^{1/2}$, from which the relation between φ and u_0 becomes evident: $\varphi = u_0$. The condition of total reflection is trivial, and we shall not discuss it here.

Now we shall deal with the other case when φ represents the fourth component of the vector potential. In this case the Klein–Gordon equation has the form

$$[(k_0 - \varphi)^2 + \Delta - m^2]\Psi = 0. \tag{1.5}$$

We again obtain the previous expressions for the wavefunction and the amplitudes r_+ and t_+, but now substitution of the function (1.2) into eqn (1.5) gives the following relation for $k'_\perp$ inside the medium:

$$k'^2_\perp = k_\perp^2 + \varphi^2 - 2k_0\varphi. \tag{1.6}$$

For small φ and non-relativistic energies this expression reduces to $k_\perp^2 - 2m\varphi$, whence the relationship is determined between φ and u_0, i.e. $u_0 = 2m\varphi$ and the condition for total reflection remains the same as before. But if the potential φ is very large, then the condition for total reflection

$$k_\perp^2 + \varphi^2 - 2k_0\varphi < 0 \tag{1.7}$$

is no longer trivial since k_0 also depends on $k_\perp$. Resolving inequality (1.7) for $k_\perp$ we find that for total reflection to occur $k_\perp^2$ must be within the interval

$$\varphi(\varphi - 2k_{0\parallel}) < k_\perp^2 < \varphi(\varphi + 2k_{0\parallel}), \tag{1.8}$$

where $k_{0\parallel} = (m^2 + k_\parallel^2)^{1/2}$. We note that if $\varphi > 2m$, then particles with small k^2 are not reflected from the barrier. To provide for total reflection at any $k_\perp$, it is necessary to make the left-hand side of (1.8) negative. This is possible only at sufficiently large $k_\parallel$. Hence it follows that if, neglecting their internal structure, we imagine the pertinent particles to be Cooper pairs in a neutron star (our equation being a scalar one), then for the neutrons to be confined in the neutron star the star must rotate at quite a high speed. But as a result of the conservation of angular momentum the axis of rotation must retain its direction in space, which means that there exist regions near the poles through which evaporation of the neutrons may take place. To prevent evaporation, the velocities of the Cooper pairs must be arranged so that at the poles $k_\perp$ are always large: $\varphi(\varphi - m) < k_\perp^2 < \varphi(\varphi + m)$. This could represent an interesting astrophysical problem. Clearly, a purely wave approach is not justified in the case of strong potentials. Field theory has to be applied, but the anomalies predicted within the wave formalism often give rise to interesting effects, also, when the formalism of quantum field theory is used.

1.4.2 The Dirac equation

Since the neutron is a spinor particle, it is described by the Dirac equation. We shall write the Dirac equation in the form

$$[k_0 + i\boldsymbol{\alpha}\nabla - \beta(m + \varphi)]\Psi = 0, \tag{1.9}$$

where $\boldsymbol{\alpha}$ and β are the Dirac matrices and Ψ is a four-component spinor, the two upper components of which describe the spinor function of a particle, whereas the lower components describe the function of an antiparticle. The amplitudes r_+ and t_+ are now matrices. In Appendix 1.2 the reflection matrix for a spinor particle is shown to be representable in the form

$$\hat{r}_+ = \frac{k_\perp(1 + \delta) - k'_\perp + i\delta\sigma_y k_\parallel}{k_\perp(1 + \delta) + k'_\perp - i\delta\sigma_y k_\parallel}, \tag{1.10}$$

where $\delta = \varphi/(k_0 + m)$, σ_y is the Pauli matrix, and $k'_\perp = (k_\perp^2 - \varphi^2 - 2m\varphi)^{1/2}$. The coordinate system is chosen as follows: the z axis is directed along the normal to the surface; the x axis along $k_\parallel$; and the y axis is chosen to be parallel to the reflecting surface and together with the x and z axes forms a right-hand Cartesian coordinate system. If the spin of the neutron is perpendicular to the scattering plane, i.e. directed along the y axis, then the amplitude $\hat{r}_+$ (1.10) is equal to

$$\hat{r}_+ = r_+ = \frac{k_\perp(1 + \delta) - k'_\perp + i\delta k_\parallel}{k_\perp(1 + \delta) + k'_\perp - i\delta k_\parallel}, \tag{1.11}$$

i.e. it is a complex number smaller than unity in absolute value. In the general case expression (1.10) can be represented in the form

$$r_+ = \frac{\sqrt{[k_\perp(1 + \delta) - k'_\perp]^2 + \delta^2 k_\parallel^2}}{\sqrt{[k_\parallel(1 + \delta) + k'_\perp]^2 + \delta^2 k_\parallel^2}} \exp(i\gamma\sigma_y), \tag{1.12}$$

i.e. besides a numerical amplitude, the reflection matrix also contains a rotation of the spin about the y axis through the angle

$$\gamma = \operatorname{arctg}\{2\delta k_\parallel k_\perp(1 + \delta)/[k_\perp^2(1 + \delta)^2 - (k'_\perp)^2 - \delta^2 k_\parallel^2]\}. \tag{1.13}$$

When $k_\perp^2 < \varphi^2 + 2m\varphi$, we have total reflection, and (1.12) assumes the form

$$\hat{r}_+ = \exp(-i\theta + i\gamma\sigma_y), \tag{1.14}$$

where

$$\theta = \mathrm{arctg}\{2k_\perp k''_\perp(1+\delta)/[k_\perp^2(1+\delta)^2 - (k''_\perp)^2 + \delta^2 k_\parallel^2]\}, \quad (k''_\perp)^2 = -(k'_\perp)^2. \tag{1.15}$$

Now let us consider the second case when the potential is the fourth component of a vector potential. Instead of (1.9), we have

$$[k_0 - \varphi + \mathrm{i}\boldsymbol{\alpha}\nabla - \beta m]\Psi = 0. \tag{1.16}$$

If this equation is solved in the same way as in Appendix 1.2 for eqn (1.9), then the amplitude of the reflected wave obtained differs from (1.10) only by the sign of δ:

$$\hat{r}_+ = \frac{k_\perp(1-\delta) - k'_\perp - \mathrm{i}\delta\sigma_y k_\parallel}{k_\perp(1-\delta) + k'_\perp + \mathrm{i}\delta\sigma_y k_\parallel}. \tag{1.17}$$

The difference seems to be insignificant. Since the potential φ in (1.10) can be considered negative, then δ in (1.10) will also be negative. However, altering the form of the equation changes the relationship between the momentum and energy. Now it assumes the form $(k_0 - \varphi)^2 = k^2 + m^2$, and therefore $k_\perp$ and $k'_\perp$ are related by expression (1.6). The condition of total reflection is then indicated in (1.8) and, if $\delta > 1$ and $k_\perp^2$ lie outside the interval indicated in (1.8), then, as follows from (1.12) when $-\delta$ is substituted for δ, the reflection amplitude becomes larger than unity. This phenomenon is known as the Klein paradox and is solved within the framework of field theory where strong potentials lead to the production of particle–antiparticle pairs.

Note that in the case of a purely scalar potential the phenomenon does not exist since the two requirements, $\delta < -1$ and $(k'_\perp)^2 > 0$, are contradictory. It is also noteworthy that a strong negative scalar potential can, in contradiction to the non-relativistic case, reflect a particle totally.

The mass occurring in the formulae obtained can be set equal to zero, and then they will describe neutrino reflection. This, however, will not concern a real neutrino. A real neutrino possesses a certain helicity and its spin always points opposite to its momentum. But, the helicity in the formulae is not conserved under reflection. Actually, these formulae involve four-component spinors, and a massless neutrino is described by a two-component spinor. So, attempts to describe the reflection of a two-component neutrino from a scalar potential turn out to be inconsistent since the requirement of conservation of helicity is not compatible with the solution of the equation.

The issue of neutrino reflection is extremely interesting in connection with the possibility of probing the Earth with the aid of neutrino beams, but we will have to set this question aside because a study of the potential for Dirac, Majorana, and Weyl neutrinos, especially if the possibility of their masses differing from zero is taken into account, would lead us too far away from the scope of this book. We shall only point out that the storage of neutrinos without taking account of their helicity was dealt with by * Bel'aev (1978).

In non-relativistic physics, the reflection of particles from a barrier is related to the problem of confining a two-dimensional film of electrons and protons on a surface of liquid helium. Here the repulsive potential competes with attraction due to the image forces, resulting in the formation of bound surface states (* Shikin 1979; * Morrow *et al.* 1981).

Atomic particles can be stored in magnetic bottles just like UCN. Moreover, it is even easier to deal with them than neutrons, because the magnetic interaction of atomic particles is three orders of magnitude greater than that of UCN. Atoms do not disappear as do neutrons due to beta-decay, so it is possible to attain a much higher density of atomic particles than for UCN. Many interesting experiments could be performed with a cold and dense localized cloud of polarized atoms in a magnetic field. See for example * Koelman *et al.* (1988), * Masuhara *et al.* (1988) and * Silvera and Walraven (1986).

Now is a suitable point to consider the reflection of photons since the nature of neutron optics and that of photon optics are very much alike.

1.5 SIMILARITY TO OPTICS. THE REFLECTION OF PHOTONS

The reflection of the neutron from matter can also be described in terms of the refractive index, i.e. in the same way as in ordinary optics (Frank 1972). The index of refraction for neutrons, $n = k'/k = (1 - u_0/k^2)^{1/2}$, was introduced by * Fermi and Zinn (1947) to explain the total reflection of thermal neutrons. For neutrons the law of refraction at a boundary of a medium is of the same form, $\sin\theta = n\sin\theta'$, as for light ($\theta$ and θ' are the angle of incidence and the angle of refraction) since this law is a consequence of the continuity condition of the tangential components of the wavevector $k_{\parallel}$. However, the index of refraction itself represents a phenomenological quantity. Its physical meaning is clarified only upon introduction of the interaction between matter and radiation. In the case of UCN it became clear immediately after the first publications (Vladimirskii 1960) that it is more convenient to describe the reflection of a neutron from a medium with the aid of a potential, instead of the index of refraction. On the other hand, optics may also in principle be described in terms of a potential. Indeed, the well-known Fresnel formulae (see, for example, * Born and Wolf (1973)) for the amplitude of the reflected wave $r_{\perp}$ (the light is polarized normally to the plane of incidence) and $r_{\parallel}$ (the light polarization is parallel to the plane of incidence) can be written in a form similar to (1.3):

$$r_{\perp} = \frac{k_{\perp} - k'_{\perp}}{k_{\perp} + k'_{\perp}}, \qquad r_{\parallel} = \frac{n^2 k_{\perp} - k'_{\perp}}{n^2 k_{\perp} + k'_{\perp}}, \tag{1.18}$$

where

$$k'_{\perp} = \sqrt{k_{\perp}^2 - (1 - n^2)k^2}. \tag{1.19}$$

Comparing (1.19) with the UCN wavevector in the medium, $k'_\perp = (k_\perp^2 - u_0)^{1/2}$, we see that $(1 - n^2)k^2$ assumes the role of the potential u_0. Thus, for instance, total reflection at all angles of incidence is possible only when $n^2 < 0$. The quantity n^2 is related to the dielectric constant ε, $n^2 = \varepsilon$, and within a certain range of frequencies it may be negative. For example, in the case of metals reflecting ultraviolet waves and X-rays when the frequencies ω of the waves are essentially higher than atomic frequencies but lower than a certain frequency $\omega_p = (4\pi N_0 e^2/m)^{1/2}$ known as the plasma frequency, the dielectric constant is $\varepsilon = 1 - \omega_p^2/\omega^2 < 0$ (see *R'asanov 1984, p. 20). In the optical frequency range negative values of ε occur in the region of anomalous dispersion.

The reflection of optical waves from metals is described by a complex dielectric constant (see * R'asanov 1984, p. 75):

$$\varepsilon = 1 + \frac{4\pi i}{\omega}\sigma(\omega), \tag{1.20}$$

where $\sigma(\omega)$ is the conductivity. The conductivity $\sigma(\omega)$ is extremely high, and therefore n^2 and the potential $(1 - n^2)k^2 = -4\pi i\sigma(\omega)k/c$ are characterized by a very large imaginary part.

The imaginary part of the potential is responsible for absorption. In the case of neutrons, $\text{Im}\, u_0 = 4\pi N_0 \text{Im}\, b$ and $\text{Im}\, b = k\sigma_t/4\pi$ in accordance with the optical theorem (here σ_t is the total interaction cross-section of the neutron with a nucleus). From Appendix 1.1 it follows that the greater the value of $\text{Im}\, u_0$, the lower the coefficient of reflection. On the other hand, we know that metal glitter is explained precisely by a high coefficient of reflection, and the stronger the absorption, the stronger the reflection. This phenomenon is called 'metallic reflection'. Utilizing the example of UCN it can readily be explained: as follows from (1.3), the reflection amplitude is always close to -1 if $k_\perp \ll |k'_\perp|$. But $|(k_\perp^2 - u_0)^{1/2}| = |k'_\perp|$ can be significantly larger than $k_\perp$ only if $k_\perp^2 \ll |u_0|$, independently of whether the potential u_0 is real, complex, or even completely imaginary. The metallic reflection of neutrons is most spectacular when considered for materials with a real part of u_0 equal to zero and even less than zero. For these materials no potential barrier seems to exist, but nevertheless, when $k_\perp^2 \ll \text{Im}\, u_0$, a significant specular reflection should be observed. When $k_\perp^2 \gg |u_0|$, reflection occurs as well, but it is insignificant, although in this case it is also greater, the larger the imaginary part of the potential.

The reflection of electromagnetic waves from dielectrics is characterized by $n^2 > 1$. This means that dielectrics attract electromagnetic waves. The dielectric constant ε is expressed through the polarizability α as follows: $\varepsilon \approx 1 + 4\pi N_0\alpha$. Hence the potential of the interaction of a photon with a medium is higher, the greater the number of atoms per unit volume, N_0, and so the photon is not only attracted by the medium but also makes the medium contract, which manifests itself in such phenomena as electrostriction and self-focusing. By analogy with neutrons, one can also consider the bound state of a photon, i.e. the problem of confining photons in a limited volume (∗ Ignatovich 1980). One may quite easily imagine a ball lightning to be a cluster of photons which, owing to total internal reflection, is confined inside a gas cloud that is more dense than the surrounding atmosphere, and the cloud itself retains its higher density as a result of the electrostrictive forces of the confined photons themselves. A cluster like that may be built up, for instance, by a laser discharge in a dense gas cloud arising from a local explosion. But the rest is silence.

APPENDICES

Appendix 1.1 Reflection of UCN from a perfect wall

$$\text{a) } [\Delta - u_0\theta(z \geqslant 0) - E]\Psi(\mathbf{r}) = 0, \quad \text{b) } E = k^2. \tag{1}$$

$$\text{a) } \Psi(\mathbf{r}) = \exp(ik_\parallel r_\parallel)\mathcal{H}_+(k_\perp, z), \quad \text{b) } k_\perp = \sqrt{E - k_\parallel^2}. \tag{2}$$

$$\mathcal{H}_+(k_\perp, z) = \theta(z < 0)[\exp(ik_\perp z) + r_+(k_\perp)\exp(-ik_\perp z)] + \theta(z \geqslant 0)t_+(k_\perp)\exp(ik'_\perp z). \tag{3}$$

$$k'_\perp = \begin{cases} \sqrt{k_\perp^2 - u_0}, & k_\perp^2 > u_0, \\ ik''_\perp = i\sqrt{u_0 - k_\perp^2}, & k_\perp^2 < u_0. \end{cases} \tag{4}$$

$$\text{a) } 1 + r_+ = t_+, \quad \text{b) } ik_\perp(1 - r_+) = ik'_\perp t_+. \tag{5}$$

$$\text{a) } r_+ = (k_\perp - k'_\perp)/(k_\perp + k'_\perp), \quad \text{b) } t_+ = 2k_\perp/(k_\perp + k'_\perp). \tag{6}$$

$$\text{a) } k_\perp^2 < u_0, \quad \text{b) } r_+ = (k_\perp - ik''_\perp)/(k_\perp + ik''_\perp) = \exp(-2i\varphi). \tag{7}$$

$$\varphi = \operatorname{arctg}(k''_\perp/k_\perp) = \arccos(k_\perp/\sqrt{u_0}). \tag{8}$$

$$\text{a) } u_0 = u'_0 - iu''_0, \quad \text{b) } u''_0 \ll u'_0, \quad \text{c) } u''_0 \ll (k''_\perp)^2. \tag{9}$$

$$\text{a) } \varphi = \varphi' - i\varphi'', \quad \text{b) } \varphi'' = u''_0(d\varphi/du_0)_{u''_0 = 0}. \tag{10}$$

$$\text{a) } |r_+|^2 = \exp(-4\varphi'') \approx 1 - 4\varphi'', \quad \text{b) } \mu = 1 - |r_+|^2 \approx 4\varphi''. \tag{11}$$

$$\text{a) } \mu = 2\eta k_\perp/k''_\perp, \quad \text{b) } \eta = u''_0/u'_0 = b''/b'. \tag{12}$$

$$\text{a) } u_0 = 4\pi N_0 b, \quad \text{b) } b = b' - ib''. \tag{13}$$

Upon being divided by $\hbar^2/2m$, the stationary Schrödinger equation reduces to form (1). We shall represent its solution in the form (2a) and (3) where θ is a step function equal to unity when the inequality indicated in the argument is satisfied, or to zero otherwise. The parameters $k_\perp$ and $k'_\perp$ are indicated in (2b) and (4). The reflection, r_+, and the transmission, t_+, amplitudes satisfy eqns (5) following from the matching function (3) and its derivative at $z = 0$. The solutions of eqn (5) are equal to (6). When (7a) is satisfied, the reflection amplitude assumes form (7b) where the phase angle equals (8). If the potential is complex, (9a), and inequalities (9b) and (9c) hold, then the imaginary part of phase (10a) can be determined with the aid of expression (10b). In this case it turns out to be small, and for the reflection coefficient expansion (11a) is valid. The loss coefficient is defined in (11b). Substituting (10b) into it, with account taken of (8), we obtain expression (12a) in which is introduced the reduced loss coefficient indicated in (12b). The latter equality follows from definition (13a) and from (13b).

Appendix 1.2 Reflection of a Dirac particle from a scalar barrier

$$[k_0 + \mathrm{i}\boldsymbol{\alpha}\nabla - \beta(m + \varphi)]\Psi(\mathbf{r}) = 0. \tag{1}$$

$$\text{a) } \boldsymbol{\alpha} = \begin{pmatrix} 0 & \boldsymbol{\sigma} \\ \boldsymbol{\sigma} & 0 \end{pmatrix}, \quad \text{b) } \beta = \begin{pmatrix} I & 0 \\ 0 & -I \end{pmatrix}, \quad \text{c) } \Psi = \begin{pmatrix} \xi \\ \chi \end{pmatrix}. \tag{2}$$

$$\sigma_x = \begin{pmatrix} 0 & 1 \\ 1 & 0 \end{pmatrix}, \quad \sigma_y = \begin{pmatrix} 0 & -\mathrm{i} \\ \mathrm{i} & 0 \end{pmatrix}, \quad \sigma_z = \begin{pmatrix} 1 & 0 \\ 0 & -1 \end{pmatrix}, \quad I = \begin{pmatrix} 1 & 0 \\ 0 & 1 \end{pmatrix}. \tag{3}$$

$$\text{a) } (k_0 - m - \varphi)\xi = -\mathrm{i}\boldsymbol{\sigma}\nabla\chi, \quad \text{b) } (k_0 + m + \varphi)\chi = -\mathrm{i}\boldsymbol{\sigma}\nabla\xi. \tag{4}$$

$$\text{a) } \chi = -\mathrm{i}\boldsymbol{\sigma}\nabla\xi/(k_0 + m + \varphi), \quad \text{b) } \Psi = \exp(\mathrm{i}\mathbf{k}_\parallel\mathbf{r}_\parallel)\mathcal{H}(z). \tag{5}$$

$$\mathcal{H}(z) = \left[\mathrm{e}^{\mathrm{i}k_\perp z}\begin{pmatrix} \xi \\ (\boldsymbol{\sigma}\mathbf{k}_+)\xi/(k_0 + m) \end{pmatrix} + \mathrm{e}^{-\mathrm{i}k_\perp z}\begin{pmatrix} \hat{r}_+\xi \\ (\boldsymbol{\sigma}\mathbf{k}_-)\hat{r}_+\xi/(k_0 + m) \end{pmatrix}\right]\theta(z < 0)$$
$$+ \mathrm{e}^{\mathrm{i}k'_\perp z}\begin{pmatrix} \hat{t}_+\xi \\ (\boldsymbol{\sigma}\mathbf{k}'_+)\hat{t}_+\xi/(k_0 + m + \varphi) \end{pmatrix}\theta(z \geqslant 0). \tag{6}$$

$$\text{a) } 1 + \hat{r}_+ = \hat{t}_+, \quad \text{b) } (1 + \delta)[(\boldsymbol{\sigma}\mathbf{k}_+) + (\boldsymbol{\sigma}\mathbf{k}_-)\hat{r}_+] = (\boldsymbol{\sigma}\mathbf{k}'_+)\hat{t}_+. \tag{7}$$

$$\text{a) } \mathbf{k}_\pm = (\mathbf{k}_\parallel, \pm\mathbf{k}_\perp), \quad \text{b) } \mathbf{k}'_\pm = (\mathbf{k}_\parallel, \pm k'_\perp), \quad \text{c) } \delta = \varphi/(k_0 + m). \tag{8}$$

$$\text{a) } k_0^2 = k^2 + m^2 = (k')^2 + (m + \varphi)^2, \quad \text{b) } k'_\perp = \sqrt{k_\perp^2 - 2m\varphi - \varphi^2}. \tag{9}$$

$$[(1 + \delta)(\boldsymbol{\sigma}\mathbf{k}_-) - (\boldsymbol{\sigma}\mathbf{k}'_+)]\hat{r}_+ = (\boldsymbol{\sigma}\mathbf{k}'_+) - (1 + \delta)(\boldsymbol{\sigma}\mathbf{k}_+). \tag{10}$$

$$[(1 + \delta)\sigma_z\mathbf{k}_\perp + \sigma_z\mathbf{k}'_\perp - \delta\boldsymbol{\sigma}_\parallel\mathbf{k}_\parallel]\hat{r}_+ = (1 + \delta)\sigma_z\mathbf{k}_\perp - \sigma_z\mathbf{k}'_\perp + \delta\boldsymbol{\sigma}_\parallel\mathbf{k}_\parallel. \tag{11}$$

$$\text{a) } \boldsymbol{\sigma}_\parallel = (\sigma_x, \sigma_y), \quad \text{b) } \sigma_z\boldsymbol{\sigma}_\parallel = \mathrm{i}\boldsymbol{\sigma}'_\parallel = \mathrm{i}(\sigma_y, -\sigma_x). \tag{12}$$

$$\text{a) } \hat{r}_+ = \frac{(1 + \delta)\mathbf{k}_\perp - \mathbf{k}'_\perp + \mathrm{i}\delta\boldsymbol{\sigma}'_\parallel\mathbf{k}_\parallel}{(1 + \delta)\mathbf{k}_\perp + \mathbf{k}'_\perp - \mathrm{i}\delta\boldsymbol{\sigma}'_\parallel\mathbf{k}_\parallel}, \quad \text{b) } \hat{t}_+ = \frac{2(1 + \delta)\mathbf{k}_\perp}{(1 + \delta)\mathbf{k}_\perp + \mathbf{k}'_\perp - \mathrm{i}\delta\boldsymbol{\sigma}'_\parallel\mathbf{k}_\parallel}. \tag{13}$$

$$\text{a) } \mathbf{k}_\parallel = (k_x, 0), \quad \text{b) } \boldsymbol{\sigma}'_\parallel\mathbf{k}_\parallel = \sigma_y k_x. \tag{14}$$

We shall choose the four-row matrices in the Dirac equation (1) in the form (2a, b) with the two-dimensional matrices defined in (3). Substituting spinor (2c) into (1) we obtain the set of equations (4). The solution of eqn (4b) equals (5a). Taking into account (5a) and (2c) we write the solution of (1) in the form (5b) and (6). The two-dimensional reflection and transmission matrices, $\hat{r}_+$ and $\hat{t}_+$, respectively, satisfy the set of equations (7) that follows from the matching at $z = 0$ of both components of the spinor (6). The wavevectors occurring in (6) and (7) are defined in (8a, b), whereas the parameter δ is defined in (8c). From the law of energy conservation (9a) follows (9b). Substituting (7a) into (7b) we obtain eqn (10) which is conveniently rewritten in form (11), where notation (12) is adopted. Multiplying both sides of eqn (11) from the left by σ_z and taking advantage of relations (12b) for the Pauli matrices we obtain the solution of eqn (11) in the form (13a), and then substituting it into (7a) we find expression (13b). If (14a) is valid, then (14b) is satisfied and expression (13a) is reduced to the form indicated in the text.

2

The production of UCN

The production of UCN should more appropriately be termed the extraction of UCN since in the main they are only extracted from the total thermal flow of neutrons coming from the surface of the moderator of the reactor. However, even if there are UCN present in this flow, they cannot enter the neutron guide since the channel for the UCN is separated from the moderator by a wall that cannot be penetrated by UCN. This is done for technological reasons, in particular for creating a vacuum in the channel. For regeneration of UCN an additional moderator is placed inside the neutron guide; henceforth this moderator will be termed a converter. In this chapter, we shall deal with the slowing down of neutrons in the converter and also discuss the properties of the converter from the standpoint of UCN production. In some cases the regeneration of UCN is performed by directly slowing down the neutrons from higher energies mechanically or with the aid of external fields. These methods will be dealt with at the end of this chapter.

2.1 THE FRACTION OF UCN IN THE MAXWELLIAN SPECTRUM OF THERMAL NEUTRONS

Let us now determine the fraction of UCN in the total thermal spectrum of neutrons coming from the moderator. We shall consider the spectrum to be Maxwellian and the limiting energy of the moderator to be negligible. The flux density of neutrons of energy k^2 (in most cases we shall omit the factor $\hbar^2/2m$) pertaining to the Maxwellian spectrum of temperature T_n is given by

$$d\Phi_M(T_n, k) = \Phi_0(k d\mathbf{k}/2\pi T_n^2)e^{-k^2/T_n}, \qquad d\mathbf{k} = k^2 dk d\Omega, \tag{2.1}$$

where Φ_0 is the total isotropic flux density. The flux of neutrons coming from the surface of the moderator is proportional to the cosine of the angle between the direction of an outgoing neutron and the normal vector to the surface and is, therefore, non-isotropic. The fraction of UCN in this flux is (Fig. 2.1)

$$\Phi_U/\Phi_0 = \int_0^{k_{lim}} (k_\perp/k) d\Phi_M(T_n, k) = \tfrac{1}{8}(E_{lim}/T_n)^2, \tag{2.2}$$

where $k_\perp$ is the component of the UCN momentum normal to the surface

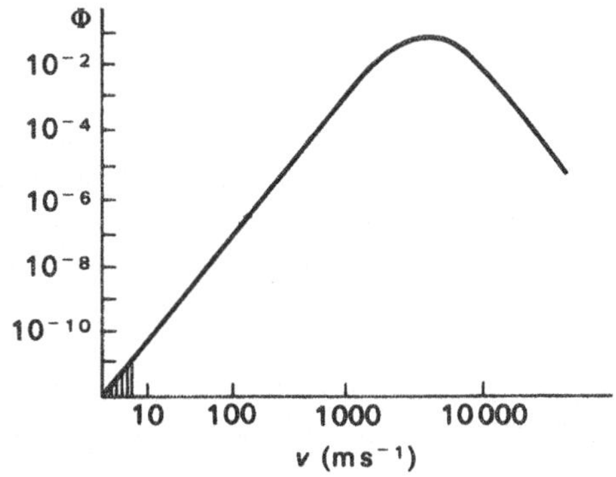

Fig. 2.1 The spectrum of the thermal neutron flux and the fraction (the shaded part) of UCN in it.

of the moderator, and the potential u_0 of the moderator is put equal to zero. Assuming $T_n = 300\ \text{K} \approx 25\ \text{meV}$ and $E_{\text{lim}} \approx 170\ \text{neV}$, we obtain $\Phi_U/\Phi_0 \approx 5 \times 10^{-12}$. When $\Phi_0 = 10^{13}\ \text{cm}^{-2}\,\text{s}^{-1}$, which is characteristic of a conventional research reactor, the UCN flux density is $\Phi_U = 50\ \text{cm}^{-2}\,\text{s}^{-1}$. In practice, the fluxes were even smaller.

We note that if the limiting energy of the moderator, E_m, cannot be neglected, then the fraction of UCN in the thermal flux is smaller than shown in (2.2) and, when $E_m > E_{\text{lim}}$, it actually vanishes.

2.2 THE THEORY OF A CONVERTER

Inside a converter the incident neutrons undergo inelastic scattering. If a neutron loses nearly all its energy as a result of an inelastic collision with a nucleus, it then becomes an ultracold neutron. The number of UCN of energy k_U^2 that are produced in the converter per unit time and per unit volume, $dn_U(k_U)$, is proportional to the incident neutron flux $d\Phi(T_n, k)$ of energy k^2, to the cooling differential cross-section $d\sigma_-(T_c, k \to k_U)$ (T_c is the temperature of the converter), and to the number of nuclei per unit volume, N_0:

$$dn_U(k_U) = d\Phi(T_n, k) N_0 d\sigma_-(T_c, k \to k_U). \tag{2.3}$$

The lifetime of the UCN inside the converter is extremely short owing to collisions with nuclei leading to their capture and to inelastic scattering involving a gain in energy (heating). The capture cross-section $\sigma_a(k_U)$ and the heating cross-section $\sigma_+(T_c, k_U)$, and consequently the total loss cross-section $\sigma_l(T_c, k_U) = \sigma_a(k_U) + \sigma_+(T_c, k_U)$, increase with decreasing k_U proportionally to $1/k_U$ and in the region of UCN energies achieve very large values. Therefore the only UCN which will leave the converter are those produced at a distance from the surface of the converter not larger than $l_0 = 1/N_0\sigma_1 \approx 1$ mm. Hence it follows that a layer of material of width l_0, and even of width $l_0/2$, may serve as a converter.

The UCN produced in the converter travel isotropically in all directions, and if the width of the converter is of the order of magnitude of $l_0/2$ and a UCN reflector is placed at the rear surface of the converter, then, in principle, all the produced UCN can enter the neutron guide. If, on the other hand, the width of the converter is larger than l_0, then only half of all the UCN produced within a layer of width l_0 will leave it because the other half travel away from the output surface deeper into the converter. The probability for UCN produced at a depth x to leave the converter is given by

$$W_{\text{out}} = \theta(k_\perp > 0)\exp[-xN_0\sigma_1(T_c, k_U)k_U/k_{U\perp}], \tag{2.4}$$

where k_U and $k_{U\perp}$ are, respectively, the total momentum and its component normal to the output surface inside the converter.

If the converter has a potential u_c, then in (2.4) it is necessary to take account of reflection at the interface, as well as the change in energy of the neutron leaving the converter, $(k'_U)^2 = u_c + k_U^2$. For both of these effects to be insignificant it is desirable to choose the material of the converter to have a small or negative value of u_c. This condition is satisfied, for instance, by water, liquid hydrogen, and gases. In the case of vertical and inclined channels the potential u_c may be quite high because the neutrons leaving the converter must be even more energetic to reach the upper end of the channel. When the neutrons are elevated by 1 m in the gravitational field, they are cooled by 10^{-7} eV and the potential barrier does not affect the UCN spectrum. Henceforth, we shall therefore neglect u_c.

Now consider a thick converter, the width of which can be considered infinite compared with l_0, but sufficiently thin for the effects of thermalization of the initial spectrum i.e. the influence of the converter on the spectrum, $d\Phi(T_n, k)$ to be neglected. The UCN flux arriving from a unit area of the surface of a converter of that sort, taking into account (2.3) and (2.4), can be represented in the form

$$\begin{aligned}\Phi_U &= N_0 \int_0^\infty d\Phi(T_n, \mathbf{k}) \int_0^{k_{\text{lim}}} d\sigma_-(T_c, \mathbf{k} \to \mathbf{k}_U) \int_0^\infty dx \\ &\quad \times \exp[-xN_0\sigma_1(T_c, k_U)k_U/k_{U\perp}] \\ &= \int_0^\infty d\Phi(T_n, \mathbf{k}) \int^{k_{\text{lim}}} k_{U\perp} d\sigma_-(T_c, \mathbf{k} \to \mathbf{k}_U)/k_U\sigma_1(T_c, k_U), \\ k'^2_U &= k_U^2 + u_c.\end{aligned} \tag{2.5}$$

where the limits of two integrals correspond to integrations over k and k_U respectively.

In Appendix 2.1 it is shown that expression (2.5) can be reduced to the form

$$\Phi_U = (\Phi_0/8)(k_{lim}^2/T_n)^2 \bar{\sigma}_-(T_n, T_c, k_{lim})/\sigma_l(T_c, k_{lim}), \tag{2.6}$$

where $\Phi_0 = \int_0^\infty \mathrm{d}\Phi(T_n, k)$, and

$$\bar{\sigma}_-(T_n, T_c, k_U) = \int \sigma'_-(T_c, \mathbf{k} \to \mathbf{k}_U)\exp(-k^2/T_n)\mathrm{d}\mathbf{k} \tag{2.7}$$

represents the integral over the Maxwellian spectrum of the 'derivative' cooling cross-section which is expressed in a simple manner through the double-differential cross-section of inelastic scattering:

$$\sigma'_-(T_c, \mathbf{k} \to \mathbf{k}') = \mathrm{d}\sigma_-(T_c, \mathbf{k} \to \mathbf{k}')/\mathrm{d}\mathbf{k}' = (\hbar^2/mk')\mathrm{d}^2\sigma(\mathbf{k} \to \mathbf{k}')/\mathrm{d}\Omega \mathrm{d}E', \tag{2.8}$$

where the energy E' has the conventional dimensionality.

The inelastic scattering cross-section must be such that, when equilibrium occurs, i.e. when $T_n = T_c$, the number of transitions from k to k' per unit time is the same as the number of transitions in the opposite direction. In other words, the cross-section must satisfy the principle of detailed balance

$$\mathrm{d}\Phi(T, \mathbf{k})\mathrm{d}\sigma(T, \mathbf{k} \to \mathbf{k}') = \mathrm{d}\Phi(T, \mathbf{k}')\mathrm{d}\sigma(T, \mathbf{k}' \to \mathbf{k}), \tag{2.9}$$

and therefore the cooling differential cross-section in (2.5) can be expressed in terms of the heating differential cross-section, $\mathrm{d}\sigma_+(T_c, k_U \to k)$, and (2.6) can be reduced to the form

$$\Phi_U = (\Phi_0/8)(E_{lim}/T_n)^2 \bar{\sigma}_+(T_n, T_c, k_{lim})/\sigma_l(T_c, k_{lim}), \tag{2.10}$$

where $\bar{\sigma}_+$ is the heating effective cross-section

$$\bar{\sigma}_+(T_n, T_c, k_U) = \int \exp[-k^2(1/T_n - 1/T_c)]\mathrm{d}\sigma_+(T_c, k_U \to k). \tag{2.11}$$

In equilibrium, when $T_n = T_c$, from (2.10) and (2.11) it follows that

$$\Phi_U = \frac{1}{8}\Phi_0\left(\frac{E_{lim}}{T_n}\right)^2 \frac{\sigma_+(T_c, k_{lim})}{\sigma_a(k_{lim}) + \sigma_+(T_c, k_{lim})}. \tag{2.12}$$

If the cross-sections in (2.12) are extrapolated to the thermal energies by the law $1/v$, they may, then, be taken not only at the point k_{lim}, but also at other values of k, for instance at $k = T_c^{1/2}$ or at $k_{300} = T_{300}^{1/2}$, where $T_{300} = 300$ K.

From the formulae obtained one can determine the most favourable conditions for UCN generation. First of all, it is clear that the smaller the capture cross-section σ_a, the greater the produced UCN flux. From this point of view the most favourable material is ^{4}He, for which $\sigma_a = 0$. In this case, therefore, the output of UCN in equilibrium coincides with the total number of UCN (2.2) in the Maxwellian spectrum, and the lower the temperature T_n, the higher the UCN flux intensity. The capture cross-section for all substances, with the exception of helium, differs from zero, and therefore the equilibrium output of UCN (2.12) may exhibit a maximum at a certain temperature $T_n = T_c = T_m$, below which it decreases if $\sigma_+(T_c)$ falls off more rapidly than T_c^2, as T_c decreases.

In the non-equilibrium case, when $T_n \neq T_c$, it is possible to follow separately the dependences of the UCN outputs on T_n and T_c for fixed T_c and T_n, respectively. From (2.10) and (2.11) it follows that as $T_n \to 0$, the UCN output increases owing to the factor T_n^{-2}, whereas it decreases because of the exponential factor in integral (2.11). Therefore the output of UCN can be expected to exhibit a maximum at a certain temperature $T_n = T_{nm}$. The integral (2.11) contains only such k that are comparable with the values present in the excitation spectrum of the material of the converter, so the maximum T_{nm} is higher, the harder the excitation spectrum.

At a fixed temperature T_n the dependence of the UCN output on T_c is determined by the excitation spectrum and the ratio σ_a/σ_+. When the cross-section $\sigma_+(T_c)$ is greater than σ_a, the decrease of T_c is accompanied by an increase of the UCN output, because when the converter is being cooled the numerator in (2.10) decreases more slowly than the denominator. But when the cross-section σ_+ becomes smaller than σ_a, the denominator stops decreasing, while the numerator still continues to decrease with T_c. Thus, the UCN output exhibits a maximum at a certain temperature T_{cm}.

Let us consider the example of a converter with two energy levels, the ground level and an excited one, separated by k_1^2. Henceforth we shall call such systems two-level systems. The corresponding cross-section can be represented as follows ($k_U^2 \ll k_1^2$ and $k_U^2 \ll T_c$):

$$d\sigma_+ = \sigma_0 \delta(k^2 - k_U^2 - k_1^2)\exp(-k_1^2/T_c)d\mathbf{k}/2\pi k_U. \tag{2.13}$$

Substituting it into (2.10) and taking account of (2.11) we obtain

$$\Phi_U \sim T_n^{-2} \frac{\exp(-k_1^2/T_n)}{\gamma + \exp(-k_1^2/T_c)}, \qquad \gamma = \sigma_a(k_U)k_U/\sigma_0 k_1. \tag{2.14}$$

In this case the numerator remains constant as T_c decreases down to zero, while the denominator stops changing when the temperature

$$T_{c1} \approx -k_1^2/\ln\gamma \tag{2.15}$$

is achieved. The temperature T_{nm} corresponding to the maximum and determined from (2.14) is

$$T_{nm} = k_1^2/2. \tag{2.16}$$

Thus, both (2.15) and (2.16) reveal that the temperature characteristic of the case under consideration increases with k_1^2, i.e. it tends to become higher the harder the excitation spectrum in the converter. To achieve the maximum number of UCN, it is desirable to lower T_c.

The limit temperature (2.15), down to which it makes sense to lower T_c, is lower the softer the k_1^2 spectrum and the smaller the cross-section σ_a when

$\gamma < 1$. When $\gamma > 1$, the flux Φ_U, as follows from (2.14), depends very weakly on the temperature T_c.

When the spectrum of excitations k_1^2 is not too hard, it is also desirable to lower the temperature of the neutron spectrum, T_n. But this can be done only by essentially increasing the mass of the converter, which can be achieved in practice only using hydrogen, with a required volume that is not very large. Such a converter has been installed for example at the Leningrad Institute of Nuclear Physics, Gatchina.

Now we can formulate the general requirements that must be satisfied by a converter. A converter should (i) possess a large inelastic scattering cross-section; (ii) have a small capture cross-section; (iii) exhibit a hard excitation spectrum; (iv) be stable with respect to radiation (a converter is placed inside the neutron guide in the vicinity of the reactor core); (v) be homogeneous (otherwise scattering on the inhomogeneities will hinder the UCN from leaving the converter); and finally (vi) have a low potential u_c in the case of horizontal channels.

Strictly speaking, a converter with potential u_c also produces sub-limiting neutrons of energy $E < u_c$, but these neutrons are created only in a thin surface layer of a width smaller than the depth $1/k_U''$ penetrated by UCN undergoing total reflection. This quantity is five orders of magnitude smaller than the escape depth $l = 1/N_0\sigma_1$. So the fraction of sub-limiting neutrons present in the total UCN flux from the converter is also five orders of magnitude smaller. However, if a beam of thermal neutrons is incident on the walls of a hermetically sealed cavity, the density of UCN in the cavity will ultimately become Maxwellian since the sub-limiting neutrons produced are progressively accumulated in the cavity.

In practice, different converters are compared with each other by their relative efficiencies. The following quantity is called the efficiency of a converter A at temperatures T_n and T_c with respect to the converter B at the reference temperatures $T_n = T_c = T_r$:

$$G_{AB}(T_n, T_c, T_r) = \Phi_U(T_n, T_c, \mathrm{A})/\Phi_U(T_r, T_r, \mathrm{B}). \tag{2.17}$$

If the limiting energies of the converters are neglected, then

$$G_{AB}(T_n, T_c, T_r) = (T_r/T_n)^2\bar{\sigma}_+(\mathrm{A}, T_n, T_c)\sigma_1(\mathrm{B}, T_r)/\sigma_+(\mathrm{B}, T_r)\sigma_1(\mathrm{A}, T_c). \tag{2.18}$$

The UCN intensity is measured not at the surface of the converter, but at the other end of the neutron guide. The total UCN flux at the detector can be represented in the form

$$J_U = W_{tr}\Phi_U S_1, \tag{2.19}$$

where S_1 is the area of the converter surface, and W_{tr} is the transmission of the neutron guide which partly depends also on the height of the converter's potential barrier. Thus, the measured efficiency is the ratio

$$G_{AB}(T_n, T_c, T_r) = J_U(T_n, T_c, \mathrm{A})/J_U(T_r, T_r, \mathrm{B}). \tag{2.20}$$

If u_c is the same for the converters A and B, then the transmission of the neutron guide drops out, for measurements performed in the same channel, and we obtain the relative efficiency (for instance, the temperature dependence of the UCN yield from one and the same converter) which is represented in the form

$$G(T_n, T_c) = (C/T_n^2)\bar{\sigma}_+(T_n, T_c)/[\sigma_a + \sigma_+(T_c)], \tag{2.21}$$

where the cross-section $\bar{\sigma}_+(T_n, T_c)$ is defined in (2.11) and the cross-section $\sigma_+(T_c)$ is

$$\sigma_+(T_c) = \int d\sigma_+(T_c, k_U \to k), \tag{2.22}$$

whereas the differential cross-section $d\sigma_+$ is related to the conventional double-differential cross-section $d^2\sigma/d\Omega dE$ by (2.8).

2.3 SOLID-STATE CONVERTERS

The simplest solid-state converter is an aluminium plate 5 mm thick cooled by flowing water. The limiting energy of aluminium is comparatively low, $E_{Al} = 53$ neV. Aluminium is very convenient from a technological point of view; it is often utilized to make windows for UCN detectors. The absence in the UCN spectrum of neutrons with energies $E < E_{Al}$ does not significantly influence the experiments since the detector windows cut off these neutrons anyway, if no special measures are adopted. The capture cross-section of aluminium is not large (0.13 barn in the thermal region (Gurevich and Tarasov 1965)), but the inelastic scattering cross-section is also not large. For $T_c = T_n = 400$ K, $\sigma_+/(\sigma_a + \sigma_+) = 0.19$. Therefore, the UCN fluxes from an aluminium converter are not high (Groshev *et al.* 1974). There is one defect of aluminium that makes the interpretation of experiments utilizing it difficult: always formed on its surface is an Al_2O_3 film of uncontrolled thickness, the limiting energy of which is essentially higher than E_{Al}. Usually, the film is relatively thin (10–20 Å) and, owing to the tunnelling effect, it does not hinder the passage of UCN through the aluminium, but under extreme conditions it may become significantly thicker, and then the oxide film will determine precisely the interaction of the UCN with aluminium.

A somewhat higher flux can be achieved with a magnesium converter, but to avoid sputtering of magnesium, it has to be placed inside a thin-walled (0.3 mm) aluminium case.

UCN fluxes of about an order of magnitude higher than that achieved with aluminium are obtained with zirconium hydride and polyethylene. But both these materials are to a certain extent unstable with respect to radiation, and they give off hydrogen which contaminates the walls of the neutron guide. In this respect, polyethylene is more dangerous, and for this reason polyethylene is utilized only in low-flux sources of thermal neutrons; in the case of zirconium hydride the emission of hydrogen exhausts the converter and worsens its properties with time. However, the zirconium hydride is stable to some extent; at any rate it is more stable than polyethylene, and it yields a relatively high output of UCN owing to the presence of hydrogen. Therefore, zirconium hydride is the most effective converter at normal room temperatures (Kosvintsev *et al.* 1977*c*).

In the case when cryogenic equipment is available freeze-deposited ice can be used as a converter (Akhmetov *et al.* 1974*b*). The output of UCN from the frozen water is about 23 times higher at 80 K than from aluminium at room temperature.

A good converter for the vertical channel is a cooled piece of beryllium. It is stable to radiation, has a very small capture cross-section, and quite a hard spectrum. When it is cooled from room temperature down to 30 K, the output of UCN from it is enhanced by a factor of 8 (Egorov *et al.* 1974). It is interesting to note that of three samples of beryllium, quasi-single crystal, melted, and moulded, the moulded beryllium possessed the smallest elastic scattering cross-section (Altarev *et al.* 1976). This means that moulded beryllium is the most homogeneous and the most suitable as a converter.

Calculations of solid-state converters (Golikov *et al.* 1973) are based on the one-phonon incoherent approximation.

Coherent inelastic scattering also contributes to the production of UCN. The mechanism is the same as in the two-level systems, (2.13). It will be dealt with somewhat later, but for the moment we shall neglect interference effects and take into account the assertion (* Turchin 1963) that in this case the error in determining the total inelastic scattering cross-section does not exceed 30%.

The multiphonon cross-section does not alter qualitatively the non-coherent one-phonon cross-section, and neglecting it leads to a relative error that does not exceed 15% (Altarev *et al.* 1976).

In the approximation indicated above we have

$$d\sigma_+(T_c, k_{\lim} \to k) = \sigma'_+(T_c, \omega_{\lim} \to \omega)d\omega, \qquad \omega = k^2, \tag{2.23}$$

where for $\omega_{\lim} \ll \omega$

$$\sigma'_+(T_c, \omega_{\lim} \to \omega) = (\sigma_{el}/M)g(\omega)n(\omega/T_c)\sqrt{\omega}/k_{\lim}, \tag{2.24}$$

$$\sigma_{el} = 4\pi(b^2_{coh} + b^2_{inc}); \qquad M = M_N/m, \quad n(x) = (e^x - 1)^{-1}; \tag{2.25}$$

$g(\omega)$ is the phonon spectrum normalized to unity and M_N is the mass of the nucleus. Substituting (2.23) into (2.22) and (2.11), and then applying (2.21), we obtain

$$G(T_n, T_c) = \frac{C}{T_n^2}\,\frac{\int \exp[-\omega(1/T_n - 1/T_c)]g(\omega)n(\omega/T_c)d\omega\sqrt{\omega/T_{300}}}{\gamma_{300} + \int g(\omega)n(\omega/T_c)d\omega\sqrt{\omega/T_{300}}}, \tag{2.26}$$

where $T_{300} = k_{300}^2$ is the neutron energy corresponding to the temperature 300 K and $\gamma_{300} = M\sigma_a(k_{300})/\sigma_{el}$. The constant C is chosen to be such that for $T_n = T_c = 300$ K the G factor equals a given number, for example unity. From (2.26) it follows that the smaller the value of γ_{300}, the higher the UCN output from the converter, provided all the other conditions are the same. The value of γ_{300} equals 1.3×10^{-4} for atomic deuterium, 3.9×10^{-3} for atomic hydrogen, 1.0×10^{-2} for beryllium, 1.1×10^{-2} for graphite, 0.5 for magnesium, and 3.8 for aluminium.

The spectrum $g(\omega)$ in (2.26) must be taken from experimental data, as is done for the numerical calculations of converters (Golikov *et al.* 1973). Here we shall adopt a model expression for the spectrum. This will allow analytical estimation of the efficiencies of various converters. Comparison with experimental data shows that these estimations correctly reveal the main trends both qualitatively and quantitatively. We shall consider two kinds of spectra: the Einstein and the generalized Debye spectra.

For the Einstein spectrum, $g(\omega) = \delta(\omega - \omega_1)$, we obtain from (2.26) a result close to that for the two-level converter, (2.13). Coherent scattering yields an expression that is approximately the same.

Now we shall adopt the generalized Debye model, i.e. we shall assume the spectrum

$$g(\omega) = \exp(-\omega/\theta)\omega^2/2\theta^3, \tag{2.27}$$

which is close to the Debye spectrum (θ is the Debye temperature) but defined for all ω. This spectrum permits representation of (2.26) in the simple analytic form (see Appendix 2.2)

$$G(T_n, T_c) = \frac{C}{T_n^2} \frac{T_c^{7/2}\zeta(7/2, T_c + T_c/T_n)}{\bar{\gamma}_\theta + T_c^{7/2}\zeta(7/2, 1 + T_c)}, \tag{2.28}$$

where $\bar{\gamma}_\theta \approx 0.6\gamma_{300}(T_{300}/\theta)^{1/2}$, $T_{300} = 300$ K, the temperatures T_n and T_c are measured in terms of θ, and $\zeta(s, x)$ is the Riemann function. In the case of small $T_c \ll 1$ and $T_c \ll T_n$, expression (2.28) can be simplified further:

$$G(T_n, T_c) = CT_n^{3/2}/(T_n/\theta + 1)^{7/2}(\bar{\gamma}_\theta + T_c^{7/2}). \tag{2.29}$$

We see that as $T_c \to 0$ the output of UCN tends towards a finite limit. Taking into account coherent effects will not change the behaviour of $G(T_n, T_c)$ qualitatively and can alter the numerical result by not more than 30%.

For small $\bar{\gamma}_\theta$ the output of UCN is enhanced as T_c decreases from room temperature, which is in good agreement with experimental results (Groshev *et al.* 1971). An exception is aluminium, for which $\theta = 416$ K and $\bar{\gamma}_\theta = 1.2 > (1 + 1/T_n)^{-7/2}$. In this case the output of UCN should not increase as T_c decreases, which is also confirmed by observations (Groshev *et al.* 1971). For low T_c one can readily find from (2.29) the temperature T_n corresponding to the maximum output of UCN: $T_{nm} = 3\theta/4$.

Evaluation of converters was performed by Golikov *et al.* (1973) including experimentally determined spectra $g(\omega)$ or spectra calculated on the basis of some dynamic solid-state model, the parameters of which were determined by using measured quantities, for example heat capacities

$$C = (\mathrm{d}/\mathrm{d}T)\int \omega g(\omega) n(\omega/T)\mathrm{d}\omega. \tag{2.30}$$

In all channels involving sufficiently high fluxes of thermal neutrons, a decrease with time was observed in the UCN output from solid-state converters (Egorov *et al.* 1974). After a year of operation of a channel the UCN output may become two–three times lower. In the case of a zirconium hydride converter the fall of the UCN output can be explained by exhaustion of the hydrogen in the sample. To explain the decrease of the UCN output from a beryllium converter two hypotheses were proposed. One presumes the accumulation in the converter of ^{6}Li produced as a result of a chain of radioactive transformations, and the other concerns the accumulation of inhomogeneities. Naturally, the inhomogeneities are produced more rapidly when the flux density of thermal neutrons is high.

2.4 GAS CONVERTERS

A gas converter (Akhmetov *et al.* 1976) is a cooled vessel (Fig. 2.2) with an aluminium window. A gas is convenient in that it is readily changed and is controllable. Obviously, as the pressure of the gas in the converter increases,

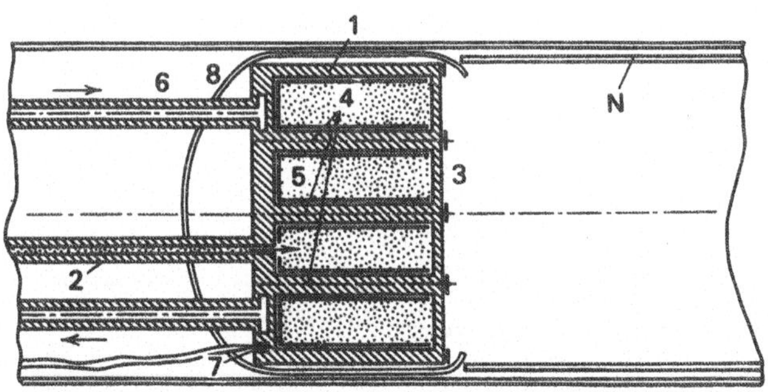

Fig. 2.2 A converter vessel for studying the output of UCN from liquids and gases, installed in the tangent channel of the WWR reactor (Alma-Ata) (Akhmetov *et al.* (1976): 1, cylindrical aluminium vessel 90 mm long and 175 mm in diameter; 2, tube for supplying gas or liquid to the converter; 3, aluminium window 0.25 mm thick for the exit of UCN towards the neutron guide; 4, supports permitting an increase of the pressure in the converter up to 5 atm; 5, copper foil 50 μm thick (it reflects the UCN produced inside the vessel and travelling away from the window, and thus the pressure at which saturation of the output of the UCN occurs is decreased by a factor of 2); 6, cooling tubes; 7, thermocouple; 8, spring positioning the converter inside the neutron guide.

the UCN yield must increase, achieving saturation when the mean free path of the UCN in the gas becomes twice the depth of the vessel. The dependences of the UCN output on the gas pressure presented in Fig. 2.3 are in good agreement with theoretical estimations.

The theory of UCN production in a monoatomic gas is the most simple one. In Appendix 2.3 it is shown that the differential cross-section of inelastic scattering on a monoatomic gas is

$$d\sigma_+(T_c, k_U \to k) \approx \sigma_{el}\sqrt{\frac{M}{4\pi T_c}}\exp\left(-\frac{M\omega}{4\mu^2 T_c}\right)\frac{d\omega}{k_U},$$

$$\omega = k^2, \qquad \sigma_{el} = 4\pi b^2, \qquad \mu = M/(M+1). \tag{2.31}$$

Substituting this cross-section into (2.11) and (2.22) and then taking advantage of (2.21) we obtain for the G factor the expression

$$G(T_n, T_c) = \frac{C}{T_n^2}\frac{\sqrt{T_c/T_{300}}}{(\alpha + T_c/T_n)(\gamma_{300} + \sqrt{T_c/T_{300}})}, \tag{2.32}$$

where $T_{300} = 300\,\text{K}$, $\alpha = (M-1)^2/4M$, and

$$\gamma_{300} = \sqrt{\pi M}\,\sigma_a(k_{300})/2\mu^2\sigma_{el}, \qquad k_{300} = \sqrt{T_{300}}. \tag{2.33}$$

Expression (2.32) makes it easy to follow the temperature dependence of the

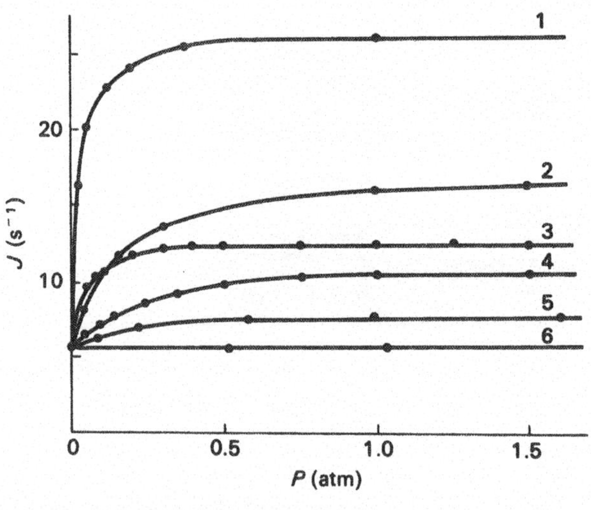

Fig. 2.3 Dependence of the UCN counting rate on the pressure and temperature of the gases in the converter (Fig. 2.2) (Akhmetov *et al.* 1976): 1, hydrogen at 80 K; 2, deuterium at 80 K; 3, hydrogen at 300 K; 4, deuterium at 300 K; 5, air at 300 K; 6, ^{4}He at 300 K.

UCN output. In the case of helium, when σ_a and consequently γ_{300} equal zero, the UCN output should increase monotonously as T_c decreases and achieve its limit value at $T_c \approx T_n$. In the case of monoatomic hydrogen ($M = 1$ and $\alpha = 0$), for $T_c \geqslant \gamma_{300}^2 T_{300}$ the UCN output increases, as T_c falls off, proportionally to $1/T_c$ and, further, proportionally to $1/T_c^{1/2}$. For other gases the UCN output exhibits a maximum at a certain temperaure T_{cm} and then decreases down to zero, as $T_c \to 0$. This decrease is characteristic of converters with monoatomic gases. It is readily understood from a physical point of view. Indeed, if the molecules of the gas are nearly motionless, then owing to the difference between their masses the incident neutron cannot transfer all its energy to the molecules and become an ultracold neutron for purely kinematic reasons: even a head-on collision leads to the velocity of the neutron, v, decreasing merely by the quantity $v/M \ll v$.

In the case of molecular gases the theory of UCN production turns out to be somewhat more complicated. Inelastic scattering of a neutron on a molecule is accompanied not only by an exchange of momentum and energy but also by intramolecular energy transitions. Specifically, we shall consider the diatomic gases hydrogen and deuterium. A molecule may have vibrational and rotational excited states. We shall not deal with the vibrational states since the characteristic energies of these states are significantly higher than the energy of a thermal neutron. The energies of rotational levels are $E = (\hbar^2/2I)L(L + 1)$, where I is the moment of inertia of the molecule. The distance between the two lowest hydrogen levels is 15 meV $\approx$ 200 K, and in the case of deuterium it is 7 meV $\approx$ 100 K. The proton has spin $\frac{1}{2}$, i.e. it is a fermion, and therefore the states of a molecule with total spin $S = 0$, called para-states, may exhibit rotational excitations only with even L. Ortho-states have $S = 1$ and so their orbital quantum number L on excited rotational levels can only assume odd values.

The deuteron has spin 1, i.e. it is a boson, so the states of a molecule with total spin $S = 0, 2$, called ortho-states, can have rotational excitations only with even L, whereas para-states ($S = 1$) can have excitations only with odd L.

Imagine the temperature of the converter to be sufficiently small to restrict our consideration to the two low-lying rotational levels with $L = 0$ and 1, the distance between them being denoted by Δ. The expression for the G factor can be reduced to the form

$$G(T_n, T_c) = C\frac{P_0[\bar{\sigma}_-(0 \to 0) + \bar{\sigma}_-(0 \to 1)] + P_1\bar{\sigma}_-(1 \to 1)}{\sigma_a + P_0\sigma_+(0 \to 0) + P_1[\sigma_+(1 \to 0) + \sigma_+(1 \to 1)]}, \tag{2.34}$$

where P_0, P_1 are the populations of the states with $L = 0$, 1, respectively, $\bar{\sigma}_-(0 \to 0)$ and $\bar{\sigma}_-(1 \to 1)$ are the cooling cross-sections without transition between levels of the molecule, $\bar{\sigma}_-(0 \to 1)$ is the cooling cross-section with excitation of the molecule. The heating cross-sections are denoted in a similar manner. For reasons of simplicity the dependences of the cross-sections upon T_n and T_c are not indicated.

We shall take advantage of the detailed balance principle to express the cooling cross-section through the heating cross-section (Appendix 2.4). Then, for the G factor, we will have

$$G(T_n, T_c) = \frac{C}{T_n^2} \frac{P_0[\bar{\sigma}_+(0 \to 0) + \bar{\sigma}_-(0 \to 1)] + P_1 \bar{\sigma}_+(1 \to 1)}{\sigma_a + P_0 \sigma_+(0 \to 0) + P_1[\sigma_+(1 \to 0) + \sigma_+(1 \to 1)]}. \tag{2.35}$$

If the gas is in thermodynamic equilibrium, then $P_1/P_0 = \exp(-\Delta/T_c)$. In general, it can also be prepared with another distribution, P_i, if collisions between the molecules do not lead to rapid relaxation of the distribution towards thermodynamic equilibrium. For example, hydrogen can be prepared in such a way that only states with even L are populated. In this case it is called para-hydrogen.

Consider our model to be such that $P_0 = 1$ and $P_1 = 0$. Then

$$G(T_n, T_c) = \frac{C}{T_n^2} \frac{\bar{\sigma}_+(T_n, T_c, 0 \to 0) + \bar{\sigma}_-(T_n, T_c, 0 \to 1)}{\sigma_a + \sigma_+(T_c, 0 \to 0)}. \tag{2.36}$$

Unlike the expression (2.32) for a monoatomic gas, we have here an additional term in the numerator. When $T_c \to 0$, it does not decrease, provided T_n is sufficiently high to excite a molecule. Thus, the G factor differs from (2.32) by the factor $1 + F$, where $F = \bar{\sigma}_-(T_n, T_c, 0 \to 1)/\bar{\sigma}_+(T_n, T_c, 0 \to 0)$ increases rapidly as T_c decreases, as a result of which at low T_c the UCN output becomes the same as in the case of a two-level converter. We see that here, unlike the case of a monoatomic gas, an incident neutron may with a finite probability turn into an ultracold neutron even when it collides with a molecule at rest, provided its energy is higher than Δ.

Inelastic scattering cross-sections always contain an elastic factor $\sigma_{LL'}$ which is different for different transitions (see Appendix 2.4), and, as follows from Appendix 2.5, the factor F can be represented in the form

$$F = q \exp(-\Delta/T_n)\sqrt{\Delta/T_c}\,\sigma_{01}/\sigma_{00}, \tag{2.37}$$

where for $T_c \ll T_n$ the quantity q is close to unity (≈ 0.6 for hydrogen and ≈ 0.4 for deuterium). The cross-sections $\sigma_{LL'}$ are presented in Appendix 2.4. In the case of hydrogen $\sigma_{01} \equiv \sigma_{po}$, the cross-section for the para–ortho transition, and $\sigma_{00} \equiv \sigma_{pp}$, the cross-section for the para–para transition, we have the following:

$$\sigma_{po} = 4\sigma_{inc}, \qquad \sigma_{pp} = 4\sigma_{coh}. \tag{2.38}$$

In the case of deuterium

$$\sigma_{op} \equiv \sigma_{01} = (3/2)\sigma_{inc}, \qquad \sigma_{oo} \equiv \sigma_{00} = 4\sigma_{coh} + (5/2)\sigma_{inc}, \tag{2.39}$$

where σ_{coh} and σ_{inc} are, respectively, the coherent and incoherent elastic scattering cross-sections on a bound proton or the deuteron. In the case of hydrogen and deuterium, respectively, we have $\sigma_{01}/\sigma_{00} = 45$ and 0.1; $\gamma_{300} = 0.22$ and 0.0023; $\alpha = 1/8$ and 9/16. Hence one can see that gaseous para-hydrogen in the ground state is always more efficient from the point of view of UCN production than ortho-deuterium, provided the neutron temperature is sufficiently high for para–ortho transitions to occur in the hydrogen. If $T_n \ll \Delta$, then, owing to the absence of para–ortho transitions, para-hydrogen is inferior to deuterium, since the capture cross-section of the latter is small. As T_c falls below the temperature of condensation, coherent inelastic scattering processes on the liquid enter the game, and deuterium also becomes more efficient for UCN production. But this will be dealt with in the next section.

Experimental data are in good agreement with theoretical calculations. As T_c falls from 300 to 80 K the UCN output from para-hydrogen becomes about 18 times higher (Akhmetov *et al.* 1977*a*), while the UCN output from deuterium increases by a factor of 2. Normal hydrogen, i.e. hydrogen where the levels are populated in accordance with thermodynamic equilibrium at room temperature, is significantly inferior to para-hydrogen, since inelastic processes may occur in the ortho-state with a cross-section σ_{oo} of approximately the same order of magnitude as the cross-section σ_{op}. Cooling normal hydrogen from 300 down to 80 K leads to the UCN output becoming only 2.5 times higher.

Experiments with para-hydrogen (Akhmetov *et al.* 1977*a*) reveal that the UCN output from it decreases with time. The time constant characteristic of the rate with which the UCN output falls depends on the reactor power and on the pressure of para-hydrogen in the converter. The observed decrease in the UCN production is explained by reconversion of the para-hydrogen to the state of thermodynamic equilibrium. Hence it is evident that UCN can be utilized for investigating reconversion processes. To eliminate the influence of the reconversion on the UCN output one can make the para-hydrogen flow through the converter continuously.

2.5 LIQUID AND FREEZE-DEPOSITED CONVERTERS

One of the most efficient of the existing sources of UCN is the liquid hydrogen converter at the Leningrad Institute of Nuclear Physics, Gatchina (Altarev *et al.* 1986*a*). This converter is quite an intricate piece of machinery.

A simple liquid converter is represented by water in an aluminium ampoule. Such a converter yields 6 times more UCN than aluminium. The aluminium window of the converter partially absorbs the UCN produced in the water. At low temperatures this can be avoided by freeze-depositing the water on to the window from the outside. If this is done, the output of UCN from water at $T_c \approx 80\,K$ is 25 times higher than the UCN output from aluminium.

Interesting effects were observed when heavy water was frozen on the converter surface. The limiting velocity of heavy ice is quite high: $v_{lim} \approx 5.5\,m\,s^{-1}$. Therefore, when a copper neutron guide with $v_{lim} \approx 5.7\,m\,s^{-1}$ is used, a converter of D_2O generates UCN within an extremely narrow energy interval. In the literature information is available (*Delseme and Wegener 1970) that indicates that the density of D_2O is higher in thin layers of ice than in thick ones. If this is actually so, then thin layers of heavy ice should not generate any UCN capable of travelling along a copper neutron guide. Freeze deposition of D_2O on to zirconium hydride reveals (Fig. 2.4) that a thin layer of ice screens the zirconium hydride almost completely and does not take part in the generation of UCN. As the ice layer becomes thicker, an enhancement of the UCN flux is observed, which testifies in favour of the generation of UCN in D_2O, when the density of the ice becomes normal, as its thickness increases. Hence it is evident that the structure of a coating of heavy ice can be studied using the UCN themselves. Such experiments are discussed by Morosov *et al.* (1988*b*).

A freeze-deposited converter with a thin film of hydrogen or deuterium frozen directly on the walls of the trap was considered by Golub and Böning (1983) and Yu *et al.* (1986). It has been shown (Yu *et al.* 1986) that the highest attainable UCN density is independent of the thickness of the film and amounts to $10^{-9}\Phi$ in the case of a deuterium film at $T_c = 5\,K$ and $T_n = 30\,K$.

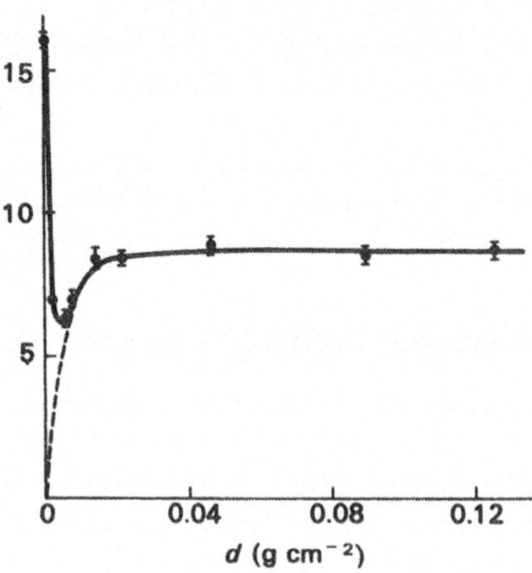

Fig. 2.4 Dependence of the UCN counting rate on the thickness of the layer of heavy water deposited on zirconium hydride at a temperature of 80 K (Akhmetov *et al.* 1977*b*).

The limit density is calculated from the relation $d\rho_U(v)\, V/t_f = d\Phi_U S$, where do is the differential density of neutrons with velocity v, V is the volume. S is the surface of the walls of the trap, t_f the free time of flight between the walls is $t_f = 4V/Sv$, and it is assumed (in the case of a thick film, for instance) that each collision of a neutron with the wall leads to its being lost. From the above relation the limit density is readily shown to be $\rho_U = 4\int d\Phi_U/v$. For the purposes of estimation one can assume the limit density inside a vessel, the walls of which play the role of converter, to be the following (see Appendices 2.1 and 2.2):

$$\rho_U = 5 \times 10^{12}\, G(T_n, T_c, T_{300})\Phi_0/v_{\lim}, \qquad (2.40)$$

where the G factor is normalized to unity at $T_n = T_c = T_{300} = 300$ K, and Φ_0 is the density of the thermal flux incident upon the trap. It is notable that (2.40) holds for converter films of arbitrary thicknesses, even in the case of sub-limiting conversion in the walls of the trap itself. In this case the rate at which the trap is filled falls, but then the time t_f should be replaced by τ_l which is larger than the time of flight between two collisions with the walls. Of course, if the filling time, which is equal to the loss lifetime τ_l, becomes comparable with or higher than the time t_β of the neutron β-decay, then the limit density becomes smaller than (2.40) by a factor of approximately t_β/τ_l. Notably, this may be utilized for measuring the lifetime of the neutron itself, t_β.

From a physical point of view liquid helium is one of the most interesting converters. Before we deal with liquid helium, however, it will be useful to consider the difference between the generation of UCN in liquids and in gases. The main peculiarity of liquids is related to the coherent inelastic scattering cross-section. At low temperatures the G factor is represented in the form

$$G(T_n, T_c) = (C/T_n^2)[\bar{\sigma}_+^{\text{inc}}(T_n, T_c) + \bar{\sigma}_+^{\text{coh}}(T_n, T_c)]/[\sigma_a + \sigma_+^{\text{inc}}(T_c)]. \qquad (2.41)$$

Coherent processes are characterized by certain laws. Inelastic scattering occurs on certain quasi-particles (collective excitations) of momentum q and energy $\Omega(q)$, the dependence of which on q is termed a dispersion relation. For example, in the case of acoustic waves this relation assumes the form $\Omega(q) = cq$, where c is the velocity of sound in the medium. In the case of inelastic scattering on quasi-particles the laws of energy and momentum conservation must be obeyed, while in the case of incoherent scattering momentum conservation plays no essential role. The differential inelastic scattering cross-section can be written in the general case as follows:

$$d\sigma_+^{\text{coh}}(T_c, k_U \to k) = \sigma_{\text{el}}^{\text{coh}}\delta(\boldsymbol{\kappa} - \mathbf{q})\delta(\omega - \Omega(q))\exp[-\Omega(q)/T_c]\,d\mathbf{q}S(q)d\mathbf{k}/k_0, \qquad (2.42)$$

where $S(q)$ is the structure factor, $\boldsymbol{\kappa} = \mathbf{k} - \mathbf{k}_U \approx \mathbf{k}$, $\omega = k^2 - k_U^2 \approx k^2$, and

σ_{el}^{coh} is the coherent elastic scattering cross-section. If, now, the inelastic scattering cross-sections (2.22) and (2.11) are calculated, taking into account (2.42), then the first one turns out to be proportional to $\exp(-\Delta/T_c)$, while the second is proportional to $\exp(-\Delta/T_n)$, where $\Delta = k^2$ is determined from the solution of the equation $k^2 = \Omega(k)$. In the case of acoustic waves we have $\Delta = c^2$ (in conventional units $\Delta = 2mc^2$), corresponding to quite a high energy. When $T_n \geqslant \Delta \geqslant T_c$, the cross-section σ_+^{coh} is negligible, while $\bar{\sigma}_+^{coh}(T_n, T_c)$ is not small, i.e. the coherent inelastic scattering cross-section only influences the numerator of the G factor. It is equivalent to the scattering cross-section involving excitation of a discrete level in para-hydrogen, as indicated in (2.36), where the first summand in the numerator of (2.36) and the cross-section $\sigma_+(T_c)$ in the denominator correspond to incoherent inelastic scattering. Liquid ^{4}He is remarkable in that no absorption and no incoherent scattering are observed in it (*Cowley and Woods 1971). Therefore, with the aid of ^{4}He it is possible, in principle, to generate any amount of UCN. But for this purpose a sufficiently large amount of superfluid helium must be placed near the reactor core under conditions of a high heat release. The helium must be superfluid since, otherwise, thermal fluctuations will occur in it, the scattering on which is of a quasi-elastic nature, but the quasi-elastic scattering cross-section is not governed by the law of momentum conservation, and so it plays the part of the incoherent scattering cross-section.

At present attempts to create a converter with superfluid helium situated in the near vicinity of the active core of the reactor are being made in Japan (Yoshiki *et al.* 1986, 1987), but in the meantime a helium source (termed 'superthermal') is being investigated only in extracted thermal neutron beams.

As we already know, the idea of using helium for generating UCN is due to Ya.B. Zel'dovich (Zel'dovich 1959). Evidently, difficulties were to arise in positioning the necessary amount of liquid helium near the reactor, so the idea was discarded. In 1975, however, the idea was revived, when it was demonstrated (Golub and Pendlebury 1975) that a sufficiently high number of UCN can also be obtained in an extracted beam of thermal neutrons. If the vessel for storage of UCN is filled with helium and exposed to a beam of thermal neutrons, then an arbitrary amount of UCN can be accumulated in the vessel, the actual number of UCN being limited only by the intensity of the beam and by the Fermi statistics. If the neutrons themselves did not decay and were not absorbed in the walls of the vessel (regretfully, with the exception of helium, there are no other substances in which absolutely no absorption of neutrons occurs), the beam intensity would play no essential role, since sooner or later the vessel would be filled with the maximum amount of UCN corresponding to limits imposed by the Fermi statistics. If this were the case, the generated UCN density would exceed the density of UCN in the incident beam. The following question arises: are not the principles of thermodynamics violated in that, even when the neutrons are in equilibrium with the medium, the fraction of UCN in the neutron spectrum cannot exceed that in the Maxwellian spectrum? The answer is: no, this is not so. Naturally, if the heat were not diverted from the helium, then the excitations formed in it after the thermal neutrons are cooled would accumulate at the same rate as the UCN, and sooner or later inverse processes would arise. But if the heat is diverted, these excitations are destroyed by the condenser, and the UCN are thus generated as a result of the work done by an external source and no violation of the principles of thermodynamics occurs.

It must be noted that, although there is no incoherent scattering in liquid helium, inelastic scattering in UCN does, nevertheless, occur. It is due to the existence of more than one quasi-particle, for these processes are not restricted rigorously by the conservation of momentum (Golub 1979a). Thus, several quasi-particles of low energies may lead to the heating of UCN, but as the temperature decreases the role of these processes also decreases rapidly. Thus, for example, at $T = 1$ K two-phonon processes limit the lifetime of the UCN in helium to 130 s, and to 4000 s at 0.6 K. Admixtures of ^{3}He present in the ^{4}He also accelerate the loss of UCN. For such admixtures not to influence the lifetime the atomic concentration of ^{3}He must not exceed 10^{-12}. Helium of such a degree of purity is quite attainable.

The density of UCN in a vessel filled with helium in a beam of thermal neutrons of total flux density Φ_0 for a lifetime τ may be estimated by the quantity

$$\rho_U = N_0 \int d\Phi(T_n k) d\sigma_-(T_c, k \to k_U)\tau. \tag{2.43}$$

Taking into account the principle of detailed balance (2.8) we express $d\sigma_-$ through $d\sigma_+$, and obtain

$$\rho_U = \tfrac{2}{3}\Phi_0 N_0 \tau[(E_w^{3/2} - E_{He}^{3/2})/T_n^2]\bar{\sigma}_+(T_n, T_c, k_U)k_U, \tag{2.44}$$

where E_w and E_{He} are the limiting energies of the walls of the vessel and of the helium itself, respectively, while the cross-section $\bar{\sigma}_+$ is obtained by substitution of (2.42) into (2.11). Upon performing the indicated substitution we arrive at the following expression:

$$\rho_U = \tfrac{2}{3} N_0 \sigma_{el} \Phi_0 \tau S(\sqrt{\Delta})\left(\frac{\Delta}{T_n}\right)^2 e^{-\Delta/T_n} \frac{(E_w^{3/2} - E_{He}^{3/2})/\Delta^{3/2}}{|1 + d\Omega/d\Delta|},$$

$$d\Omega/d\Delta \approx \tfrac{1}{2}. \tag{2.45}$$

In the case of helium the quantity $\Delta \approx 12$ K. The structure factor $S(\Delta^{1/2})$, as follows from measurements, can be estimated to be $S(\Delta^{1/2}) \approx 0.2$. If the spectrum of thermal neutrons is assumed to have the most favourable temperature $T_n = \Delta/2$ (at this point ρ_U exhibits a maximum), then, taking into account the fact that $\sigma_{el} = 1.1$ barn and $N_0 \approx 1.5 \times 10^{22}\ \mathrm{cm}^{-3}$, we obtain (Golub and Pendlebury 1977)

$$\rho_U \approx 10^{-9}\Phi_0\tau. \tag{2.46}$$

For a flux $\Phi_0 = 10^9\ \mathrm{cm}^{-2}\,\mathrm{s}^{-1}$ and $\tau = 100\ \mathrm{s}$ we obtain a density $\rho_U \approx 100\ \mathrm{cm}^{-3}$. (Steyerl *et al.* 1986).

The first investigation of the generation of UCN in helium was performed using the layout of Fig. 2.5. The experimental results are presented in Fig. 2.6.

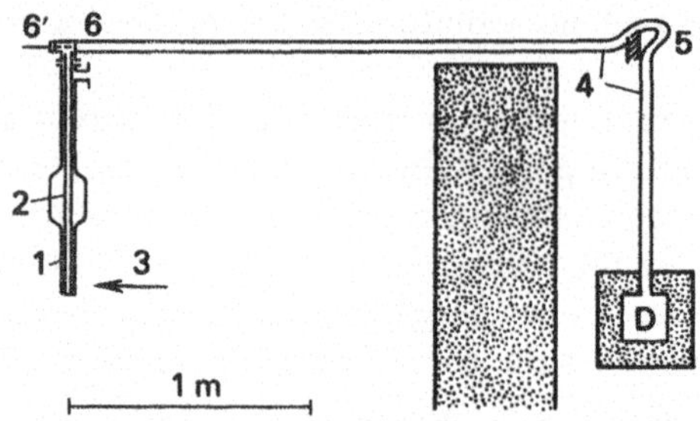

Fig. 2.5 Experimental layout for measuring the output of UCN from superfluid ^{4}He (Ageron *et al.* 1978*b*): 1, cryostat; 2, stainless steel tube 45 mm in diameter; 3, incident beam of cold neutrons (the beam cross-section is $13 \times 5\ \mathrm{cm}^2$, $\bar{\lambda} \approx 10$ Å); 4, neutron guide of electropolished stainless steel 67 mm in diameter; 5, filter representing a part of the neutron guide with four rectangular bends; 6, stainless steel piston, closed in position 6 and open in position 6'. (When the piston was closed, the counting rate of the detector fell to a background level $0.5\ \mathrm{s}^{-1}$.)

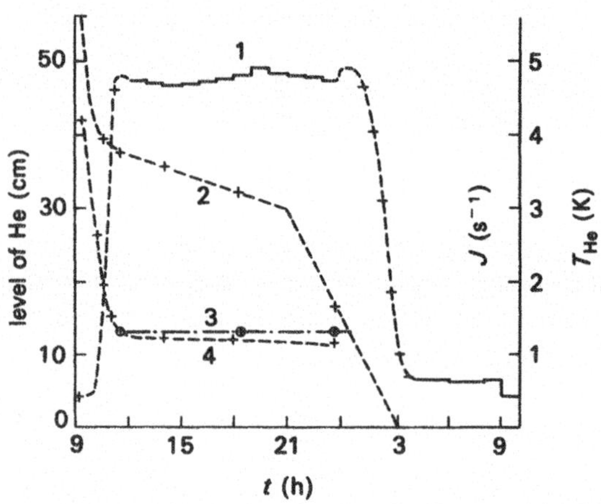

Fig. 2.6 Results of investigation (Ageron *et al.* 1978*b*) of the UCN output from helium: 1, counting rate of UCN as function of time; 2, position of the helium level; 3, UCN counting rate when piston is closed; 4, helium temperature. The UCN counting rate remained constant all the time the helium was superfluid, and its level is higher than the level of the incident neutron beam.

In this experiment the maximum counting rate of the detector amounted to $3.5\ s^{-1}$, which, if the transmission coefficient of the neutron guide is taken into account, is equal to 0.17 (Ageron *et al.* 1978*b*), corresponding to the UCN production rate of approximately $20\ s^{-1}$ within a volume of $207\ cm^3$.

In the second experiment (Jewell *et al.* 1981) the possibility was checked of storing UCN in a trap with helium at a temperature of 1.1–1.2 K. The storage time in a trap made of stainless steel was $\sim$ 10 s. Only in the third experiment (Golub *et al.* 1983) performed at 0.49 K was the production successful, in a beryllium trap, of UCN with the indicated density of 20 UCN/cm^3, and the storage time of $\sim$ 100 s was achieved.

In all the later experiments (Kilvington *et al.* 1987), (Yoshiki *et al.* 1987, 1988) the rate of UCN production was in agreement with the expected value, but difficulties with the extraction of UCN hindered this method. Generation with the help of a cold deuterium source and Steyerl turbine is therefore more effective, where the UCN density of $100\ n\,cm^{-3}$ can be achieved (Steyerl *et al.* 1986). Nevertheless, the generation of UCN in liquid helium is very useful, because it gives a good new possibility to study the interaction of neutrons with collective liquid helium excitations. In Kilvington *et al.* (1987) the rate of UCN production was estimated by measuring the intensity of their up scattering.

To conclude this discussion on converters, we present a composite table which compares the efficiency of various substances for the generation of UCN (Table 2.1). Unfortunately, different converters were investigated at different channels. If one and the same converter were measured at the same temperature at different channels, it should be possible, by comparing the counting rates for this single converter, to judge the other converters. On the other hand, such a comparison could in certain cases not be performed at all. For example, the output from beryllium in a vertical channel cannot in any way be compared with the yield from polyethylene or aluminium measured only in horizontal channels. In this case the yield from beryllium was normalized to the theoretical value (Golikov *et al.* 1973), calculated without taking into account the limiting energy of the material of the converter.

Calculations of converters have also been performed in other works. In Chapter 9 tables are presented from Golikov *et al.* (1973) and Akhmetov *et al.* (1974*a*, *b*, 1976). Utsuro and Hetzelt (1977) calculated the yield from liquid hydrogen and deuterium and measured the output of neutrons with velocities $v \approx 10\ m\,s^{-1}$ when interaction with a beam of thermal neutrons occurred. Their results are in quite good agreement with theory. In the Leningrad Institute of Nuclear Physics (Gatchina) extensive studies (Altarev *et al.* 1980*b*) were performed, as a result of which the dependences of the UCN yields from hydrogen and deuterium were determined, as shown in Fig. 2.7.

TABLE 2.1 Comparison of UCN yields from various converters

Substance	T_c (K)	G factor	Reference
$(CH_2)_n$	300	1.0	Golikov *et al.* (1973)
	130	2.2	Golikov *et al.* (1973)
	90	4.0	Golikov *et al.* (1973)
$ZrH_{1.9}$	300	1.0	Golikov *et al.* (1973)
			Groshev *et al.* (1974)
	130	2.3	Golikov *et al.* (1973)
$ZrH_{1.6}$	300	0.9	Groshev *et al.* (1974)
	80	1.5	Groshev *et al.* (1974)
Mg^{*a}	300	0.5	Golikov *et al.* (1973)
	300	0.14	Groshev *et al.* (1974)
			Akhmetov *et al.* (1976)
H_2O^*	300	0.35	Akhmetov *et al.* (1976)
	80	3.5	Akhmetov *et al.* (1976)
H_2^*	300	0.35	Akhmetov *et al.* (1976)
			Altarev *et al.* (1980*b*)
H_2^*	80	0.7	Akhmetov *et al.* (1976)
H_2^*	20	8.75	Altarev *et al.* (1980*b*)
D_2^*	300	0.28	Akhmetov *et al.* (1976)
D_2^*	80	0.42	Akhmetov *et al.* (1976)
D_2O	80	0.25	Akhmetov *et al.* (1976)
Be	300	1.0	Golikov *et al.* (1973)
			Abov *et al.* (1981*b*)
			Egorov *et al.* (1974)
	160	3.5	Abov *et al.* (1981)
	40	8.0	Egorov *et al.* (1974)
H_2^* (para)[b]	300	0.35	Akhmetov *et al.* (1976), (1977*a*)
H_2^* (para)[a]	80	1.7	Akhmetov *et al.* (1977*a*)

[a] The substances marked by asterisks are inserted in ampoules of aluminium

[b] For substances with more than one reference normalization is performed to make comparison convenient.

2.6 PRODUCTION OF UCN BY REFLECTION OF NEUTRONS FROM MOVING MIRRORS

If a mirror is moving with a velocity c and the velocity of a neutron along the direction of motion of the mirror is $v_\perp$, then the component of the relative velocity normal to the surface of the mirror will be equal to $v_\perp - c$. Since in the reference system connected with the mirror the normal component of the neutron velocity changes sign, if total reflection occurs, the component of the velocity $v_\perp$ in the laboratory system becomes, upon reflection,

$$v_{U\perp} = v_\perp - 2(v_\perp - c) = 2c - v_\perp. \tag{2.47}$$

Total reflection is possible either when $v_\perp - c < v_m$ (the limiting velocity of the mirror) (Groshev *et al.* 1974), or when $v_\perp - c$ satisfies the condition of

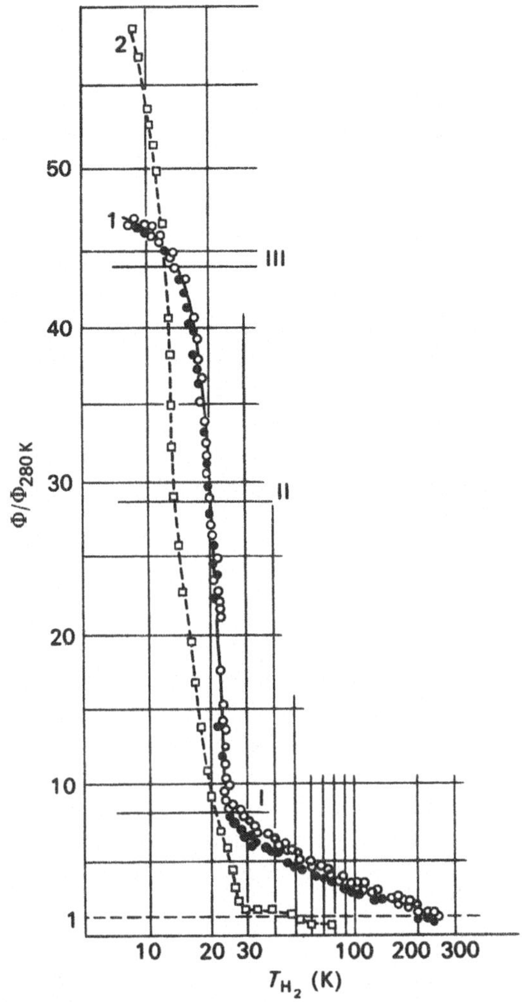

Fig. 2.7 The output of UCN as a function of: 1, the hydrogen temperature (○ and ● are experimental points obtained when the source is cooled and heated, respectively); 2, the deuterium temperature (□ are experimental points obtained when the source is cooled). The levels I, II, and III correspond to the starting point of condensation, the total transition from gas to liquid, and crystallization, respectively. At 280 K the pressure of hydrogen is 3.2 atm, corresponding to the UCN output at a 90% level of saturation. In calculations T_n was assumed to be equal to 300 K (Altarev *et al.*, 1980*b*).

Bragg reflection (Steenstrup and Buras 1978). In the first case a single reflection from the runaway mirror allows reduction of the neutron velocity by not more than $2V_m$, i.e. by not more than $\sim 16\ \mathrm{m\,s^{-1}}$ (if the mirror is made of ^{58}Ni). To slow down the neutrons to even lower energies multiple scattering must be applied. For example, the motion of the mirror may be programmed so that it slows down after each reflection of a neutron, so the neutron can once again catch up with the mirror and undergo reflection. Other versions are also feasible, naturally.

A simpler version consists of making use of a mirror curved in the form of a trough (Fig. 2.8), instead of a plane mirror. The neutron will then slide along the mirror, in the reference system connected with it, each time impinging upon the mirror's surface at an angle greater than the critical angle and thus undergoing total reflection. On the other hand, in the laboratory system this will seem as if the neutron undergoes a single reflection from the mirror. This method of slowing down neutrons was realized by A. Steyerl (1975). A beam of thermal neutrons with an intensity of 10^{13} cm^{-2} s^{-1} was directed from the reactor towards a turbine with its blades curved in the form of a trough. The linear velocity of the blades at the periphery was 25 m s^{-1}, and neutrons with an initial velocity of ~50 m s^{-1} became ultracold. The beam of UCN obtained in this manner had a cross-section of 10 × 20 cm^2 and the flux was ~240 s^{-1}.

In another version realized by Kashukeev *et al.* (1975*b*) the blades of the turbine are flat, but the neutron beam impinges upon the blade at a grazing angle, towards the axis of rotation. In this case, in a rotating reference system connected with the blade a centrifugal force acts on the neutron, so the neutron slows down as it approaches the axis of rotation. Since the centrifugal force can be described by the potential $\varphi = -m\Omega^2 r^2/2$, where Ω is the angular rotation frequency and r is the distance from the centre, the neutron kinetic energy will decrease by the quantity $\Delta E = m\Omega^2 R^2/2$ as the neutron travels over a distance R equal to the radius of the turbine. In the experimental set-up being considered the initial neutron velocity was ~70 m s^{-1}. The UCN produced were gathered in the neutron guide situated

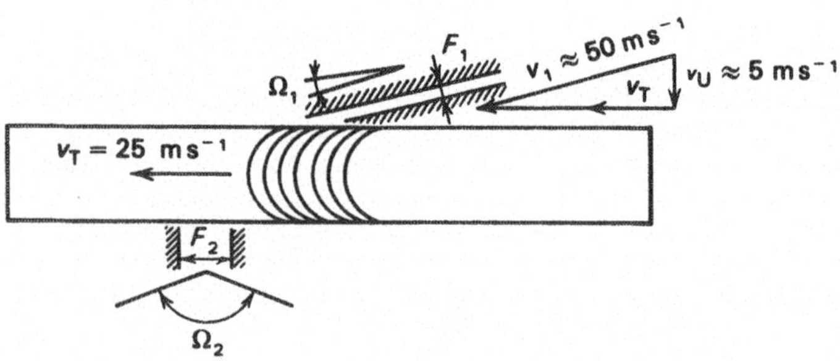

Fig. 2.8 Form of the blades of the Steyerl turbine (Steyerl 1975) (top view). Geometry of the input neutron guide of cross section F_1 and solid angle Ω_1 and of the outgoing neutron guide of cross-section F_2 and solid angle for the UCN Ω_2. The velocity transformation is shown.

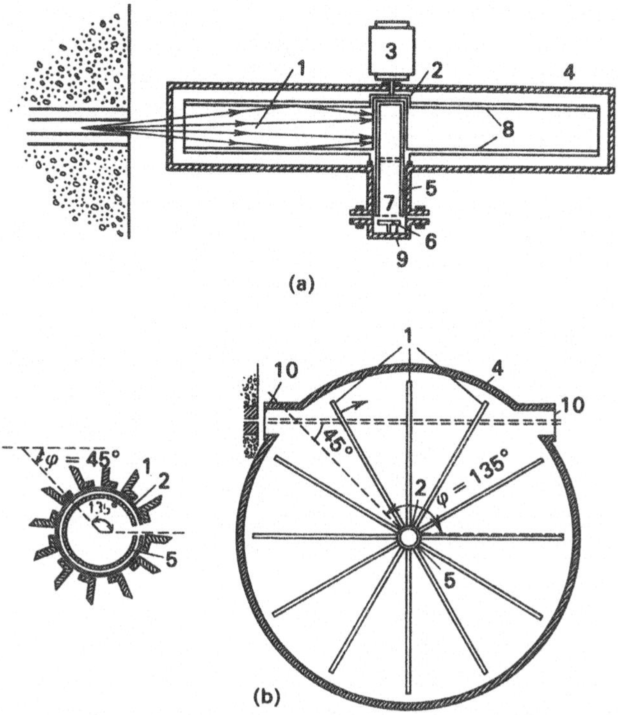

Fig. 2.9 Kashukeev turbine (Kashukeev *et al.*, 1975*a*, *b*). (a) Vertical view in section. (b) Horizontal view in section: 1, mobile neutron reflectors (blades) of length 1.25 m and width 12 cm; 2, socket where the UCN are gathered; 3, electric motor; 4, steel jacket: 5, UCN channel 5 cm in diameter and 18 cm in length; 6, foil (shutter of detector); 7 and 10, aluminium membranes; 8, lateral reflectors; 9, leads of power supply.

on the axis of the turbine (Fig. 2.9). In the case of a detector 40 cm^2 in area a counting rate of 4 s^{-1} was achieved for a thermal flux from the reactor of 10^{12} $cm^{-2}\,s^{-1}$.

In Japan also a mechanical generator of UCN is constructed (Utsuro *et al.* 1988*b*) that is similar to the Steyerl turbine but involves a certain improvement of the reflecting mirror. The surface of the mirror is covered with alternating layers of two materials with strongly differing potentials, such as, for example, nickel ($u_0 > 0$) and titanium ($u_0 < 0$). In all, about 170 layers are deposited with thicknesses gradually decreasing with distance from the surface. The alternating layers of variable thickness form something like an artificial crystal with a complicated mosaic structure which enhances E_{lim} up to 15 $m\,s^{-1}$ as a result of Bragg reflection from the superstructure. Earlier a mirror of this type was proposed by Mezei (∗Mezei 1976). A similar structure was considered by A.V. Antonov *et al.* (1981*a*), (1986*c*, *h*) consisting of several groups, each of which represented a periodic set of alternating layers. At present investigations of such supermirrors are being carried out at many institutes. Presented in Fig. 2.10 is the dependence on energy of the reflection coefficient of a neutron from a supermirror constructed in Japan (Utsuro *et al.* 1982, 1988*b*). With the aid of a turbine with such a mirror a density flux of $0.6\ cm^{-2}\ s^{-1}$ of neutrons with velocities between 0 and $15\ m\ s^{-1}$ is achieved for a thermal flux from the reactor equal to $10^{12}\ cm^{-2}\ s^{-1}$. It is hoped that with a liquid deuterium cold source, the UCN flux density of $10^2\ n\ cm^{-2}\ s^{-1}$ and total flux $10^3\ n\ s^{-1}$ will be achieved.

By making use of Bragg reflection from crystals it is possible to slow down neutrons with comparatively high velocities. This method of generating UCN was studied by Monte Carlo calculations by Steenstrup and Buras (1978), Dombeck *et al.* (1979), and Majkrzak and Malik (1980) and realized at the Argonne National Laboratory in the USA (Brun *et al.* 1980). The mirror was chosen to be of thermic mica characterized by Bragg reflection of neutrons with velocities close to $200\ m\ s^{-1}$. Aluminium wedges were applied to make the mosaic structure of the crystal more complex and to extend the range of reflected velocities.

The beam of thermal neutrons with velocities of about $400\ m\ s^{-1}$ impinged upon a crystal placed at the end of a rotating shaft (Fig. 2.11) and then, having been transformed into UCN, was directed through the neutron guide (nickel, 30 cm long, $\phi = 9.5$ cm) towards the detector. Upon performing the necessary corrections of the experimental results, about five neutrons were found (Brun *et al.* 1980) to impinge on the detector during each 140 μs burst

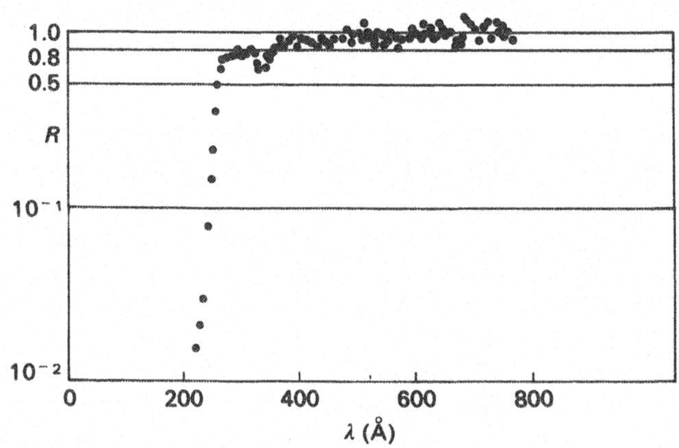

Fig. 2.10 Experimentally measured reflection coefficient from a supermirror with alternating layers of nickel (85 layers) and titanium (84 layers) of gradually decreasing thickness (Utsuro *et al.* 1982).

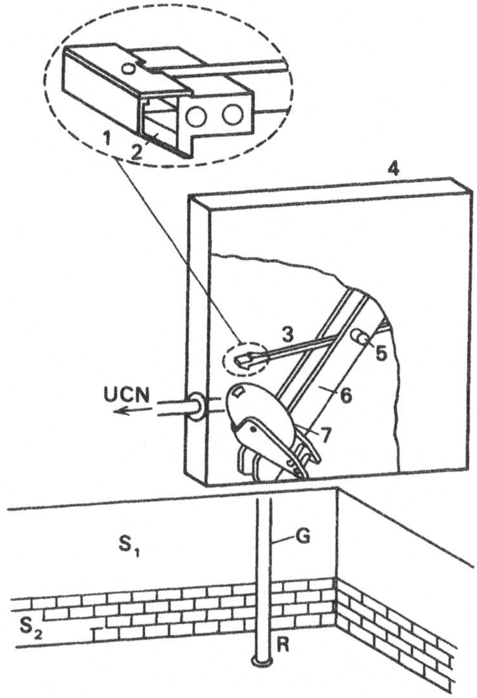

Fig. 2.11 Plan of installation for obtaining UCN by reflection of neutrons from a moving crystal (Brun *et al.* 1980): 1, crystal holder; 2, mica with aluminium wedges; 3, rotor; 4, vacuum chamber; 5, rotor of motor; 6, internal supports; 7, rotating disk with slit. G, channel for thermal neutrons; S_2, lead and concrete shielding; S_1, concrete.

occurring with a frequency of 15 Hz when the pulsed thermal neutron flux was $10^{13}\,\mathrm{cm^{-2}\,s^{-1}}$. In the future, it will be possible, with the aid of a mechanical generator, to achieve record neutron densities in traps amounting to 500–700 UCN/cm^3.

A disadvantage of a mechanical generator is that it represents quite a complicated device in comparison with a converter (its turbines are approximately 2 m in diameter and rotate with a frequency of about 1000 rev min^{-1}). On the other hand, one advantage is that the UCN generated in the turbine do not have to be transported in a long neutron guide, where significant losses are possible as a result of multiple collisions with the walls. Neutrons reflected from the blades also undergo several collisions; thus, for instance, the total number of collisions with the blades in the Steyerl turbine amounts to 10. It is essentially smaller than in a neutron guide, but, since the neutrons in the turbine are not ultracold, each non-mirror reflection may lead to significant losses. Therefore, at present it cannot be stated definitely which method—the mechanical one or the one employing converters—is preferable for the production of UCN.

At the outset the mechanical generator looked attractive because it seemed (Antonov *et al.* 1969*a*) possible with its aid to transfer all the neutrons from a given interval of the spectrum involving high velocities to the UCN region. The reasoning was as follows. The flux of thermal neutrons of velocity $v_\perp$ in the direction towards the mirror may be represented in the form $d\Phi_T = f(v_\perp)v_\perp dv_\perp$, and since $|dv_{U\perp}| = |dv_\perp|$ when reflection from the mirror occurs, in accordance with Antonov *et al.* (1969*a*), the flux of UCN reflected from the mirror is $f(v_\perp)v_\perp dv_{U\perp} = d\Phi_T$, which is about $v_\perp/v_{0\perp}$ times greater than the UCN flux $d\Phi_U = f(v_{U\perp})v_{U\perp}dv_{U\perp}$ in the primary (i.e. Maxwellian) spectrum.

Actually, such arguments are not valid. The total neutron flux after slowing down by the runaway mirror is not $f(v)v_\perp dv_{U\perp}$, but $f(v)v_{U\perp}dv_{U\perp}$. At first glance this seems contradictory. Indeed, in the laboratory reference frame the flux of primary neutrons through a fixed plane parallel to the moving mirror and coincident with it at time $t = 0$ is $J_1 = f(v)v_\perp dv_{U\perp}$, while the flux of reflected neutrons is $J_2 = f(v)v_{U\perp}dv_{U\perp}$, where $v_{U\perp} = v_\perp - 2c$. The flux is therefore not conserved and in the time interval t the number of neutrons $\Delta N = t(J_1 - J_2) = 2ctf(v)$ seems to be lost. But by that time the mirror runs away from the plane under consideration through the distance $l = ct$, and these neutrons turn out to be in the volume of space swept out by the mirror. They are therefore not lost at all, but fill the space in front of the mirror with the density $2f(v)$.

In the Maxwellian spectrum, when v is high then $f(v)$ is small because of the presence of the exponential factor $\exp(-E/T_n)$. Therefore, mechanical generators are most efficiently applied for transforming low-energy neutrons into UCN and have no advantages over converters. This, however, is only true for the Maxwellian spectrum. If, on the other hand, the spectrum is narrower and the primary beam is well collimated, then it may turn out to be more efficient to transform such a beam to the region of UCN with the aid of a mechanical generator, rather than use a converter.

2.7 SLOWING DOWN NEUTRONS IN EXTERNAL FIELDS

For producing UCN it is also possible to utilize the gravitational and inhomogeneous magnetic fields. Elevation of a neutron in the gravitational field by 1 m lowers its kinetic energy by 100 neV. In the existing vertical and inclined channels a neutron is raised to heights between 5 (Egorov *et al.* 1974) and 14 m (Steyerl 1969). The neutrons, which at the top are ultracold, have an energy at the bottom which exceeds significantly the limiting energy even of such a material as beryllium. The number of neutrons generated at the bottom is essentially higher than the amount of UCN in the Maxwellian spectrum. However, only a small part of the neutrons within a narrow solid angle arrives at the top. Moreover, neutrons occurring in this solid angle must undergo mirror reflection from the walls of the neutron guide, otherwise they may pass through the walls when undergoing the next collision. Therefore, the walls of the vertical channels have to be well polished. The highest intensity beam of UCN was obtained by utilizing a vertical channel (Steyerl *et al.* 1986).

In practice, inhomogeneous magnetic fields are not utilized for slowing down neutrons, since a field of about 2 T is required to cool a neutron by 0.1 μeV. In the literature, however, three methods of producing UCN with the aid of external magnetic fields are discussed. The first has been dealt with in Chapter 1 (Drabkin and Zhitnikov 1960).

The second method (Namiot 1973) consists of making the neutrons pass, for example, through solid atomic deuterium polarized, say, dynamically in an external field B. A neutron, the spin of which is directed opposite to the field, may undergo incoherent scattering on the deuterium nucleus and reverse its spin direction. At the same time spin flip of the deuterium nucleus will also occur. This process is accompanied by a decrease of the neutron energy by the quantity $(2\mu + \mu_N)B$, where μ_N is the magnetic moment of the nucleus. Backward scattering does not occur, since the deuterium polarization is assumed to be 100%. For the process to occur repeatedly a resonance radio-frequency field has to be imposed that can reverse the flipped neutron spins back to the initial direction. In this way the whole initial neutron spectrum can be cooled down to the UCN region. For this purpose a sufficiently large volume of deuterium is needed, and it is necessary, upon having cooled the deuterium, to suppress the inelastic phonon scattering to a sufficient extent.

The third method (*Summhammer *et al.* 1986) is a cooling of neutrons with a travelling magnetic wave. The idea is simple. Consider a pulse source, that gives a neutron cluster moving with some average speed v_0. This cluster will disperse continuously in space due to different velocities of the neutrons. Suppose that a magnetic well moving with the speed v_0 is created in the centre of mass of this cluster. Then in the centre of mass reference system some neutrons with energy lower then the potential barrier of magnetic walls will be trapped in the well.

If the magnetic well starts to decelerate, then the trapped neutrons also start to decelerate, and when the well stops, the trapped neutrons become ultracold. This is the essence of the third method. A better theoretical analysis of it is obtained with the help of the equivalence principle but we shall not dwell on this subject, since the similar problem, the transportation of UCN in a moving bottle, will be met in chapter 4.

2.8 ACCUMULATION OF UCN AT PULSED REACTORS

The density of the UCN accumulated in a trap is proportional to the thermal neutron flux density. At pulsed reactors the impulse density is significantly higher than the average, and a natural desire arises to enhance the density of the accumulated neutrons up to the impulse density. In principle, this can be done with the aid of shutters which open up the UCN source during bursts and cover it between the bursts (Shapiro 1972*a*). Then the neutrons, which during the time in between the pulses are capable of returning to the source (for instance, to the converter) and of being lost in it, are reflected from the shutter and are lost only at the moment when the shutter is open. Dynamic equilibrium is established when the number of UCN entering the trap during the burst equals the sum of UCN leaving it and of those lost in the trap in between the bursts. If a neutron is confined in the trap for a sufficiently long time, then dynamic equilibrium arises when the density of accumulated UCN achieves the impulse value (Stoica and Strelkov, 1978*a*; Novopoltzev and Pokotilovskii, 1980*a*; Shapiro 1972*a*). If, on the other hand, the losses of UCN due to their leaving the trap when the shutter is open are small compared with other losses, then the utilization of shutters cannot essentially enhance the neutron density in the trap.

The simplest shutter is the mechanical one. It was utilized in the Argonne laboratory (Brun *et al.* 1980) to obtain UCN by reflection from a moving crystal and at the IIN pulsed aperiodic reactor (Antonov *et al.* 1976*a*), where the trap was brought close to the reactor during the burst and whence the converter was covered by the shutter and the trap was transported to the experimental hall. Other types of shutters are also feasible. We shall now list them.

1. Imagine a converter with a high limiting velocity (e.g. beryllium) placed within a horizontal channel and travelling towards a flux of thermal neutrons. In the laboratory reference frame part of the neutrons produced inside the converter turns out to be ultracold. If, when the burst has come to an end, the converter is brought to a stop, it then becomes a reflector of UCN (Pokotilovskii 1980; Stoica and Jipa 1984).

2. Impulse storage is possible with the aid of a vessel falling down on the UCN source and stopping abruptly. During the motion the velocity of the UCN relative to the vessel is high, and therefore the walls of the vessel are transparent. When the vessel stops, the neutrons happening to be inside it turn out to be blocked (Stoica and Strelkov 1978*a*). It may be that this method will yield the highest number of UCN stored in a trap, when a reactor with a compact zone (e.g. TRIGA) is utilized together with a cold moderator. For extracting the UCN confined in the trap it suffices to accelerate the trap to a high velocity together with the neutrons inside it and then once again to

stop it abruptly (Stoica and Jipa 1984). The UCN would then be shaken out from the trap.

3. Impulse storage is also possible with the aid, for example, of para-hydrogen. Consider the storage vessel to be filled with para-hydrogen. At the moment of the burst the para-hydrogen represents a converter of UCN which are produced throughout the whole volume. If, after the burst, the para-hydrogen is taken away rapidly (the storage vessel is connected to a large evacuated cavity), then the UCN density in the vessel will immediately become equal to the impulse density.

Other methods, considered below, are related to the utilization of magnetic fields and are based on the interaction of a neutron with matter in the presence of a magnetic field B described by the potential

$$u = \frac{\hbar^2}{2m} 4\pi N_0 b - \mu B,$$

the value of which is greater for neutrons with spins directed along the field and smaller for neutrons polarized in the opposite direction.

4. Now consider a non-magnetic converter with a high-energy E_{lim}. Let a magnetic field be switched on during the reactor burst. Its magnitude is chosen such that the total potential for neutrons with spins opposite to the field inside the converter equals zero. After the burst the magnetic field is switched off, and the UCN produced with spins opposite to the field will now be reflected from the converter (Golikov and Taran 1975). In this case depolarization of the neutrons in between the bursts does not lead to their loss.

5. The simplest magnetic shutter is a ferromagnetic film which is magnetized along a certain direction during the burst and in the opposite direction between the burst (Pokotilovskii *et al.* 1978). In this case, however, the neutrons must not undergo depolarization between the bursts.

6. Instead of remagnetizing the film, a spin flipper may be established in sequence with it and switched on only during the burst (Pokotilovskii *et al.* 1976). Here depolarization is also important.

7. The film may also be completely demagnetized between the bursts by heating it to temperatures higher than the Curie temperature (Nikitenko and Taran 1978). In this case the depolarization does not affect the accumulation of UCN.

8. Finally, use can be made of multilayer systems. Consider, for example, two films which at the moment of the burst are magnetized in parallel and antiparallel between the bursts. One of the films may be remagnetized either with the aid of foil with current separating the films (Pokotilovskii *et al.*

1976), or by utilizing the difference between the coercive forces of the films (Novopoltzev and Pokotilovskii, 1979). It is also possible to use more than two films. This method was subjected to experimental verification (Novopoltzev and Pokotilovskii 1978, 1979) which revealed that in the remagnetized state confinement is not complete. This means that the neutron may undergo partial depolarization in the space between the films. Experiments have also revealed that the ratio of the UCN flux passing through a set of films when the shutter is open to that passing through when the shutter is closed increases as the number of films increases.

9. Yet another system can be considered, namely polarized deuterium deposited on a beryllium surface and, also, a strong magnetic field present in the system. Then, neutrons undergoing scattering on the deuterium nuclei involving a spin flip cool down and turn out to be confined. If no depolarization occurs, then polarized neutrons accumulate in the trap on the upper magnetic level, i.e. population inversion arises. Under certain conditions a laser effect may arise in such a system, which can serve as a model of a pulsar.

Now let us clarify under which conditions it is sensible to use shutters. Let the duration of the burst be τ, the repetition period of the impulses be θ, and the storage time of the UCN in the trap be T, when the converter is closed. Dynamic equilibrium is established when

$$S(\rho_{imp}v/4)\tau = S(\rho v/4)\tau + \rho V(\theta/T),$$

where S is the converter area, V is the volume of the trap, ρ is the density of UCN in the trap, and ρ_{imp} is the density of UCN in the impulse. Hence it immediately follows that the maximum density of the accumulated UCN equals

$$\rho = \frac{\rho_{imp}}{1 + \theta T_1/\tau T},$$

where $T_1 = 4V/Sv$ is the time the UCN flow out from the trap to the converter. If the lifetime T is large compared with $\theta T_1/\tau$, then the maximum density of accumulated neutrons equals the impulse value. If, on the other hand, $T \approx T_1$, then $\rho = (\tau/\theta)\rho_{imp}$ and we obtain the density averaged over the whole period. Thus, a shutter does not always lead to an enhancement of the density of UCN. Shutters, however, can be utilized for the interruption of the UCN beam, for example in a correlation spectrometer or in other applied problems.

Now let us determine the maximum density of UCN that can be obtained under ideal conditions. This density is determined by Fermi statistics. If the limiting wavevector is k_{lim}, then the maximum density will be $\rho_{max} = 2(4\pi/3)k_{lim}^3(2\pi)^{-3}$ and, when $k_{lim} \approx 10^6\ cm^{-1}$, it amounts to $\sim 10^{-3}$ of the density of the atmosphere, i.e. $\sim 3 \times 10^{16}\ cm^{-3}$. In astrophysics, at densities of the order of magnitude of the nuclear density essentially higher values of ρ_{max} are possible, as was mentioned in Chapter 1.

2.9 CONCLUSION

Practically all the methods of obtaining UCN that can be realized at present have been considered in the preceding sections. Interestingly, many of them had already been described in the very first papers on UCN. Some of them underwent further development, but, at any rate, the number of possible ways of obtaining UCN increased with time. Hopefully, in the future methods will be invented which will allow the intensity of the available sources of UCN to be increased by at least an order of magnitude. However, obtaining UCN does not represent the complete solution of the problem. The UCN produced have to be transported far away from the sources of background to the experimental hall, and therefore the problems of UCN transportation still remain. Before tackling these issues we shall deal with the methods of detecting UCN and with the spectrometry of UCN.

APPENDICES

Appendix 2.1 Theory of UCN generation in a homogeneous converter

$$\mathrm{d}\Phi_U(\mathbf{k}_U) = \Phi_0 N_0 \overline{\mathrm{d}\sigma_-}(T_n, T_c, \mathbf{k}_U) \int_0^L \mathrm{d}x \exp[-xN_0\sigma_1(T_c, k_U)k_U/k_{U\perp}]. \quad (1)$$

$$\text{a) } \mathrm{d}\sigma_- = \sigma'_- \mathrm{d}\mathbf{k}_U, \quad \text{b) } \overline{\mathrm{d}\sigma_-}(T_n, T_c, \mathbf{k}_U) = \int \mathrm{d}\Phi(T_n, \mathbf{k})\mathrm{d}\sigma_-(T_c, \mathbf{k} \to \mathbf{k}_U)/\Phi_0. \quad (2)$$

$$\text{a) } \mathrm{d}\Phi(T, \mathbf{k}) = \Phi_0(k\mathrm{d}\mathbf{k}/2\pi T^2)\mathrm{e}^{-k^2/T}, \quad \text{b) } \mathrm{d}\mathbf{k} = k^2 \mathrm{d}k\mathrm{d}\Omega, \quad \text{c) } \mathrm{d}\Omega = \mathrm{d}\phi \mathrm{d}\cos\theta. \quad (3)$$

$$\sigma'(T, \mathbf{k} \to \mathbf{k}') = \mathrm{d}\sigma(T, \mathbf{k} \to \mathbf{k}')/\mathrm{d}\mathbf{k} = (\hbar^2/mk)\mathrm{d}^2\sigma(T, \mathbf{k} \to \mathbf{k}')/\mathrm{d}\Omega\mathrm{d}E. \quad (4)$$

$$\text{a) } LN_0\sigma_1 \ll 1, \quad \text{b) } \mathrm{d}\Phi_U(\mathbf{k}_U) = \Phi_0 N_0 L\overline{\mathrm{d}\sigma_-}(T_n, T_c, \mathbf{k}_U), \quad \text{c) } \Phi_U = \int \mathrm{d}\Phi_U(\mathbf{k}_U). \quad (5)$$

$$\text{a) } LN_0\sigma_1 \gg 1, \quad \text{b) } \mathrm{d}\Phi_U(\mathbf{k}_U) = k_{U\perp}\Phi_0\overline{\mathrm{d}\sigma_-}(T_n, T_c, \mathbf{k}_U)/k_U\sigma_1(T_c, k_U). \quad (6)$$

$$\text{a) } \mathrm{d}\Phi(T, \mathbf{k})\mathrm{d}\sigma_-(T, \mathbf{k} \to \mathbf{k}_U) = \mathrm{d}\Phi(T, \mathbf{k}_U)\mathrm{d}\sigma_+(T, \mathbf{k}_U \to \mathbf{k}), \quad \text{b) } \mathrm{d}\sigma_+ = \sigma'_+ \mathrm{d}\mathbf{k}. \quad (7)$$

$$\text{a) } \mathrm{d}\sigma_-(T, \mathbf{k} \to \mathbf{k}_U) = (k_U \mathrm{d}\mathbf{k}_U/k\mathrm{d}\mathbf{k})\exp(\omega/T)\mathrm{d}\sigma_+(T, \mathbf{k}_U \to \mathbf{k}),$$

$$\text{b) } \omega = k^2 - k_U^2 \approx k^2. \quad (8)$$

$$\mathrm{d}\Phi_U(\mathbf{k}_U) = (\Phi_0/2\pi T_n^2)k_{U\perp}\mathrm{d}\mathbf{k}_U\bar{\sigma}_+(T_n, T_c, \mathbf{k}_U)/\sigma_1(T_c, k_U). \quad (9)$$

$$\bar{\sigma}_+(T_n, T_c, k_U) = \int \sigma'_+(T_c, \mathbf{k}_U \to \mathbf{k})\exp(\omega/T_c - k^2/T_n)d\mathbf{k}$$
$$= \int \exp(k^2/T_c - k^2/T_n)d\sigma_+(T_c, \mathbf{k}_U \to \mathbf{k}). \quad (10)$$

$$\text{a) } k_U^2 \ll T_n, T_c, \quad \text{b) } \sigma_{+,1} \propto 1/k_U, \quad \text{c) } k_U\sigma(T, k_U) = k_{lim}\sigma(T, k_{lim}). \quad (11)$$

$$\Phi_U(T_n, T_c) = (\Phi_0/8)(k_{lim}^2/T_n)^2\bar{\sigma}_+(T_n, T_c, k_{lim})/\sigma_1(T_c, k_{lim}). \quad (12)$$

$$\text{a) } G(T_n, T_c) = (C/T_n^2)\bar{\sigma}_+(T_n, T_c, k_T)/\sigma_1(T_c, k_T), \quad \text{b) } k_T = \sqrt{T} = \sqrt{300}\text{ K}. \quad (13)$$

$$\text{a) } d\rho_U(\mathbf{k}_U)V/t = d\Phi_U(\mathbf{k}_U)S, \quad \text{b) } t = 4V/Sk_U, \quad \text{c) } \rho_U = \int d\rho_U(\mathbf{k}_U). \quad (14)$$

$$\rho_U = (16/3)\Phi_U(T_n, T_c)/k_{lim} = (2E_{lim}^{3/2}/3T_n^2)[\bar{\sigma}_+(T_n, T_c, k_T)/\sigma_1(T_c, k_T)]\Phi_0. \quad (15)$$

The differential UCN flux density is defined by (1), where $d\bar{\sigma}_-$ is the differential cooling (down-scattering) cross-section (2a) averaged over the thermal neutron spectrum (3a), as shown in (2b). $k_{U\perp}$ is the component of the UCN wavevector normal to the exit converter surface; L is the thickness of the converter. The derivative cross-section σ' can be expressed with the aid of the well-known double-differential cross-section, as shown in (4). Elements of the wavevector space and angles are defined in (3b) and (3c), respectively. For a thin converter, when (5a) is satisfied, the integration in (1) yields (5b). The total UCN flux density is defined in (5c). In the case of a thick converter, when (6a) is satisfied, the integration in (1) leads to (6b). Taking into account the detailed balance (7) the down-scattering cross-section can be substituted by the up-scattering (heating) cross-section $d\sigma_+$, as shown in (8a). The energy gained by the UCN in inelastic scattering is defined in (8b). Substituting (7b) into (8a), then (8a) and (3a) into (2b), and, finally, (2b) into (6b) we obtain (9), where the effective up-scattering cross-section defined in (10) is introduced. The relations (11) are usually satisfied for UCN. Substituting (9) into (6b) we arrive at the main expression (12), taking into account (11c). Using (11b) and (11c) we can write (12) in the form (13), where C is a normalization constant and $T = 300$ K is a thermal point. Expression (14a) can be written for the UCN density in a bottle with walls covered with a thick layer of converter, where t is the time of flight (14b) between two subsequent collisions with the walls, V is the volume of the bottle, and S represents the surface area of the walls. Substituting (9) into (14a) we obtain the integral density (14c) in the form (15).

Appendix 2.2 UCN generation in a solid-state converter

$$\text{a) } d\sigma_+(T, \mathbf{k}_U) = \sigma_{el}(\omega^{1/2}/k_U M)g(\omega)n(\omega/T)d\omega, \quad \text{b) } n(x) = 1/(\exp x - 1). \quad (1)$$

$$\text{a) } \sigma_{el} = \sigma_{el}^{coh} + \sigma_{el}^{inc}, \quad \text{b) } g(\omega) = \exp\left(-\frac{\omega}{\theta}\right)\frac{\omega^2}{2\theta^3}, \quad \text{c) } k_\theta = \sqrt{\theta}, \quad \text{d) } \bar{T} = \frac{T}{\theta}. \quad (2)$$

a) $\sigma_+(T, k_U) = (\sigma_{el}k_\theta/2Mk_U)\bar{T}^{7/2}\Gamma(7/2)\zeta(7/2, 1 + \bar{T})$,

b) $\Gamma(7/2) = 15\sqrt{\pi}/8$. (3)

$$\zeta(s, x) = \int_0^\infty \exp(-xy)y^{s-1}\,dy/[1 - \exp(-y)]\Gamma(s) = \sum_{n=0}^\infty (n + x)^{-s}. \quad (4)$$

$$\bar{\sigma}_+(T_n, T_c, k_U) = \int \exp(\omega/T_c - \omega/T_n)\,d\sigma_+(T_c, \mathbf{k}_U)$$
$$= (\sigma_{el}k_\theta/2Mk_U)(T_c/\theta)^{7/2}\Gamma(7/2)\zeta(7/2, T_c/\theta + T_c/T_n). \quad (5)$$

a) $\zeta(s, x) = x^{-s} + \zeta(s, 1 + x)$, b) $\zeta(s, 1 + x) = \zeta(s) - xs\zeta(s + 1) + \ldots$. (6)

a) $\zeta(s) = \sum_{n=1}^\infty n^{-s}$, b) $\zeta(7/2) \approx 1.13$, c) $\zeta(9/2) \approx 1.05$. (7)

a) $T_c/\theta \ll 1$, b) $G(T_n, T_c) = C\bar{T}_n^{3/2}/(\bar{T}_n + 1)^{7/2}(\gamma + \xi\bar{T}_c^{7/2}) + O(\bar{T}_c^{9/2})$. (8)

a) $\xi = \zeta(7/2)[1 - \gamma(1 + 1/T_n)^{7/2}]$, b) $\bar{T}_n \approx 1$, c) $\gamma \ll 1$, d) $\xi \approx 1$. (9)

a) $\gamma = (16/15\pi^{1/2})(k_{lim}/k_\theta)M\sigma_a(k_{lim})/\sigma_{el}$, b) $T_{nm} = 3\theta/4$. (10)

a) $\bar{T}_n \ll 1, \bar{T}_c \ll 1$, b) $\rho_U = 5\sqrt{\pi}(E_{lim}/\theta)^2\bar{T}_n^{3/2}[\sigma_{el}/8Mv_{lim}\sigma_a(k_\theta)]\Phi_0$. (11)

$$d\sigma_+^{coh}(T, \mathbf{k}_U \to \mathbf{k}) = \sigma_{el}^{coh}\sum_{i,\tau}(\mathbf{e}_i\mathbf{k})^2(d\mathbf{k}/2\pi k_U k^2)\int n[\omega_i(\mathbf{q})/T]\delta(\mathbf{k} + \tau - \mathbf{q})$$
$$\times\, \delta[k^2 - \omega_i(\mathbf{q})]\,d\mathbf{q} = \sigma_{el}^{coh}\sum_j S_j(\mathbf{k}, T)\delta(k^2 - E_j)\,d\mathbf{k}/2\pi k_U. \quad (12)$$

$$S_j(\mathbf{k}, T) = \sum_{i,\tau}(\mathbf{e}_i\mathbf{k}/k)^2 n(E_j/T)/|1 - \nabla_{k^2}\omega_i(\mathbf{k} + \tau)|_{k^2 = E_j}, \quad E = k^2 = \omega(\mathbf{k} + \tau). \quad (13)$$

The differential up-scattering cross-section in the incoherent one-phonon approximation neglecting the Debye–Waller factor can be represented in the form (1a), the Bose factor being shown in (1b) and the elastic factor in (2a). We choose a phonon density normalized to unity in the form (2b) with the parameter θ equivalent to the Debye temperature. The up-scattering cross-section obtained upon integration of (1a) with the parameter defined in (2c) and the reduced temperatures defined in (2d) is shown in (3a), where $\Gamma(x)$ is the Euler Γ function, and its value for $x = 7/2$ is given in (3b); $\zeta(s, x)$ in (4) is the generalized Riemann function. Substituting (1a) into eqn (10) of Appendix 2.1 and taking into account (7b) we obtain (5). From definition (4) it follows that the equality (6a) and the power expansion (6b) are valid. Here the Riemann function (7a) is introduced. Its values for $s = 7/2$ and $s = 9/2$ are given in (7b) and (7c), respectively. Substituting (3a) and (5) into (13a) of Appendix 2.1 we obtain the G factor which in approximation (8a) can be reduced to the form (8b), the parameters ξ and γ being defined in (9a) and (10a), respectively. ξ can be approximated by (9d), when (9b) and (9c) hold. We see that the UCN yield plotted against T_n exhibits a maximum at the point (10b). Substituting (3a) and (5) into eqn (15) of Appendix 2.1 we obtain (11). The coherent differential up-scattering cross-section can be represented in the form (12). The dynamical structure factor occurring in expression (12) is defined in (13), and E_l is one of the solutions of the equation $E = k^2 = \omega(\mathbf{k} + \boldsymbol{\tau})$.

Appendix 2.3 UCN generation in a monoatomic gas converter

$$\text{a) } d\sigma_+(T_c, \mathbf{k}_U \to \mathbf{k}) = \sigma_{el} S_+(T_c, \boldsymbol{\kappa}, \omega) d\mathbf{k}/k_U, \quad \text{b) } \sigma_{el} = 4\pi b^2. \tag{1}$$

$$\begin{aligned} S_+(T_c, \boldsymbol{\kappa}, \omega) &= \frac{1}{2\pi} \int \delta(\boldsymbol{\kappa} + \mathbf{p} - \mathbf{p}_0) \delta(\omega + p^2/M - p_0^2/M) \\ &\quad \times \exp(-p_0^2/MT_c) d^3p d^3p_0/(\pi M T_c)^{3/2} \\ &= \frac{1}{2\pi} \int \delta\left(\omega + \frac{\kappa^2}{M} - \frac{2\kappa}{M} p_\kappa\right) e^{-p_\kappa^2/MT_c} \frac{dp_\kappa}{(\pi M T_c)^{1/2}} \\ &= \frac{1}{2\pi\kappa} \sqrt{\frac{M}{4\pi T_c}} \exp\left[-\left(\omega + \frac{\kappa^2}{M}\right)^2 \frac{M}{4T_c\kappa^2}\right]. \end{aligned} \tag{2}$$

$$\text{a) } \boldsymbol{\kappa} = \mathbf{k} - \mathbf{k}_U \approx \mathbf{k}, \quad \text{b) } \omega = k^2 - k_U^2 \approx k^2, \quad \text{c) } \mu = M/(M+1). \tag{3}$$

$$d\sigma_+(T_c, \mathbf{k}_U \to \mathbf{k}) \approx \sigma_{el}(M/4\pi T_c)^{1/2} \exp(-M\omega/4\mu^2 T_c) d\omega/k_U. \tag{4}$$

$$\begin{aligned} \bar{\sigma}_+(T_n, T_c, k_U) &= \int \exp(\omega/T_c - \omega/T_n) d\sigma_+(T_c, \mathbf{k}_U \to \mathbf{k}) \\ &= \sigma_{el}\sqrt{MT_c/4\pi k_U^2(\alpha + T_c/T_n)^2}. \end{aligned} \tag{5}$$

$$\begin{aligned} &\text{a) } \alpha = (M-1)^2/4M, \\ &\text{b) } \bar{\sigma}_+(T_n = T_c, k_U) = \sigma_+(T_c, k_U) = 2\mu^2 \sigma_{el}\sqrt{T_c/\pi M k_U^2}. \end{aligned} \tag{6}$$

$$\begin{aligned} \Phi_U(T_n, T_c) &= \frac{C}{T_n^2} \frac{\bar{\sigma}_+(T_n, T_c, k_U)}{\sigma_+(k_U) + \sigma_+(T_c, k_U)} \\ &= \frac{C_1}{T_n^2} \frac{(T_c/T_{300})^{1/2}}{(\alpha + T_c/T_n)[\gamma_{300} + (T_c/T_{300})^{1/2}]}. \end{aligned} \tag{7}$$

$$\text{a) } \gamma_{300} = (\pi M)^{1/2} \sigma_a(k_{300})/2\mu^2\sigma_{el}, \quad \text{b) } k_{300} = T_{300}^{1/2}, \quad \text{c) } C_1 = CM/4\mu^2. \tag{8}$$

$$\bar{\sigma}_+(T_n, T_c, k_U) = \int \exp(-\omega/T_n) d\sigma_-(T_c, \mathbf{k} \to \mathbf{k}_U) k d\mathbf{k}/k_U d\mathbf{k}_U. \tag{9}$$

$$\begin{aligned} d\sigma_-(T_c, \mathbf{k} \to \mathbf{k}_U) &= \sigma_{el} S_-(T_c, \boldsymbol{\kappa}, \omega) d\mathbf{k}_U/k \\ &= \sigma_{el} \frac{1}{2\pi\kappa} \sqrt{\frac{M}{4\pi T_c}} \exp\left[-\left(\omega - \frac{\kappa^2}{M}\right)^2 \frac{M}{4T_c\kappa^2}\right]. \end{aligned} \tag{10}$$

The differential up-scattering cross-section is presented in (1a), the elastic factor being defined in (1b) and the scattering function in (2). The correctness of this function is checked in the limits as $MT_c \to 0$ and $M \to \infty$, where $S_+ \to \delta(\omega)/2\pi$ and $\int d\sigma_+ = 4\pi b^2$. The wavevector and energy transferred to the neutron are determined and approximated for low UCN energies in (3a) and (3b), respectively. Taking into account the approximations (3a) and (3b) as well as expression (3c) for the effective mass and substituting (2) into (1a) we obtain (4). Substituting this into eqn (10) of Appendix 2.1 we obtain (5), while the parameter α is defined in (6a). Substituting $T_n = T_c$ into (5) we obtain (6b). Substituting (5) and (6b) into eqn (12) of Appendix 2.1 we obtain the gain factor in the form (7) where the normalization constant C_1 is given together with other parameters in (8). Sometimes it is more expedient to make use of (9) and (10), instead of (5) and (2).

Appendix 2.4 UCN generation in a diatomic gas converter

$$G(T_n, T_c) = (C/T_n^2)\bar{\sigma}_+(T_n, T_c, \bar{P}, k_{\lim})/[\sigma_a(k_{\lim}) + \sigma_+(T_c, \bar{P}, k_{\lim})]. \tag{1}$$

$$\sigma_+(T_c, \bar{P}, k_{\lim}) = \sum_{L, L'} \bar{P}_L \int d\sigma_+(T, L \to L', \mathbf{k}_{\lim} \to \mathbf{k}). \tag{2}$$

$$\bar{\sigma}_+(T_n, T_c, \bar{P}, k_U) = \sum_{L, L'} \bar{P}_L \bar{\sigma}_+(T_n, T_c, L \to L', k_U). \tag{3}$$

$$\bar{\sigma}_+(T_n, T_c, L \to L', k_U) = \int \exp(-k^2/T_n) d\sigma_-(T_c, L \to L', \mathbf{k} \to \mathbf{k}_U) k d\mathbf{k}/k_U d\mathbf{k}_U. \tag{4}$$

$$d\sigma_+(T, L \to L', \mathbf{k}_U \to \mathbf{k}) = \frac{2d\mathbf{k}}{k_U} \int \delta(\boldsymbol{\kappa} + \mathbf{q} - \mathbf{q}')\delta\left(\omega + \omega_{L'L} + \frac{q^2}{M} - \frac{q'^2}{M}\right) \times d\tilde{P}(T, \mathbf{q}') Q_{LL'}^2 d\mathbf{q}. \tag{5}$$

$$d\sigma_-(T, L \to L', \mathbf{k} \to \mathbf{k}_U) = \frac{2d\mathbf{k}_U}{k} \int \delta(\boldsymbol{\kappa} - \mathbf{q} + \mathbf{q}')\delta\left(\omega - \omega_{L'L} - \frac{q^2}{M} + \frac{q'^2}{M}\right) \times d\tilde{P}(T, \mathbf{q}') Q_{LL'}^2 d\mathbf{q}. \tag{6}$$

a) $\omega = k^2 - k_U^2 \simeq k^2$, b) $\boldsymbol{\kappa} = \mathbf{k} - \mathbf{k}_U \simeq \mathbf{k}$, c) $\omega_{ab} = E_a - E_b$,

d) $E_L = \Delta L(L+1)/2$. (7)

a) $d\tilde{P}(T, \mathbf{q}) = \exp(-q^2/MT) d\mathbf{q}/(\pi MT)^{3/2}$, b) $Q_{LL'}^2 \equiv B_{LL'}^2 b_{LL'}^2$. (8)

a) $B_{LL'}^2 = \sum_{S, S'} P_S^L |\langle S' | b_L^{L'} | S \rangle|^2 = \sum_S P_S^L \langle S | |b_L^{L'}|^2 | S \rangle$,

b) $b_L^{L'} \equiv b_1 + (-1)^{L+L'} b_2$. (9)

a) $b_i = b_{coh} + \boldsymbol{\sigma}\mathbf{j}_i b_{inc}$, b) $\mathbf{j}_i = \mathbf{S}_i/\sqrt{S_i(S_i+1)}$, c) $j_J(x) \approx x^J/(2J+1)!!$. (10)

$$b_{LL'}^2 = |\langle L' | \exp(i\boldsymbol{\kappa}\mathbf{R}) | L \rangle|^2 = \sum_{m, m'} \left| \sum_J j_J(\kappa d/2) A_{LL'}^J(m, m') \right|^2 \Big/ (2L+1). \tag{11}$$

$$\exp(\mathrm{i}\mathbf{kr}) = 4\pi \sum_{L,\, m} i^L j_L(kr)\, Y^*_{Lm}(\hat{\mathbf{r}})\, Y_{Lm}(\hat{\mathbf{k}}) = \sum_L i^L(2L+1) j_L(kr) p_L(\hat{\mathbf{k}}\hat{\mathbf{r}}). \tag{12}$$

$$\text{a) } A^J_{LL'}(m, m') = i^J(2J+1)\int Y_{lm}(\Omega) p_J(\cos\vartheta) Y_{l'm'}(\Omega)\mathrm{d}\Omega, \quad \text{b) } \cos\vartheta = \hat{\mathbf{k}}\hat{\mathbf{r}}. \tag{13}$$

$$H_2\text{:}\quad \text{a) } L \to S, \quad \text{b) } L + L' = 2n, \quad \text{c) } B^2_{LL'} = 4b^2_{\mathrm{coh}} + b^2_{\mathrm{inc}} S(S+1)/S_1(S_1+1). \tag{14}$$

$$\text{a) } L = 2n, L' = 2n+1, \quad \text{b) } B_{LL'} = b^2_{\mathrm{inc}}[4 - S(S+1)/S_1(S_1+1)]. \tag{15}$$

$$\mathrm{d}\sigma_+(T, L \to L', \mathbf{k}_\mathrm{U} \to \mathbf{k}) = \sigma_{LL'} b^2_{LL'}(\kappa) S_+(T, \boldsymbol{\kappa}, \omega + \omega_{L'L})\mathrm{d}\mathbf{k}/k_\mathrm{U}. \tag{16}$$

$$\text{a) } \sigma_{LL'} = 4\pi B^2_{LL'},$$

$$\text{b) } S_+(T_\mathrm{c}, \boldsymbol{\kappa}, \omega) = \frac{1}{2\pi\kappa}\sqrt{\frac{M}{4\pi T_\mathrm{c}}}\exp\left[-\left(\omega + \frac{\kappa^2}{M}\right)^2 \frac{M}{4T_\mathrm{c}\kappa^2}\right]. \tag{17}$$

$$\mathrm{d}\sigma_-(T, L \to L', \mathbf{k} \to \mathbf{k}_\mathrm{U}) = \sigma_{LL'} b^2_{LL'}(\kappa) S_-(T, \boldsymbol{\kappa}, \omega - \omega_{L'L})\mathrm{d}\mathbf{k}_\mathrm{U}/k. \tag{18}$$

$$S_-(T_\mathrm{c}, \kappa, \omega) = (1/2\pi\kappa)\sqrt{M/4\pi T_\mathrm{c}}\exp[-(\omega - \kappa^2/M)^2 M/4T_\mathrm{c}\kappa^2]. \tag{19}$$

$$H_2\text{:}\quad \text{a) } \sigma_{\mathrm{pp}} = \sigma_{00} = 4\sigma_{\mathrm{coh}}, \quad \text{b) } \sigma_{\mathrm{oo}} = \sigma_{11} = 4\sigma_{\mathrm{coh}} + (8/3)\sigma_{\mathrm{inc}},$$
$$\text{c) } \sigma_{\mathrm{po}} = \sigma_{01} = 4\sigma_{\mathrm{inc}}, \quad \text{d) } \sigma_{\mathrm{op}} = \sigma_{10} = (4/3)\sigma_{\mathrm{inc}}. \tag{20}$$

$$D_2\text{:}\quad \text{a) } \sigma_{\mathrm{po}} = \sigma_{10} = 3\sigma_{\mathrm{inc}}, \quad \text{b) } \sigma_{\mathrm{oo}} = \sigma_{00} = 4\sigma_{\mathrm{coh}} + (5/2)\sigma_{\mathrm{inc}},$$
$$\text{c) } \sigma_{\mathrm{op}} = \sigma_{01} = (3/2)\sigma_{\mathrm{inc}}, \quad \text{d) } \sigma_{\mathrm{pp}} = \sigma_{11} = 4\sigma_{\mathrm{coh}} + \sigma_{\mathrm{inc}}. \tag{21}$$

The total yield of UCN from a polyatomic gas, neglecting vibrational states, is given by (1), where $\bar{P}$ is the distribution of rotational states of the gas molecules. Up scattering is represented by expression (2) and effective up scattering by expressions (3) and (4); the differential up and down scattering are given by (5) and (6), respectively, and the notations are explained in (7)–(11). The rotational energy levels of a molecule are given by (7d), where L is the angular momentum. We assume the distribution of translational degrees of freedom of the gas molecules to be Maxwellian, eqn (8a). To calculate the spin part (9a) of the square of the matrix element averaged over the initial states and summed over the final states we have introduced the sum (9b) of the amplitudes (10a) of the individual ith nuclei, separating explicitly the coherent and incoherent parts. The σ in (10a) denote the Pauli matrices related to the neutron spin, and j is the spin operator of the nucleus S normalized to unity in accordance with (10b). To calculate the coordinate part of the square of the matrix element (11) we made use of the plane wave expansion (12), where p_L denote Legendre polynomials and j_L spherical Bessel functions with the asymptotics (10c) for small x. The angular part entering (11) is defined in (13). In the case of hydrogen molecules a single total momentum of an individual molecule, (14a), corresponds to each angular momentum, so the calculation of (9a) is straightforward. Here P^L_S stands for averaging over the projections of the spin uniquely related to the given angular momentum L. The results of calculations for transitions between rotational states with the same parity, (14b), and those with different parities, (15a), are presented in (14c) and (15b), respectively. In the case of deuterium these expressions must be averaged over different spin states, since (14a) is no longer valid. The differential up-scattering cross-section is reduced, upon evaluation of the integral in (5), to (16) where the elastic part and the scattering function are defined in (17a) and (17b), respectively. The differential down-scattering cross-section (6) is reduced to (18) with the scattering function (19). The elastic cross-sections for different transitions between para- and ortho-states are shown in (20) for hydrogen and in (21) for deuterium. The subscript 01 in (20c), for instance, indicates that the initial state (0) has an even angular momentum L, while the final state (1) has an odd angular momentum.

Appendix 2.5 UCN generation in a cold diatomic gas converter

$$\text{a) } \bar{P}_L = \delta_{L0}, \quad \text{b) } \bar{\sigma}_+(T_n, T_c, \bar{P}, k_U) = \sum_L \bar{\sigma}_+(T_n, T_c, 0 \to L, k_U). \tag{1}$$

$$\mathrm{d}\sigma_+(T, L \to 0, \mathbf{k}_U \to \mathbf{k}) = \sigma_{L0}\, j_L^2(kd/2) S_+(T_c, \boldsymbol{\kappa}, \omega + \omega_{0L})\mathrm{d}\mathbf{k}/k_U. \tag{2}$$

$$\text{a) } \sigma_{LL'} = 4\pi B_{LL'}^2, \quad \text{b) } S_+(T_c, \boldsymbol{\kappa}, \omega) = \frac{1}{2\pi\kappa}\sqrt{\frac{M}{4\pi T_c}}\exp\left[-\left(\omega + \frac{\kappa^2}{M}\right)^2 \frac{M}{4T_c\kappa^2}\right]. \tag{3}$$

$$\bar{\sigma}_+(T_n, T_c, 0 \to L, k_U) = \sigma_{0L}\int j_L^2(kd/2)\exp(-\omega/T_n) \times (2L+1)S_-(T, \boldsymbol{\kappa}, \omega + \omega_{0L})\mathrm{d}\mathbf{k}/k_U, \tag{4}$$

$$S_-(T_c, \boldsymbol{\kappa}, \omega) = (1/2\pi\kappa)\sqrt{M/4\pi T_c}\exp[-(\omega - \kappa^2/M)^2 M/4T_c\kappa^2]. \tag{5}$$

$$\text{a) } \omega \approx \kappa^2 \approx k^2, \quad \text{b) } L = 0, 1, \quad \text{c) } T_c \ll T_n. \tag{6}$$

$$\text{a) } j_l(x) = x^l\left(\frac{\mathrm{d}}{x\mathrm{d}x}\right)^l\left(\frac{\sin x}{x}\right), \quad \text{b) } j_0(x) = \frac{\sin x}{x}, \quad \text{c) } j_1(x) = \frac{x\cos x - \sin x}{x^2}. \tag{7}$$

$$f_0(x^2) = j_0^2(x) = (1 - \cos 2x)/2x^2 = 2\sum(-4x^2)^n/(2n+2)! \approx 1 - x^2/3. \tag{8}$$

$$f_1(x^2) \equiv 9j_1^2(x)/x^2 = \sum(-4x^2)^n 72(n+1)/(n+3)(2n+4)! \simeq 1 - x^2/5. \tag{9}$$

$$\text{a) } S_+(T_c, k, \omega) = (1/2\pi k)\sqrt{M/4\pi T_c}\exp(-M\omega/4\mu^2 T_c), \quad \text{b) } \mu = M/(M+1). \tag{10}$$

$$\text{a) } \sigma_+(T_c, 0, k_U) = \int \mathrm{d}\sigma_+(T_c, 0 \to 0, \mathbf{k}_U \to \mathbf{k}) = A\sigma_{00}I_0, \quad \text{b) } A = (2/k_U d^2)\sqrt{M/\pi T_c}. \tag{11}$$

$$\text{a) } I_0 = I_0(\eta) = \int \exp(-\eta y) f_0(y)\mathrm{d}y \approx (1/\eta)(1 - 1/3\eta), \quad \text{b) } \eta = M/\mu^2 T_c d^2. \tag{12}$$

$$\text{a) } \bar{\sigma}_+(0, k_U) = \bar{\sigma}_+(T_n, T_c, 0 \to 0, k_U) = A\sigma_{00}\bar{I}_0, \quad \text{b) } \bar{I}_0 = I_0(\bar{\eta}). \tag{13}$$

$$\text{a) } \bar{\eta} = (M/\bar{\mu}^2 T_c + 4/T_n)/d^2, \quad \text{b) } \bar{\mu} = M/(M-1). \tag{14}$$

$$\bar{\sigma}_+(1, k_U) = \bar{\sigma}_+(T_n, T_c, 0 \to 1, k_U) = A\sigma_{01} I_1/3. \tag{15}$$

$$\text{a) } I_1 = \int \exp(-\bar{\eta}x + \alpha - \delta/x) x f_1(x)\mathrm{d}x, \quad \text{b) } \alpha = M\Delta/2\bar{\mu}T_c,$$

$$\text{c) } \delta = (\Delta d/4)^2 M/T_c. \tag{16}$$

$$\text{a) } s \equiv \sqrt{\delta\bar{\eta}} = \tfrac{1}{2}\alpha\sqrt{1 + 4\bar{\mu}^2 T_c/MT_n}, \quad \text{b) } \bar{\eta}x = sy, \quad \text{c) } y + 1/y = 2(z+1). \tag{17}$$

$$I_1 = (s/\bar{\eta})^2 \int_0^\infty \exp(-2sz - 2s + \alpha)[y_1^2 f_1(y_1) + y_2^2 f_1(y_2)]\mathrm{d}z/\sqrt{z^2 + 2z}. \tag{18}$$

$$\text{a) } y_{1,2} = z + 1 \pm (z^2 + 2z)^{1/2}, \quad \text{b) } I_1 \simeq \sqrt{\pi/s}(s/\bar{\eta})^2 \exp(-\Delta/T_n). \tag{19}$$

$$G(T_n, T_c) = (C/T_n^2)[\bar{\sigma}_+(T_n, T_c, 0) + \bar{\sigma}_+(T_n, T_c, 1)]/[\sigma_\alpha + \sigma_+(T_c, 0)]. \tag{20}$$

$$G(T_n, T_c) = G_1(T_n, T_c)[1 + g\exp(-\Delta/T_n)\sigma_{01}/\sigma_{00}\bar{T}_c^{1/2}]. \tag{21}$$

$$g = aM\sqrt{\pi/(M-1)}\bar{\Delta}^{3/2}/6[1 + (T_c/T_n)4M/(M-1)^2]^{1/4}$$

$$\simeq \alpha M\bar{\Delta}^{3/2}[\pi/36(M-1)]^{1/2}. \tag{22}$$

$$\text{a) } a \equiv T_{300}d^2/4 = 1.7, \quad \text{b) } \bar{T} = T/T_{300}, \quad \text{c) } \bar{\Delta} = \Delta/T_{300}. \tag{23}$$

$$\text{a) } \Delta_H \simeq 0.6, \quad \text{b) } g_H \simeq 0.5, \quad \text{c) } \Delta_D \simeq 0.3, \quad \text{d) } g_D \simeq 0.2. \tag{24}$$

For the distribution (1a) the sum (3) of Appendix 2.4 reduces to (1b). The differential up-scattering cross-section is given by (2), with the notation of (3). The effective up-scattering cross-section is presented in (4) and (5). We shall adopt the approximations (6). The spherical Bessel functions given in (7) are utilized for defining the functions (8) and (9) equal to 1 for $x \to 0$. Substitution of (10a) into (2) for $L = 0$ reduces it to the form (11a), the coefficient A and the integral I_0 being defined in (11b) and (12a, b). Using (5), expression (4) can be reduced to the form (13), (14), in the case of $L = 0$, and to the form (15), (16), when $L = 1$. Changing the variables in the integral (16) in accordance with (17) transforms it to the form (18), with the notation of (19a). Using (6c) this integral can be approximated in accordance with (19b). Substituting the cross-sections into the gain function (20) we obtain (21) where the first factor is represented by function (7) of Appendix 2.3, and the factor g is given by (22); the parameters in (22) are defined in (23) and the relation (6c) is taken into account. The parameter of rotational energy levels is given by (24a) and (24c) for hydrogen and deuterium, respectively, so the numerical value of the factor given by (22) equals (24b) or (24d), respectively.

3

Detectors and spectrometers of UCN

3.1 DETECTORS

Detectors of UCN are based on the same principle as conventional neutron counters, namely they contain a substance called the radiator, the nuclei of which absorb neutrons and subsequently decay, emitting charged particles. The charged particles are then registered either by their tracks (track detectors), or by light flashes (scintillation detectors), or with the aid of electric discharges (proportional counters and solid-state detectors). The most commonly used radiators are ^{3}He (^{3}He + n = p(0.57 MeV) + t(0.19 MeV); ^{6}Li (6Li + n = t(2.7 MeV) + α(2.0 MeV)); ^{10}B (^{10}B + n = α(1.5 MeV) + ^{7}Li(0.9 MeV)); and ^{235}U, which upon capturing a neutron disintegrates into charged fragments.

However, detectors of UCN do exhibit certain features of their own: a small amount of radiator matter is sufficient for them, but it is necessary to make the windows transparent. Indeed, the cross-section of absorption by nuclei, being inversely proportional to the neutron velocity, becomes extremely large in the UCN energy region, and therefore neutrons penetrating the radiator are captured very effectively. This means, on the one hand, that for the registration of UCN a small amount of radiator matter is sufficient, and, on the other hand, that a small amount of radiator matter permits the elimination of the background arising from neutrons of higher energies, in particular, of thermal energies.

However, like any substance, the radiator, or the window of the detector, exhibits a potential barrier with respect to the neutrons, and therefore UCN of energies lower than the barrier are practically unable to penetrate the radiator, so the probability of their being captured turns out to be small. The potential barrier also reduces the detection efficiency for neutrons with energies somewhat higher than the barrier.

There are several ways of overcoming this obstacle. (i) One can decrease the coherent scattering amplitude and, consequently, the height of the potential barrier by an appropriate choice of chemical and isotopic composition of the radiator. (ii) Before registration the neutrons can be accelerated in the gravitational field (by lowering the detector). (iii) The neutrons may be heated to thermal energies (for instance, with the aid of polyethylene) and then registered by thermal neutron counters. (iv) The detector can be made to move towards the neutrons with a velocity greater than v_{lim}. This motion may be realized either in the oscillatory (Antonov *et al.* 1976*b*) or the rotational mode (Kaipov *et al.* 1979; Novopoltzev and Pokotilovskii 1980*b*).

The latter method was verified experimentally (Stoica and Strelkov 1978*a*) by making UCN pass through a rotating goffered foil (Figs 3.1 and 3.2). Two peculiarities of the obtained curves are noteworthy. The first is the presence of a maximum. It is due to the neutrons being reflected from the rear side of the fold upon passing through the foil only once. By covering the appropriate sides with polyethylene this maximum can be reduced. The second peculiarity is that at high angular velocities the transmission coefficient becomes constant and turns out to be less than unity. This fact is readily understood if one takes into account that the probability for a neutron to be knocked out of the

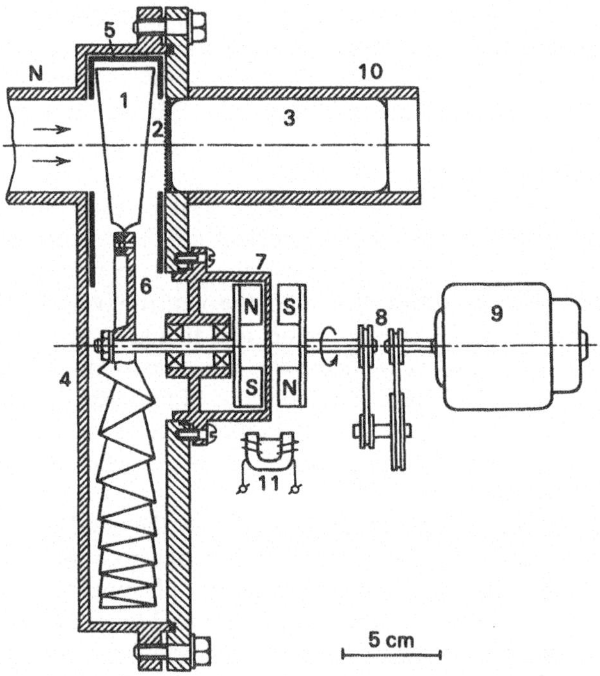

Fig. 3.1 Experimental set-up for measuring the transmission of neutrons through a rotating goffered disk (Stoica and Strelkov (1978*a*): 1, goffered foil (with an angle of 90° between the planes); 2, scintillation detector (LiOH + ZnS(Ag)); 3, photomultiplier; 4, vacuum chamber; 5, absorber (polyethylene) preventing percolation of UCN around the rotating foil; 6, disk for fastening the goffered foil; 7, magnetic dynamic clutch; 8, gearing system; 9, electric motor; 10, permalloy screen; 11, magnetic pick-up of rotation velocity.

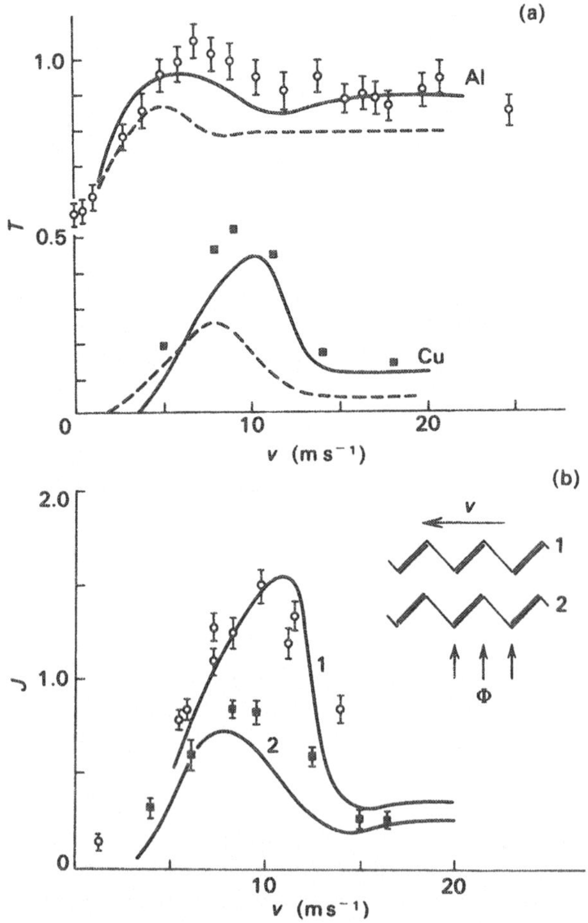

Fig. 3.2 Transmission of UCN through aluminium and copper foils as a function of the velocity of motion of the folds. (a) Without absorbers. The solid lines represent a theoretical calculation under the assumption that neutrons travel in a direction perpendicular to the disk, while the dashed lines correspond to an isotropic angular distribution. In both cases an ideal Maxellian spectrum was assumed. The maxima on the curves correspond to the case when the neutrons pass through the first face of the fold and are reflected from the second. The foil was 100 μm thick. (b) In the presence of an absorber (polyethylene) glued to the front side, 1, and to the rear side, 2, of the second face of the fold. In case 2 reflection from the second face is suppressed, and the maximum decreases significantly (the foil is of copper) (Stoica and Strelkov 1978*a*).

beam may be represented in the form $tN_0v\sigma(v)$, where t is the time the neutron is within the zone of influence of the goffers, N_0 is the average number of atoms per unit volume in this zone, v is the linear velocity of the goffered foils, and $\sigma(v)$ is the loss cross-section which is proportional to $1/v$.

The first detectors of UCN were scintillation detectors. At present gas proportional counters are used most often. In the main, we shall deal with these two types of detectors. The others will only be considered from a general point of view.

3.1.1 Scintillation detectors

Scintillation counters are the most convenient to use, but those that have been utilized up to now in UCN physics essentially possess the defect of exhibiting properties that differ significantly from sample to sample. Note was made of this in the first experiments, where the substance $LiOH.H_2O$ was utilized as a radiator of the detector with an isotopic composition of $0.9^6Li + 0.1^7Li$. It should have a low potential barrier and reflect practically no neutrons. However, the counting rates of different counters were not the same. This could indicate some reflection arising uncontrollably, and indeed such a reflection was actually observed later (Novopoltsev and Pokotilovskii 1980*b*).

One of the reasons for the reflection being enhanced may be the evaporation of crystalline water from the radiator since measurements are usually performed in a vacuum of 10^{-3}–10^{-4} mm Hg; another reason may be the inhomogeneity of the radiator film. That inhomogeneities exist can be confirmed by the effect observed with a LiF radiator (Roy 1976). This radiator has quite a high limiting velocity of 5.1 m s^{-1}; however, neutrons of lower velocities were also registered with its aid by a time-of-flight spectrometer. This could be explained by the presence of pores and tunnels in the radiator, as a result of which the slow neutrons could penetrate into the radiator and be captured there. This assumption was verified experimentally. A 1 μm layer of LiF vacuum deposited on a glass surface transmits neutrons with velocities lower than 5.1 m s^{-1}, while a thin LiF monocrystal does not let them pass. A similar anomalous transmission was also observed in boron nitrite, ^{10}BN, and in lithium glass (Bates 1978).

Usually the radiator is deposited directly on to the scintillator (ZnS(Ag), for instance), which itself represents a fine crystalline powder. This intensifies the problems of inhomogeneities. If the radiator is deposited on a thin support, for example of aluminium, the amount of inhomogeneities decreases.

Figure 3.3 shows the pulse-height spectrum obtained using a scintillation counter of UCN with a $^6Li_{0.9}{}^7Li_{0.1}OH$ (180 μg cm^{-2}) radiator deposited on an aluminium support (60 μm) and a ZnS(Ag) scintillator (8 mg cm^{-2}). In this detector the thicknesses of the radiator and the scintillator were chosen to be optimal (Novopoltzev and Pokotilovskii 1980*b*).

Reflection of UCN from the radiator can be avoided if the detector is made in the form of a rotating profiled disk (Fig. 3.4). Measurements have revealed (see Section 3.3) that such a counter exhibits an efficiency of ~ 1 throughout

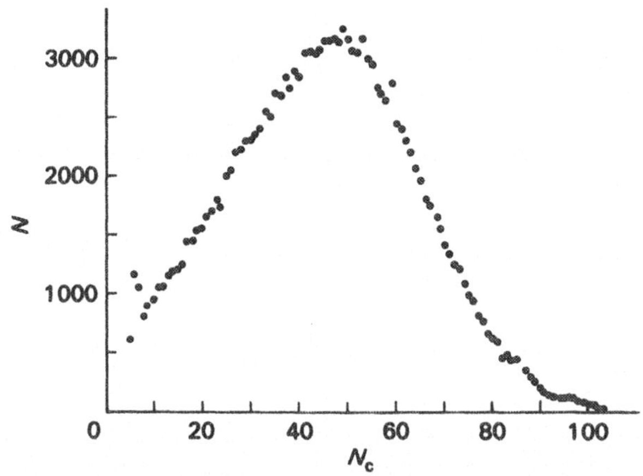

Fig. 3.3 Pulse-height spectrum from a scintillation counter in a beam of UCN. The thickness of the $^6Li_{0.9}\,^7Li_{0.1}OH$ (180 mg cm^{-2} on an Al support 60 μm thick) and that of the ZnS (Ag) scintillator (8 mg cm^{-2}) were chosen to be optimal (Novopoltzev and Pokotilovskii 1980*b*).

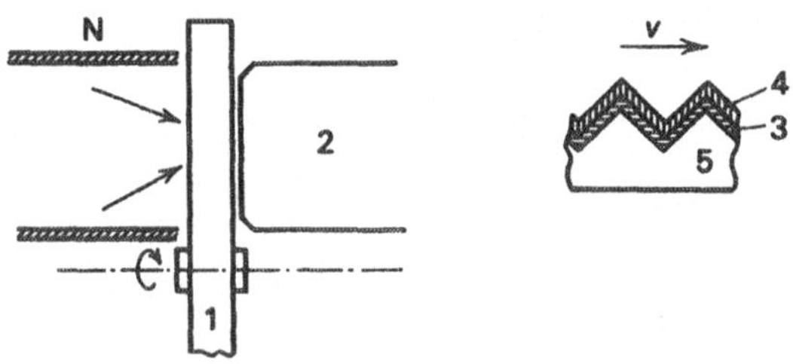

Fig. 3.4 Scheme of a scintillation detector on a rotating profiled disk of plexiglas: 1, rotating disk, the profile of the left-hand side of which is shown in the picture on the right; 2, photomultiplier; 3, scintillator (paraterphenyl); 4, LiOH radiator; 5, disk profile (Novopoltzev and Pokotilovskii 1980*b*).

the entire range of UCN energies, and if it were not for the relative complexity of its construction, it could be considered close to being perfect.

When working in the vicinity of the active zone of the reactor, or under conditions of strong activation, it is necessary to have a detector that is insensitive to β- and γ-rays. Such properties are exhibited by a detector with a radiator which releases a large amount of energy at each neutron capture event, for example ^{238}U ($Q = 160$ MeV). The coherent scattering amplitude of uranium is high, so it is necessary to dissolve it in a substance with negative amplitude, for instance titanium. Dissolution is achieved by thermodiffusion (Antonov *et al.* 1974*c*). A scintillation detector with such a radiator and a

ZnS(Ag) scintillator was investigated by Antonov *et al.* (1974*e*). Its efficiency was estimated to be ~15%.

Antonov *et al.* (1974*d*) studied a detector consisting of a cylinder made of quartz glass, one end of which is covered with a uranium–titanium radiator. Gaseous xenon was used as the scintillator. The fragments from the fission of the uranium gave rise to scintillation of the xenon, which was transferred with the aid of *p-p*-diphenylstilbene to a range convenient for the photomultiplier. The efficiency of such a detector was estimated to be ~25%.

3.1.2 A gas proportional counter (Groshev *et al.* 1974)

The gas proportional counter known as DUNya (detector of ultracold neutrons) has a simple construction and readily reproducible parameters and is reliable in operation (Fig. 3.5(a)). The detector consists of a copper or

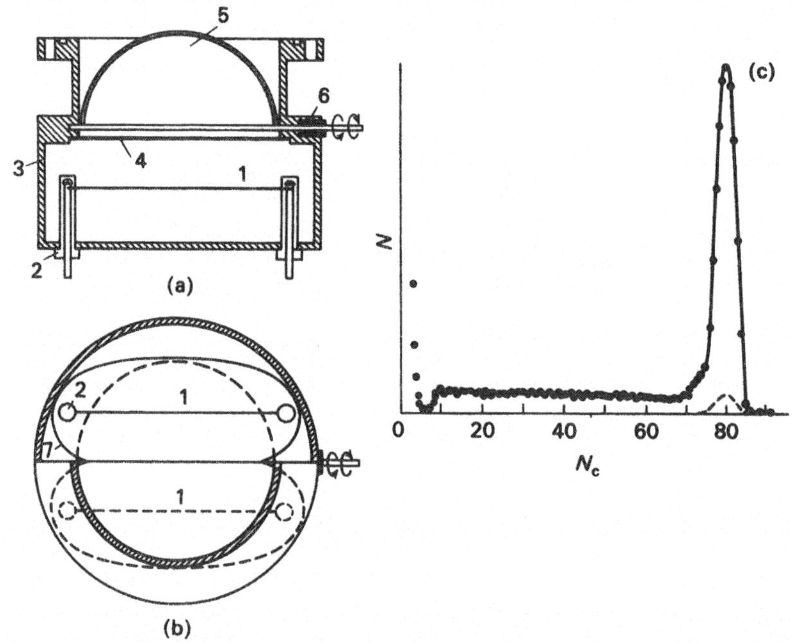

Fig. 3.5 Gas proportional detector of UCN, DUNya (Groshev *et al.* 1974): (a) vertical section; (b) horizontal plan. 1, wire (anode) 15 μm in diameter; 2, insulator; 3, body (cathode) with 17 cm internal diameter; 4, aluminium window; 5, 10 μm copper shutter; 6, gasket; 7, cathode profile; (c) pulse-height spectrum from the detector. The background amounts to about 30% of the total area under the curve and corresponds to the capture of a neutron in the vicinity of the wall; the spectrum was obtained in a UCN beam; the dashed line indicates the spectrum obtained with a closed shutter.

stainless steel cylinder divided along the diameter into two equal parts, each of which has its own anode wire. Thus, in a single body there are combined two independent detectors (single-chamber and many-chamber detectors also exist). One end of the cylinder is covered with 60 μm thick aluminium foil of total area 60 cm^2 serving as the window of the detector. In front of the window there is a copper shutter. It is capable of closing one or the other half of the window leading to one of the respective chambers of the detector. We note that the UCN flux density was in many cases defined as the ratio of the UCN counting rate to the total area of the detector window.

The gas filling the counter is a mixture of 5 mm Hg each of ^{3}He, CO_2, and argon at atmospheric pressure. The partial pressures of the individual gases were optimized. The amount of argon was chosen so as to minimize the number of charged particles that did not succeed in losing their energy through ionization (the wall effect). The amount of argon, however, must not be too large, otherwise the processes of UCN heating on its nuclei will compete with the capture processes in helium. The probability of heating in the mixture amounts to $\sim$4%. As usual, the CO_2 admixture serves to quench the discharge.

The profile of the walls inside the chambers of the detector was chosen to provide for the best possible equalization of electric field gradients along the wire. The pulse-height spectrum of DUNya is shown in Fig. 3.5(c). It exhibits a single narrow peak corresponding to the total ionization of the proton and the triton and occurring far from the noise signals of small amplitudes (compare Fig. 3.3).

If no measures are taken to enhance its transparency, the transmission of the window of DUNya, averaged over the initial part of the Maxwellian spectrum and over the isotropic angular distribution (the relevant formulae are presented in Appendix 3.2), only amounts to 40% when the upper end of the UCN spectrum coincides with the limiting energy of copper. To enhance window transparency the detector is lowered by $\sim$50 cm, so that the energies of all the neutrons become higher than the limiting energy of aluminium. One may also goffer the walls of the detector and make it rotate.

In proportional counters other gases are also used as radiators, for example BF_3. The release of energy in this gas is higher than in helium, and the ionization provided by a single particle is sometimes sufficient to achieve signals that are significantly higher than the noise. If the partial pressure of ^{10}B is large, then the UCN are mainly captured near the wall. In this case, the greater the energy of a neutron, the farther from the wall it is captured and the higher the signal. This makes it possible to measure the UCN spectrum.

Anikolenko *et al.* (1973*a*) described a solid-state counter in which ^{3}He serves as the radiator, but the charged particles are registered by two surface-barrier solid-state detectors situated in the ^{3}He at 0.5 cm from each other. This counter exhibits a comparatively low efficiency of $\sim$13%.

3.1.3 Other types of detectors

An activation detector was used by Robson and Winfeld (1972). This detector consists of an 88% Mn–12% Ni alloy foil 0.13 mm thick. Absorption of neutrons by the manganese nuclei induces β-activity which is measured by a scintillation counter of β-particles. The efficiency of this detector is estimated to be 29%.

A track detector consists of a support of mica, or glass, or some other transparent dielectric, on to which the radiator is deposited, for instance a uranium–titanium radiator (Antonov *et al.* 1974*b*). The fission fragments leave tracks in the support (in Antonov *et al.* (1974*b*) mica served as the support) that, when the mica is treated with hydrofluoric acid, can readily be counted in a microscope. The sensitivity threshold of mica relative to the charge of the registered particle corresponds to ^{20}Ne nuclei, and so this detector is insensitive to the α-, β-, and γ-background. Its efficiency as measured in a beam of UCN was 50%. A track detector provides a clear image of the distribution of neutrons in a beam, and it will most likely be widely applied in experiments when the neutron image of an object is required (see, for instance, Kashukeev and Chikov (1979*a*)). A neutron image may also be obtained in another way, for example with the aid of a detector involving a radiator, a scintillator, and a photographic film (Bates 1981).

It is now appropriate to consider the measurement of the spectral efficiency of a detector. Before we do so it is necessary to become familiar with the methods of measuring UCN spectra.

3.2 SPECTROMETERS

The region of UCN in the ideal Maxwellian spectrum (see the shaded part of Fig. 2.1) can be blown up and represented by the straight line 1 in Fig. 3.6. The flux density within the interval dE is

$$d\Phi = \Phi'(E)dE, \qquad \Phi'(E) = E\theta(E < E_{\text{lim}}).$$

When the converter has a potential barrier u_c, the spectrum must be cut off at energies less than u_c. Thus, an ideal UCN spectrum is trapeziform: $\Phi'(E) = E\theta(u_c < E < E_{\text{lim}})$. A real spectrum (for definiteness we shall consider the spectrum at the end of a long horizontal channel) can qualitatively be represented by the curve in Fig. 3.6. The lower part of the spectrum lacks neutrons as a result of reflection from the interface of the converter. The upper part, on the one hand, lacks neutrons because losses along the neutron guide due to collisions with the walls are higher when their energy is higher (see Appendix 1.1), but on the other hand it is enriched with over-limiting neutrons which have a certain probability of percolating through all the filters that happen to be in their way. Now we shall see what is actually

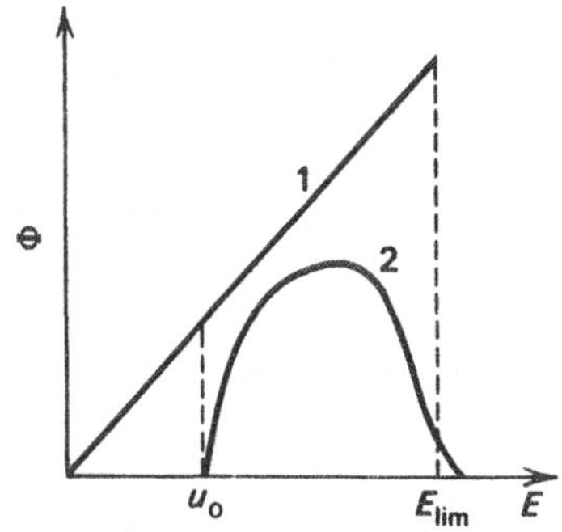

Fig. 3.6 UCN spectrum: 1, part of the ideal Maxwellian spectrum. When the converter has a limiting energy u_c, the spectrum is cut off below u_c; 2, qualitative shape of real spectrum.

obtained by measurements, and we shall consider the methods of measuring spectra.

3.2.1 Time-of-flight spectrometers

The most reliable information is obtained with time-of-flight spectrometers (Fig. 3.3) which, in principle, operate nearly in the same way as the corresponding devices applied in the region of thermal neutrons. In most cases they are utilized within the energy range extending far beyond UCN energies. These spectrometers may have vertical or horizontal time-of-flight bases. In the case of a vertical base it is necessary to take into account the influence of gravity on the time-of-flight spectrum. The time-of-flight base for the UCN themselves may be relatively short. In Novopoltsev and Pokotilovskii (1981) it varied from 15 to 100 cm. It is interesting to see whether the time-of-flight spectrum changes when the length of the base changes. Since a neutron travels inside the neutron guide, each of its collisions with the wall will lead with a certain probability to non-specular reflection. The higher this probability, the stronger the variation of the time-of-flight spectrum must be as the length of the flight base increases. However, no changes within the standard experimental uncertainties were observed.

To derive the spectrum from the time-of-flight measurements, it is necessary to collimate the neutron beam. Indeed, the spectrum of a neutron flux is described by the expression

$$d\Phi = v_{\parallel} dv_{\parallel} \int f(v^2) d^2 v_{\perp},$$

where the integral is taken over the velocity components $v_{\perp}$, perpendicular to the flight base, and the time-of-flight spectrum actually determines the distribution of the neutron velocities $v_{\parallel}$ along the flight base. The scope of spectrometry consists in determining the function $f(v^2)$ which, for simplicity, we shall consider to be isotropic: $f(v^2) = f(v_\parallel^2 + v_\perp^2)$. Collimation of the beam signifies that $v_\perp \ll v_\parallel$, i.e. $f(v^2) \approx f(v_\parallel^2)$ and

$$d\Phi \approx f(v_\parallel^2)v_\parallel dv_\parallel \int d^2 v_\perp = Cf(v_\parallel^2)v_\parallel dv_\parallel,$$

where C is a constant. When there is no collimation the components $v_\perp$ are not small compared with $v_\parallel$, and to find the function $f(v^2)$ from the time-of-flight spectra may turn out to be quite a difficult task.

In the spectrometry of UCN, as well as in the spectrometry of thermal neutrons, various devices can be applied to enhance intensity and the signal-to-noise ratio. Fourier spectrometry (*Poyry *et al.* 1975), of which the correlation technique can be considered a special case, is becoming more and more popular nowadays. The essence of Fourier spectrometry is as follows. In the way of the neutrons a chopper wheel is established with periodically situated peripheral slots. The rotation frequency of the wheel changes gradually. At each given moment of time the detector, by registering all the neutrons arriving, actually measures the Fourier component of the time-of-flight spectrum. Since the frequency of the Fourier component is known at each moment of time, by multiplying a periodic function of this frequency by the measured Fourier component and performing a summation over all the frequencies we can obtain the time-of-flight spectrum being measured (see Appendix 3.1). Multiplication and summation are performed automatically by a computer during the experiment, one of the most difficult tasks in the case of measurements with thermal neutrons being to provide the appropriate computer speed. In the case of UCN these problems are significantly less complicated, and therefore Fourier spectrometry will, most likely, continue to develop here. So far, however, no work has been carried out in this field, probably because Fourier spectrometry is itself a novel technique.

The correlation technique has been known for a long time, although it is now clear that it only represents a special case, and not the most favourable one, of Fourier spectrometry. Indeed, Fourier spectrometry applies periodic functions, which permits the simultaneous transmission of a neutron beam through several periodically situated slots of the chopper wheel. However, the correlation technique makes use of non-periodic functions (see Appendix 3.1) and therefore the neutron beam is transmitted through a single slot. Thus, for a given resolution Fourier spectrometry has an advantage in intensity. In UCN physics the correlation technique has been given successful trials, and with its aid not only has the spectrum of the primary neutron beam been measured, but also a resonance phenomenon such as the transmission of UCN through a multilayer system (Novopoltzev and Pokotilovskii 1981). Although correlation spectrometry is inferior to Fourier spectrometry, in certain cases it nevertheless exhibits, in comparison with the conventional time-of-flight technique, the advantage of enhancing the utilization of the primary beam. For UCN with their low intensity this advantage may prove to be decisive.

The layout of a time-of-flight correlation spectrometer is presented in Fig. 3.7 (Novopoltzev and Pokotilovskii 1981). The flight base is a segment of a specular neutron guide and the detector is a scintillation counter made in the form of a rotating profiled disk (see Fig. 3.4). The latter is extremely important, since such a detector reflects no neutrons and does not, therefore, deform the spectrum (Fig. 3.8).

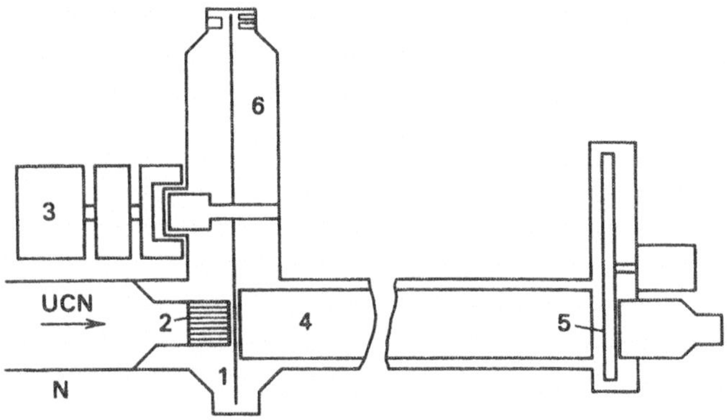

Fig. 3.7 Layout of correlation spectrometer (Novopoltzev and Pokotilovskii 1981): 1, chopper with $5 \times 50\ \mathrm{mm}^2$ window consists of steel bands 0.2 mm thick extended between the disks and the external rings. The period of rotation is 127 ms, the channel length is 1 ms, the duty cycle is 0.5; 2, multislit polyethylene $5 \times 20\ \mathrm{mm}^2$ collimator; 3, system for rotation of chopper; 4, specular neutron guide with $25 \times 50\ \mathrm{mm}^2$ cross-section, made of float glass spray-coated with a thin layer of nickel; 5, the detector (Fig. 3.4); 6, vacuum case.

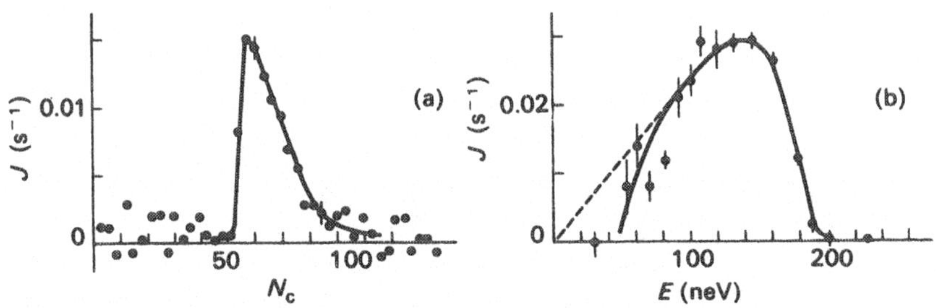

Fig. 3.8 Differential UCN spectrum at the end of the neutron guide at the SM-2 reactor (Dimitrovgrad), measured with a correlation spectrometer with a time-of-flight base of 31 cm: (a) from the time-of-flight spectrum; (b) upon recalculation taking into account the energy spectrum (Novopoltzev and Pokotilovskii 1981).

3.2.2 Gravitational spectrometers

In the spectrometry of UCN a most simple spectrometer (a crank) (Fig. 3.9) is often utilized. It actually consists of a filter that cuts off from the detector neutrons of energy E lower than the height of the gravitational barrier mgH. By measuring the counting rate of the detector as a function of the height of the crank H we determine the total number of neutrons with energies higher than mgH. The resulting curve (Fig. 3.10) is known as the integral spectrum. When the crank is lower by H, the neutrons inside it acquire an additional energy of mgH, and those neutrons with energies higher than $E_{\text{lim}} - mgH$, where E_{lim} corresponds to the potential barrier of the walls of the neutron guide, turn out to be over-limiting and are able to leave the neutron guide through its walls. In this case the detector counts the total number of neutrons with energies lower than $E_{\text{lim}} - mgH$.

The integral spectrum shown in Fig. 3.10 was measured with a horizontal channel (Kosvintsev *et al.* 1977*a*) partitioned by an aluminium foil. For this reason the UCN spectrum contains no neutrons with energies lower than the limiting energy of aluminium, E_{Al}. Correspondingly, the integral spectrum exhibits a plateau extending from zero to E_{Al}. The solid line in Fig. 3.10 is the integral spectrum calculated for the interval of the perfect Maxwellian spectrum $\Phi'(E) = E\theta(E_{\text{Al}} < E < E_{\text{Cu}})$ (the upper limit equals E_{Cu}, because

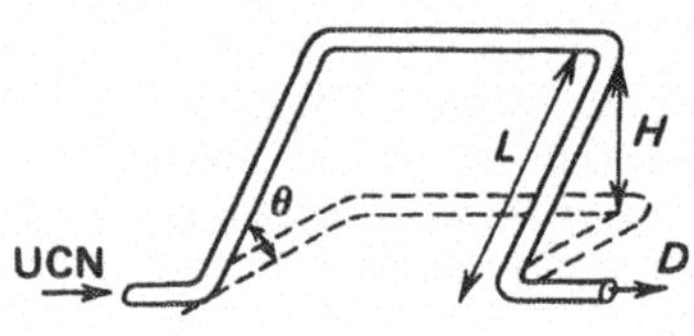

Fig. 3.9 Integral gravitational spectrometer in the form of a crank.

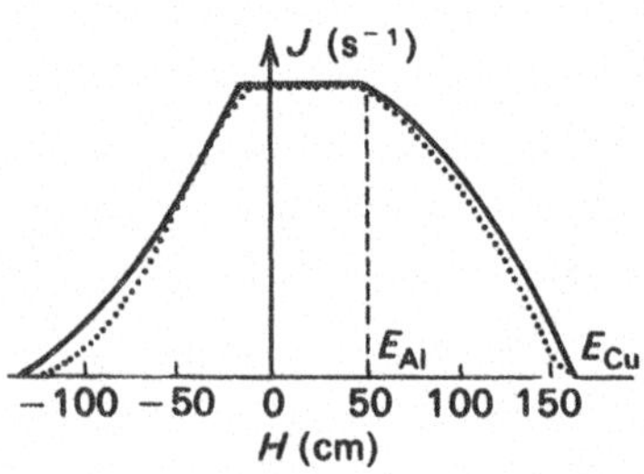

Fig. 3.10 Integral UCN spectrum measured at the output end of the horizontal channel of the SM-2 reactor (Dimitrovgrad) with the aid of the spectrometer presented in Fig. 3.9. The solid line corresponds to the theoretical calculation, and the points are experimental (Kosvintsev *et al.* 1977*a*).

in the above case the neutron guide was made of copper). The plateau in Fig. 3.10 also extends a little into the region of negative height H. This is because the material, of which the crank is constructed (stainless steel), has a somewhat higher limiting velocity than the main neutron guide.

Integral spectra obtained with the aid of a gravitational spectrometer usually have a form similar to the one shown in Fig. 3.10 which is in qualitative agreement with the theoretically expected spectrum (see Fig. 3.6). Sometimes, however, anomalies are encountered. Twice spectra were obtained which exhibited two plateaux separated by a small but well-identified sharp drop (Fig. 3.11) (Kosvintsev *et al.* 1980*d*). These spectra were obtained with the horizontal channel of the SM-2 reactor (Dimitrovgrad) involving a zirconium hydride converter after the operation of the channel for a long period (1.5 years). The neutron guide was separated from the head of the channel by an aluminium foil, and the drop occurred near the energy E_{Al}. This drop corresponds to the narrow monoline near E_{Al} in the differential spectrum. It is interesting to note that the spectrum presented in Fig. 3.8, which was measured by applying the correlation method, also exhibits an indication of the existence of such a monoline. Its source still remains unclear. The only reasonable explanation could be the sub-limiting generation of UCN beyond the aluminium window. But since this phenomenon was not studied and was not observed further, the problem still remains.

In some cases a rise in the integral curve, instead of a plateau, is observed when the crank is raised. This phenomenon may be due to the presence in the spectrum of a significant fraction of over-limiting neutrons, for which passage to the detector is facilitated when the crank is elevated.

The normal form of the integral spectrum, in particular the presence of a plateau in it, seems at first glance to confirm our expectation of the spectrometer operation. A more detailed investigation, however, has revealed an-

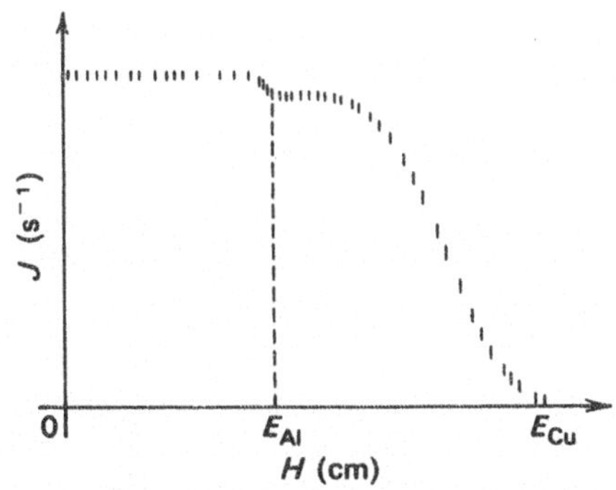

Fig. 3.11 Integral UCN spectrum measured at the SM-2 reactor after the channel was operated for 1.5 years (Kosvintsev *et al.* 1980*d*).

other peculiarity which of course must be taken into account. To understand it, imagine that at the entrance to the spectrometer (see Fig. 3.9) there are incident monoenergetic neutrons of energy E_0, i.e. their spectrum has the form

$$d\Phi = Cv\delta(E - E_0)d^3v,$$

where C is a normalizing constant. In accordance with the picture above the integral spectrum in this case should represent a step exhibiting a sharp drop at $H = E_0/mg$. However, as the crank is raised to the height H, the neutron density in the upper bar decreases, for otherwise it would be possible to cool all the neutrons at the bottom by decreasing the energy by mgH; but, in accordance with the Liouville theorem, this is impossible. Let us clarify the above statement.

Assume the neutrons to be moving upwards along a vertical neutron guide with perfectly smooth walls. Then the flux of neutrons arriving at the height H is

$$\Phi(H) = \int v_z d^3v C\delta(E - E_0)\theta(v_z^2 > 2gH), \tag{3.1}$$

where the θ function is unity if the inequality in its argument is satisfied, and zero otherwise. We shall denote v' by $(v^2 - 2gH)^{1/2}$. Since $v_z dv_z = v'_z dv'_z$, and the components perpendicular to the z axis do not change, expression (3.1) can be represented in the form

$$\Phi(H) = \int v'_z d^3v' C\delta(E' - E_0 + mgH), \qquad E' = E - mgH. \tag{3.2}$$

Note that, as follows from (3.2), the phase density $C\delta(E - E_0)$ at the height H is equal to the phase density at zero height (just as it should be, in accordance with the Liouville theorem), but the volume in the phase space which the particles are allowed to occupy at the height H is smaller than the volume at lower levels ($v_0'^3 < v_0^3$), and therefore the spatial density $\rho = (4\pi/3)Cv_0'^3$ is also lower. Note also that, in accordance with (3.2), if the angular distribution at the bottom is isotropic, it is also isotropic at the top. Hence it follows that, if the detector were in the upper bar of the crank, the flux incident upon it would be $v\rho/4$ and, thus, the counting rate of the detector would decrease with enhancement of the height, even if $H < E_0/mg$. So, if all the neutrons coming over the bar are registered by the detector after the crank, then the integral spectrum of the spectrometer in Fig. 3.9 should exhibit no plateau at all, yet the experiment reveals that it actually exists.

The above fact may be understood if the flux through the upper bar is assumed to exceed the number of neutrons captured by the detector, as a result of which UCN are accumulated in the vicinity of the detector and their density becomes proportional to the density at the input and is independent of the height of the crank within a wide range of values.

This assertion may be verified as follows. Let us partition the upper bar of the crank using a diaphragm. Then the flux of UCN toward the detector will decrease and the detector will be able to register the neutrons arriving before they are accumulated. When accumulation of UCN in the bar of the crank occurs, quite intense fluxes will exist both from the left to the right and vice versa, and it is only the small difference between these two fluxes that the detector registers. In the presence of the diaphragm no accumulation occurs and no flux is directed away from the detector, so the integral spectrum will acquire a form corresponding to the detector situated right inside the upper part of the crank. If another diaphragm is now set up just in front of the detector, then the reflection from the detector will be enhanced and the detection rate will decrease.

As a result, conditions for accumulation will arise once again, and the shape of the integral spectrum measured by the spectrometer will change (Fig. 3.12). Thus, the integral spectrum curve produced by the detector depends significantly on the conditions of measurement. For this reason one must distinguish between the integral spectrum obtained in the flow mode and the one obtained in the accumulation mode. In the flow mode the counting rate of the detector must decrease steadily as H increases. For example, the point E_{Al} of the spectrum in Fig. 3.10 is singular only to the extent that the fall in the counting rate of the detector becomes more rapid

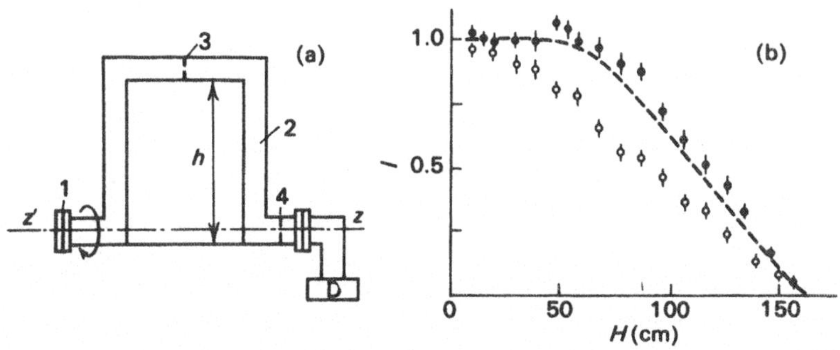

Fig. 3.12 Change of shape of the integral spectrum when passing from the flow mode to the accumulation mode. (a) Layout of experiment by Kosvintsev *et al.* (1977*a*): 1, flange opening to the main neutron guide; 2, crank of the spectrometer; 3, internal diaphragm of the crank; 4, diaphragm in front of the detector. (b) Integral spectra measured with (upper points) and without (lower points) diaphragm 4.

with the enhancement of H. For an interval of an ideal Maxwellian spectrum it is also possible to perform analytical computations.

Consider a spectrum of the form

$$\mathrm{d}\Phi = CE\mathrm{d}E\theta(E_{\mathrm{Al}} < E < E_{\mathrm{Cu}}). \tag{3.3}$$

The counting rate of the detector in the flow mode will then be proportional to the following, as the upper part of the crank is raised to the height H:

$$(E_{\mathrm{Cu}} - mgH)^2 - [\max(E_{\mathrm{Al}} - mgH, 0)]^2, \tag{3.4}$$

while in the accumulation mode it is proportional to

$$E_{\mathrm{Cu}}^2 - [\max(E_{\mathrm{Al}}, mgH)]^2.$$

In a real experiment one never knows in which mode the gravitational spectrometer actually operates (what is absolutely clear is that this mode is always intermediate); thus, there always exists some uncertainty in the interpretation of integral spectra. We shall sometimes encounter below accumulation similar to the one just dealt with, and it always has to be taken into account.

All the above may give the impression that a gravitational spectrometer is of no use for the measurement of spectra, since interpretation of the results is too difficult. Recently, however, significant progress was achieved in the theory of a gravitational spectrometer (Ignatovich and Nikitenko 1988). When the problem of diffusion of UCN in a gravitational field was solved (Ignatovich 1987*a*) it became possible to follow analytically the dependence of the spectrometer's readout both on the spectrum being measured and on the actual construction of the spectrometer. Figure 3.13 demonstrates the computed integral curve for various parameters of the spectrometer in the case of a truncated Maxwellian spectrum (see Fig. 3.6) at the input. It can be seen that the theory reflects all the peculiarities of the integral curves that are encountered in the experiment. The main points of the theory of a gravitational spectrometer will be presented in the next chapter, since it is necessary first to introduce the concept of diffusion of UCN along neutron guides.

A somewhat different kind of integral spectrometer is a storage vessel into which a polyethylene piston is inserted that can slide from the top (Fig. 3.14). If the height of the piston is h, then only neutrons with energies from zero up to mgh can accumulate in the vessel. The operation of this spectrometer is similar to that of the crank that is being lowered. The similarity becomes even more complete if the neutrons are made to pass through the vessel in a continuous flow (Kosvintsev *et al.* 1977*a*).

The spectrometer can be made into a differential spectrometer if, for a given position of the piston, one alters the position of the vessel itself, H. In this case the resolution of the spectrometer is $\Delta E = mgh$. If, contrariwise, it

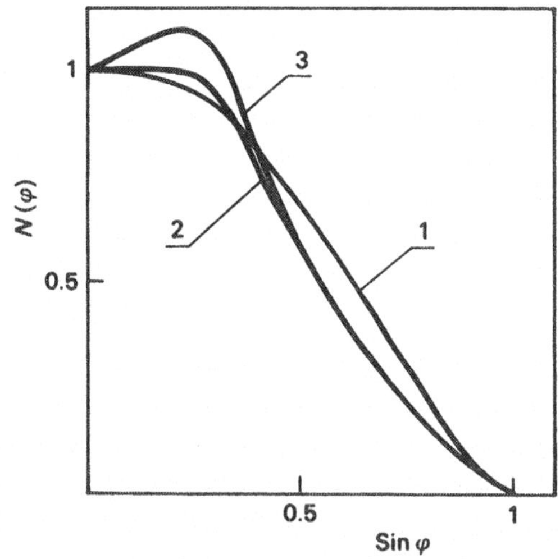

Fig. 3.13 Spectral curves calculated (Ignatovich and Nikitenko 1987*b*) for various lengths of the sections of the crank of the gravitational spectrometer. 1, curve without plateau (flow mode version); 2, curve with plateau (accumulation mode); 3, curve with peak for the spectrometer with strong absorption; φ is the turn angle of the crank from the horizontal plane. All curves are normalized to unity for the horizontal position.

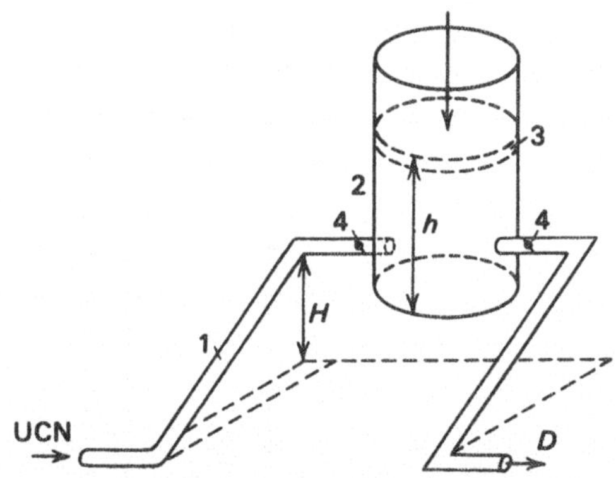

Fig. 3.14 Schematic picture of a differential spectrometer involving a crank, 1; storage vessel, 2; and polyethylene piston, 3. The valves, 4, allow filling and emptying of the vessel. When the valves are closed, the neutrons are kept in the vessel for some time, so that those neutrons with an energy higher than mgH have time to be absorbed in the polyethylene. When the valves are open, the spectrometer operates in the flow mode. The integral spectrum of neutrons present in the vessel is measured for a given height H and for various positions of the piston h.

operates in the continuous flow mode, its resolution turns out to be somewhat worse than mgh, since neutrons with energies higher than mgh can pass through it without being completely absorbed in the polyethylene during their stay in the vessel (see, for example, Kosvintsev *et al.* (1980*d*) and Dvoretskii (1976*a*)). A spectrometer that operates in a similar manner consists of two cranks connected in series (see Fig. 3.9), one of which is lowered (say, by H_1) while the other is raised (by H_2). Neutrons of energies within the limits $mgH_2 < E < E_{\text{lim}} - mgH_1$ can pass through such a system. By varying H_1 and H_2 so that $H_1 + H_2$ remains constant, one can measure the distribution of neutrons throughout the entire spectrum with a resolution $\Delta E = E_{\text{lim}} - mg(H_1 + H_2)$. Such a spectrometer was investigated by Akhmetov *et al.* (1976). The differential spectrum obtained is in good agreement with the expected one and with the spectrum shown in Fig. 3.8.

A somewhat improved differential spectrometer may be represented by the construction shown in Fig. 3.15. Here both a crank and a storage chamber are utilized. The resolution of this spectrometer is $\Delta E = E_B - mgH$, where E_B is the limiting energy of the storage chamber, and the advantage of this construction as compared with the spectrometer shown in Fig. 3.14 is that, when the storage vessel is lowered, the number of neutrons accumulated in it should be higher. However, this spectrometer has not yet been tested.

3.2.3 Other types of spectrometers

Other types of spectrometers differ from the gravitational spectrometer shown in Fig. 3.9 only in that, instead of the gravitational barrier, other types of barriers are used. An example may be provided by the magnetic spectrometer described by Kosvintsev *et al.* (1980*b*) (Fig. 3.16). The magnetic field can be varied continuously and no complicated mechanical displacements are

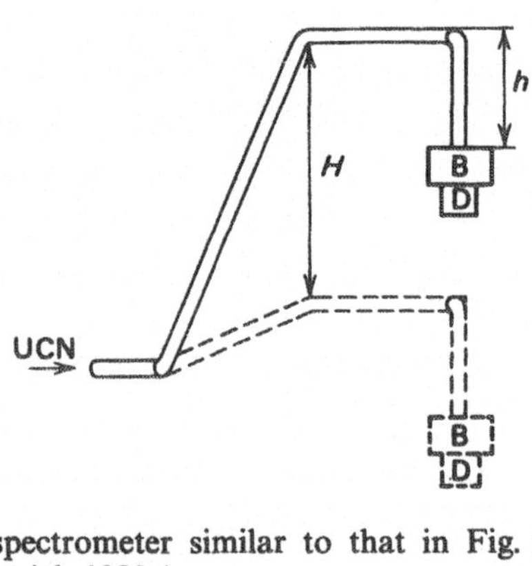

Fig. 3.15 Differential spectrometer similar to that in Fig. 3.14, but with lowered storage vessel B (Ignatovich 1980*a*).

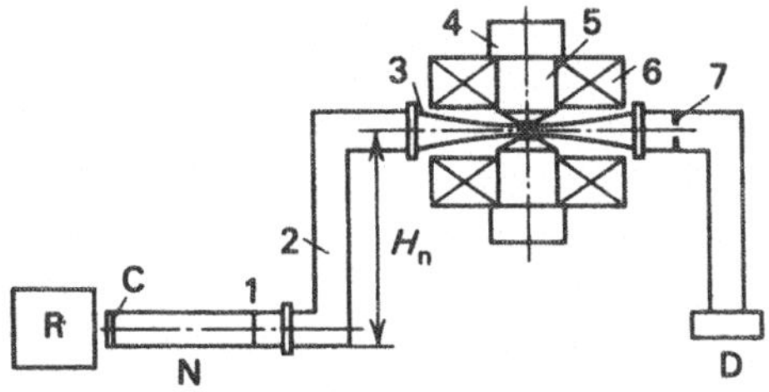

Fig. 3.16 Magnetic integral spectrometer (Kosvintsev *et al.* 1980*b*): 1, aluminium membrane; 2, turnable crank; 3, neutron guide of the spectrometer inside the magnet (electropolished copper of cross-section $1 \times 8\ \mathrm{cm}^2$); 4, yoke of magnet; 5, poles of the electromagnet (allowing establishment of fields up to 2.6 T); 6, magnet coil; 7, diaphragm.

required. The magnetic field, however, acts as a barrier only for neutrons of a single polarization. This makes it necessary either to decrease the neutron intensity by a factor of 2, making use of only polarized neutrons, or to deal with a comparatively high background. It is still not clear how depolarization processes influence the results yielded by the spectrometer. Evidently, the magnetic spectrometer can serve as a good tool for studying depolarization processes. The integral spectra measured are in reasonable agreement with theoretical expectations.

To measure integral spectra one can also make use of films with potential barriers of various heights. If the films are situated directly at the detector, then such a spectrometer will operate in the flow mode, while if it is not close to the detector the accumulation mode is realized. Measurement of the shutter difference, which is utilized for the estimation of the total number of UCN in the neutron flux, is also a measurement of the spectrum. It is necessary, however, to bear in mind that the shutter difference provides a correct idea of the spectrum only for neutrons inside the storage vessel or at the output of a long neutron guide with rough walls or with a large number of bends. In the direct beam the shutter difference yields an enhanced number of neutrons with energies lower than the potential barrier of the shutter, u_{sh}, since for the neutrons to be reflected it is necessary that the energy of their normal motion be less than u_{sh}, while their total energy may be higher.

A spectrometer may combine the various elements considered above. For example, in the magnetic spectrometer both magnetic and gravitational fields are used simultaneously. One may also use the gravitational barrier and the potential of a foil simultaneously.

We have not yet mentioned the spectrometers used for specific experiments, for example the time focusing spectrometer FOTOF by Gahler and Golub (1984). They can have very good resolution. But it is better to refer to them in connection with particular experiments. In some of these spectrometers properties of free fall parabola are used. For example, if you have well collimated UCN beam directed at 45° with respect to the horizon, then particles with different velocities will fall at different distances from the source. So, their distribution along the horizontal line gives a spectrum of UCN in the source (see Kawabata *et al.* (1986), Utsuro (1983) and Steyerl *et al.* (1983)).

Now we turn to the issue of measuring the spectral efficiency of detectors.

3.3 MEASURING THE SPECTRAL EFFICIENCY OF DETECTORS

First of all it is necessary to come to an agreement on the definition of the efficiency of a detector. The point is that in the case of UCN the probability for the detector to register a neutron depends on the actual experimental set-up. Thus, for example, if the detector is situated in the neutron beam so that each individual neutron may undergo interaction with it only once, then a certain registration probability is achieved, while if the detector is attached to the output of the neutron guide, the neutron, upon leaving the detector once, still has the possibility of interacting with it several times, so the registration probability will be enhanced. Finally, if the detector is attached to the output of the storage chamber, the neutron, with its sufficiently long lifetime and high energy, will be registered sooner or later with a probability equal to unity. To avoid uncertainty in what follows, the efficiency of the detector will be understood to be the registration probability of a neutron interacting with the detector a single time. To what extent this concept of efficiency conforms to the registration efficiency in each experiment is up to readers to decide.

Evidently, the absolute registration efficiency can only be determined theoretically. In the case of the DUNya detector, for example, it is made up of the probability that UCN penetrate the window, $\bar{T}$, and the absorption probability inside the chamber. The probability of penetration through the window depends on the angular distribution of the incident neutrons. Since the angular distribution, generally speaking, is unknown, it is taken to be isotropic in theoretical computations. It turns out that the probability of a neutron penetrating into the chamber through an aluminium window 100 μm thick increases rapidly from zero at $E = E_{Al}$ to $\bar{T} = 60\%$ at $E = E_{Cu}$. When the neutron travels along a path strictly perpendicular to the window, the probability of penetration through a thin aluminium window for $E = E_{Cu}$ amounts to 97%. If the angular distribution is assumed to be isotropic and the form of the spectrum is considered identical to that part of the Maxwellian spectrum between the limiting energies of aluminium and copper, the penetration probability averaged over the entire spectrum only amounts to 40%. To enhance the efficiency of DUNya the detector must be lowered so that the neutrons being registered by it will be accelerated by gravity and will penetrate the window with a significantly higher probability. Truly, there may arise another problem here: the accelerated neutrons may leave the neutron guide just before the detector, if the angle of incidence on its walls is smaller than the angle of total reflection.

Having entered the chamber a neutron should be absorbed with a probability close to unity, since the amount of ^3He is quite sufficient for this to happen. However, the neutron may, with a certain probability, be heated on nuclei of the inert argon gas, too. In the case of DUNya the probability is 4%. Thus, the total efficiency of the DUNya detector may be estimated to be $0.96\bar{T}$, where $\bar{T}$ is the probability of penetration through the window (see Appendix 3.2).

The efficiency of a rotating scintillation counter with a profiled surface is mainly determined by the probability of scintillations occurring and being registered, since the energy barrier of the radiator and the probability of heating (in the case of LiF it amounts to some parts of a per cent) may be neglected. Certainly, as the energy of the incident neutron increases, its velocity with respect to the moving material of the radiator may turn out to be very high, and then the probability of capture may diminish slightly. Experiments reveal that such an effect actually does occur, but on the whole the efficiency of such a detector may be considered close to unity and constant within the entire range of UCN energies.

Only the relative efficiency can be measured; however, here also certain difficulties are encountered. For example, if two detectors are connected, in turn, to the output of the neutron guide, the first one having a higher efficiency and the second a lower one as a result of reflection, the efficiency of the second detector measured by the counting rate with respect to the first detector will be enhanced owing to the accumulation effect. If, on the other hand, the detectors are connected to the neutron guide simultaneously, then the neutrons reflected from the second detector will be registered by the first detector, as a result of which the relative efficiency of the second detector may turn out to be underestimated. However, if the two detectors are absolutely symmetrically connected in parallel, then reflection from one of them effectively enhances the total UCN flux, so, since the enhancement of the flux is the same for both detectors, the ratio between their counting rates really does provide information on the relative efficiency for a single interaction of a neutron with the counters. In the absence of symmetry the ratio between the counting rates may differ slightly from the relative efficiency. At any rate, the above arguments are only of a qualitative nature, since none of the effects has been investigated thoroughly enough.

Measurement of the relative efficiency of various detectors was performed with the experimental layout shown in Fig. 3.17 (Novopoltsev and Pokotilovskii 1980*b*). The UCN flux was measured in parallel by two detectors, one of which served as the monitor, and its efficiency was taken to

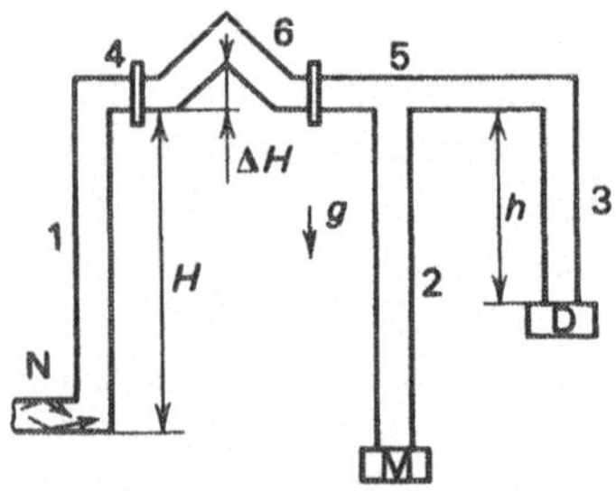

Fig. 3.17 Experimental layout for measurement of the relative efficiencies of various detectors in dependence on the UCN energy. M, monitor (DUNya); D, detector under investigation; 1, 2, 3, neutron guides for slowing and accelerating neutrons in the gravitational field; 4, horizontal part of neutron guide; 5, flux splitter; 6, turning crank. The primary UCN spectrum is determined by the interval $E_1 < E < E_2$. Along section 5 the spectrum is within the limits $0 < E < E_2 - mgH$ (H is chosen to be such that $mgH > E_1$) if the crank 6 is lowered, and within the limits $mg\Delta H < E < E_2 - mgH$ if crank 6 is raised vertically. Near the monitor the spectrum exhibits the limits $mgH < E < E_2$ when the crank 6 is horizontal, and the limits $mg(H + \Delta H) < E < E_2$ when crank 6 is vertical. In the vicinity of the detector being investigated the boundaries of the spectrum, under the same conditions, are $mgh < E < E_2 - mg(H - h)$ and $mg(h + \Delta H) < E < E_2 - mg(H - h)$, respectively. The efficiency of the detector D for neutrons of energy mgh within the interval $mg\Delta H$ is determined as the ratio of the difference between the counting rates of the detector for a lowered and raised crank, respectively, to the corresponding difference between the counting rates of the monitor (Novopoltzev and Pokotilovskii 1980*b*).

be unity. A narrow spectral line was supplied to the input of the system (to be more precise, difference measurements were carried out) (see Fig. 3.17). The monitor, which was the DUNya detector, was lowered to such a level that the neutrons arriving at it had an energy sufficient to penetrate the aluminium window with a probability close to unity. On the other hand, the level (to which the detector being investigated was lowered) varied, resulting in the energy of the neutrons incident on it also varying. The relative efficiency of the detector was determined by the ratio of its counting rate to the counting rate of the monitor.

Presented in Fig. 3.18 are the results of the measurement of the relative spectral efficiency of the DUNya under investigation with a 100 μm aluminium window. The experimental points follow the theoretical curve closely. The decreasing efficiency with the UCN energy is mainly due to reflection from the window. This is verified by Fig. 3.19. It can be seen that as the counting rate of the detector decreases the counting rate of the monitor rises, but the total counting rate remains constant.

The results in Fig. 3.20 similar to those of Fig. 3.19, except that they represent a rotating profiled scintillator. In this case the counting rate of the

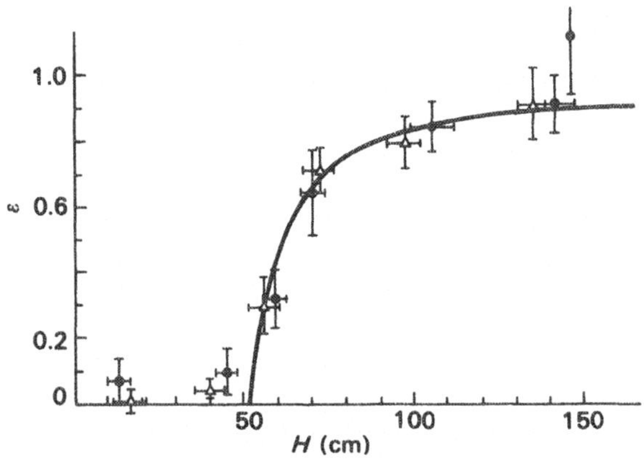

Fig. 3.18 Spectral efficiency of the DUNya detector. The solid line corresponds to the theoretical computation of the transmission through the window for an isotropic angular distribution of the UCN (Novopoltzev and Pokotilovskii 1980*b*).

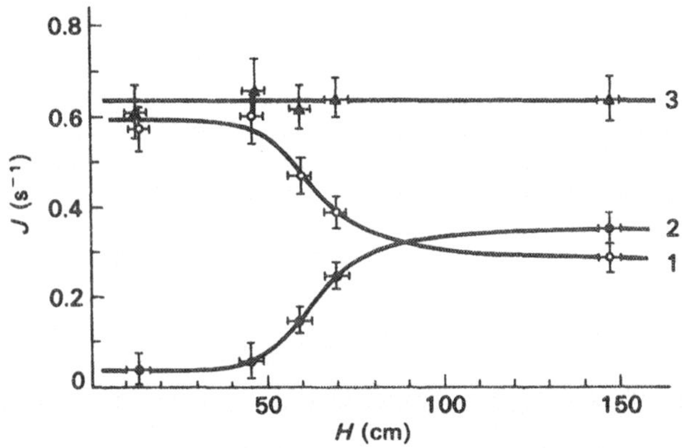

Fig. 3.19 Change in counting rate of the monitor 1, the detector under investigation 2 (DUNya), and their sum, 3, as a function of the UCN energy $E = mgH$. The reduction of the counting rate of DUNya together with the enhancement of the counting rate of the monitor with the decrease of E points to reflection from the window of the detector (Novopoltzev and Pokotilovskii 1980*b*).

detector is nearly independent of the UCN energy, and it only falls slightly as the energy increases (this effect was mentioned above). The counting rate of the monitor remains absolutely intact here, i.e. the small decrease in the efficiency of the detector at higher energies is not related to the reflection. This is a clear indication that a scintillation detector with a rotating radiator is the most efficient detector of UCN.

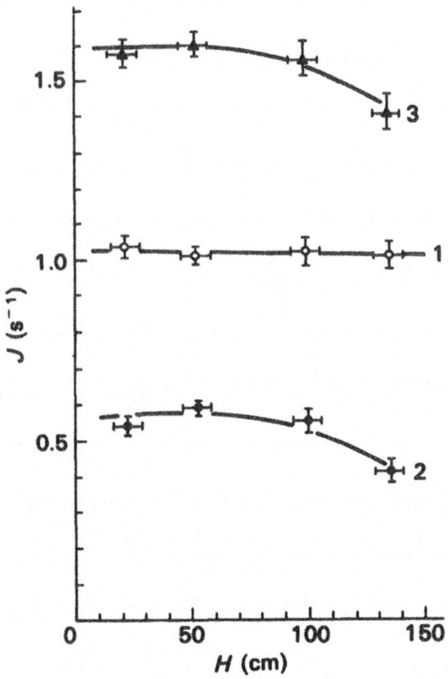

Fig. 3.20 Curves similar to those in Fig. 3.19 for the case when the detector depicted in Fig. 3.4 is investigated. The counting rate of the detector and, hence, its efficiency fall somewhat as the energy increases. As the counting rate of the monitor is constant, this indicates that the above fall is not related to reflection (Novopoltzev and Pokotilovskii 1980*b*).

APPENDICES

Appendix 3.1 Fourier and pseudorandom neutron spectroscopy

$$\text{a) } dN_d(t) = dt[I \int W(\tau)p(t - a\tau)T_c(t - \tau)d\tau + \Phi], \quad \text{b) } a = L_{sd}/L_{cd}. \tag{1}$$

$$\text{a) } \int_0^{t_m} W(t)dt = 1, \quad \text{b) } p(t) = T_{ps}(t\Omega), \quad \text{c) } \int_0^{t_m} p(t)dt = \tau_p. \tag{2}$$

$$\text{a) } p(t) = T_{SS}(t) = 1, \quad \text{b) } I_{ps} = QI_{SS}, \quad \text{c) } Q = t_i/\tau_p, \quad \text{d) } t_i = 2\pi/\Omega. \tag{3}$$

$$\text{a) } p(t - \tau)p(t - t_c) = \Delta(t_c - \tau),$$

$$\text{b) } \int \Delta_{ps}(t)g(t)dt \simeq \langle g(0)\rangle\tau_p, \quad \text{c) } \Delta_{SS} = 1. \tag{4}$$

$$\text{a) } S(t_c) = \int dN_d(t)p(t - at_c)f(t - t_c),$$

$$\text{b) } \int_0^T T_c(t - \tau)f(t - t_c)dt = TF(t_c - \tau). \tag{5}$$

$$\text{a) } F(t_c - \tau) = \alpha\delta(t_c - \tau) + A_0, \quad \text{b) } \int \delta(t - t')g(t')dt' = g(t)r. \tag{6}$$

$$\text{a) } A_0 \simeq \alpha \simeq 1, \quad \text{b) } r \ll \tau_p, \quad \text{c) } t_m < t_i. \tag{7}$$

$$S(t_c) = TI \int W(\tau)F(t_c - \tau)\Delta[a(t_c - \tau)]d\tau + \int \Phi f(t - t_c)p(t - at_c)dt. \tag{8}$$

$$\text{a) } \int_0^T f(t - t_c)p(t - at_c)dt = nA_f\tau_p, \quad \text{b) } A_f \simeq 1, \quad \text{c) } n = T/t_i. \tag{9}$$

$$\int W(t)\Delta(at)dt = \langle W(0)\rangle_a \tau_p/a. \tag{10}$$

$$S(t_c) = TI[\alpha r W(t_c) + A_0\tau_p\langle W(t_c)\rangle_a/a] + \Phi A_f\tau_p n. \tag{11}$$

$$\text{a) } N_c = \alpha r TI W(t_c), \quad \text{b) } \Delta N_c = \sqrt{S(t_c)}, \quad \text{c) } \Delta N_c/N_c = k. \tag{12}$$

$$\text{a) } \rho = N_c/[S(t_c) - N_c], \quad \text{b) } \rho \simeq rW(t_c)/\{\int W(\tau)\Delta[a(t_c - \tau)]d\tau + \tau_p\Phi/It_i\}$$
$$\simeq (ra/\tau_p)[W(t_c)/\langle W(t_c)\rangle_a][1 + \Phi a/I\langle W(t_c)\rangle_a t_i]^{-1}. \tag{13}$$

$$T = S(t_c)T/N_c^2k^2$$
$$= [r\alpha W(t_c) + A_0\tau_p\langle W(t_c)\rangle_a/a + \Phi A_f\tau_p/It_i]/I[\alpha r W(t_c)k]^2. \tag{14}$$

$$T \simeq (\tau_p/ra)[\langle W(t_c)\rangle_a/W(t_c)][1 + a\Phi/t_i I\langle W(t_c)\rangle_a]/\alpha r I W(t_c)k^2. \tag{15}$$

$$\text{a) } T_{pS}/T_{SS} = (\tau_p/t_i)^2, \quad \text{b) } T_{pS}/T_{SS} = (\tau_p/t_m)^2, \quad \text{c) } T_{pS}/T_{SS} = (\tau_p/t_m)^3. \tag{16}$$

$$\text{a) } I_{pR} = qI_{pF}, \quad \text{b) } q \ll 1, \quad \text{c) } T_{pF}/T_{pR} = q, \quad \text{d) } T_{pF}/T_{pR} = q^2. \tag{17}$$

$$\text{a) } T_c(t) = T_c(t\omega), \quad \text{b) } \omega = \omega(t), \quad \text{c) } dt = T(dt/d\omega)d\omega/T = Tg(\omega)d\omega. \tag{18}$$

$$TF(t_c - \tau) = \int_0^T T_c(t - \tau)f(t - t_c)dt$$
$$= T\int T_c[(t - \tau)\omega]f[(t - t_c)\omega]g(\omega)d\omega. \tag{19}$$

$$\text{a) } T_c(t_i) = \beta_i, \quad \text{b) } \sum \beta_i\beta_{i+j} = \alpha\delta_{0j} + A_0, \quad \text{c) } \alpha = m(1 - c)/n. \tag{20}$$

$$\text{a) } A_0 = mc/n, \quad \text{b) } c = (m - 1)/(n - 1). \tag{21}$$

The number of counts of a detector occurring during the time interval dt at time t equals (1a), where I is the total effective intensity of the neutron source, i.e. the count of the detector per unit time when the chopper is completely open, W is the spectrum in the time of flight τ from the chopper to the detector, T_c is the transmission of the chopper, p is the reduced intensity of the source, which can also be considered the transmission of a stationary source of average intensity I. The quantity a equals (1b), where L_{sd} is the flight path from the source to the detector, L_{cd} is the path from the chopper to the detector; thus $a\tau$ is the time of flight from the source to the detector. Φ in (1a) represents the background, assumed to be constant. The spectrum W is assumed normalized in accordance with (2a), where t_m is the maximum time of flight present in the spectrum investigated. In the case of a periodic pulsed source of frequency Ω the reduced intensity p may be represented by some function (2b). We determine the pulse width by the equality (2c). In the case of a stationary source the reduced intensity p is equal to (3a). The efficiencies of pulsed and stationary sources are conveniently compared for identical numbers of neutrons registered by the detector and averaged over unit time, when the chopper is open, so the intensity of the pulsed source I_{ps} is assumed to be Q times higher than the intensity I_{SS} of the stationary source (3b), where Q equals (3c) and t_i is the period of the pulsed source (3d). We shall introduce function (4a), where t_c is the time of flight from the chopper to the detector corresponding to the channel number in the experiment. In the case of a pulsed source, in the absence of recycling both in τ and in t_c this function exhibits a well-defined sharp peak of unit height and width τ_p and, therefore, the approximation (4b) holds for any smooth function g, where $\langle g(0) \rangle$ is the value of the function in the vicinity of 0 averaged over the time interval τ_p. Expression (4c) is valid for stationary sources. To derive the spectrum sought, W, from the experimental data expression (1a) is multiplied by a specially chosen function f and the integral (5a) is computed. The function f must be such to enable relation (5b) to be satisfied, where T is the total measurement time, and F can be represented in the form (6a), α being the modulation amplitude of the chopper transmission, A_0 is the mean transmission of the chopper, and δ is a function with a sharp peak of unit height and width r, for which (6b) holds for any smooth function g. The quantity r is the resolution and is determined by the maximum rotation frequency of the chopper. We shall assume below that the relations (7) are valid. Substituting (1a) into (5a) we obtain, taking into account (5b), expression (8). The integral in the second term of (8) can be represented in the form (9a), where A_f is the amplitude of the function f, the value of which is close to unity (relation (9b)), and n is the ratio (9c). With the aid of relation (4b) we obtain equality (10), where $\langle W(0) \rangle_a$ is the value of the function in the vicinity of 0 averaged over the time interval τ_p/a. Substituting (9) and (10) into (5a) we obtain (11). The useful count is defined in (12a), the statistical uncertainty is defined by (12b), and the notation for the experimental precision is introduced in (12c), while the notation for the signal-to-noise ratio is given by (13a). Substituting (12a) and (11) into (12c) and using (10) we obtain (13b). For a given k the total measurement time in the case of a pulsed source is given by expression (14) which, taking into account (7), is rewritten in the form (15). In the case of a stationary source one must substitute t_i for τ_p in (15). The ratio of the total measurement time for a pulsed source, T_{pS}, to that for a stationary source, T_{SS}, is indicated in (16a), where relations (3b) and (3c) are taken into account and the background in (15) is assumed to be small. If the maximum time of flight is smaller than the period ($t_m < t_i$), then (16a) transforms into (16b). If the background is high, (16b) transforms into (16c). In Fourier spectroscopy the slits of the chopper are of the same width and are situated at equal distances from each other, so the neutrons can be transmitted through several slits simultaneously, while in pseudorandom correlation spectroscopy the beam can only be let to pass through a single slit at a time. For this reason the beam intensity utilized in pseudorandom correlation spectrometry with a pulsed source, I_{pR}, is lower than the intensity I_{pF} used in Fourier spectrometry with a pulsed source (eqns (17a, b)). Correspondingly, the ratio of the total measurement times is equal to (17c) when Φ is small; when Φ is large, it is (17d). In the case of Fourier spectrometry the transmission of the chopper is a periodic function, (18a), of the frequency ω. The chopper rotates with acceleration and therefore the frequency depends on the time, (18b). Equation (18c) follows from (18b), where the frequency distribution density $g(\omega)$ is introduced, determined by the last equality in (18c). Applying (18c) we rewrite relation (5b) in the form (19). To obtain an expression for the function F in the form (6a) it is necessary to take f as a periodic function similar to the transmission of the chopper but shifted in phase by the time t_c corresponding to the channel number, as shown in (19). In the case of pseudorandom spectrometry the transmission of the chopper at time t is characterized by a discrete number β (eqn (20a)), equal either to zero or to unity. The sequence of units is chosen to be such that relation (20b) is satisfied, where δ is the Kronecker symbol, the constants α and A_0 are indicated in (20c) and (21a) respectively, and c is the fraction of the area of the chopper open to the beam (eqn (21b)): m is the total number of open holes situated at the periphery of the chopper and n is the total number of cells of areas equal to the area of a single hole, which can all occupy the periphery of the chopper.

Appendix 3.2 Transmission of UCN through a film of non-uniform thickness

$$\text{a) } A = t_+ t_- \exp(ik'_\perp l)/[1 - r_-^2 \exp(2ik' l)], \qquad \text{b) } t_+ = 2k_\perp/(k_\perp + k'_\perp). \tag{1}$$

$$\text{a) } t_- = 2k'_\perp/(k_\perp + k'_\perp), \quad \text{b) } r_- = (k'_\perp - k_\perp)/(k'_\perp + k_\perp),$$
$$\text{c) } k'_\perp = \sqrt{k_\perp^2 - u_1}\,. \tag{2}$$

$$\text{a) } l = l_0 + x, \quad \text{b) } T(k_\perp) = \int_{-a}^{a} |A|^2 \, dx/2a = \int_0^{2\pi} |A|^2 \, d\varphi/2\pi. \tag{3}$$

$$\text{a) } \varphi = 2k'_\perp l, \quad \text{b) } \int_0^{2\pi} d\varphi/2\pi(\beta + \gamma\cos\varphi) = 1/\sqrt{\beta^2 - \gamma^2}, \quad \text{c) } \beta > \gamma. \tag{4}$$

$$T(k_\perp) = |t_+ t_-|^2/(1 - |r_-|^4) = (1 - |r_-|^2)/(1 + |r_-|^2) = 2k_\perp k'_\perp/(k_\perp^2 + k_\perp'^2). \tag{5}$$

$$\text{a) } \bar{T} = \int T(k_\perp)\cos\theta \, d\Omega/\pi = \int_{\sqrt{u_1}}^{k} 2T(k_\perp)k_\perp dk_\perp/k^2, \quad \text{b) } k_\perp = \sqrt{u_1}\,\text{ch}\,\psi. \tag{6}$$

$$\text{a) } k = \sqrt{u_1}\, y, \quad \text{b) } \sqrt{k^2 - u_1} = \sqrt{u_1}\, y', \quad \text{c) } y' = \sqrt{y^2 - 1}\,. \tag{7}$$

$$\text{a) } \bar{T} = y'/y - (1/2y^2)\,\mathrm{arctg}(2yy'), \quad \text{b) } \mathfrak{D} = \bar{\bar{T}} = \int_{\sqrt{u_1}}^{\sqrt{u_2}} \bar{T} 4k^3\,\mathrm{d}k/u_2^2. \tag{8}$$

$$\text{a) } \mathfrak{D} = 0.5\{(2-\alpha)[\sqrt{1-\alpha} - \alpha\,\mathrm{arctg}(2\sqrt{1-\alpha}/\alpha)] + \alpha^2\,\mathrm{arch}(1/\sqrt{\alpha}),$$

$$\text{b) } \alpha = u_1/u_2. \tag{9}$$

$$T = |t_+t_-|^2 \exp(-u_1'' l_0/k_\perp')/[1 - r_-^4 \exp(-2u_1'' l_0/k_\perp')]. \tag{10}$$

$$\mathfrak{D} = 8 \int_{\sqrt{u_1}}^{\sqrt{u_2}} \frac{k^3\,\mathrm{d}k}{u_2^2} \int_{\sqrt{u_1}}^{k} \frac{k_\perp\,\mathrm{d}k_\perp}{k^2} T(k_\perp) = 4 \int_{\sqrt{u_1}}^{\sqrt{u_2}} (u_2 - k_\perp^2)\, T(k_\perp) \frac{k_\perp\,\mathrm{d}k_\perp}{u_2^2}. \tag{11}$$

$$\text{a) } \mathfrak{D} = 4\alpha^2 \int_0^s (1 - \alpha^2 - \alpha^2 x^2)\, x T(x)\,\mathrm{d}x, \quad \text{b) } s = \sqrt{\alpha^{-2} - 1}. \tag{12}$$

$$T(x) = (1 - r^2)^2 \exp(-\beta/x)[1 - r^4 \exp(-2\beta/x)]^{-1}. \tag{13}$$

$$\text{a) } r = (\sqrt{1+x^2} - x)/(\sqrt{1+x^2} + x), \quad \text{b) } \beta = l_0 u_1''/\sqrt{u_1'}. \tag{14}$$

The amplitude of a wave transmitted through a film of thickness l and of potential u_1 is given by (1a) where (1b) and (2a) are the amplitudes of transmission through the interface from vacuum into the matter and in the opposite direction respectively, (2b) is corresponding reflection amplitude and (2c) is the normal component of the wave vector inside the matter. Let the thickness of the film be (3a), where l_0 is the mean value; x varies within the range $(-a, +a)$, and a is such, that $k'a \gg 1$. Then the film transmission coefficient can be represented by (3b), where averaging over the thickness is replaced by averaging over the phase (4a). Taking into account relation (4b) with parameters satisfying (4c), we reduce (3b) to the form (5). We average it over the angle of incidence, as shown in (6a). After substitution of (6b) and (7a)–(7c) the integral (6a) is reduced to (8a). Averaging it over energy in accordance with (8b), leads to (9a) with the notation (9b). When the imaginary part u_1'' of potential u_1 is present, expression (5) should be replaced with (10). Then, the integral (8b) takes the form (11) and can be reduced to (12a) with the upper limit of the integral shown in (12b) and the transmission coefficient (10) represented by (13) with the notation (14). The integral (12a) can be calculated only numerically. In the case of $\alpha = 0.315$, $l = 60\ \mu$m and $\beta = 0.062$ we get $\mathfrak{D} = 0.36$. If the UCN spectrum begins at $\sqrt{u_1}$, then $\mathfrak{D} = 0.39$, which is $(1 - \alpha^2)^{-1}$ larger than before.

4

Transporting UCN

This chapter deals in the main with neutron guides, and only at the very end will the issue be touched upon of how neutrons are transported in a sealed vessel.

Neutron guides serve to extract the neutrons from the reactor core to the experimental hall, beyond the biological shield, and to separate low-energy neutrons from the direct beam. The intensity of the UCN themselves, however, is then attenuated. Obviously the attenuation is more significant when the walls of the neutron guide are rougher and more contaminated by hydrogen and when the vacuum inside it is poor. As far as the vacuum is concerned, operating experience is in quite good agreement with theoretical expectations and reveals that, at pressures lower than 10^{-3} mm Hg, residual gases actually do not hinder the passage of the UCN. Nevertheless, the vacuum is quite controllable. The contamination of the walls, however, and the degree of roughness are much more difficult to control.

Usually, to remove contamination, neutron guides are rinsed with degreasing solutions and distilled water and then polished either mechanically, chemically or electrochemically. Mechanical polishing gives a surface a specular appearance in visible light, but neutron guides polished mechanically usually transmit UCN quite poorly. The point is that the condition for specular reflection of light does not coincide with the same condition for UCN, the wavelength of which is of about an order of magnitude smaller. Therefore, a surface that is specular and, thus, mirror like may turn out to be rough for UCN.

In practice, it appears that electrochemical polishing yields smooth surfaces from the point of view of reflecting UCN. But neutron guides represent quite long tubes and the electropolishing of these tubes requires a special technology. Such a technology was developed at the Leningrad Institute of Nuclear Physics (Gatchina) and it can be applied not only for the construction of experimental equipment, but also in industry.

The best neutron guides at present are obtained by the following technology. A layer (of about 100 nm) of ^{58}Ni is vacuum deposited on to the surface of float glass, and a layer of copper is further deposited. The layer of copper is then galvanized to a thickness of $\simeq 1$ mm. The resulting sheet is separated from the glass and then the sheets are joined. The neutron guides thus obtained have a rectangular cross-section and their inner surface covered with ^{58}Ni exhibits both a high limiting energy and a high degree of perfection.

There are problems in the physics of UCN that do not require neutron guides with high-quality surfaces for their solution. Moreover, the fact that the guides are rough may play a positive role in segregating neutrons of higher energies from the UCN beam. The roughness leads to the non-specular reflection of neutrons of higher energies and to their leaving the neutron guide. Sometimes labyrinths are actually set up inside the neutron guides, playing the part of additional roughnesses.

In studying the transportation of UCN it is important to understand how the neutrons propagate along a neutron guide. The problems that arise here have much in common with the diffusion problems of molecular gases in pipes. UCN, however, differ from a molecular gas, first, in that they decay and can be absorbed by the walls, whereas the number of molecules in a gas is always conserved, and, secondly, in that the molecules of a gas mainly undergo diffuse reflection from the walls (with a probability not less than 0.9) and are readily thermalized, whereas UCN can, to a large extent, undergo specular and elastic scattering from the walls. Transportation of UCN, and their storage, likewise, depends ultimately on their interaction with the walls. In transporting UCN, however, numerous regularities are revealed that are not apparent when the UCN are stored, such as the dependence of the transmission on the length of the neutron guide and on the neutron wavelength; the variation of the transmission when the neutron guide is filled with a gas; the angular distributions and their relation with the length of the neutron guide, the quality of its surface, and the neutron spectrum; the density distribution of UCN along the neutron guide; non-stationary processes; the propagation of thermal neutrons; and many other things. Unfortunately, all the studies of these regularities are still at a rudimentary level. We still do not know how to relate unambiguously the parameters of a neutron guide with the measured characteristics. There is still no clear understanding about the variation of parameters to which various characteristics are sensitive. Therefore, the only possible thing for us to do is to present the empirical material concerning the transportation of UCN and those theoretical models which may help to interpret the experimental data. To what extent a certain model corresponds to reality is up to readers to decide. The author, however, hopes they will be able to make use of the material expounded here for actually computing the transmission of neutron guides, for instance in the case of thermal neutrons.

4.1 METHODS OF EXPERIMENTAL INVESTIGATION OF NEUTRON GUIDES

The transmission of a neutron guide is the ratio of the UCN flux at the output to the flux at the input. From this definition it follows that direct measurement of the transmission of the entire system from the converter to the detector is practically impossible, since at the converter the background of neutrons of higher energies is so high that it is impossible to separate from it those neutrons that pertain to the UCN flux, i.e. the UCN flux at the input cannot be determined.

Usually, the transmissions of small segments of the neutron guide are measured. To this end a segment of length ΔL is connected to a system of length L, which has a UCN flux at its output equal to $J(L)$, so that at the output of the total system a flux $J(L + \Delta L)$ is obtained and the transmission of the segment is

$$W_{tr}(\Delta L) = J(L + \Delta L)/J(L) < 1.$$

Clearly, such a method cannot, generally speaking, provide a quantitative idea of the actual transmission of a segment in the presence of non-specular reflection from the walls, since the transmission depends strongly on the resistance of the entire preceding neutron guide of length L. To understand this, we must turn to the analogous case of electrical circuits. Suppose the small resistance to be studied, r, is connected in series in a circuit with an electromotive force $\mathscr{E}$ and a resistance R. Then, if the transmission of the segment studied is understood to be represented by the ratio of the currents $I_2 = \mathscr{E}/(R + r)$ and $I_1 = \mathscr{E}/R$, we obtain the quantity $I_2/I_1 = R/(R + r)$, which tends to unity as R increases. The losses, here, depend only on the ratio r/R. Naturally, the analogy with the electric current cannot be complete, since the UCN flux along the neutron guide is not conserved, as a result of losses, while the current is conserved; but for relatively short neutron guides this analogy is quite plausible. In spite of the indicated fault, the above method is applied by experimenters because it is simple and makes possible judgement of the quality of a particular neutron guide.

The transmission of the segment of length ΔL of a neutron guide may also be measured in an indirect way (Kosvintsev *et al.* 1984*b*). For this purpose a small outlet of cross-section $s \ll S$ is made at a point L, where S is the cross-section of the segment of interest, and the flux through it, $j(\mathrm{L})$, is measured. If the losses in the outlet can be neglected and the neutron angular distribution at point L can be considered isotropic, the transmission of the segment ΔL can then be estimated by the quantity

$$W_{tr}(\Delta L) = J(\mathrm{L} + \Delta L)s/j(\mathrm{L})S.$$

Measurement of the flux in the outlet, in this case, is equivalent to measurement of potential difference in electric circuits. In this way it is possible to measure the transmission not only of horizontal channels, but also of vertical and inclined ones if the spectral efficiency of the detector is calibrated first.

Obviously, the transmission of a neutron guide depends not only on its length, but also on its diameter $d = 2\rho$. However, if the influence of gravity is neglected (this may be done in the case of horizontal neutron guides, when *mgd* is essentially less than the neutron energy, *E*), the transmission of a neutron guide of length *L* and radius ρ depends only on the ratio L/ρ. Therefore, if not stated otherwise, we shall give below the length of a neutron guide in terms of the radius.

In one of the first experiments on UCN (Groshev *et al.* 1971) an exponential dependence of $W_{tr}(L)$ was obtained for a chemically polished copper neutron guide (Fig. 4.1). For a neutron guide of length $L = 100$ the transmission amounts to about 35%. A neutron guide of non-elaborated stainless steel of length 100, as measured by Kosvintsev *et al.* (1976*b*) (Fig. 4.2), exhibits a transmission about 3%, but when it is electropolished the transmission rises to 40%. At the Leningrad Institute of Nuclear Physics a transmission of 70% (Egorov *et al.* 1974) was obtained with a neutron guide of nickel evaporated on polished glass having a cross-section of $8 \times 6\ cm^2$ and a length of 3 m (which was equivalent to $L \approx 100$).

In experiments (Kosvintsev *et al.* 1984*a*) with copper and stainless steel neutron guides the exponential fall of the transmission was also observed as a function of the length. A decrease by a factor of e occurs at lengths $L \approx 100$. The investigation of various methods of polishing neutron guides revealed that a neutron guide made of stainless steel has an advantage in transmission as compared with a copper neutron guide, and the best transmissions were exhibited by stainless steel neutron guides that underwent electropolishing for 10 min in H_3PO_4 at 30°C, when the current density was $10\ A\,dm^{-2}$.

Measurement of the transmission only permits the determination of the exponent characterizing the attenuation of the neutron beam as the length of the neutron guide increases. But this exponent depends on two parameters: the loss coefficient μ and the probability of diffuse reflection *G* when UCN undergo collisions with the walls of the neutron guide. To determine the values of both quantities, a measurement was performed (Groshev *et al.* 1971)

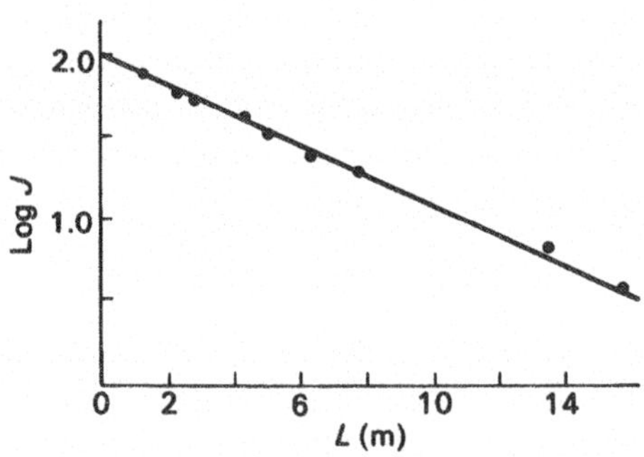

Fig. 4.1 Dependence of UCN counting rate *J*(*L*) on the length of a neutron segment attached to the UCN channel 5.3 m long. The neutron guide represents a copper pipe, polished chemically, with an internal diameter of 9.4 cm (Groshev *et al.* 1971).

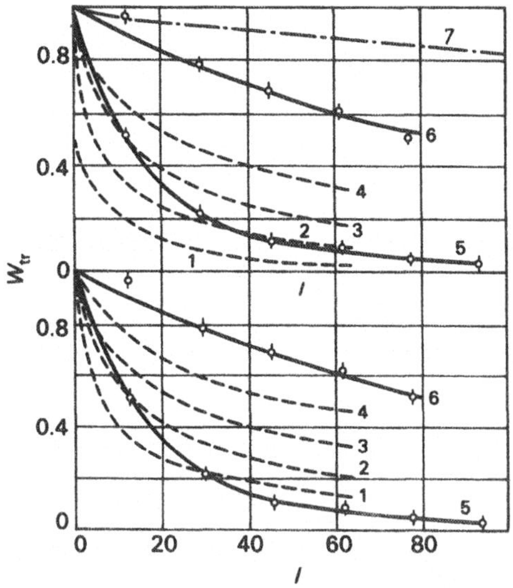

Fig. 4.2 Transmission of straight neutron guides of stainless steel (Kosvintsev *et al.* (1976*b*): 1–4, theoretical curves computed by the Monte Carlo method (Berceanu and Ignatovich 1973) with the reflection indicatrix (4.8) and the parameters $G = 1.0, 0.5, 0.25, 0.125$, respectively. The upper curves are computed for $A(\Omega) = 1$ and the lower ones for $A(\Omega) = \cos\theta$; 5, experimental curve for a non-polished neutron guide; 6, experimental curve for electropolished neutron guide with internal diameter equal to 36 mm; 7, theoretical curve for the transmission of a perfect specular neutron guide with a loss coefficient $\mu = 10^{-3}$. The reduced length of the neutron guide is plotted along the horizontal axis.

of the transmission of a diaphragmed neutron guide depending on the area of the diaphragm. Such a measurement allows the pre-exponential coefficient of the transmission of the non-diaphragmed neutron guide to be factored out and, thus, one more independent quantity relating μ and G to be found. When this is done both μ and G can be readily determined independently. Measurements have revealed that a value of μ of 10^{-3} is an order of magnitude larger than the expected theoretical value, whereas the probability of diffuse reflection, G, amounts to 18%.

The quality of treatment of the walls of a neutron guide influences not only the transmission, but also the angular distribution of neutrons leaving the exit orifice. The angular distribution at the output may, naturally, also depend on the initial angular distribution of the neutrons at the input. However, its influence is less significant, the more diffuse-like the reflection of UCN from the walls and the longer the neutron guide. In this connection it is reasonable to follow the variation of the angular distribution of the outgoing neutrons with an increase in the length of the neutron guide (Fig. 4.3), and in the presence of bends in the neutron guide (Fig. 4.4).

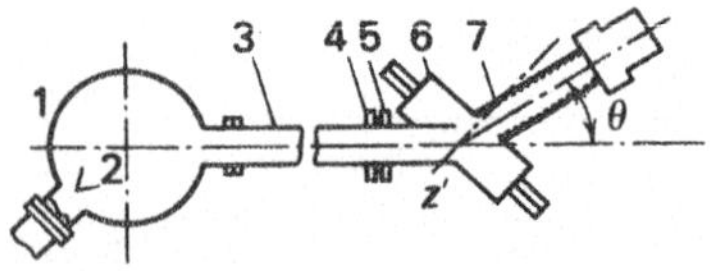

Fig. 4.3 Experimental set-up for measuring angular distributions: 1, spherical copper cavity for providing an isotropic distribution at the input of the neutron guide (the diameter of the sphere is 32 cm); 2, copper scattering funnel preventing the UCN from passing directly from the neutron guide into tube 3; 3, pipe 88 mm in diameter; 4, rotatable flange; 5, connecting cylinder; 6, turning box; 7, collimator with ribs 4 mm in height and polyethylene to avoid the reflection of neutrons in the collimator from the walls (Kosvintsev *et al.* 1976*b*).

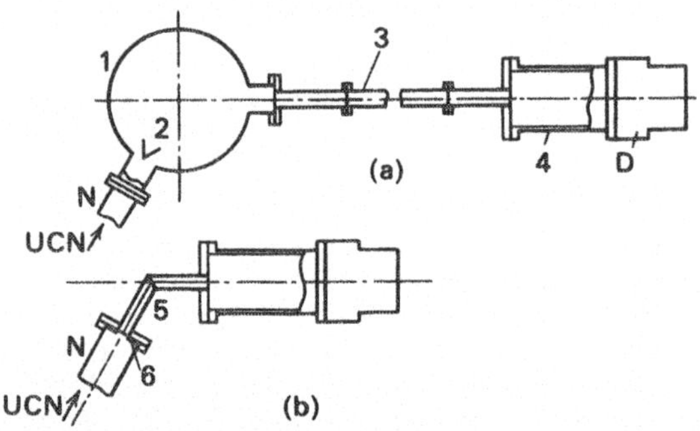

Fig. 4.4 Experimental layout for transmission measurements of (a) straight pipes and (b) elbows (Kosvintsev *et al.* 1976*b*): 1, 2, the same as in Fig. 4.3; 3, set of straight pipes 36 mm in diameter (chosen sufficiently small for the outgoing UCN flux, through the neutron guide being studied, not to affect strongly the neutron density inside the sphere); 4, transitional volume 94 mm in diameter and 25 cm long, which diminishes the influence of reflection from the detector on the UCN flux; 5, an elbow; 6, polyethylene.

The results of experiments are presented in Figs 4.5 and 4.6. Shown in Fig. 4.6 is the dependence of the transmission of a bend in the neutron guide on its angle for arms of given lengths. The lengths of the arms may influence the transmission of the bend quite significantly: the longer the arms, the less the transmission of the neutron guide changes with the angle of the bend. Only in the case of perfect specular reflection from the walls does the length of an arm have no influence on the resistance of a bend. In this connection it is interesting to note that, as shown by Kosvintsev *et al.* (1984*b*), the transmission depends only on the total length of the arms and is independent of their ratio.

Ageron *et al.* (1978*b*) determined the transmission of a neutron guide system in a different way. It was measured as follows. The storage bottle was

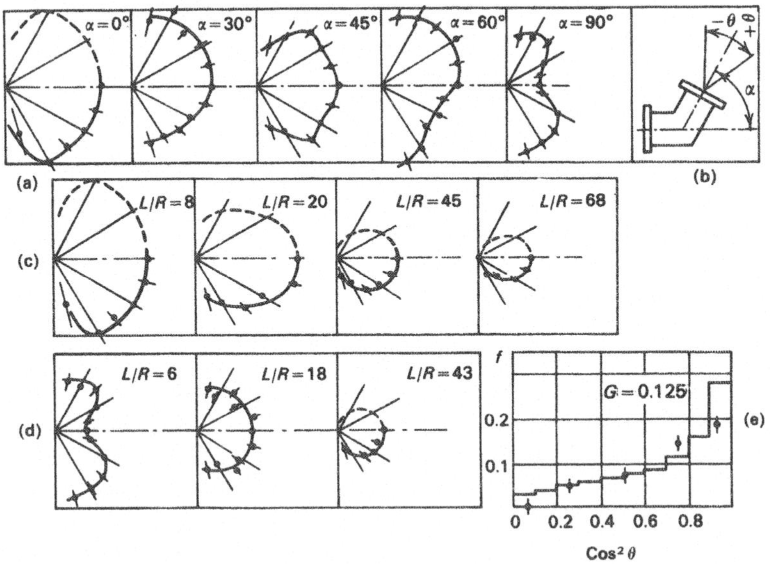

Fig. 4.5 Angular distributions in the UCN flux: (a) angular distributions after the elbow; (b) layout for measuring angular distributions after an elbow of the angle α; (c) alignment of initially isotropic UCN flux; (d) change in angular distributions as the length of the tube increases, when the initial distribution is formed by the elbow; (e) comparison of experimental and theoretical limit angular distributions (Kosvintsev *et al.* 1977*c*).

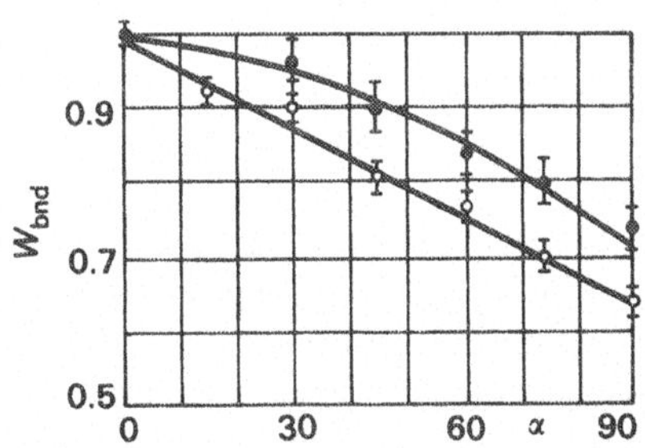

Fig. 4.6 Dependence of the transmission of an elbow upon the angle. The upper curve is obtained for an isotropic distribution at the input, and the lower one for the angular distribution formed by a long neutron guide (Kosvintsev *et al.* 1977*c*).

filled for 5 s with neutrons, which were kept for 2 s in the closed bottle; then, in 13 s the remaining number of neutrons in the bottle was counted once by a detector connected directly to the bottle and another time when the neutron guide system being studied was put between the detector and the bottle. The ratio of the counting rates was 0.17 and was considered to be the transmission of the neutron guide. This method differs from the preceding one for determining the transmission. Obviously, if there were no losses of UCN due to heating, absorption, and their decays, and if the time during which the neutrons remaining in the vessel are registered were sufficiently long, implying that all the neutrons are registered, the transmission would always be equal to unity, since in the long run all the neutrons leave the vessel and impinge upon the detector. So, instead of the transmission of the neutron guide, what is measured here is the relationship between the storage time τ, the outflow time τ_{ext}, and the measurement time τ_{meas}. To derive the transmission of the neutron guide from this relationship is not a simple task.

Indeed, the UCN gas flux per unit surface area is $nv/4$, where n is the number of neutrons per unit volume and v is the velocity. The number of neutrons leaving an opening of area s during the time dt is $nvs\mathrm{d}t/4 = N\mathrm{d}t(sv/4V)$, where N is the total number of neutrons in the vessel and V is its volume. Thus, the emptying time is $\tau_{emp} = 4V/sv$. If the storage time of UCN in the vessel is τ, then the number of neutrons registered by the detector at t within the interval $\mathrm{d}t$ is $N\exp(-t/\tau - t/\tau_{emp})\mathrm{d}t/\tau_{emp}$, and the total number of neutrons registered by a non-reflecting detector situated directly at the exit opening is $N\tau/(\tau + \tau_{emp})$. If the exit opening is connected with the detector through a neutron guide, which reflects neutrons with a probability R and transmits them with a probability T, then the outflow time τ'_{emp} will differ from τ_{emp} determined as $\tau'_{emp} = \tau_{emp}(1 - R) + \tau_0$, where τ_0 is the emptying time of the neutron guide itself. If the measurement time is sufficiently long, the detector connected with the vessel containing UCN through the neutron guide will count a number of neutrons that may be estimated roughly by the quantity $NT\tau/[\tau + \tau_{emp}(1 - R) + \tau_0]$. Hence it follows that comparison of the two readings of the detector does not yield sufficient information for deducing the transmission of the neutron guide, T.

The nature of the motion of a neutron inside a neutron guide can be studied by filling the neutron guide with a gas and measuring the dependence on pressure of the UCN flux at the output (Fig. 4.7). The most convenient gas for performing this measurement is ^{4}He, since it is a noble gas, i.e. it does not react chemically with the walls, is not absorbed by surfaces, and is an atomic gas, i.e. the interaction of a neutron with it is described by an extremely simple theory. The curve depicted in Fig. 4.7 has been termed the helium curve. Its form is readily explained qualitatively. Moreover, one can predict that the more diffuse-like the reflection from the walls, the more complicated and long is the trajectory of a neutron and the higher the probability of a collision with an atom of the gas. In other words, as the surface becomes rougher, the drop of the helium curve becomes steeper. Further, it is obvious that when the losses in collisions of a neutron with the walls increase, the helium curve decreases slowly with pressure, since in this case only those

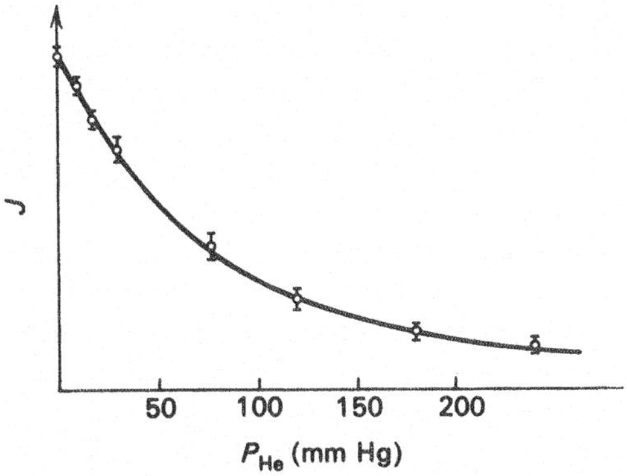

Fig. 4.7 Helium curve. Dependence of the counting rate of a UCN detector situated at the end of the neutron guide on the helium pressure in the neutron guide (Groshev *et al.* 1971).

UCN that exhibit the shortest trajectories reach the exit opening. Correspondingly, these neutrons are the least likely to undergo collisions with the gas. It is interesting to note that since the UCN energies correspond to a temperature of 1 mK, at all higher temperatures the velocities of the ^{4}He atoms are significantly higher than the neutron velocity, so the interaction of a neutron with the gas may be represented as the impingement of atoms upon a neutron at rest. If we denote the number of atoms per unit volume by N_0, the velocity of an atom by v, and the interaction cross-section with a neutron by $\sigma(v)$ (the relative velocity is practically equal to the velocity of the atoms), then the number of collisions with a neutron per unit time can be written in the form

$$1/\tau_{col} = N_0 v\sigma(v), \tag{4.1}$$

where τ_{col} is the lifetime of a neutron before it collides with an atom of the gas. From the equation of state for an ideal gas, $p = N_0 k_B T$, we may express N_0 in terms of the pressure p and introduce the parameter $p\tau$. Then

$$1/\tau_{col} = p/p\tau, \qquad p\tau = k_B T/v\sigma(v), \tag{4.2}$$

where $p\tau$ has the dimensions of mm Hg s. Formulae (4.1) and (4.2) hold for any gas. In the case of helium, computation of the cross-section is very simple. At a temperature of 300 K the parameter $p\tau$ for helium turns out to be equal to 343 mm Hg s.

Another way of studying the propagation of UCN along a neutron guide is the investigation of non-stationary processes. For instance, we may follow the variation with time of the counting rate of a detector at the exit opening after the reactor has been shut down for an instant, or the access of thermal neutrons to the channel has been interrupted by a shiber. In this case, obviously, the longer neutrons are counted after their generation in the converter has been stopped, the more entangled are their trajectories inside the neutron guide, i.e. the more diffuse-like is the reflection from the walls and the smaller are the losses in collisions from the walls. Robson (1976*a*) measured the time of the UCN outflow from a neutron guide 5.3 m long when the reactor was abruptly stopped. It turned out to be about 4 s. In the case of absolute specular reflection from the walls and a neutron velocity of 4–5 $\mathrm{m\,s^{-1}}$ this time should be $2L/v \approx 2$–3 s.

Non-stationary experiments (Egorov *et al.* 1974) of another kind were carried out at the Leningrad Institute of Nuclear Physics. Here the neutron guide was partitioned at a certain distance from the detector by a reflecting shutter. At a certain moment the shutter was opened and the change of the counting rate of the detector was observed as a function of time (Fig. 4.8). The counting rate of the detector might be expected to rise monotonically when the shutter is opened, until it reaches a stationary value for any ratio of the total length of the neutron guide to the length of the section between the shutter and the detector. In the case of perfect specular reflection from the walls this would actually be true. When the reflection is diffuse, however, the picture becomes somewhat more complicated and the dependence of the counting rate on time may exhibit bumps such as resonance peaks, as can be seen in Fig. 4.8. These bumps may be explained qualitatively by taking advantage, as an example, of the flow of a molecular gas which undergoes totally diffuse reflection from the walls. When the shutter is closed, the entire neutron guide is uniformly filled with the gas and there is significantly more

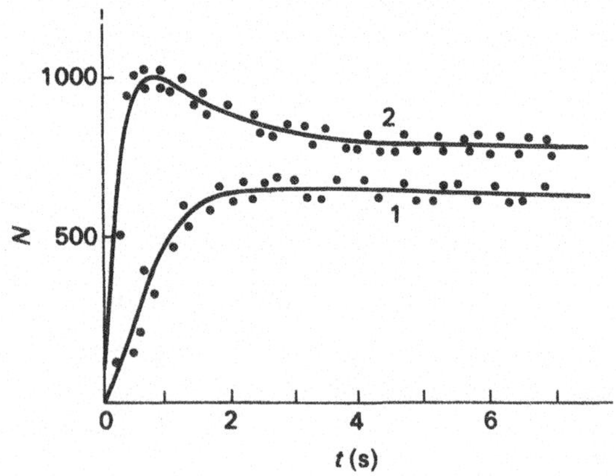

Fig. 4.8 Experimental data on non-stationary diffusion (Egorov *et al.* 1974). At a certain distance from the detector the neutron guide is partitioned by a shutter which is opened at a given moment, whereupon the counting rate of the detector is measured as a function of time. The smaller the resistance of the section of the neutron guide adjacent to the detector, the clearer is the subsequent maximum.

gas than the amount contained in the same volume when a stationary flow takes place. When the shutter is opened, the excess gas flows into the newly opened section, and if its amount exceeds the amount necessary for filling the supplementary section to the point when a stationary condition is established, an excess counting rate is observed in the detector. This phenomenon is similar to the impulse discharge of a capacitor.

Obviously, this excess will be higher, relatively, when the length of the neutron guide from the converter to the shutter increases, when the additional section is short, and when the reflection of a neutron from the walls of the main neutron guide is more diffuse-like.

Recently, a further way of studying neutron guides has been developed; it was revealed in the course of totally different investigations (Golikov *et al.* 1985*b*). The essence of the experiments is explained in Fig. 4.9. In one of the arms of a branched neutron guide there is a sample. The detector is put in the other arm, and from the counting rate of the detector the absorption in the sample is determined. For such measurements it is sufficient, in principle, to calibrate the experimental equipment with the aid of known samples, and it is not at all necessary to know how the UCN propagate along the neutron guide. But the problem may also be inverted, so that we might derive, from the calibration curves, parameters characterizing the propagation of UCN along neutron guides.

We shall now consider how and by which parameters it is possible to describe transportation processes.

4.2 MATHEMATICAL SIMULATION

The most straightforward method of calculating the transmission of a neutron guide is to perform numerical experiments, similar to the real experiments. In numerical experiments neutron trajectories are generated and

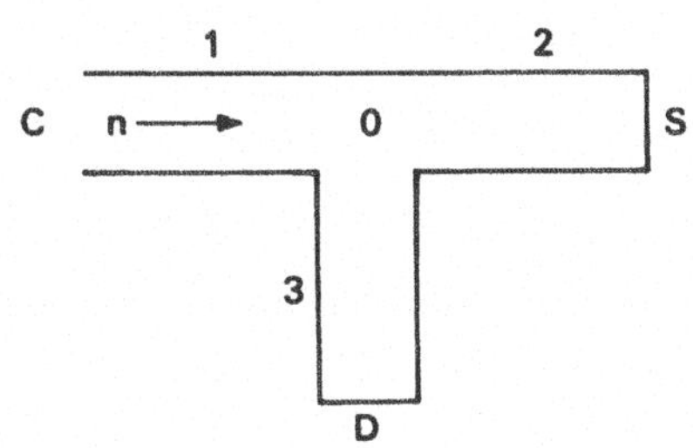

Fig. 4.9 Experimental layout involving a branched neutron guide. C, source of UCN; n, neutrons, 0, branching point; S, absorbing sample; D, detector. The detector and absorber can change places. The branches 1, 2, and 3 may contain bends, diaphragms, and be situated at an arbitrary angle to the force of gravity (Golikov *et al.* 1985*b*).

the fraction of the total number of simulated neutrons which reach the exit opening is counted. This fraction precisely determines the transmission of the neutron guide. The outgoing neutrons can be classified according to the directions of their flights, and thus it is possible to obtain their angular distribution. For a comparison between numerical experiments and real ones it is necessary to define certain parameters, for example the probability of losses occurring in collisions with the wall and the reflection indicatrix. It is necessary to know the coordinate distribution of the incident neutrons at the input opening and their angular distribution at the entrance to the neutron guide. Since no information on these two distributions is available, the coordinate distribution is taken to be uniform over the input opening cross-section, and the angular distribution is taken to be isotropic within the forward hemisphere.

An essential point is the choice of the reflection indicatrix $W_{ref}(\Omega_0, \Omega)$ which characterizes the probability for a neutron incident on a surface from the direction Ω_0 to be reflected in the direction Ω. At first sight it may seem that the possibilities for simulation here are infinite. Actually, the indicatrix $W_{ref}(\Omega_0, \Omega)$ has to satisfy certain constraints, so no contradictions should arise with thermodynamics. We shall only deal with elastic collisions of a neutron with the wall, and for the present we shall neglect losses occurring in collisions. The condition to be imposed on the indicatrix may be formulated as follows. If the neutron guide is filled with a homogeneous and isotropic UCN gas, then collisions with the walls must not violate the isotropy of the gas and must not give rise to any fluxes whatever, either in the coordinate or in the momentum space. The formulated requirement signifies that the number of neutrons reflected per unit time from direction Ω_0 in direction Ω must be equal to the number of neutrons reflected from direction Ω in direction Ω_0. This is simply the condition of detailed balance:

$$\cos\theta_0 \mathrm{d}\Omega_0\, W_{ref}(\Omega_0, \Omega)\mathrm{d}\Omega = \cos\theta\, \mathrm{d}\Omega\, W_{ref}(\Omega, \Omega_0)\, \mathrm{d}\Omega_0, \tag{4.3}$$

where θ_0 and θ are the angles between the normal to the reflecting surface and the directions Ω_0 and Ω, respectively. We therefore have $\mathrm{d}\Omega = \mathrm{d}\cos\theta\, \mathrm{d}\varphi$, $\mathrm{d}\Omega_0 = \mathrm{d}\cos\theta_0\, \mathrm{d}\varphi_0$, and φ and φ_0 are the corresponding azimuthal angles. Condition (4.3) can be satisfied only by functions of the form

$$W_{ref}(\Omega_0, \Omega) = \tilde{F}(\Omega_0, \Omega)\cos\theta,$$

where $\tilde{F}(\Omega_0, \Omega)$ is an arbitrary function that is symmetric in the argument. Besides condition (4.3), the indicatrix $W_{ref}(\Omega_0, \Omega)$ must also satisfy the normalization condition

$$\int W_{ref}(\Omega_0, \Omega)\mathrm{d}\Omega = 1. \tag{4.4}$$

Let us consider some examples. In the case of a purely diffuse reflection $\tilde{F} = 1/\pi$ and $W_{ref}(\Omega_0, \Omega) = \cos\theta/\pi$, i.e. the reflection is independent of the direction along which the particle impinges upon the surface. In the case of ideal specular reflection

$$\tilde{F}(\Omega_0, \Omega) = \delta(\Omega - \Omega_0)/\sqrt{\cos\theta_0 \cos\theta} = \delta(\cos\theta - \cos\theta_0)\delta(\varphi - \varphi_0)/\cos\theta.$$

Here Ω represents the specular direction relative to Ω_0. In the more general case the reflection indicatrix can be represented in the form of the sum of a specular term and a non-specular term:

$$W_{\text{ref}}(\Omega_0, \Omega) = [1 - F(\Omega)]\delta(\Omega - \Omega_0) + F(\Omega_0, \Omega)\cos\theta, \qquad (4.5)$$

where $F(\Omega_0, \Omega)$ is a function symmetric in the argument, and $F(\Omega)$ is related to $F(\Omega_0, \Omega)$ by the normalization condition (4.4).

When the neutron is reflected (generally speaking, it may be of arbitrary energy up to thermal energies and even higher) from a rough surface the non-specular reflection indicatrix $F(\Omega_0, \Omega)$, as demonstrated by Steyerl (1972*c*) (the derivation will be presented in Chapter 6), is

$$F(\Omega_0, \Omega) = (\zeta^2 l^2 k^2 u_0 / 8\pi k_{0\perp} k_\perp)|t_+(k_{0\perp}) t_+(k_\perp)|^2 \exp(-\kappa_\parallel^2 l^2/2), \qquad (4.6)$$

where $k_\perp = k\cos\theta$, $k_{0\perp} = k\cos\theta_0$, k is the wavevector of the neutron, $\kappa_\parallel^2 = (\mathbf{k}_\parallel - \mathbf{k}_{0\parallel})^2 = k^2(\sin^2\theta_0 + \sin^2\theta - 2\sin\theta_0\sin\theta\cos\varphi)$ is the momentum transfer which is parallel to the surface, ζ is the root-mean-square height, and l is the mean transverse dimension of the roughnesses (to be more precise, the correlation length along the plane). The coefficients $t_+(k_\perp)$ are given in Appendix 1.1, and for $k_\perp^2$, $k_{0\perp}^2$ less than the potential u_0 of the reflecting material $|t_+(k_\perp)|^2 = 4k_\perp^2/u_0$, i.e. expression (4.6) reduces to the form

$$F(\Omega_0, \Omega) = (2/\pi)\zeta^2 l^2 k^4 \cos\theta_0 \cos\theta \exp(-l^2\kappa_\parallel^2/2). \qquad (4.7)$$

Sometimes indicatrixes of the following simpler form are utilized:

$$W_{\text{ref}}(\Omega_0, \Omega) = [1 - GA(\Omega)]\delta(\Omega - \Omega_0) + cGA(\Omega_0)A(\Omega)\cos\theta, \qquad (4.8)$$

where G is a non-specularity parameter. The smaller it is, the higher the fraction of specular reflection. For $A(\Omega) = 1$ we have so-called mixed diffuse–specular reflection. The normalization condition (4.4) leads to $c = 1/\pi$. In this case the fraction of specular reflection, $1 - G$, is independent of the angle of incidence, and the non-specular reflection turns out to be totally isotropic and also independent of the direction Ω_0. When $A(\Omega) = \cos\theta$, the normalization condition (4.4) leads to $c = 3/2\pi$. In this case the properties of the indicatrix are closer to reality since the fraction of specular reflection, $1 - G\cos\theta_0$, increases with the angle of incidence, as is often observed in optics.

Simulation of trajectories by the Monte Carlo method is carried out as follows. First of all, the coordinate at the entrance opening where a trajectory begins is determined. The coordinate distribution is assumed to be homogeneous, so the starting point of the trajectory may happen to be within any interval $d^2\rho$ of the input opening with the same probability. Since $d^2\rho = d\rho^2 d\varphi_\rho/2$, where ρ is the distance from the centre of the entrance and φ_ρ is the azimuthal angle, for choosing ρ and φ_ρ two random numbers are obtained with the aid of a random number generator distributed uniformly within the interval from 0 to 1. One of these numbers is equated to ρ^2 for determining ρ, while the other one is equated to $\varphi_\rho/2\pi$.

The direction of flight of the neutron is then determined. Since in the case of an isotropic angular distribution the flux through a given elementary area $d^2\rho$ of the entrance opening is proportional to $\cos\vartheta_z d\Omega = d\cos^2\vartheta_z d\varphi_z/2$ (ϑ_z is the angle between the axis of the neutron guide and the direction of flight, and φ_z is the azimuthal angle read off from the radius vector ρ at the point being considered), then, as before, two random numbers are generated, one of which is equated to $\cos^2\vartheta_z$ and the other to $\varphi_z/2\pi$. Sometimes it is surprising to see an additional cosine in the angular distribution. In the case of isotropy the number of neutrons travelling in the direction Ω within the interval $d\Omega$ should seem to be proportional to $d\Omega = d\cos\vartheta_z d\varphi_z$. However, since we are interested in the flux through a given elementary area, the amount of neutrons passing through an element $d^2\rho$ is proportional to the projection of the area of $d^2\rho$ seen from the direction Ω. It is thus that the additional factor $\cos\vartheta$ arises.

Having determined the coordinates ρ, φ_ρ, and the direction of flight, ϑ_z and φ_z, of the neutron entering the neutron guide, we then apply the formulae derived in Appendix 4.1 for a cylindrical neutron guide and determine the coordinates at which the neutron collides with the wall, as well as the angles Ω_0. Further, a random number is once again generated, and, if it turns out to be less than the loss coefficient μ, then the trajectory of the neutron is completed and generation of the next trajectory is initiated. If, on the other hand, this random number turns out to be greater than μ, then the trajectory continues to be built up. The next random number permits the determination of whether the reflection will be specular (if the number is less than $1 - F(\Omega_0)$), in accordance with the reflection indicatrix (4.5), or non-specular (if it is greater than $1 - F(\Omega_0)$). When the reflection is non-specular, two more random numbers are generated that are equated to the normalized function

$$F(\Omega_0, \Omega)\cos\theta\, d\Omega \Big/ \int_{2\pi} F(\Omega_0, \Omega)\cos\theta\, d\Omega,$$

from which new reflection angles Ω are found. (This procedure becomes somewhat more complicated if the integrals cannot be found analytically and is simplified if the angles θ and φ can be separated.)

Monte Carlo calculations of neutron guides were performed by Berceanu and Ignatovich (1973), Brown *et al.* (1975), Kügler *et al.* (1980), and others. Berceanu and Ignatovich (1973) studied the transmission of straight neutron guides neglecting absorption and for a reflection indicatrix (4.8) with

$A(\Omega) = 1$ and $A(\Omega) = \cos\theta$. The parameter G was chosen to be equal to 1, 0.5, 0.25, and 0.125 (see Chapter 9). Computations were performed for neutron guides of lengths 1, 2, 4, 8, 16, 32, and 64. Instead of being isotropic, the angular distribution of the outgoing neutrons was observed to be slightly elongated along the axis of the neutron guide. This fact is readily explained qualitatively as follows. In the presence of non-specular reflection, and when the flux is stationary, there is established along the neutron guide a certain UCN density distribution which decreases monotonically from the beginning towards the end of the neutron guide. Since the trajectories of the neutrons leaving the exit opening in the direction $\mathbf{\Omega}$ start from an element of the surface of the wall lying on the intersection of the straight line $\mathbf{\Omega}$ with the surface of the neutron guide, their number increases when the flux increases towards the wall at this site, i.e. the UCN density at the given cross-section of the neutron guide increases. And, since the angle between the direction of flight of the outgoing neutrons and the axis decreases when the corresponding surface element of the wall is closer to the entrance opening, where the neutron density is higher, there are more neutrons travelling at smaller angles to the axis than at large angles to the axis. If the probability of specular reflection increases with the angle of incidence of the UCN to the wall (for instance, for $A(\Omega) = \cos\theta$), then the angular distribution at the exit exhibits a bump within a narrow cone of angles close to the axis, corresponding to those neutrons which as a result of purely specular reflection have arrived without losses from the entrance to the output opening. As the neutron guide becomes longer, the range of angles within which passage of neutrons is possible shrinks and the corresponding bump becomes more narrow. On the other hand, since the amount of neutrons passing through the neutron guide at other angles falls monotonically as the length increases, the relative height of the bump rises monotonically. If the probability of non-specular reflection is independent of the angle of incidence ($A(\Omega) = 1$), then the angular distribution at the output, starting from a certain length of the neutron guide, reaches the asymptotic form and no longer depends on the length. Experiments, the results of which are presented in Fig. 4.5, reveal that the angular distribution actually does develop gradually as the length of the neutron guide increases, and it turns out to be elongated in the forward direction instead of being isotropic. However, as the resolution is insufficient, it is not possible to establish whether or not the angular distribution exhibits a narrow bump.

It must be pointed out, however, that in Monte Carlo computations the neutron is considered not to be subject to gravity, so it may travel in a direction parallel to the axis for an infinitely long time. Actually, the gravitational field does bend the neutron trajectory, and so there are no neutrons at all travelling at very small angles to the axis of the neutron guide. Correspondingly, there can be no bump tending to be infinitely narrow on the angular distribution of the outgoing UCN, i.e. at some time or other, whatever the law of reflection from the walls, the angular distribution will acquire an asymptotic form that will further be independent of the length of the neutron guide. In the experiment by Kosvintsev *et al.* (1976*b*) (see Fig. 4.5) the asymptotic angular distribution at the output of the neutron guide, made of electropolished stainless steel, was found to be described by the expression $4\cos^3 \vartheta_z \mathrm{d}\cos \vartheta_z$, where ϑ_z was the angle between the axis of the neutron guide and the direction of flight of the outgoing neutrons. Contrary to this, an isotropic distribution is of the form $2\cos \vartheta_z \mathrm{d}\cos \vartheta_z$.

Brown *et al.* (1975) performed Monte Carlo calculations utilizing the reflection indicatrix (4.5) with the function $F(\Omega_0, \Omega)$ defined in (4.7). The quantities $c = k^2 l^2/2$ and $f = 4k^4\zeta^2 l^2/3$ served as parameters. It was found that the angular distribution at the exit for $L = 64$, $f = 0.503$, $c = 2.193$ could be represented in the form $(0.8 + 2.4\cos^2 \vartheta_z)\cos \vartheta_z \mathrm{d}\cos \vartheta_z$. When the neutron wavelength is 750 Å, the chosen parameters f and c correspond to the roughnesses $\zeta = 35$ Å and $l = 250$ Å, which in accordance with the observations of Steyerl (1972*c*) are typical for electropolished copper. Note that the attenuation of a beam of UCN by the neutron guide caused by the reflection indicatrix (4.5) and (4.7) is less significant, the greater the parameter c in the exponent, since, then, the cone of angles is narrower about the specular direction into which neutrons, having undergone non-specular reflection, may be scattered, i.e. the non-specular reflection is nearly specular.

Mysenholder (1976) applied the Monte Carlo method to study the transmission of bends of a neutron guide. The resistance of a bend was determined through the ratio of the transmission of a bent neutron guide to the transmission of a straight neutron guide of the same length. In the case of purely specular reflection and a loss coefficient of 0.02 this author found that bends of 45° and 90° attenuate the UCN flux by 4% and 22%, respectively, if the arms of the elbows are $L_1 = L_2 = 40$. The experiments of Kosvintsev *et al.* (1976*b*) made it possible to establish the empirical dependence of the transmission of the bend of a neutron guide, made of electropolished stainless steel, on its angle α for $L_1 = L_2 = 5$. If the angular distribution at the entrance to the elbow is isotropic, then the transmission of the bend may be described by the expression $W_{\mathrm{bnd}} = 0.072 + 0.28\cos\alpha$; if, on the other hand, it is not isotropic, but as formed by a long neutron guide, $W_{\mathrm{bnd}} = 1 - 0.23\alpha$. Note that the resistance of a bend is smaller when the fraction of non-specular UCN reflection from the walls is high. In the case of purely diffuse reflection the transmission is no longer affected by the existence of bends when $L_1 = L_2 = 2$, as shown by the calculations of *Davis (1960).

Monte Carlo calculations were also performed for neutrons of energies somewhat higher than UCN energies. Thus, for instance, Kügler *et al.* (1980) studied the transmission of smooth bends for neutrons with velocities up to 25 m s^{-1}. (These neutrons were directed toward a toroidal magnetic trap.) The spectrum of the outgoing neutrons that succeeded in passing was measured by their time of flight. The calculation is in agreement with experiment if, for example, for a velocity of 10 m s^{-1} the loss coefficient for each collision with the wall is assumed to be equal to 3%. Naturally, this loss coefficient depends mainly on non-specular reflection.

In the experiment of Kügler *et al.* (1980) an interesting peculiarity was observed, in that if the curved part of a neutron guide is partitioned into a series of narrow parallel channels that are also curved (Fig. 4.10), the total

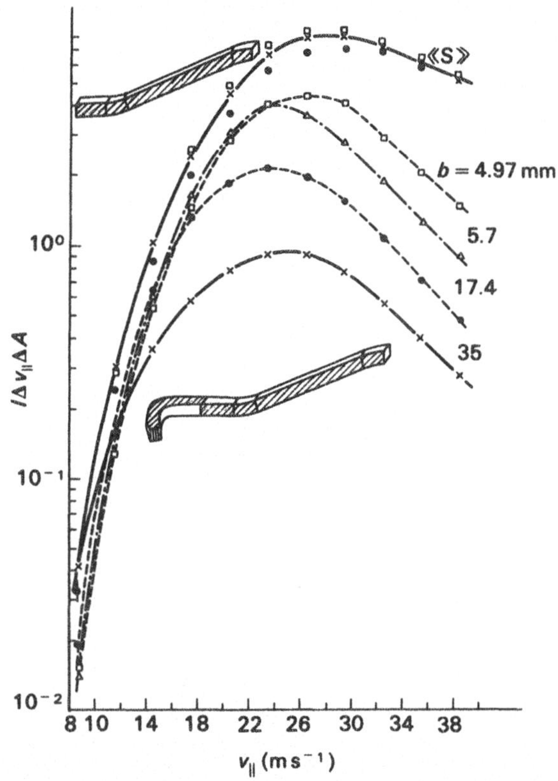

Fig. 4.10 Time-of-flight spectrum at the output of the neutron guide shown in the lower part of the picture for different widths of the slitlike sections into which the turn to the left through 105° is partitioned. S indicates the spectrum at the exit from the section of the neutron guide depicted at the top of the picture. Various points on this spectrum correspond to measurements separated in time. The neutron intensity within the middle range of velocities increases slightly with time (during an interval of 2–3 days), which may indicate that the surfaces are gradually liberated from contaminations containing hydrogen at room temperature and a vacuum of 10^{-3} mm Hg. The neutron guide is made of glass plates and of plexiglas plates which are subsequently coated with nickel. The neutron guide shown at the top of the picture has two smooth bends of 22.5° curvature with a radius of 25 cm and two straight section 644 and 176 mm long (Kügler *et al.* 1980).

transmission increases as the channels are made narrower. This phenomenon may be explained as follows. When the channel is narrow, a neutron travels a small distance between two collisions with the walls and along which the wall cannot make a significant turn, so the grazing angle and consequently the probability of non-specular reflection and the probability of losses always turn out to be small. Figure 4.10 testifies that over-limiting neutrons may, with a high probability, surmount bends, even if the radius of their curvature is relatively small. This is also confirmed by Fig. 4.11 (Ageron *et al.* 1978*a*), from which it follows that even four 90° bends do not make it possible to clear the beam of UCN completely of faster neutrons.

4.3 THEORETICAL MODELS

Monte Carlo methods are inconvenient for fitting parameters to experimental data. From this point of view they are less advantageous than analytic methods, although the latter often require approximations, the precision of

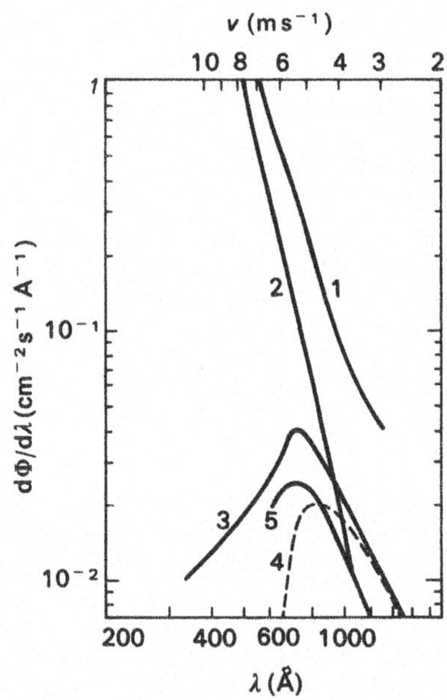

Fig. 4.11 UCN spectra measured at the inclined channel at ILL (Grenoble) (Ageron *et al.* 1978*a*): 1, time-of-flight spectrum for a non-collimated neutron beam; 2, time-of-flight spectrum for a collimated neutron beam; 3, time-of-flight spectrum measured after a filter representing a horizontal neutron guide system of electropolished stainless steel with four sharp 90° bends; 4, calculated time-of-flight UCN spectrum, which at large wavelengths asymptotically approaches the experimentally measured spectrum; 5, spectrum measured with the aid of a gravitational spectrometer (Π-formed crank).

which is not always sufficiently well established. We shall consider several theoretical transport models which allow the derivation of analytic expressions for the transmission of a neutron guide.

4.3.1 Specular propagation of UCN

The simplest version is specular propagation along a straight cylindrical neutron guide. The losses occurring in collisions with the wall can be described by a phenomenological parameter, μ. Then the total transmission of a neutron guide of length L can be represented in the form:

$$W_{tr}(L) = \overline{(1-\mu)^{n(\Omega, \boldsymbol{\rho})}}, \tag{4.9}$$

where $n(\Omega, \boldsymbol{\rho})$ is the number of collisions with the walls of the neutron guide of a neutron incoming through point $\boldsymbol{\rho}$ of the input opening in the direction Ω. The bar over the expression in (4.9) signifies averaging over a uniform distribution in $\boldsymbol{\rho}$ and an isotropic distribution in Ω. Obviously, $n(\Omega, \boldsymbol{\rho}) = L/\zeta(\Omega, \boldsymbol{\rho})$, where $\zeta(\Omega, \boldsymbol{\rho})$ is the UCN path length along the neutron guide in between two successive collisions. The mean value of $1/\zeta(\Omega, \boldsymbol{\rho})$ is calculated in Appendix 4.1 and equals 2. If μL is small, then expression (4.9) may be expanded in powers of μ, the expansion being limited to the linear term. As a result, we obtain $W_{tr}(L) = 1 - 2\mu L$. For large values of μL expression (4.9) can be represented in the form

$$W_{tr}(L) = \exp(-2\mu L).$$

If the surface of the wall is not too rough and the reflection indicatrix is of the form (4.8) with a small G (for simplicity we shall assume $A(\Omega) = 1$), then, if the length of the neutron guide is small ($GL \ll 1$), a neutron can be considered to be reflected with a probability of $G/2$ in the backward direction each time it impinges upon the wall and is, thus, lost. Then $G/2$ must be substituted for μ in (4.9), and the resulting transmission of a quite short neutron guide is of the form $W_{tr}(L) = 1 - GL$. Comparison with experiments reveals that in the case of good treatment (e.g. polished glass spray coated with nickel (Egorov *et al.* 1974)) a value of $G \approx 0.3\%$ can be obtained. In the case of electropolished stainless steel (see Fig. 4.2), for $L = 100$ we have the transmission $W_{tr}(L) = 1 - 0.006L$; hence it follows that $G \approx 0.6\%$. If GL is considered to be large and the formula $\exp(-GL)$ is adopted for the transmission, then the nonspecularity in a chemically polished copper neutron guide (see Fig. 4.1) amounts to $G \approx 1\%$, while in a non-treated stainless steel neutron guide $G \approx 4\%$ (Fig. 4.2).

However, for large L this approach is not justified, since the neutrons undergoing backward reflection may, in a long neutron guide, be reflected non-specularly in the forward direction, which has to be taken into account.

We shall, therefore, consider another model—that which was the first to appear in papers on UCN (Groshev *et al.* 1971)—namely, the model of neutron diffusion. From the outset it was clear that, unlike thermal neutrons, UCN can diffuse along a neutron guide like a rarefied (molecular) gas. In the case of a molecular gas diffusion processes have long been studied relatively well. Moreover, the fraction of diffuse reflection has been established to exceed 90% in the collision of an atom with a wall. Therefore, it seems reasonable to start by considering purely diffuse reflection from walls.

4.3.2 The Clausing equation (*Clausing 1971)

The most precise formulation of the problem for this case was given by Clausing. He wrote down an integral equation for the probability of an atom entering a straight pipe through one opening and leaving it through the other opening, considering the angular distribution at the entrance to be isotropic (see Appendix 4.2). He also proposed an approximate and quite cumbersome solution of this problem, which was successfully verified experimentally (*Levenson *et al.* 1960) for lengths $L \leqslant 6$ and by Monte Carlo numerical calculations (*Davis 1960) for lengths $L \leqslant 5$. The Clausing approximate analytic solution differs little from the more simple expression

$$W_{tr}(L) = 1/(1 + 3L/8), \tag{4.10}$$

obtained by Dushman (*Clausing 1971), and at large L the two solutions coincide asymptotically. The maximum difference corresponds to $L \approx 1$ and amounts to 10%. Expression (4.10) is readily obtained from the diffusion equation, into which, unlike the Clausing equation, it is easy to introduce partial specularity and losses in the neutron guide. We shall, therefore, turn directly to the diffusion equation.

4.3.3 The diffusion equation

In the diffusion model the neutron density $n(z, t)$ is introduced, where t is the time and z is the coordinate along the neutron guide. The flux $J(z, t)$ is determined through the density gradient and the diffusion coefficient D:

$$J(z, t) = -D\mathrm{d}n(z, t)/\mathrm{d}z. \tag{4.11}$$

In the stationary case the density is independent of time, and in the absence of losses inside the neutron guide the current along the neutron guide must be constant. Hence we obtain the equation

$$\mathrm{d}J(z)/\mathrm{d}z = -D\mathrm{d}^2 n(z)/\mathrm{d}z^2 = 0, \tag{4.12}$$

which has the obvious solution

$$n(z) = a + bz. \tag{4.13}$$

If losses exist, then we may introduce the neutron lifetime τ, the time until a neutron is lost in the neutron guide, and then, instead of (4.11), we obtain the equation

$$-Dd^2n(z)/dz^2 = -n(z)/\tau, \tag{4.14}$$

the solution of which also has a simple form:

$$n(z) = a\exp(-z/L_D) + b\exp(z/L_D), \tag{4.15}$$

where the constant $L_D = (D\tau)^{1/2}$ is termed the diffusion length.

In a non-stationary case the equation becomes even more complicated:

$$\partial n(z, t)/\partial t = Dd^2n(z, t)/dz^2 - n(z, t)/\tau.$$

Let us return, however, to eqn (4.14) and the solution (4.15). This solution involves two constants, a and b, the determination of which requires formulation of boundary conditions. Usually, the one-dimensional diffusion problem is considered on the finite interval $0 \leqslant z \leqslant L$, and for determining the constants a and b in (4.15) it is sufficient to define only two boundary conditions at the two ends. For example, we may give the external fluxes incident on both openings (Ignatovich and Terekhov 1978).

We shall go another way, which seems to be more fruitful. At input and output openings we shall define two conditions which have quite a clear physical meaning: the continuity conditions for the density and flux. This will permit us not only to solve the simplest problem of determining the transmission of a neutron guide of finite length, but also to find the transmission of significantly more complicated branched systems, for instance, such as the one shown in Fig. 4.9.

To demonstrate the essential features of the method we shall consider a semi-infinite neutron guide directed along the axis $z \geqslant 0$. If the UCN density inside the neutron guide is created by the neutrons incident on its entrance opening from the outside, then the density inside the guide can only fall exponentially with the distance from the input opening. In other words, the solution of (4.15) in this case has the form

$$n(z) = a\exp(-z/L_D), \tag{4.16}$$

i.e. it involves a single arbitrary constant. We shall denote the density of the neutrons impinging upon the input opening from the outside by n_0, and the density of the reflected neutrons by rn_0, while the constant a in (4.16) will be denoted by tn_0. The quantities r and t, respectively, represent the reflection and transmission coefficients of the input opening.

The total neutron density outside the neutron guide equals $(1 + r)n_0$. Obviously the density inside the guide at the point $z = 0$ must be the same. From this continuity condition for the density at the input opening follows the first equation for determining r and t:

$$1 + r = t. \tag{4.17}$$

The second equation follows from continuity of the flux. The incident flux is $J_0 = n_0 v/2$ and the internal flux is $-D\mathrm{d}n(z=0)/\mathrm{d}z = tn_0 D/L_\mathrm{D}$, while the reflected flux may be set equal to $\alpha r J_0$, where the new constant α is introduced and which represents a quantity dependent on the angular distribution of the reflected neutrons. In the simplest diffusion theory the angular distribution is considered nearly isotropic. Therefore, we shall assume the reflected neutrons to be distributed isotropically in the backward hemisphere and we shall set $\alpha = 1$. Then, from the continuity condition for the flux follows the second equation for determining r and t:

$$\text{a)}\ 1 - r = qt, \qquad \text{b)}\ q = 2D/vL_\mathrm{D}. \tag{4.18}$$

The solution of eqns (4.17) and (4.18a) has the form

$$\text{a)}\ r = (1-q)/(1+q), \qquad \text{b)}\ t = 2q/(1+q). \tag{4.19}$$

If $\alpha \neq 1$, then αq must be substituted for q in expressions (4.19).

For a semi-infinite neutron guide the problem is, thus, solved. Now let us find the reflection and transmission of a finite-length neutron guide. In principle, we may directly define the solution in the form (4.15) and, with the aid of four matching conditions at the two points $z = 0$ and $z = L$ leading to a set of four equations, find both the coefficients a and b and the transmission and reflection coefficients T and R of a finite-length neutron guide. We shall not, however, take this way; instead we shall take advantage of the elegant method of recursive relations, to be often applied later. Let us demonstrate this method by applying it to our problem.

4.3.4 The transmission and reflection of a finite-length neutron guide. The recursion relations method

Let us choose two cross-sections 1 and 2 of the semi-infinite neutron guide at points z_1 and z_2 such that $z_2 - z_1 = L$. We denote the neutron density at these points by n_1 and n_2. Since the density, in accordance with (4.16), decreases exponentially along the guide, then

$$n_2 = en_1 \tag{4.20}$$

where e is given by

$$e = \exp(-L/L_\mathrm{D}). \tag{4.21}$$

The neutron density n is conveniently represented as the sum $n = \vec{n} + \overleftarrow{n}$, where an arrow shows the direction of the neutron velocity projection on to the neutron guide axis (for definiteness we admit the density to decrease from left to right). Let us find $\vec{n}_2$. First, we imagine it to be known. Then we can write down $\overleftarrow{n}_2$ as

$$\overleftarrow{n}_2 = r\vec{n}_2, \tag{4.22}$$

where r is the reflection coefficient of the entrance orifice of the semi-infinite part of the neutron guide to the right of the point z_2. For $\overleftarrow{n}_1$ we can write two expressions, one of which is similar to (4.22):

$$\overleftarrow{n}_1 = r\vec{n}_1 = \vec{R}\vec{n}_1 + \overleftarrow{T}\overleftarrow{n}_2. \tag{4.23}$$

Here the reflection coefficient from left to right, $\vec{R}$, and the transmission coefficient from right to left, $\overleftarrow{T}$, are introduced. In our case reflection and transmission in both directions are the same, so we shall use the notation R and T without arrows. The component $\vec{n}_2$ of the density n_2 can be represented in the form

$$\vec{n}_2 = R\overleftarrow{n}_2 + T\vec{n}_1. \tag{4.24}$$

Here the first term is due to reflection of the component $\overleftarrow{n}_2$ from the neutron guide section of length L, and the second to the transmission of the component $\vec{n}_1$ through the same section.

Substituting (4.22) into (4.23) and (4.24) we obtain the recursion relations:

$$\text{a) } r\vec{n}_1 = R\vec{n}_1 + Tr\vec{n}_2, \qquad \text{b) } \vec{n}_2 = Rr\vec{n}_2 + T\vec{n}_1. \tag{4.25}$$

From (4.25b) it follows that

$$\vec{n}_2 = (1 - Rr)^{-1}T\vec{n}_1. \tag{4.26}$$

Substituting (4.20) into (4.25) we obtain two equations to determine R and T:

$$re = Re + Tr, \tag{4.27}$$

$$e = Rr + T. \tag{4.28}$$

The solution of these equations is

$$R = r(1 - e^2)/(1 - r^2e^2), \tag{4.29}$$

$$T = e(1 - r^2)/(1 - r^2e^2). \tag{4.30}$$

Note that when the losses decrease, eqn (4.14) transforms into (4.12) and (4.30) transforms into $T = 1/(1 + Lv/4D)$, which coincides with (4.10) when $D = 2v/3$.

4.3.5 Diffusion of UCN in composite and branched neutron guides

Consider a neutron guide consisting of two sections, 1 and 2. The reflection R_i and the transmission T_i of the ith section ($i = 1, 2$) are given by expressions (4.29) and (4.30) with appropriate parameters. The question is how to find the reflection R_{12} and the transmission T_{12} of the whole system. To answer this question it is necessary to take advantage of relations (4.25b) and (4.26), substituting R_i and T_i for R and T, and R_2 for r. In the relation (4.25a) the coefficient r on the left-hand side of the equation is to be replaced by R_{12}. The transmission of the whole system can be represented as

$$T_{12} = T_2 \vec{n}_2.$$

Substituting (4.26) here and in (5.25a) we obtain equations, the solution of which, in the new notation, is equal to

$$\text{a) } T_{12} = T_1 T_2/(1 - R_1 R_2), \qquad \text{b) } R_{12} = R_1 + T_1^2 R_2/(1 - R_1 R_2). \tag{4.31}$$

Now let us consider the chain shown in Fig. 4.9. It consists of three sections of lengths L_1, L_2, and L_3, respectively. We shall assume the density of UCN incident upon the entrance opening of the primary section (point C) to be unity. A detector is placed at the output of section 3 (point D), and a sample exhibiting an absorption probability of M is at the output of section 2 (point S). The problem is to determine the UCN counting rate J of the detector as a function of M. Clearly, J depends on the transmission T_{123} of the chain 13 in the presence of section 2.

Now we shall assume each section to be characterized by its own diffusion length $L_{Di} = (D_i \tau_i)^{1/2}$, where D_i and τ_i are the diffusion coefficient and the lifetime in the ith section, respectively. The transmission of each section are found from formulae (4.30).

If the output opening of a section is covered with a diaphragm of transmission M and reflection $\rho = 1 - M$, then the transmission T_M and the reflection R_M of the resulting system is computed by applying formulae (4.31), considering the diaphragm as the second section of a double-section neutron guide:

$$\text{a) } T_M = MT/(1 - \rho R), \qquad \text{b) } R_M = R + T^2 \rho/(1 - \rho R). \tag{4.32}$$

(Note that in measuring the transmission T_M as a function of M we can determine separately the transmission and reflection of the non-diaphragmed neutron guide and, thus, determine two of the most essential parameters for the description of diffusion: D and τ.)

For the determination of the transmission of a branched system it is necessary to find the matching conditions at the branching point 0. Owing to the continuity of density and flux the following conditions must obviously be satisfied at the branching point 0:

$$\text{a) } n_1 = n_2 = n_3, \qquad \text{b) } j_1 = j_2 + j_3. \tag{4.33}$$

Now we introduce at point 0 a fictitious vacuum gap and denote by N_i the component of neutron density due to neutrons incident on the opening of the ith section present in the gap. From condition (4.33a) it follows that

$$T_1 + (1 + R_1)N_1 = (1 + R_2)N_2 = (1 + R_3)N_3, \tag{4.34}$$

where R_i is the reflection coefficient of the ith section.

With the aid of the density N_i we may readily write the expression for the respective flux. Assuming the N_i to be distributed isotropically in the respective hemispheres, we find, for example, the resulting flux in the gap at 0 from section L_0:

$$j_1 = [T_1 - (1 - R_1)N_1]v/2. \tag{4.35}$$

Thus, condition (4.33b) leads to the equation

$$T_1 - (1 - R_1)N_1 = (1 - R_2)N_2 + (1 - R_3)N_3. \tag{4.36}$$

The solution of the set (4.34), (4.36) reduces to the form

$$\begin{aligned} &\text{a) } N_1 = c(1 - \eta_2 - \eta_3)/(1 + R_1), \quad \text{b) } N_i = c(1 + \eta_1)/(1 + R_i), \quad \text{c) } i = 1, 2 \\ &\text{d) } c = T_1/(\eta_1 + \eta_2 + \eta_3), \quad \text{e) } \eta_j = (1 - R_j)/(1 + R_j), \qquad \text{f) } j = 1, 2, 3. \end{aligned} \tag{4.37}$$

The counting rate of the detector is determined by the quantity $J_d = T_3 N_3 v/2$; its ratio to the primary flux $v/2$ incident on the entrance opening S determines the transmission $T_{1_2 1}$:

$$T_{1_2 3} = T_3 N_3 = 2T_1 T_3/(1 + R_1)(1 + R_3)(\eta_1 + \eta_2 + \eta_3). \tag{4.38}$$

Absorption at the end of section 2 can be modelled by a diaphragm the transmission of which, M, is proportional to the absorption. In this case the transmission and the reflection of section 2 must be modified in accordance with expressions (4.32). As a result, the counting rate of the detector as dependent on M will be represented in the form:

$$j_d(M) = j_d(0)(1 + bM)/(1 + aM), \tag{4.39}$$

where

$$\begin{aligned} a &= [R_3(1 - R_3)(1 + \eta_1 + \eta_2) + T^2(1 - \eta_1 - \eta_2)]/[(1 - R_3)^2 - T_3^2 \\ &\quad + (\eta_1 + \eta_2)(1 - R_3^2 + T_3^2)], \\ b &= [R_3(1 - R_3) - T_3^2]/(1 - R_3^2 + T_3^2). \end{aligned} \tag{4.40}$$

Until now we have assumed a monochromatic spectrum, but this restriction can now be abandoned. For an incident neutron spectrum $W(v)$ the counting rate of the detector is

$$J = \int W(v) j_d(v) v \, dv/2. \tag{4.41}$$

This formula may serve as the basis for comparison of theory and experiment.

Experiments were carried out at the UCN channel of the IBR-2 reactor. The channel diameter was 20 cm. A T joint was connected to the output of the channel which included a single 90° bend and three 60° bends. All channel sections and the T joint were made of electropolished copper tubes with internal diameters of 0.2 m. The channel itself and the branch of the T joint of common length equal to 13.3 m correspond to the left section in Fig. 4.9. The other two branches of the T joint with identical lengths equal to 0.34 m correspond to the sections L_2 and L_3. The sections 2 and 3 are perpendicular to each other.

A polyethylene disk with an area of S_p was used as a sample. The absorption of such a sample is $M = S_p/S$, where S is the area of the cross-section of the neutron guide: $S = 314\ \text{cm}^2$.

The quantity that is measured experimentally is

$$\gamma = J(0)/J(M), \tag{4.42}$$

where J is the counting rate of the detector determined in accordance with expression (4.41). When $M = 0$, no absorber is present in the place of the sample, and the corresponding branch of the T joint is just closed tightly at the output opening by a wall reflecting UCN.

Two calibration curves (eqn (4.42)) are presented in Fig. 4.12 corresponding to two different positions of the sample. Curve 1 corresponds to the case when the absorber is situated at the output of section 2, and curve 2 to the absorber at the output of section 3. These curves are fully described by four parameters corresponding to three different diffusion lengths L_{Di} and to the common loss coefficient μ. The loss coefficient determines the lifetime τ in eqn (4.14):

$$\text{a) } \tau = l_f/\mu v, \qquad \text{b) } \mu = 2\eta(\arcsin y - y\sqrt{1-y^2})/y^2, \tag{4.43}$$

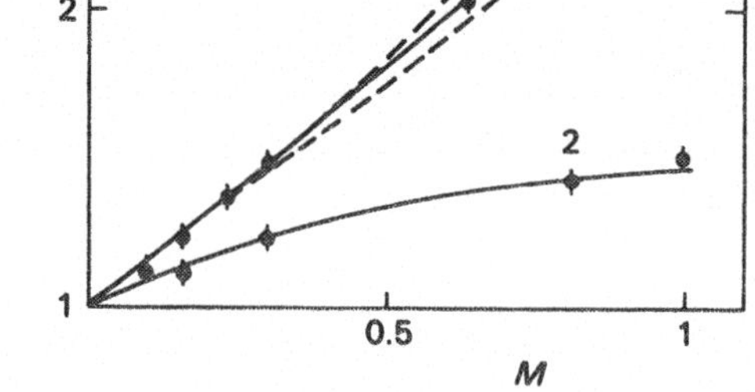

Fig. 4.12 Dependence of γ on the absorption probability by the sample. The solid lines 1, 2 are the result of the fit indicated in the text. The dashed lines 3, 4 are the result of calculations with L_{D3}^2 differing from the optimal value by $\pm$ 15%, respectively.

where l_f is the mean free path between two collisions with the walls, the coefficient μ represents expression (12a) (given in Appendix 1.1) averaged over the angles, $y = v/v_{Cu}$, and v_{Cu} is the limiting velocity of copper, $5.67\ m\,s^{-1}$. The reduced loss coefficient η was chosen to be equal to 2×10^{-3}, a value close to the experimentally measured one for the material of the neutron guide subjected to treatment according to the given technology. Simultaneous fitting of parameters L_{Di} for both curves reveals that $L_{D1}^2 : L_{D2}^2 : L_{D3}^2 = 12.0:3.5:0.55$. We can see that with the use of four parameters a good description of the entire set of experimental data is achieved. In performing the fit the UCN spectrum in the neutron guide was considered to be Maxwellian, but the experimental results are also well described in the approximation of a monochromatic spectrum if we adopt, as the velocity, the velocity $\langle v \rangle$ averaged over the whole spectrum. The mean velocity can be also considered as a parameter to be fitted.

The good agreement of the fitted values with the experimental data cannot be considered as proof of the validity of the simple diffusion theory. Moreover, the factors pointing to the inconsistency of the developed theory can be immediately indicated. When the sample and the detector are transposed, the counting rate of the latter changes by a factor of 1.4. This fact is the main testimony against elementary diffusion theory, since, in accordance with expression (4.38), the counting rate of the detector should not depend on the mutual positions of the detector and sample, if the branches L_2 and L_3 are identical. The defect of the elementary diffusion theory, which correctly predicts an exponential decrease of the transmission of a neutron guide (Groshev *et al.* 1971) as its length increases ($L_D = 100$), is that in no way does it take account of angular distributions which most likely play an essential role in the situation under consideration.

We shall, therefore, turn to the microscopic diffusion theory which provides information on angular distributions and makes possible computation of the diffusion coefficient that, in elementary diffusion theory, represents a phenomenological parameter. In doing so it is necessary to choose the law governing neutron reflection from a wall.

The reflection indicatrix determines the UCN angular distribution in the neutron guide and the features of the random walk of a neutron in it. The diffusion coefficient can be determined either with the aid of random walk theory (see Appendix 4.3), or with the aid of microscopic kinetic diffusion theory. We shall deal in more detail with the latter. In principle, a theory can be developed for any arbitrary indicatrix, but to simplify our task we shall limit our consideration here and later on to expression (4.8) with $A(\Omega) = 1$ or $\cos\vartheta$. For simplicity we shall also consider the coefficient μ to be independent of the angle at which the neutron impinges upon the wall and of its velocity, if not stated otherwise.

4.3.6 Microscopic description of neutron diffusion

Consider an infinitely long straight cylindrical neutron guide and denote the coordinate along its axis by z. Let the number of neutrons per unit volume at point z, in the vicinity of the wall and travelling in direction Ω, be $n_w(z, \Omega)\mathrm{d}\Omega$. The number of neutrons within the interval of angles $\mathrm{d}\Omega$ that impinge per unit time upon a surface area $\mathrm{d}s_w$ of the wall of the neutron guide at point z from direction Ω is given by

$$\mathrm{d}N^+(z, \Omega) = n_w(z, \Omega)v \cos \vartheta_n \mathrm{d}s_w \mathrm{d}\Omega, \tag{4.44}$$

where v is the neutron velocity and ϑ_n is the angle between direction Ω and the normal vector to the area under consideration. The number of neutrons leaving this surface area within the range of angles $\mathrm{d}\Omega$ is

$$\mathrm{d}N(z, \Omega) = (1 - \mu)\mathrm{d}\Omega \int n_w(z, \Omega'')v \cos \vartheta'' \mathrm{d}s_w W_{\mathrm{ref}}(\Omega'', \Omega)\mathrm{d}\Omega''. \tag{4.45}$$

Upon travelling a certain distance $\zeta_{ww}(\Omega)$ along the neutron guide, the neutrons will once again collide with the wall impinging upon a section of the wall of area $\mathrm{d}s'$. We shall denote the range of solid angles within which these neutrons are located, near the area $\mathrm{d}s'$, by $\mathrm{d}\Omega'$. From the condition of detailed balance it follows directly that

$$\mathrm{d}s_w \cos \vartheta_n \mathrm{d}\Omega = \mathrm{d}s' \cos \vartheta' \mathrm{d}\Omega', \tag{4.46}$$

where ϑ' is the angle between the normal vector to the plane area $\mathrm{d}s'$ and the direction along which the neutron impinges upon this area. If the laws of motion are known, then it is not difficult to relate $\mathrm{d}\Omega$ with $\mathrm{d}s'$, ϑ_n with ϑ', and $\mathrm{d}\Omega'$ with $\mathrm{d}s_w$. For example, in the case of free motion we have:

$$\text{a) } \mathrm{d}\Omega = \mathrm{d}s' \cos \vartheta'/l^2, \qquad \text{b) } \vartheta_n = \vartheta', \qquad \text{c) } \mathrm{d}\Omega' = \mathrm{d}s_w \cos \vartheta_n/l^2,$$

where l is the distance between the surface areas along direction Ω. Now, we substitute the indicatrix (4.8) into (4.46) and express the number of neutrons incident upon $\mathrm{d}s'$ at point $z + \zeta_{ww}$ through the number of neutrons impinging upon the area $\mathrm{d}s_w$ at point z:

$$\begin{aligned}\mathrm{d}N^+(z + \zeta_{ww}, \Omega) &= n_w(z + \zeta_{ww}, \Omega)v \cos \vartheta_n \mathrm{d}s_w \mathrm{d}\Omega \\ &= (1 - \mu)\{[1 - GA(\Omega)]n_w(z, \Omega) \\ &\quad + GA(\Omega)c \int n_w(z, \Omega'')\cos \vartheta'' A(\Omega'')\mathrm{d}\Omega''\}v \cos \vartheta_n \mathrm{d}s_w \mathrm{d}\Omega.\end{aligned}$$

Hence it follows that

$$\begin{aligned}n_w(z + \zeta_{ww}, \Omega) = (1 - \mu)\{&[1 - GA(\Omega)]n_w(z, \Omega) \\ &+ GA(\Omega)c \int n_w(z, \Omega'')\cos \vartheta'' A(\Omega'')\mathrm{d}\Omega''\}.\end{aligned} \tag{4.47}$$

Far from the entrance opening the density $n_w(z, \Omega)$ achieves the asymptotic form. There it is determined by the diffusion equation and exhibits a functional dependence on z of the form (4.16). We shall denote this density by

$$n_d(z, \Omega) = C\nu_d(\Omega)\exp(-z/L_D), \tag{4.48}$$

where the constant C is determined from the boundary conditions and the function $\nu_d(\Omega)$ satisfies the normalization condition

$$c\int \cos\vartheta A(\Omega)\nu_d(\Omega)d\Omega = 1. \tag{4.49}$$

Substituting (4.48) into (4.47) we obtain, taking into account (4.49), the following:

$$\nu_d(\Omega) = (1-\mu)GA(\Omega)/\{\exp[-\zeta_{ww}(\Omega)/L_D] - (1-\mu)[1-GA(\Omega)]\}. \tag{4.50}$$

Substituting (4.50) into (4.49) we obtain the equation for determining L_D:

$$\frac{c\int \cos\vartheta_n A(\Omega)(1-\mu)GA(\Omega)}{\{\exp[-\zeta_{ww}(\Omega)/L_D]-(1-\mu)[1-GA(\Omega)]\}}\,d\Omega = 1. \tag{4.51}$$

The integral in (4.51) converges badly, since ζ_{ww} is an unlimited function. If, however, gravity is taken into account, then the function ζ_{ww} turns out to be finite and the integral converges. Assuming L_D to be a large parameter, we expand (4.51) into a series in powers of $1/L_D$ and restrict our consideration to the quadratic term. When such an approximation is applied, the integral in (4.51) converges even without gravity being taken into account.

For the case of $A = 1$ we have $c = 1/\pi$, and in the lowest order in μ we obtain

$$1/L_D^2 \approx 3G\mu/4(2-G). \tag{4.52}$$

It is interesting to note that if the diffusion coefficient is calculated on the basis of the random walk theory (see Appendix 4.3), then we obtain $D = \frac{2}{3}v(2-G)/G$. The lifetime τ equals $l/v\mu$, where l is the mean free path in between two collisions with the walls, which in the case of a cylindrical neutron guide of unit radius is equal to 2. Hence $L_D^2 = D\tau = 4(2-G)/3G\mu$, which is in good agreement with expression (4.52).

When $A = \cos\vartheta_n$ ($c = 3/2\pi$), gravity must be taken into account. If it is not, then the integral in (4.51) increases without limit as the loss coefficient μ decreases. The integral in the power expansion up to $1/L_D^2$ is prevented from diverging by the μ in the denominator of (4.51). In the case of small μ it is convenient to neglect μ in the denominator when computing the integral in (4.51). The computations then become simpler. But in this case it becomes necessary to introduce a cut-off at $\vartheta_z \approx \mu/G(1-\mu)$. Taking into account gravity also leads to an effective cut-off of the integral (4.51). This cut-off is due to the fact that the path between two collisions with the walls cannot, for purely kinematic reasons, be greater than $v(4r/g)^{1/2}$, where r is the radius of the neutron guide, v is the neutron velocity, and g is the gravitational constant. Equating this quantity to the geometric distance $2r\cos\vartheta_z$ we obtain the equivalent cut-off angle $\vartheta_z \approx (rg/v^2)^{1/2}$. Naturally, of the two cut-off parameters the greatest must be chosen. Thus, for example, if the neutron guide is filled with a gas, the losses may be great. In this case gravity may be neglected for any reflection indicatrix. Thus, computation of the integral (4.51) up to the second order in $1/L_D^2$ leads to the expression

$$1/L_D^2 \approx \mu G/4[\ln(2/u) - G/3], \tag{4.53}$$

where

$$u = \max(\sqrt{rg/v^2}, \mu/G(1-\mu)). \tag{4.54}$$

4.3.7 Angular distribution and the diffusion coefficient

Knowledge of L_D makes it possible to determine the UCN angular distribution as well as their distribution over the cross-section. Indeed, the flux density of neutrons travelling in direction Ω in the element $d\Omega$ and crossing at point z an area $d^2\rho$ near the point $\boldsymbol{\rho}$ of the cross-section (ρ is the radius) depends on the number of neutrons leaving the area ds_w at the intersection of the line Ω with the wall of the neutron guide, and situated at a distance $\zeta_{ws}(\rho, \Omega)$ from z:

$$n(z, \rho, \Omega) \cos \vartheta_z \, d\Omega_s \, d^2\rho = ds_w \, d\Omega_w \, dN_w(z - \zeta_{ws}(\rho, \Omega), \Omega), \tag{4.55}$$

where ϑ_z is the angle between direction Ω and the axis of the neutron guide z, $d\Omega_w$ is the solid angle element within which the area $d^2\rho$ is seen from the element ds_w, and $d\Omega_s$ is the solid angle element within which the area ds_w is seen from the area element of the cross-section of the neutron guide, $d^2\rho$. We shall adopt for $n(z, \rho, \Omega)$ an expression similar to (4.48):

$$n_d(z, \rho, \Omega) = \nu_d(\rho, \Omega) \exp(-z/L_D). \tag{4.56}$$

Substituting into (4.35) expressions (4.45), (4.48), and (4.55), and using (4.8) and (4.49), we obtain

$$\nu_d(\rho, \Omega) = (1-\mu)\{[1 - GA(\Omega)]\nu_d(\Omega) + GA(\Omega)\} \exp[\zeta_{ws}(\rho, \Omega)/L_D]. \tag{4.57}$$

Expression (4.57) represents the exact self-consistent solution for an asymptotic flow established along an infinitely long cylindrical neutron guide. With its aid the angular distribution averaged over the entire cross-section of the neutron guide is readily determined:

$$W_{ang}(\Omega) = \int d^2\rho n(z, \rho, \Omega) / \iint d^2\rho n(z, \rho, \Omega) \, d\Omega.$$

In Appendix 4.4 it is shown that, for $A(\Omega) = 1$,

$$W_{ang}(\Omega) = 1/4\pi + \tfrac{2}{3} \operatorname{ctg} \vartheta_z (2 - G)/\pi^2 G L_D, \tag{4.58}$$

and, when $A(\Omega) = \cos \vartheta_n$,

$$W_{ang}(\Omega) = 1/4\pi + (1/G \sin \vartheta_z - 4/3\pi) \operatorname{ctg} \vartheta_z / 2\pi L_D. \tag{4.59}$$

Knowing the angular distribution we can readily find the diffusion coefficient. Indeed, from the definition of the diffusion flux it follows that $D = -J/(\partial N/\partial z)$, where

$$\text{a) } J = v \iint d^2\rho \cos\vartheta_z n(z, \rho, \Omega)\, d\Omega, \qquad \text{b) } N = \iint d^2\rho n(z, \rho, \Omega)\, d\Omega. \tag{4.60}$$

From (4.56) it follows that $\partial N/\partial z = -N/L_D$, therefore

$$D = vL_D \int W_{ang}(\vartheta_z) \cos\vartheta_z\, d\Omega. \tag{4.61}$$

Substituting (4.58) and (4.59) here we obtain, respectively (see Appendix 4.4)

$$\text{a) } D = 2v(2-G)/3G, \qquad \text{b) } D \approx 2v\left[\frac{1}{G}\ln\left(\frac{2}{u}\right) - \frac{1}{3}\right], \tag{4.62}$$

where u is defined in (4.54). With the aid of expressions (4.62) the diffusion lengths (4.52) and (4.53) can be represented in the form $L_D^2 = 2D/\mu v$, and the angular distribution (4.58) may be written in the form

$$W_{ang}(\Omega) = [1 + (2/\pi)\sqrt{2D\mu/v}\,\mathrm{ctg}\,\vartheta_z]/4\pi. \tag{4.63}$$

4.3.8 Matching at the openings

We shall now turn to computation of the coefficients r and t indicated in (4.19) and (4.21). Here the same faults will be seen that are inherent in diffusion theory. The density and flux at the entrance opening must be defined in order to perform calculations. Once again we shall consider an infinitely long neutron guide. If at the entrance opening the outside neutron angular distribution were the same as in the asymptotic region, eqns (4.59), (4.50), then

$$\nu_d(\rho, \Omega) = \frac{(1-\mu)GA(\Omega)\exp[\zeta_{ws}(\rho, \Omega)/L_D]}{\{1 - \exp[\zeta_{ww}(\Omega)/L_D](1-\mu)[1 - GA(\Omega)]\}}. \tag{4.64}$$

The neutron densities incident upon the opening, n_0, reflected, n_r, and present inside the neutron guide, n_t, are, respectively, equal to the following:

$$n_0 = \int \nu_d(\rho, \Omega)\, d\Omega\, \theta(0 \leqslant \cos\vartheta_z \leqslant 1),$$

$$n_r = \int \nu_d(\rho, \Omega)\, d\Omega\, \theta(-1 \leqslant \cos\vartheta_z \leqslant 0), \tag{4.65}$$

$$n_t = \int \nu_d(\rho, \Omega)\, d\Omega\, \theta(-1 \leqslant \cos\vartheta_z \leqslant 1), \tag{4.66}$$

i.e. the condition

$$n_t = n_0 + n_r \tag{4.67}$$

is satisfied automatically; in a similar manner the continuity condition is satisfied for the flux. For example, when the angular distribution is of the form (4.58) we obtain

$$\text{a) } n_0 = \tfrac{1}{2} + 4(2 - G)/3\pi G L_D, \quad \text{b) } n_r = \tfrac{1}{2} - 4(2 - G)/3\pi G L_D, \quad \text{c) } n_t = 1. \tag{4.68}$$

Hence it follows that

$$\text{a) } r = (1 - q)/(1 + q), \quad \text{b) } t = 2/(1 + q), \quad \text{c) } q = 4D/\pi v L_D, \tag{4.69}$$

where D is indicated in (4.62). Similarly, in the case of (4.59) we have

$$\text{a) } n_{0,r} = \tfrac{1}{2} \pm \{[\ln(1/u) + 1]/G - 4/3\pi\}/L_D, \qquad \text{b) } n_t = 1. \tag{4.70}$$

$$q = 2\{[\ln(1/u) + 1]/G - 4/3\pi\}/L_D. \tag{4.71}$$

On the other hand, if the angular distribution is isotropic within the forward hemisphere, and the angular distribution of the reflected neutrons is likewise isotropic, then, as follows from (4.18b),

$$q = 2D/vL_D.$$

So, to conclude! The main fault of the diffusion model is the asymptotic distributions over the angles and over the coordinates in the cross-section of the neutron guide differing from the given incident neutron distributions at the entrance opening. To match these distributions is a task for the future; the necessary analytic tools have not yet been developed. This discrepancy, however, is essential only for short neutron guides of lengths smaller than the transition region, the dimensions of which can be estimated by the quantity $1 \approx r/G$. If we deal only with long neutron guides, then the transition region near the opening can be neglected, and the discrepancy between the angular distributions becomes quite acceptable. Below we shall assume the angular distribution in the diffusion model to be isotropic both inside and outside the neutron guide. The anisotropy, as follows from (4.58) and (4.59), is inversely proportional to GL_D or proportional to the quantity $(\mu/G)^{1/2}$ which, in most cases, is significantly less than unity.

4.4 DIFFUSION OF UCN IN VERTICAL NEUTRON GUIDES

The collisions of UCN with the walls of neutron guides are mainly elastic, which affects the diffusion processes in the gravitational field, since the total energy is conserved in the propagation of the neutrons along a vertical or an inclined neutron guide, instead of the mean velocity. Formula (4.11) for a diffusion flux is of no use when the gravitational field is taken into account. Indeed, when UCN are stored in a vertical hermetically sealed vessel, the equilibrium UCN distribution over the height turns out to be inhomogeneous. The density gradient turns out to differ from zero, and in accordance with (4.11) there should arise in the vessel a diffusion flux to become level with the density. But no such flux can exist since it would make the system depart from the equilibrium state. This fact proves precisely the inapplicability of (4.11).

Usually, a diffusion flux in the presence of an external field is written in the form

$$\text{a) } J(z) = -D\partial n(z)/\partial z + \bar{v}n(z), \qquad \text{b) } \bar{v} = -\mu dU(z)/dz, \tag{4.72}$$

where $\bar{v}$ is the mean velocity, μ is the mobility, and $U(z)$ is the external field potential. In the case of UCN this formula is also inapplicable, since the mean velocity depends on the height.

To obtain an expression for the diffusion flux that holds for UCN, let us see how such an expression is derived in kinetic gas theory. The number of molecules passing through a unit area at point z per unit time is given by

$$J(z) = v[n(z-\lambda) - n(z+\lambda)] = -2\lambda v\partial n(z)/\partial z, \tag{4.73}$$

where λ is the mean distance of the area from the site of the last preceding collision of molecules and $n(z)$ is the density at point z. Not only does the form of the diffusion flux follow from expression (4.73), but the expression for the diffusion coefficient is also found.

To derive the expression we are interested in we shall note that the density n depends, besides the coordinate z, also on the momentum p, which is a function of the field

$$p^2 = 2m[E - U(z)], \tag{4.74}$$

where E is the total energy of the neutron. Therefore, from (4.73),

$$\begin{aligned} j(z, p) &= v[f(z-\lambda, p) - f(z+\lambda, p)] \\ &= -2v\lambda\{\partial f(z, p)/\partial z - 2m[\partial U(z)/\partial z]\partial f(z, p)/\partial p^2\} \\ &= -D[\partial f(z, p)/\partial z - 2m^2 g\partial f(z, p)/\partial p^2], \end{aligned} \tag{4.75}$$

where $j(z, p)$ is the flux density, $f(z, p)$ is the distribution density, and g is the gravitational constant. From (4.75) it immediately follows that in a vertical vessel, where no fluxes exist, the distribution f can be written in the form $f(p^2 + 2m^2gz)$. For example, the distribution for monochromatic neutrons of total energy E is represented in the form

$$f(z, p) = f_0(E)\delta(E - p^2/2m - mgz), \tag{4.76}$$

where $\delta(x)$ is the Dirac delta function.

In the presence of a diffusion flux, expression (4.76) undergoes a little change, namely

$$f(z, p) = f_0(z, E)\delta(E - p^2/2m - mgz). \tag{4.77}$$

Substituting (4.77) into (4.75) we obtain

$$j(z, p) = -D\delta(E - p^2/2m - mgz)\partial f_0(z, E)/\partial z. \tag{4.78}$$

Integrating (4.77) and (4.78) over d^3p we obtain the density and total flux at point z:

$$n(z) = \int f(z, p)d^3p = 4\pi m\sqrt{2m(E - mgz)}f_0(z, E) = \sqrt{1 - z/z_E}n_0(z, E), \tag{4.79}$$

$$J(z) = \int j(z, p)d^3p = -D\sqrt{1 - z/z_E}\partial n_0(z, E)/\partial z, \tag{4.80}$$

$$\text{a) } z_E = E/mg, \qquad \text{b) } n_0(z, E) = 4\pi m\sqrt{2mE}\, f_0(z, E). \tag{4.81}$$

Note that the diffusion coefficient depends linearly on the velocity, so it depends also on z and can be represented in the form

$$D = \sqrt{1 - z/z_E}D_0, \tag{4.82}$$

where D_0 is the diffusion coefficient at the height $z = 0$.

Now consider the stationary diffusion equation

$$\operatorname{div} J(z, E) = -n(z, E)/\tau(z, E). \tag{4.83}$$

If for τ we utilize expression (4.43a) and approximate μ, eqn (4.43b), by a linear function of the velocity v, then, taking into account the dependence of the velocity on the height z, we obtain

$$1/\tau(z, E) = (1 - z/z_E)/\tau(0, E). \tag{4.84}$$

Changing the variable $x = 1 - z/z_E$ we reduce equation (4.83) to the form

$$(d/dx)x(d/dx)n_0(x, E) = \beta^2 x^{3/2} n_0(x, E), \tag{4.85}$$

where the parameter $\beta^2 = \eta(E/E_{Cu})^{1/2}(z_E/l_f)^2$ is introduced. (The mean free path l_f, as shown in Appendix 4.1, equals 2.) Performing one more change of variable:

$$\text{a) } y = \beta x^{\nu}/\nu, \qquad \text{b) } \nu = 5/4, \tag{4.86}$$

we reduce (4.85) to the Bessel equation, the solution of which is represented in the form (Ignatovich 1987*a*)

$$n_0(z, E) = aI_0(y) + bK_0(y), \tag{4.87}$$

where I_0 and K_0 are modified Bessel functions.

The solution obtained is similar to (4.15) but involves more complicated functions. Nevertheless, the application of recursive relations is still possible. For example, the transmission of a vertical neutron guide, as shown in Appendix 4.5, is given by

$$T_{21} = (1 - \bar{r}_2\vec{r}_2)e_{21}(1 - \vec{r}_2 e_{21}\bar{r}_1 e_{12})^{-1}, \tag{4.88}$$

where T_{21} is the transmission in the direction from the opening at the coordinate z_1 to the opening at the coordinate z_2. The direction must be specified, since the two openings are not identical. If z_1 corresponds to the lower end of the neutron guide, then the propagation functions e_{ij} in (4.88) are, respectively, equal to the following:

$$\text{a) } e_{21} = \sqrt{x_2/x_1}\, I_0(y_2)/I_0(y_1), \quad \text{b) } e_{12} = \sqrt{x_1/x_2}\, K_0(y_1)/K(y_2). \tag{4.89}$$

The quantities $\vec{r}_i$ and $\overleftarrow{r}_i$ are the reflection coefficients from a semi-infinite neutron guide above and below the section at the point z_i. They are given in Appendix 4.5. It is necessary to note that the transmission T is the ratio of the neutron density at the output to the density at the input. On the other hand, the ratio of the fluxes also involves the ratio of velocities, therefore

$$J(z_2)/J(z_1) = W = \sqrt{x_2/x_1}\, T_{21}. \tag{4.90}$$

If losses in the neutron guide can be neglected, then, assuming $z_1 = 0$ and $z_2 = L$, we obtain

$$\text{a) } W = 2x/[1 + x - (3z_{\text{E}}/4)x\ln(x)], \qquad \text{b) } x = 1 - L/z_{\text{E}}, \tag{4.91}$$

where, as we recall, all the lengths are measured in terms of the radius of the neutron guide.

If the neutron guide is inclined, then in all the preceding formulae $g\cos\varphi$ must be substituted for g, where φ is the angle between the vertical direction and the axis of the neutron guide. Thus, for $\varphi = \pi/2$ we have a horizontal neutron guide, and formula (4.91a) transforms into the Dushman formula (4.10). If the neutron guide consists of several sections, then the transmission of the whole chain is computed using the method of recursion relations. Precisely this method was applied in computing the transmission of the gravitational spectrometer mentioned in Chapter 3.

4.5 LOSSES IN A NEUTRON GUIDE

The lifetime τ entering eqn (4.14) is determined by several factors: the decay of the neutron itself (lifetime τ_β), losses in collisions with the walls (the corresponding partial time τ_1), and losses in collisions with molecules of the gas present in the neutron guide (τ_{col}, which time depends on the composition of the gas and on its pressure p). The total τ is found from the relation

$$1/\tau = 1/\tau_\beta + 1/\tau_1 + 1/\tau_{\text{col}}. \tag{4.92}$$

In a typical experimental situation τ_β (about 1000 s) is large compared with τ_1 ($10 \div 100$ s), so the term $1/\tau_\beta$ can be neglected, which will be done. The quantity τ_1 is indicated in (4.43a), and τ_{col} can be calculated by the same formula if we introduce the effective loss coefficient for the free path time between two collisions with the walls. This is done in the following way.

The velocities of UCN are negligible compared with the velocities of the gas atoms. Therefore a neutron can be considered to be at rest, and the probability μ_{col} of its interacting with the gas during the free path time $t_f = l_f/v$ between two collisions with the walls is then

$$\mu_{col} = \sigma_t t_f N_g v_g/4 = \sigma_t l_f N_g v_g/4v = \sigma_t l_f(\sqrt{3}/4)\,p/v\sqrt{k_B T_g m_g}, \quad (4.93)$$

where σ_t is the total interaction cross-section of the gas atoms with a neutron at a velocity of v_g, v_g is the velocity of the gas atom ($v_g \simeq (3k_B T_g/m_g)^{1/2}$, where m_g is the mass of the gas atom, k_B is the Boltzmann constant, T_g is the temperature), v is the neutron velocity, and N_g is the density of the gas atoms ($N_g = p/k_B T_g$, where p is the pressure). Substituting (4.93) into (4.43a) we obtain

$$\tau_{col} = (4/\sqrt{3})\sqrt{k_B T_g m_g}/p\sigma_t. \quad (4.94)$$

A somewhat different form of this expression is presented in (4.1) and (4.2).

From (4.93) it follows that, for $\sigma_p = 10$ barn, $p = 10^{-3}$ Torr, $T_g = 300$ K, and a radius of the neutron guide of 10 cm, the effective loss coefficient amounts to about 10^{-7}, which is significantly smaller than the loss coefficient arising from the interaction of the UCN with the walls (about 10^{-3}). Therefore, at such pressures the influence of the residual gas on the UCN propagation may be neglected.

On the other hand, the neutron guide can be especially filled with a gas in order to investigate its influence on the UCN propagation. The most convenient gas for this purpose, as already mentioned, is ^{4}He. When a gas is present in the neutron guide the diffusion length decreases, as follows from (4.52) for instance, since the loss coefficient μ involves μ_{col} additively. By measuring the decrease of the transmission of the neutron guide as a function of the gas pressure in it we can determine the diffusion coefficient and, consequently, the nature of the neutron reflection from the walls.

At first sight, however, a paradox arises here. Indeed, since the transmission exhibits an exponential dependence, $T \simeq \exp(-L/L_D)$, which in accordance with (4.52) may be written in the form $\exp[-aL(G\mu)^{1/2}]$ (a is a constant), then at high pressures, when the main contribution to the losses is due to the gas, this dependence acquires the form $\exp(-bLp^{1/2})$ (b is another constant). This seems quite strange if we bear in mind that, when the neutrons are transmitted directly through a region of length L occupied by the gas, the neutron intensity decreases as $\exp(-cLp)$ (c is a certain parameter), i.e. it decreases much more significantly than when the motion inside the neutron guide occurs along a complicated trajectory.

Actually, no paradox exists. Expression (4.52) is approximate and holds only to the first order in μ. The exact formula (4.51) from which L_D is found involves, instead of G, the combination $G + \mu$. Therefore, the transmission of the neutron guide decreases, when μ increases, as $\exp\{-aL[\mu(G+\mu)]^{1/2}\}$, i.e. as μ increases because of the increase in pressure of the gas, when

μ becomes greater than G this dependence transforms into the form $\exp(-aL\mu)$ or $\exp(-cLp)$.

Incidentally, the paradox is also removed if the diffusion coefficient is determined more correctly. Thus, in the case of diffuse reflection from the wall with a probability G, the diffusion coefficient is written in the form $D \propto l_f Gv/3$, where l_f, the free path between two collisions with the wall, is equal to 2ρ in a cylindrical neutron guide, where ρ is the radius of the neutron guide. But, instead of l_f there should occur a length that takes into account losses, as in studies of neutron diffusion in moderators. This leads precisely to the diffusion coefficient containing the parameter $G + \mu$, instead of the parameter G.

The same result is obtained in the theory of Frank (1976). In this theory two opposite flows are considered inside the neutron guide and equations are written for the mutual pumpback of each flow into the other. These equations actually reduce to the diffusion equation obtained earlier, but involve a modified diffusion coefficient.

4.6 NON-STATIONARY DIFFUSION

To conclude with the diffusion theories it is also necessary to consider the non-stationary problem. The non-stationary experiments carried out at the Leningrad Institute of Nuclear Physics (Egorov *et al.* 1974) serve as a good indication of the existence of diffusion inside a neutron guide, since the presence of a peak (see Fig. 4.8) cannot be explained in the case of purely specular reflection from the wall. Regretfully, the experiments were performed in such a way that no reliable quantitative information on μ and G could be derived from them. But, as follows from Fig. 4.8, in which the solid curve corresponds to the solution obtained in Appendix 4.6, the experimental data can in principle be interpreted with excellent precision. The computational difficulties consist, first, of the weak convergence of the solution represented in the form of a series at small t. Secondly, diffusion theory does not take into account the delay time before the beginning of the counting rate, because of the finite time required for the first neutrons to travel from the shutter to the detector. Upon fitting the parameters of interest we obtained for a nickel-plated copper neutron guide the values $\mu = 10^{-2}$ and $G = 0.22$, and the mean neutron velocity turned out to be at a level of $3\ \mathrm{m\,s^{-1}}$.

4.7 TRANSPORTATION OF UCN IN VESSELS

Now we shall turn to the problem of transportation of UCN in closed vessels. The problem may be formulated as follows. Imagine we have accumulated neutrons in a trap and then the trap is transported to another site. The question arises as to how many UCN are lost as a result of the transport process itself. These losses, naturally, depend on many things: the neutron spectrum in the trap, the limiting velocity of the material of the trap, geometric dimensions, acceleration, the mode of motion, etc. Experiments involving transportation of a trap have been performed (see Anikolenko *et al.* 1973*b*; Antonov *et al.* 1969*b*; and Trüsted *et al.* 1970).

To solve this problem we shall consider the loss coefficient to be zero, the trap to be transported along the x axis with its dimension along x equal to l, the UCN spectrum to be initially monoenergetic, and the neutron distribution to be homogeneous and isotropic. We neglect the gravitational field of the Earth, and we consider the motion to consist of three stages: acceleration (the acceleration is a), uniform motion, and deceleration (the acceleration is b). The solution of the problem of measuring the spectrum is elementary but extremely elegant, so it is difficult to withstand the desire to describe it. Obviously, the UCN spectrum in the trap changes only if there exists acceleration. When the motion is uniform, the neutron spectrum relative to the trap remains unaltered, in accordance with the Galilean principle of relativity. Therefore we must consider the two extreme stages of the propagation process only, and, moreover, since the direction of acceleration along or opposite the axis is irrelevant to the change of the spectrum, it is sufficient to consider only the first stage, generally speaking.

Let the distribution of UCN with respect to the velocity and the x coordinate be described by the following density, until acceleration is switched on:

$$n(v, x) = N\theta(0 < x < l)\,\frac{S(x)}{V}\,\frac{\delta(v^2 - v_0^2)}{2\pi v}, \tag{4.95}$$

where V is the volume and $S(x)$ is the cross-section of the trap at point x. The spectrum $n(v)$ and the number of neutrons present in the trap are

$$n(v) = \int n(v, x)\,\mathrm{d}x = N\delta(v^2 - v_0^2)/2\pi v, \qquad N = \int n(v)\,\mathrm{d}^3v. \tag{4.96}$$

Switching on the acceleration a is equivalent to switching on 'gravity'. When this happens, the total energy of each particle situated at a distance x from the 'bottom' of the trap changes immediately (an energy of amx is added, where m is the neutron mass). Thus, the distribution (4.95) is transformed into

$$\tilde{n}(v, x) = (N/2\pi v V)\theta(0 < x < l)\delta(v^2 - v_0^2 - 2ax)S(x). \tag{4.97}$$

This results in a broadening of the spectrum. It will now be described, in accordance with (4.96), by the expression

$$\tilde{n}(v) = \frac{N}{4\pi a V v}\,\theta(0 < v^2 - v_0^2 < 2al)\,S\!\left(\frac{v^2 - v_0^2}{2a}\right). \tag{4.98}$$

For $a = 1\ \mathrm{m\,s^{-2}}$ and $l = 40$ cm the maximum change in the neutron energy amounts to ~ 4 neV.

Note that only the switching on of the acceleration leads to a change in the spectrum. Further motion with acceleration does not alter the spectrum. The spectrum next undergoes a change when the acceleration is switched off and transition to the stage of uniform motion takes place. Interruption of the acceleration is equivalent to switching off the gravitational field and to instantaneous alteration of the total energy of the particles (the gravitational energy is subtracted). Now let us see what the total change in neutron energy is when the acceleration is switched on and off. If, when switch-on took place, a particle was at a distance x_1 from the 'bottom' and at the time of switch-off it was at a distance of x_2, then the total change in kinetic energy by the end of the stage of acceleration is equal to $am(x_1 - x_2)$. This quantity may be positive or negative, and the total width of the spectral line, upon transition to uniform motion, will amount to $2mal$. We shall now show that in the presence of roughnesses this line will be non-symmetric if the acceleration stage is sufficiently long for an equilibrium density distribution to be established in the vessel. The asymmetry of the line will be such that, on the whole, the mean neutron energy will increase somewhat upon transition to uniform motion.

Indeed, in the case of perfectly specular reflection the neutron angular distribution at the 'bottom', when acceleration is switched on, becomes non-isotropic since acceleration occurs only along the x axis and does not take place in perpendicular directions. In the presence of roughnesses isotropy is again restored (this is evident, owing to the increase of entropy), but on the other hand the UCN density turns out to be inhomogeneous, in accordance with barometric height formulae. To determine the spectrum at a distance x from the 'bottom', we must substitute $(v^2 + 2ax)^{1/2}$ for v in (4.98), i.e. the density (4.97) is transformed into

$$\tilde{\tilde{n}}(v, x) = c\frac{N}{\sqrt{v^2 + 2ax}}\frac{S[(v^2 + 2ax - v_0^2)/2a]S(x)}{V^2}$$
$$\times\ \theta(0 < v^2 + 2ax - v_0^2 < 2al)\theta(0 < x < l), \qquad (4.99)$$

where c is a normalizing constant. Clearly, the admissible phase volume is greater at the 'bottom', so the density is also higher. The mean change in the neutron energy upon transition to uniform motion is

$$\Delta E = am(\bar{x}_1 - \bar{x}_2),$$

where the bar indicates averaging over all the neutrons, or integration with the densities (4.95) and (4.99). Since $\bar{x}$ characterizes the centre of mass of the neutron gas, it is clear that in the second case it is closer to the 'bottom' than before acceleration, and therefore $\Delta E > 0$. When $S(x)$ is constant, calculation of ΔE is readily performed exactly. We leave this to readers and only present the answer here. If $2al \ll v_0^2$, then $\bar{x}_2$, up to the constant $\alpha \approx 1$, which is negligible for our purposes, must be of the following form: $\bar{x}_2 = \frac{1}{2}l(1 - \alpha al/v_0^2)$. Correspondingly, $\Delta E = (m\alpha/2)(al/v_0)^2$. After the completion of the deceleration stage with an acceleration of b the UCN spectrum in the vessel will increase somewhat by the quantity $\Delta E = (m\alpha/2)(bl/v_0)^2$ and will simultaneously become wider by $\Delta E = 2mbl$.

The case of an instantaneous change in velocity of the trap along the x axis by u must also be considered. This will lead to the v_x velocity component of all the neutrons changing by u and to the neutron spectrum relative to the trap acquiring, instead of (4.96), the form

$$n(v) = N\delta((\vec{v} - \vec{u})^2 - v_0^2)/2\pi[(\vec{v} - \vec{u})^2]^{1/2}. \tag{4.100}$$

The mean energy will, in this case, become $\overline{v^2} = v_0^2 + u^2$ and the width of the spectrum will be

$$\sqrt{\overline{(\Delta v^2)^2}} = 2v_0 u/\sqrt{3}.$$

This section concludes transportation issues, and we shall now turn to the question of why neutrons are reflected from a wall.

APPENDICES

Appendix 4.1 Geometrical relations in a cylindrical tube

$$\text{a) } \mathbf{R} = \boldsymbol{\rho} + \mathbf{z}, \quad \text{b) } \boldsymbol{\rho} = \rho(\cos\varphi_\rho, \sin\varphi_\rho, 0), \quad \text{c) } \mathbf{z} = z\mathbf{k}, \quad \text{d) } \mathbf{k} = (0, 0, 1). \tag{1}$$

$$\text{a) } \mathbf{l} = \mathbf{R} - \mathbf{R}' = \Delta\boldsymbol{\rho} + \Delta\mathbf{z}, \quad \text{b) } \Delta\boldsymbol{\rho} = \boldsymbol{\rho} - \boldsymbol{\rho}', \quad \text{c) } \Delta\mathbf{z} = \mathbf{z} - \mathbf{z}'. \tag{2}$$

$$\text{a) } \mathbf{l} = l\mathbf{e}, \quad \text{b) } \mathbf{e} = (\sin\vartheta_z\cos\varphi_z, \sin\vartheta_z\sin\varphi_z, \cos\vartheta_z). \tag{3}$$

$$\text{a) } \boldsymbol{\rho}, \mathbf{e}, \rho' \text{ are given}, \quad \text{b) } \boldsymbol{\rho}' = \boldsymbol{\rho} - \Delta\rho\mathbf{s}, \quad \text{c) } \mathbf{s} = (\cos\varphi_z, \sin\varphi_z, 0). \tag{4}$$

$$\text{a) } \Delta z = \Delta\rho\,\mathrm{ctg}\,\vartheta_z, \quad \text{b) } l = \Delta\rho/\sin\vartheta_z, \quad \text{c) } \rho'^2 = \rho^2 + (\Delta\rho)^2 - 2\Delta\rho(\boldsymbol{\rho}\mathbf{s}). \tag{5}$$

$$\text{a) } \rho' < \rho, \quad \text{b) } \Delta\rho_{1,2} = \boldsymbol{\rho}\mathbf{s} \pm \sqrt{(\boldsymbol{\rho}\mathbf{s})^2 + \rho'^2 - \rho^2}. \tag{6}$$

$$\text{a) } \Delta z = \mathrm{ctg}\,\vartheta_z[\rho\cos(\varphi_\rho - \varphi_z) \pm \sqrt{\rho'^2 - \rho^2\sin^2(\varphi_\rho - \varphi_z)}],$$

$$\text{b) } l = \Delta z/\cos\vartheta_z. \tag{7}$$

$$\text{a) } \rho' > \rho, \quad \text{b) } \Delta z = \mathrm{ctg}\,\vartheta_z[\rho\cos(\varphi_\rho - \varphi_z) + \sqrt{\rho'^2 - \rho^2\sin^2(\varphi_\rho - \varphi_z)}]. \tag{8}$$

$$\text{a) } \Delta z = \zeta_{\mathrm{ww}}(\mathbf{\Omega}) = 2\cos(\varphi_r - \varphi_z)\,\mathrm{ctg}\,\vartheta_z, \quad \text{b) } l = l_{\mathrm{f}} = 2\cos(\varphi_r - \varphi_z)/\sin\vartheta_z. \tag{9}$$

$$\Delta z = \zeta_{\mathrm{sw}}(\rho, \mathbf{\Omega}) = \mathrm{ctg}\,\vartheta_z[\rho\cos(\varphi_\rho - \varphi_z) + \sqrt{1 - \rho^2\sin^2(\varphi_\rho - \varphi_z)}]. \tag{10}$$

$$\zeta_{\mathrm{ww}}(\rho, \mathbf{\Omega}) = \zeta_{\mathrm{sw}}(\rho, \mathbf{\Omega}) - \zeta_{\mathrm{sw}}(\rho, -\mathbf{\Omega}) = 2\,\mathrm{ctg}\,\vartheta_z\sqrt{1 - \rho^2\sin^2(\varphi_\rho - \varphi_z)}]. \tag{11}$$

$$\text{a) } \int\zeta_{\mathrm{ww}}(\rho, \mathbf{\Omega})\,\mathrm{d}^2\rho/\pi = 16\,\mathrm{ctg}\,\vartheta_z/3\pi, \quad \text{b) } \int\zeta_{\mathrm{ww}}^2(\rho, \mathbf{\Omega})\,\mathrm{d}^2\rho/\pi = 3\,\mathrm{ctg}^2\,\vartheta_z. \tag{12}$$

$$\text{a)}\ \overline{\overline{\zeta_{ww}}} = \int \zeta_{ww}(\rho, \mathbf{\Omega})\,d^2\rho \cos\vartheta_z d\Omega/\pi^2 = 16/3, \quad \text{b)}\ \overline{\overline{\zeta_{ww}^2}} = 0. \tag{13}$$

$$\overline{\overline{1/\zeta_{ww}}} = \int [1/\zeta_{ww}(\rho, \mathbf{\Omega})]\,d^2\rho \cos\vartheta_z d\Omega/\pi^2 = 2. \tag{14}$$

$$\text{a)}\ \rho' = 1, \quad \text{b)}\ \vec{n} = -\boldsymbol{\rho}', \quad \text{c)}\ \cos\vartheta_n = \mathbf{en} = -\mathbf{e}\boldsymbol{\rho} + \Delta\rho \sin\vartheta_z. \tag{15}$$

$$\text{a)}\ \cos\vartheta_n = \sin\vartheta_z \sqrt{1 - \rho^2 \sin^2(\varphi_\rho - \varphi_z)}, \quad \text{b)}\ \cos\vartheta_n = \sin\vartheta_z \cos(\varphi_r - \varphi_z). \tag{16}$$

$$\text{a)}\ \bar{l}_f = \int \cos\vartheta_n l_f\, d\Omega/\pi = 2 \int_{-\pi/2}^{\pi/2} \cos^2\varphi_z d\varphi_z \int_{-1}^{1} d\cos\vartheta_z/\pi = 2, \quad \text{b)}\ \bar{\zeta}_{ww} = 0. \tag{17}$$

$$\overline{\zeta_{ww}^2} = \int \cos\vartheta_n \zeta_{ww}^2(\mathbf{\Omega})\, d\Omega/\pi = 4 \int_{-\pi/2}^{\pi/2} \cos^3\varphi_z d\varphi_z \int_0^{\pi} \cos^2\vartheta_z d\vartheta_z/\pi = 8/3. \tag{18}$$

$$\text{a)}\ \cos\vartheta_z = \sin\vartheta_n \cos\varphi_n, \quad \text{b)}\ \cos\varphi_z = \cos\vartheta_n/\sin\vartheta_z. \tag{19}$$

$$\zeta_{ww} = 2\cos\vartheta_n \cos\vartheta_z/\sin^2\vartheta_z = 2\cos\vartheta_n \sin\vartheta_n \cos\varphi_n/(1 - \sin^2\vartheta_n \cos^2\varphi_n). \tag{20}$$

$$l_f = 2\cos\vartheta_n/(1 - \sin^2\vartheta_n \cos^2\varphi_n). \tag{21}$$

The radius vector is conveniently represented in the cylindrical system of the neutron guide in the form (1a), where the vector $\boldsymbol{\rho}$, eqn (1b), is in the plane of the cross-section, φ_ρ represents its azimuthal angle with respect to a certain chosen axis, and the vector **z**, eqn (1c), is directed along the axis of the neutron guide that can be represented by the unit vector (1d). The distance between two points inside the neutron guide can be represented by the vector (2a) where the notation (2b, c) is introduced. Vector (1a) can also be written in the form (3a) with the aid of the unit vector (3b) where the index z assigned to the angles indicates that they are measured in a coordinate system with the polar axis directed along **z**. The azimuthal angle is read off with respect to the same axis as φ_ρ in (2b). Now consider a point $\boldsymbol{\rho}$ at which a neutron arrives from the direction **e**. If the distance from the axis, ρ', of the point from which the neutron leaves to arrive at the considered point $\boldsymbol{\rho}$ (4a) is given, then the vector $\boldsymbol{\rho}$ is immediately expressed by expression (4b), where $\Delta\rho$ is the length of vector (2b) and the unit vector (4c) is introduced. With the aid of $\Delta\rho$ it is possible to find the distance between two cross-sections, eqn (5a), where the points ρ and ρ' belong, and the total flight length (5b). Owing to (4a) the square of vector (4b) is known and, therefore, (5c) represents the equation for determining $\Delta\rho$. When $\rho' < \rho$, we have two solutions (6b). Substituting (1b) and (4c) into (6b) and (6b) into (5a) we obtain (7a). Expression (5b) can be written in the form (7b). In the case $\rho' > \rho$, eqn (5c) has a unique solution $\Delta\rho_1$, the substitution of which into (5a) leads to (8b). When $\rho = \rho' = 1$, expression (8b) transforms into (9a). Substitution of (9a) into (7b) yields (9b). If $\rho' = 1$ only, then (8b) transforms into (10). Now consider the sum (11) corresponding to displacement along the axis between two collisions under the condition that the line of flight passes through the given point $\boldsymbol{\rho}$. In (12), (13), and (14) are presented certain values of (11) averaged over the coordinates $\boldsymbol{\rho}$ and over the flux. To determine the analogous value averaged over the output angles, we introduce on the wall of the neutron guide (15a) the polar axis directed along the vector (15b). The projection of vector (3b) on to this axis equals (15c). Substituting, here, (1b), (3b), and $\Delta\rho_1$ from (6b), we obtain (16a). For $\rho = 1$ we obtain (16b). In (17) and (18) some averages are given over the output angles from the walls where expression (16b) is utilized for $\cos\vartheta_n$ and integration is performed in the coordinate system with the polar axis directed along z. Sometimes it is convenient to use a coordinate system with the polar axis along the external normal vector to the surface of the wall at the point from which the neutron departs. Choosing $\varphi_r = 0$ in (9) and (16b) and substituting (19a) and (19b) into (9) we obtain expressions (20) and (21).

Appendix 4.2. The Clausing equations

$$W_{tr}(l) = w_{ss}(l) + \int_0^l w_{sr}(z)w_w(z)dz. \tag{1}$$

$$w_w(z) = w_{rs}(l - z) + \int_0^l w_{rr}(|z - z'|)w_w(z')dz'. \tag{2}$$

$$\text{a) } R = \sqrt{(z/2)^2 + 1}, \quad \text{b) } h = R - z/2. \tag{3}$$

$$w_{ss}(z) = 2\pi Rh/2\pi(z + h) = (\sqrt{z^2 + 4} - z)/(\sqrt{z^2 + 4} + z). \tag{4}$$

$$w_{sr}(z) = -dw_{ss}(z)/dz = 2w_{ss}(z)/\sqrt{z^2 + 4}. \tag{5}$$

$$\text{a) } 2\pi dz\, w_{rs}(z) = \pi w_{sr}(z)dz, \quad \text{b) } w_{rs}(z) = w_{sr}(z)/2. \tag{6}$$

$$w_{rr}(z) = -\, dw_{rs}(z)/dz = w_{ss}(z)(z + 2\sqrt{z^2 + 4})/(z^2 + 4)^{3/2}. \tag{7}$$

$$J(z) = \pi\Phi_0 w_{ss}(z) + \int_0^z n(z')2\pi dz' w_{rs}(z - z') - \int_z^l n(z')2\pi dz' w_{rs}(z' - z). \tag{8}$$

$$W_{tr} = J(z)/\pi\Phi_0 = w_{ss}(z) + \int_0^z \tilde{n}(z')dz' w_{sr}(z - z') - \int_z^l \tilde{n}(z')dz' w_{sr}(z' - z). \tag{9}$$

$$\text{a) } \tilde{n}(z) = n(z)/\Phi_0 \approx a + bz/l, \quad \text{b) } l \gg 1, \quad \text{c) } W_{tr} \approx 8/3l. \tag{10}$$

The transmission of a neutron guide of length l is presented in (1), where $w_{ss}(l)$ is the probability of direct flight without collisions with the walls, $w_{sr}(z)$ is the probability of direct flight from the entrance opening and arrival at a circular belt, around the wall, of width dz, while w_w is the probability of a molecule leaving through the exit opening upon colliding with the wall at point z. For $w_w(z)$ we may write eqn (2), where w_{rs} is the probability of leaving without collisions with the walls and w_{rr} is the probability of direct flight from one ring to another. To calculate w_{ss} Clausing applied the following device: a cylinder of length l and of unit radius is circumscribed by a sphere of radius R

(eqn (3a)). Then the molecules entering the cylinder can be imagined as leaving isotropically the surface of a segment of height (3b) covering the entrance opening. The probability (4) equals the ratio of the areas of the spherical segment covering the entrance opening and of the spherical belt covering the lateral surface. The probability of hitting a belt of width dz is $w_{ss}(z + dz) - w_{ss}(z)$, from which (5) follows. From the condition of detailed balance (6a) follows (6b). Formula (7) is obtained similarly to (5). For an approximate solution of the equations we introduce the density $n(z)$ of the molecular distribution at the wall and require the flux (8) in the absence of losses to be the same for all z, where Φ_0 is the flux density for the molecules incident on the entrance opening. The transmission of the pipe is given by expression (9), where the reduced density (10a) is introduced. If the density $\tilde{n}$ is given in the form of a linear function, expression (10a), then the parameters a and b of this function are found from the condition that the difference between (8) and a constant be the least possible. We shall not present here the relevant calculations, but only indicate that, at large lengths, expression (10b), substitution of the corresponding expression into (9) leads to a transmission close to (10c).

Appendix 4.3 Determination of the diffusion coefficient by the random walk method

$$\text{a) } \left\langle \left(\sum_{i=0}^{n} z_i \right)^2 \right\rangle = 2Dt, \quad \text{b) } t = n\langle l_f \rangle / v, \quad \text{c) } D \to D/rv. \tag{1}$$

$$\text{a) } D = \lim_{n\to\infty} \left\langle \left(\sum_{i=1}^{n} z_i \right)^2 \Big/ 2\langle l_f \rangle n \right\rangle, \quad \text{b) } \langle B \rangle = \int_{2\pi} B(\Omega)\cos\theta \, d\Omega/\pi. \tag{2}$$

$$\text{a) } \langle l_f \rangle = 2, \quad \text{b) } \left\langle \sum_{ij} z_i z_j \right\rangle = \sum_{i=1}^{n} \langle z_i^2 \rangle + 2\sum_{i<j} \langle z_i z_j \rangle,$$

$$\text{c) } \langle z_i^2 \rangle = \langle z_0^2 \rangle = \langle \zeta_{ww}^2 \rangle. \tag{3}$$

$$\text{a) } z_0 = 2\sin\theta\cos\theta\cos\varphi/(1 - \sin^2\theta\cos^2\varphi), \quad \text{b) } \langle z_0^2 \rangle = 8/3. \tag{4}$$

$$\int_0^{2\pi} d\varphi \cos^2\varphi/(1 - \sin^2\theta\cos^2\varphi)^2 = \pi/\cos\theta. \tag{5}$$

$$W_{ref}(\Omega', \Omega) = [1 - gA(\Omega)]\delta(\Omega' - \Omega) + cgA(\Omega')A(\Omega)\cos\theta. \tag{6}$$

$$\text{a) } \langle z_i z_j \rangle = \langle z_i^2 [1 - gA(\Omega)]^{j-i} \rangle, \quad \text{b) } \sum_{i<j}^{n} \langle z_i z_j \rangle = \langle z_0^2 n(1 - gA)/gA \rangle. \tag{7}$$

$$\text{a) } D = \langle z_0^2 [2 - gA(\Omega)]/4gA(\Omega) \rangle, \quad \text{b) } A = 1, \quad \text{c) } D = \tfrac{2}{3}(2 - g)/g. \tag{8}$$

$$\text{a) } A(\Omega) = \cos\theta, \quad \text{b) } \langle z_0^2/A \rangle = 4\int_{\alpha}^{1} \sin^2\theta \, d\cos\theta/\cos\theta = -2 - 4\ln\alpha. \tag{9}$$

$$\text{a) } \alpha = 1/Lg, \quad \text{b) } Lg \gg 1, \quad \text{c) } D \approx (2/g)\ln Lg. \tag{10}$$

$$\text{a) } f(z,\rho,\Omega) = [n(z) - n'(z)Q(\rho,\Omega)]/4\pi^2, \quad \text{b) } D = \int_{4\pi} Q\cos\theta\,\mathrm{d}\Omega\,\mathrm{d}s/4\pi^2. \tag{11}$$

$$Q(\rho,\Omega) = \operatorname{ctg}\vartheta[\rho\cos\chi + \sqrt{1-\rho^2\sin^2\chi}\,[2 - gA(\Omega)]/gA(\Omega)]. \tag{12}$$

$$D = \int_0^{1-\alpha^2/2} 2(1/g\sin\vartheta - 4/3\pi)\operatorname{ctg}\vartheta\cos\vartheta\,\mathrm{d}\cos\vartheta \approx (2/g)\ln Lg. \tag{13}$$

Now we shall determine the coefficient D with the aid of equality (1a), valid for a one-dimensional random walk. Here z_i is the displacement along the pipe after the ith collision with the wall. The angle brackets indicate averaging over the angles after each reflection. The time t of n steps can be determined by equality (1b), where $\langle l_f \rangle$ is the mean free path determined in (17a) of Appendix 4.1. We shall search for D in the dimensionless form (1c), where v is the neutron velocity and r is the pipe radius. The diffusion coefficient is determined from (1a, b) with the aid of the limit (2a). The average of an arbitrary quantity B will be determined with the aid of (2b). Thus, for instance, (3a) holds for l_f. All the summands z_i^2 in the sum (3b) are independent and identical (eqn (3c)). As follows from Appendix 4.1, z_0 is as expressed in (4a). Substituting z_0^2 into (2b) for B, we obtain (4b). The integral (2b) is readily computed if we first integrate over φ, as indicated in (5). For the reflection indicatrix (6) the correlation between various steps can be calculated as indicated in (7a); correspondingly, the non-diagonal part of the sum (3b) reduces to the form (7b). Substituting (7b) and (3a–c) into (2a) we obtain (8a). If (8b) holds, then substituting (8b) into (8a) we obtain (8c). If (9a) holds, then averages such as (9b) encountered in (7b) do not exist since the integral diverges. At the lower limit we are constrained to introduce the cut-off (10a). When (10b) holds, the cut-off (10a) signifies that, at grazing angles smaller than α, the reflection is considered specular. In this case the integral (9b) has a finite value and the value of D can be approximated by expression (10c). If the neutron angular distribution (11a) (see Appendix 4.5) is applied, then coefficient D is determined by the integral (11b). Expression (12) holds for the angular part of the distribution. Substituting (8b) into (12) we obtain, with the aid of (11b), expression (8c). Substituting (9a) into (12), we obtain, with the cut-off (10a), expression (13), in agreement with (10c).

Appendix 4.4 Diffusion length and angular distribution

$$\text{a) } c\int v_d(\Omega)A(\Omega)\cos\vartheta_n\,\mathrm{d}\Omega = 1, \quad \text{b) } A(\Omega) = 1, \quad \text{c) } c = 1/\pi. \tag{1}$$

$$v_d(\Omega) = G(1-\mu)/\{\exp[-\zeta_{ww}(\Omega)/L] - (1-\mu)(1-G)\}. \tag{2}$$

a) $\nu_d(\Omega) = (1 - \mu/q)[1 + \zeta_{ww}/qL + \zeta_{ww}^2(2-q)/2q^2L^2]$,

b) $q = G + \mu - G\mu \approx G.$ (3)

a) $\zeta_{ww}(\Omega) = 2\operatorname{ctg}\vartheta_z\cos(\varphi_r - \varphi_z)$, b) $\cos\vartheta_n = \sin\vartheta_z\cos(\varphi_r - \varphi_z)$. (4)

$$\int(1-\mu/q)2(2-q)\operatorname{ctg}^2\vartheta\sin\vartheta\cos^3\varphi\,d\Omega/q^2L^2 = \int(\mu/q)A(\Omega)\cos\vartheta_n\,d\Omega. \quad (5)$$

$$\int\operatorname{ctg}^2\vartheta\sin\vartheta\cos^3\varphi\,d\Omega = \int_0^\pi \operatorname{ctg}^2\vartheta\sin^2\vartheta\,d\vartheta\int_{-\pi/2}^{\pi/2}\cos^3\varphi\,d\varphi = 2\pi/3. \quad (6)$$

$$1/L^2 = \tfrac{3}{4}\mu q/(2-q)(1-\mu/q) \approx \tfrac{3}{4}\mu G/(2-G). \quad (7)$$

a) $A(\Omega) = \cos\vartheta_n$, b) $q = \mu + G\cos\vartheta_n(1-\mu)$, c) $q \approx G\cos\vartheta_n$. (8)

a) $\int(\mu/q)A(\Omega)\cos\vartheta_n\,d\Omega = \pi\alpha\{1 - 2\alpha + 2\alpha^2\ln[(1+\alpha)/\alpha]\}$,

b) $\alpha = \mu/G(1-\mu)$. (9)

a) $L^{-2} \approx \mu G/2I$, b) $I = \int\operatorname{ctg}^2\vartheta_z\sin\vartheta_z\,d\vartheta_z\int\cos^2\varphi_z\,d\varphi_z(2 - G\cos\vartheta_n)/\pi$. (10)

$$\int_u^1\sqrt{1-x^2}\,dx/x + \int_0^u dx/u = -\sqrt{1-u^2}\ln u + 1 + \int_u^1 x\ln x\,dx/\sqrt{1-x^2}. \quad (11)$$

$$\int_u^1 x\ln x\,dx/\sqrt{1-x^2} = \int_0^1 x\ln x\,dx/\sqrt{1-x^2} + O(u\ln u) \approx \ln 2 - 1. \quad (12)$$

a) $I \approx -2[\ln(u/2) + G/3]$, b) $u = \max[\sqrt{rg/v^2}, \mu/G(1-\mu)]$. (13)

$$W_{ang}(\Omega) = \tilde{C}\int A(\Omega)d^2\rho\exp[\zeta_{sw}(\rho,\Omega)/L]/\{1 - (1-q)\exp[\zeta_{ww}(\Omega)/L]\}. \quad (14)$$

$$\zeta_{sw}(\rho,\Omega) = \operatorname{ctg}\vartheta_z[\rho\cos(\varphi_\rho - \varphi_z) + \sqrt{1-\rho^2\sin^2(\varphi_\rho-\varphi_z)}]. \quad (15)$$

a) $\Omega = \Omega(\rho)$, b) $\zeta_{ww}(\Omega) = 2\operatorname{ctg}\vartheta_z\sqrt{1-\rho^2\sin^2(\varphi_\rho - \varphi_z)}$. (16)

a) $A(\Omega) = 1$, b) $W_{ang}(\vartheta_z) = \tilde{C}[1 + w(2-G)\operatorname{ctg}\vartheta_z/LG]$. (17)

a) $w = \int d^2\rho\sqrt{1-\rho^2\sin^2(\varphi_\rho-\varphi_z)}/\pi = 8/3\pi$, b) $\int W_{ang}(\vartheta_z)d\Omega = 1$. (18)

a) $\tilde{C} = 1/4\pi$, b) $D = L\int W_{ang}(\vartheta_z)\cos\vartheta_z\,d\Omega = 2(2-G)/3G$. (19)

a) $A(\Omega) = \cos\vartheta_n$, b) $W_{ang}(\vartheta_z) = [1 + (2/G\sin\vartheta_z - w)\operatorname{ctg}\vartheta_z/L]/4\pi$. (20)

$$D = L\int W_{ang}(\vartheta_z)\cos\vartheta_z\,d\Omega \approx -2\ln(u/2)/G - 2/3 = I/G. \quad (21)$$

The angular distribution of neutrons reflected from the wall is normalized by the condition (1a). In the case of (1b) we have (1c) and (2). Expansion of (2) in a series in powers of $1/L$ up to $1/L^2$ has the form (3a) where the notation (3b) is adopted. Equalities (4) follow from Appendix 4.1, where φ_r is the azimuth of the radius of the segment of the wall in a cylindrical coordinate system, the axis of which is directed along the neutron guide. Upon substitution of (3a) and (4) into (1a) we reduce eqn (1a) to the form (5). The integral on the left-hand side of (5) is computed in (6). Expression (7) follows from (5) and (6). In the case of (8a) we have (8b, c). The integral on the right-hand side of (5) is computed, with the substitution of (8b) in (9a), where the notation (9b) is utilized. The solution of eqn (5) can be written in the form (10a), where the notation (10b) is introduced. The integral (10b) is reduced, upon introducing the cut-off $u = \sin\,\vartheta_z \ll 1$ for small angles ϑ_z, to the form (11). The second term on the right-hand side of (11) has the asymptotic expansion (12). Thus, (10b) reduces to the form (13a). The cut-off parameter is given by expression (13b), where g is the gravitational constant and r is the radius of the neutron guide. The second term in (13b) follows from expression (8b). For an established diffuse motion the UCN angular distribution averaged over the cross-section of the neutron guide has the form (14), where $\tilde{C}$ is a normalizing constant and $\zeta_{sw}(\rho, \Omega)$ is the distance (15) of the point ρ in the cross-section from the segment of the wall along the flight direction Ω. In calculating the distance $\zeta_{ww}(\Omega)$ between two collisions with the walls it is necessary to take into account that the path must pass through point ρ. Therefore Ω depends on ρ (eqn (16a)) and then $\zeta_{ww}(\Omega)$ is represented by expression (16b). In the case of (17a) the integral (14) reduces to the form (17b), where the parameter w is defined in (18a) and the normalization is given in (18b). Expression (19a) follows from (18b). The diffusion coefficient is determined by the integral (19b). In the case of (20a) the integral (14) reduces to the form (20b), where (19b) is taken into account. The diffusion coefficient is computed in (21) with the aid of the same cut-off at small ϑ_z, as applied in calculating the integral (10b).

Appendix 4.5 Transmission of a vertical neutron guide

$$\text{a) } x = 1 - \frac{z}{z_E}, \quad \text{b) } z_E = \frac{E}{mg}, \quad \text{c) } n(z) = \sqrt{x}\,\tilde{n}, \quad \text{d) } \frac{d}{dx} x \frac{d}{dx} \tilde{n} = \beta^2 x^{3/2} \tilde{n}. \tag{1}$$

$$\text{a) } \beta^2 = 4\eta\sqrt{E/E_{\text{lim}}}\,(z_E^2/\bar{l}_f\bar{l}_d). \quad \text{b) } \bar{l}_f = 2, \quad \text{c) } \bar{l}_d = 2(2 - G)/G. \tag{2}$$

$$\text{a) } y = \delta x^{\nu}, \quad \text{b) } \tilde{n} = \tilde{n}(y), \quad \text{c) } \nu = 5/4, \quad \text{d) } \delta = \beta/\nu. \tag{3}$$

$$\text{a) } \tilde{n}'' + \tilde{n}'/y - \tilde{n} = 0, \quad \text{b) } \tilde{n} = aI_0(y) + bK_0(y),$$

$$\text{c) } I'_0 = I_1, \quad \text{d) } K'_0 = -K_1. \tag{4}$$

$$\text{a) } n_{\text{out}} = (1 + \vec{r}_1)\vec{n}_0, \quad \text{b) } n_{\text{in}} = \vec{t}_1\vec{n}_0\sqrt{x/x_1}\,I_0(y)/I_0(y_1). \tag{5}$$

$$\vec{n}_0 v_1(1 - \vec{r}_1)/2 = -\vec{t}_1\vec{n}_0 D_1(d/dz)I_0(y)/I_0(y_1)_{z=z_1}. \tag{6}$$

$$\text{a) } y_i = y(x_i), \quad \text{b) } v_i = \sqrt{2mEx_i}, \quad \text{c) } D_i = l_d v_i/3, \quad \text{d) } x_i = 1 - z_i/z_E. \tag{7}$$

$$\text{a) } df(y)/dz = (dy/dz)f'(y), \quad \text{b) } y_i = \tfrac{4}{3}\sqrt{\eta l_d/l_f}(x_i E/E_{\text{lim}})^{1/4}. \tag{8}$$

a) $1 + \vec{r}_1 = \vec{t}_1$, b) $1 - \vec{r}_1 = \vec{q}_1 \vec{t}_1$, c) $\vec{q}_i = \gamma_i I_1(y_i)/I_0(y_i)$. (9)

a) $n_{\text{out}} = (1 + \overleftarrow{r}_1)\overleftarrow{n}_0$, b) $n_{\text{in}} = \overleftarrow{t}_1 \overleftarrow{n}_0 \sqrt{x/x_1}\, K_0(y)/K_0(y_1)$. (10)

$\overleftarrow{n}_0 v_1(1 - \overleftarrow{r}_1)/2 = \overleftarrow{t}_1 \overleftarrow{n}_0 D_1 (\mathrm{d}/\mathrm{d}z) K_0(y)/K_0(y_1)_{z=z_1}$. (11)

a) $1 + \overleftarrow{r}_1 = \overleftarrow{t}_1$, b) $1 - \overleftarrow{r}_1 = \overleftarrow{q}_1 \overleftarrow{t}_1$, c) $\overleftarrow{q}_i = \gamma_i K_1(y_i)/K_0(y_i)$. (12)

a) $\overset{\rightleftarrows}{r}_i = (1 - \overset{\rightleftarrows}{q}_i)/(1 + \overset{\rightleftarrows}{q}_i)$, b) $\overset{\rightleftarrows}{t}_i = 2/(1 + \overset{\rightleftarrows}{q}_i)$, c) $i = 1, 2$. (13)

a) $\vec{n}_2 = T_{21} \vec{n}_1 + \overleftarrow{R}_2 \overleftarrow{n}_2$, b) $\overleftarrow{n}_1 = \vec{R}_1 \vec{n}_1 + T_{12} \overleftarrow{n}_2$, c) $\overleftarrow{n}_2 = \vec{r}_2 \vec{n}_2$. (14)

a) $\overleftarrow{n}_1 = \vec{r}_1 \vec{n}_1$, b) $\vec{n}_2 = e_{21} \vec{n}_1$, c) $e_{21} = \sqrt{x_2/x_1}\, I_0(y_2)/I_0(y_1)$. (15)

a) $e_{21} = T_{21} + \overleftarrow{R}_2 \vec{r}_2 e_{21}$, b) $\vec{r}_1 = \vec{R}_1 + T_{12} \vec{r}_2 e_{21}$. (16)

a) $\overleftarrow{n}_1 = T_{12} \overleftarrow{n}_2 + \vec{R}_1 \vec{n}_1$, b) $\vec{n}_2 = \overleftarrow{R}_2 \overleftarrow{n}_2 + T_{21} \vec{n}_1$, c) $\vec{n}_1 = \overleftarrow{r}_1 \overleftarrow{n}_1$. (17)

a) $\vec{n}_2 = \overleftarrow{r}_2 \overleftarrow{n}_2$, b) $\overleftarrow{n}_1 = e_{12} \overleftarrow{n}_2$, c) $e_{12} = \sqrt{x_1/x_2}\, K_0(y_1)/K_0(y_2)$. (18)

a) $e_{12} = T_{12} + \vec{R}_1 \overleftarrow{r}_1 e_{12}$, b) $\overleftarrow{r}_2 = \overleftarrow{R}_2 + T_{21} \overleftarrow{r}_1 e_{12}$. (19)

a) $\vec{R}_1 = (\vec{r}_1 - e_{12} \vec{r}_2 e_{21})/Q_{12}$, b) $T_{12} = (1 - \vec{r}_1 \overleftarrow{r}_1) e_{12}/Q_{12}$. (20)

a) $\overleftarrow{R}_2 = (\overleftarrow{r}_2 - e_{21} \overleftarrow{r}_1 e_{12})/Q_{21}$, b) $T_{21} = (1 - \overleftarrow{r}_2 \vec{r}_2) e_{21}/Q_{21}$. (21)

a) $Q_{12} = 1 - \overleftarrow{r}_1 e_{12} \vec{r}_2 e_{21}$, b) $Q_{21} = 1 - \vec{r}_2 e_{21} \overleftarrow{r}_1 e_{12}$. (22)

a) $W_{21} = J_2/J_1 = \sqrt{x_2/x_1}\, T_{21}$, b) $W_{12} = J_1/J_2 = \sqrt{x_1/x_2}\, T_{12}$. (23)

Upon performing the change of variables (1a), where the parameter (1b) is introduced, the equation for the density represented in form (1c) acquires the form (1d). The parameter β^2 indicated in (2a) involves the mean free path between two collisions with the walls, eqn (2b), and the transport length l_d which has the form (2c) in the case when, upon colliding with the wall, a neutron undergoes diffuse reflection with a probability G. If the change of variables (3a) is performed and $\tilde{n}$ considered to be a function of y, eqn (3b), then for η and δ chosen in accordance with (3c, d) eqn (1d) reduces to the form (4a). Its solution is represented in the form of the superposition (4b) of modified Bessel functions exhibiting the properties (4c, d). Now consider an infinitely high neutron guide ($z \geqslant 0$), the entrance opening of which is situated at the point $z = z_1 = 0$. Assume the neutrons incident on the entrance opening from the outside to be isotropically distributed in the forward hemisphere and to have a density $\overleftarrow{n}_0$. The total outside density is (5a), taking reflection into account, and the inside density is represented in the form (5b). Note that the density decreases going from the entrance opening along the neutron guide, as it should. The continuity condition for the flux at the entrance opening is written in the form (6), under the assumption that the reflected neutrons are also distributed isotropically within their hemisphere. The parameters y_1, v_1, D_1, and x_1 are found from relations (7) by the substitution $z_i = z_1$. In calculating the derivative in (6) we utilize the rule for differentiating a complex function and introduce the parameter (8b), whence the continuity conditions for the density and flux can be reduced to the form (9). Now consider a semi-infinite neutron guide going down the point z_1. In this case, after similar considerations we obtain expressions (10), (11) analogous to (5), (6). Equations (9) acquire the form (12). The solutions of eqns (9a, b) and (12a, b) are shown in (13) for the point z_i as the coordinate of the entrance of the semi-infinite neutron guide going upwards or downwards. To determine the reflection and transmission of a vertical neutron guide of finite height $L = z_2 - z_1$, let us consider an infinite guide with a diffusion flux going upwards in it. Imagine two cross-sections 1, 2 at points $z_1 < z_2$. The densities of neutrons going up and down (the arrows $\rightarrow$ and $\leftarrow$, respectively) in these two cross-sections are related by (14) and (15). Substitution of (14c) and (15a, b) into (14a, b) reduces the latter to the form (16). If the diffusion flux inside the guide is going downwards, the relations (14)–(16) are transformed into analogous ones, (17)–(19). The solution of eqns (16b) and (19a) is given in (20), and of (16a) and (19b) in (21), where the notation (22) is used. The upward and downward transmission as the ratio of the outgoing to incident fluxes is represented by (23a) and (23b) respectively.

Appendix 4.6 Non-stationary diffusion processes

$$\text{a) } t/\tau \to t, \quad \text{b) } z/l_D = y, \quad \text{c) } n(z)v/4J_0 = f(y,t), \quad \text{d) } \alpha = 2D/vl_D. \tag{1}$$

$$\text{a) } \partial_t f = \partial_y^2 f - f, \quad \text{b) } f(y,t) = f_0(y) + g(y,t)\exp(-t). \tag{2}$$

$$\text{a) } f_0(y) = \frac{\text{sh}(L-y) + \alpha\,\text{ch}(L-y)}{2\alpha\,\text{ch}\,L + (1+\alpha^2)\text{sh}\,L}, \quad \text{b) } f(y,0) = \frac{\theta(0<y<y_1)\text{ch}(y_1-y)}{\text{ch}\,y_1 + \alpha\,\text{sh}\,y_1}. \tag{3}$$

$$\text{a) } g(y,t) = \sum b_n f_n(y)\exp(-\lambda_n^2 t), \quad \text{b) } b_n = \int [f(y,0) - f_0(y)] f_n(y)\text{d}y. \tag{4}$$

$$\text{a) } \text{d}_y^2 f_n(y) + \lambda_n^2 f_n(y) = 0, \quad \text{b) } f_n(0) - \alpha f_n'(y) = 0, \quad \text{c) } f_n(L) + \alpha f_n'(L) = 0. \tag{5}$$

$$\text{a) } f_n(y) = N_n(\sin\lambda_n y + \alpha\lambda_n\cos\lambda_n y), \quad \text{b) } \text{tg}\,\lambda_n L = 2a\lambda_n/(\alpha^2\lambda_n^2 - 1). \tag{6}$$

$$\text{a) } N_n^2 = 2/L(1 + 2\alpha + \alpha^2\lambda_n^2), \quad \text{b) } \int f_n(y)\text{d}y = -f_n'(y)/\lambda_n^2. \tag{7}$$

$$\int g(y) f_n(y)\text{d}y = -g(y)f_n'(y)/\lambda_n^2 + g'(y)f_n(y)/\lambda_n^2 - \int g''(y)f_n(y)\text{d}y/\lambda_n^2. \tag{8}$$

$$\text{a) } f''(y,0) = f(y,0), \quad \text{b) } f_0''(y) = f_0(y),$$

$$\text{c) } f(0,0) - \alpha f'(0,0) = 1, \quad \text{d) } f_0(L) + \alpha f_0'(L) = 0. \tag{9}$$

$$\text{a) } b_n = -f_n'(y_1)/(1+\lambda_n^2), \quad \text{b) } J(L,t) \sim \partial_y f(y,t)_{y=L}. \tag{10}$$

$$Q(t) = J(L,t)/J(L,\infty) = 1 - \exp(-t)\sum c_n(-1)^n\exp(-\lambda_n^2 t). \tag{11}$$

$$c_n = [2\alpha\,\text{ch}\,L + (1+\alpha^2)\text{sh}\,L]N_n b_n \lambda_n \sim \exp(L - y_1). \tag{12}$$

$$Q(t) = 1 + \exp(L - y_1)\exp(-t)[a_1 - a_2\exp(-\lambda_2^2 t)]. \tag{13}$$

$$\text{a) } a_2 = a_1 + \exp(y_1 - L) > a_1 > 0, \quad \text{b) } \lambda_2^2 \gg 1. \tag{14}$$

$$\text{a) } f(y,t) = g(y,t)\exp(-t), \quad \text{b) } f(y,0) = f_0(y). \tag{15}$$

$$b_n = \int f_0(y) f_n(y) \mathrm{d}y = f_n(y) f_0'(y) - f_n'(y) f_0(y)|_0^L/(1 + \lambda_n^2)$$
$$= f_n'(0)/(1 + \lambda_n^2) = N_n\lambda_n/(1 + \lambda_n^2). \tag{16}$$

$$J(L,t) \approx \exp(-t)\sum f_n'(L) b_n \exp(-\lambda_n^2 t) \sim \exp[-t(1 + \lambda_1^2)]. \tag{17}$$

$$\text{a) } \lambda_1 \approx \pi/2L = \pi l_D/2l, \quad \text{b) } \tau_{ext} = 4l^2\tau/(4l^2 + \pi^2 D_t\tau). \tag{18}$$

$$\text{a) } l_D \gtrsim l, \quad \text{b) } \tau_{exp} = 4l^2/\pi^2 D. \tag{19}$$

Let us introduce the dimensionless variables (1a–c) and the parameter (1d). The diffusion equation written in terms of the variables (1a–c) will be of the form (2a). We shall search for its solution in the form (2b), where the stationary part of the solution is defined in (3a) and the initial solution is (3b). We assume the neutron guide to be partitioned by a perfectly reflecting shutter at the point y_1, before the initial moment of time. In this case the form of function (3b) also follows from the stationary diffusion equations. The non-stationary part of solution (2b) is represented in the form of the series (4a) with coefficients (4b) and eigenfunctions satisfying equation (5a) with eigenvalues λ_n^2 and boundary conditions (5b, c). The solution of eqn (5a) is written in the form (6a), the eigenvalues of which are determined from the transcendental equation (6b), while the normalizing factor is determined from expression (7a). Equation (5a) can also be rewritten in the form (7b), and therefore integrals as in (4) are conveniently represented in the form (8), upon double integration by parts. The functions being integrated in (4b) satisfy conditions (9). Making use of (8), this allows us to reduce integration to the solution of an algebraic equation, which leads to (10a). The flux at the end of the neutron guide is determined by expression (10b). It is convenient to reduce it to the stationary flux and to consider the reduced flux (11), where for brevity notation (12) is introduced. The appearance of a maximum in function (11) can be understood by taking advantage of the example of (13). If the coefficients $a_{1,2}$ satisfy (14a) and (14b) holds, then expression (13) may turn out to exceed unity, and the maximum will be high when a_1 is large, and narrow when τ is small. In addition, the closer y_1 is to L, the higher the maximum.

It is also interesting to consider another problem. Consider the reactor to be switched off immediately upon the establishment of a stationary flux; the solution of eqn (2a) must then be sought in the form (15). The coefficients of the series (4a) are determined in (15) and the flux is given by (17). The flux decreases exponentially over a long period of time, and the flowing out time can then be determined in accordance with (18) and (19).

5

The foundation for the interaction potential of a neutron with matter

Ultracold neutrons can undergo total reflection from most materials, i.e. most substances repel the neutrons. At first sight this seems quite improbable since each individual nucleus attracts neutrons. To dispel the seeming contradiction, it is necessary to analyse the main ideas relevant to the interaction of a neutron with matter. Naturally, we cannot go into details of nuclear structure or of the theory of strong interactions here, but, on the other hand, we will not restrict the scope of our discussion to the region of UCN since the issues considered here, for instance repulsion by a solid body, are characteristic of slow neutrons within a wide range of energies: from UCN up to thermal, and even higher.

To start with, we recall that from an experimental point of view the interaction of slow neutrons with nuclei is described in terms of a single, although complex, parameter b which we shall call the scattering amplitude and treat as a phenomenological quantity. Scattering of slow neutrons on a single fixed nucleus is isotropic, and the problem consists of determining the scattering from an aggregate of nuclei representing a solid body. The scattering on an aggregate of nuclei can be determined without taking advantage of the concept of a potential. We shall show, however, that the scattering proceeds as if matter interacts with a neutron by means of a potential, $u_0 = 4\pi N_0 b$, where N_0 is the number of atoms per unit volume. This is what represents the foundation of the potential, which is the purpose of this chapter.

Naturally, it is practically impossible to describe scattering from the surface of a solid body when the atoms in it are distributed arbitrarily, so we are compelled to restrict ourselves to the consideration of certain mathematically soluble models, i.e. totally random and perfect crystalline bodies. The investigation of crystals is performed by the method of recursion relations. It then turns out to be possible to treat, in a novel way, a number of subtle effects following from dynamic diffraction theory, which is of interest to a large number of physicists—those who deal with UCN and those who do not.

5.1 THE FERMI PSEUDOPOTENTIAL

In quantum mechanics scattering is described by a wavefunction which can be split into two parts:

$$\Psi = \Psi_0 + \Psi_{\text{scat}}, \tag{5.1}$$

where Ψ_0 represents the incident wave which we shall consider to be a plane wave, $\Psi_0 = \exp(\mathrm{i}\mathbf{k}_0\mathbf{r})$, with the wavevector $\mathbf{k}_0$, and Ψ_{scat} is the scattered wave which at distances r from the scatterer, large compared with the dimensions of the scatterer itself, is represented by an outgoing spherical wave, $\exp(\mathrm{i}kr)/r$, where $k = |\mathbf{k}_0|$. If the scattering nucleus is motionless and fixed at the point $\mathbf{r}$, then the total wavefunction of the neutron can be written in the form

$$\Psi = \exp(\mathrm{i}\mathbf{k}_0\mathbf{r}) - (b/|\mathbf{r} - \mathbf{r}'|)\exp(\mathrm{i}\mathbf{k}_0\mathbf{r}')\exp(\mathrm{i}k|\mathbf{r} - \mathbf{r}'|). \tag{5.2}$$

Expression (5.2) can be considered a solution of the Schrödinger equation with some potential created by the nucleus. This potential is not determined uniquely. For instance, it may be represented in the form of a potential well of finite radius, as in nuclear physics, but it may also be represented in the form of a potential barrier. Naturally, the amplitudes b of two differing potentials are, strictly speaking, different; thus, their energy dependences differ from each other. But if only the region of low energies is dealt with, where the amplitude can be considered constant, then this constant can be obtained with the aid of a large variety of potentials.

The most convenient one is the potential

$$u(\mathbf{r}) = (\hbar^2/2m)4\pi b\delta(\mathbf{r} - \mathbf{r}'), \tag{5.3}$$

known as the Fermi pseudopotential (*Fermi and Zinn 1947). With this potential the amplitude b is obtained to the first order of perturbation theory. The sign of potential (5.3) coincides with the sign of the amplitude, and for $b > 0$, which is characteristic of most substances, the potential turns out to be repulsive. (Generally speaking, the amplitude b is a complex number, but for the present we shall neglect its imaginary part.)

The potential of an aggregate of nuclei can be represented as a sum of potentials (5.3) (below we shall omit the factor $(\hbar^2/2m)$:

$$u(\mathbf{r}) = 4\pi\sum b_i\delta(\mathbf{r} - \mathbf{r}_i). \tag{5.4}$$

If all the amplitudes b_i are identical and equal to b, then, passing from the sum in (5.4) to the integral form, for example when the wavelength is significantly greater than the distance between the scatterers, we obtain

$$u(\mathbf{r}) = 4\pi b\int N_0 \mathrm{d}^3r'\delta(\mathbf{r} - \mathbf{r}') = u_0\theta(\mathbf{r} \in V), \tag{5.5}$$

where $u_0 = 4\pi N_0 b$, and the θ function is unity when r is inside the volume V occupied by the material, and zero otherwise. The derivation of (5.5) is not the foundation of the potential u_0, but it allows us to understand why the interaction potential of a neutron with matter is assigned the value u_0. However, if $b > 0$ then $u_0 > 0$, and the material repels neutrons. Actually, instead of the interaction potential (5.3) of the neutron with a single nucleus, we could take advantage of the real nuclear potential $v(r - r') < 0$, which leads to the same wavefunction (5.2) and also takes into account all kinds of corrections. In such a case the procedure (5.5) would lead to the negative potential

$$u(\mathbf{r}) = \sum_i v(\mathbf{r} - \mathbf{r}_i) = N_0 \int \mathrm{d}^3 r' v(\mathbf{r} - \mathbf{r}') < 0, \tag{5.6}$$

signifying attraction of the neutron by the substance. The contradiction between (5.5) and (5.6) points to the necessity for a rigorous foundation of the potential. Note that expression (5.3) is not at all suitable to serve as the potential. Indeed, function (5.2) does not satisfy the Schrödinger equation

$$[-\Delta - k^2 + 4\pi b_0 \delta(\mathbf{r} - \mathbf{r}')]\Psi(\mathbf{r}) = 0$$

but is the solution of the inhomogeneous equation

$$(-\Delta - k^2)\Psi(\mathbf{r}) = -4\pi b \delta(\mathbf{r} - \mathbf{r}')\Psi_0(\mathbf{r}) = 0.$$

The potential which leads to function (5.2) can be written, for example, in the form $4\pi b_0 \delta(\mathbf{r} - \mathbf{r}')\hat{\Pi}(\mathbf{r}')$, where we have introduced the operator $\hat{\Pi}$, acting on the wavefunction as follows:

$$\hat{\Pi}(\mathbf{r}')\Psi(\mathbf{r}) = (\mathrm{d}/\mathrm{d}|\mathbf{r} - \mathbf{r}'|)|\mathbf{r} - \mathbf{r}'| \int \mathrm{d}\Omega_{\mathbf{r}-\mathbf{r}'} \Psi(\mathbf{r}). \tag{5.7}$$

Here integration is performed over angles about the point $\mathbf{r}'$. The solution of the Schrödinger equation

$$[-\Delta - k^2 + 4\pi b_0 \delta(\mathbf{r} - \mathbf{r}')\Pi(\mathbf{r}')]\Psi(\mathbf{r}) = 0 \tag{5.8}$$

is written in the form

$$\Psi(\mathbf{r}) = \exp(\mathrm{i}\mathbf{k}_0\mathbf{r}) - C(b_0/|\mathbf{r} - \mathbf{r}'|)\exp(\mathrm{i}k|\mathbf{r} - \mathbf{r}'|), \tag{5.9}$$

where the constant C represents the value of expression (5.7) at the point $\mathbf{r} = \mathbf{r}'$. Substitution of (5.9) into (5.7) yields the following equation for C:

$$C = \exp(\mathrm{i}\mathbf{k}_0\mathbf{r}') - \mathrm{i}kb_0 C,$$

the solution of which is

$$C = \exp(\mathrm{i}\mathbf{k}_0\mathbf{r}')/(1 + \mathrm{i}kb_0).$$

Thus, the wavefunction (5.2) is a solution of eqn (5.8), if the amplitude b is of the form

$$b = b_0/(1 + \mathrm{i}kb_0). \tag{5.10}$$

In the absence of absorption the quantity b_0 is real and is called the scattering length. The amplitude (5.10) satisfies the unitarity condition and the optical theorem, according to which the imaginary part of b is equal to $k\sigma_t/4\pi$, where σ_t is the total interaction cross-section of a neutron with the nucleus. A rigorous foundation for the interaction (5.3) of a neutron with a pointlike nucleus is given in the book by *Albeverio *et al.* (1988).

For the foundation of the interaction potential of a neutron with an aggregate of nuclei we must, starting from (5.2) and without resorting to the concept of potential, find the wavefunction of the neutron in the presence of numerous scatterers, solve the inverse problem, and then determine the potential and demonstrate it to be equal, precisely, to u_0.

In the presence of many scatterers Ψ_{scat} can be written in the general form

$$\Psi_{\text{scat}} = \hat{R}(\mathbf{k} \to \mathbf{k}')\Psi_0(\mathbf{k}), \tag{5.11}$$

where $\hat{R}$ is the scattering operator. If we are able to calculate Ψ_{scat}, then we can also find the potential leading to such a scattered wave. The wavefunction (5.1) must satisfy the Schrödinger equation

$$[\Delta + k^2 - u(\mathbf{r})]\Psi = 0. \tag{5.12}$$

Substituting (5.1) into (5.12) we obtain for Ψ_{scat} the equation

$$[\Delta + k^2 - u(r)]\Psi_{\text{scat}} = u(r)\Psi_0(k),$$

the solution of which is written with the aid of the Green function G

$$\Psi_{\text{scat}} = \hat{G} \oplus u(\mathbf{r})\Psi_0(\mathbf{k}). \tag{5.13}$$

Comparing (5.13) and (5.11) we obtain the expression

$$\hat{R} = \hat{G} \oplus u(\mathbf{r}),$$

which can be considered the equation for the potential $u(r)$. Naturally, if we make use of the exact wavefunction (5.1) and (5.11), then solution of the inverse scattering problem will yield the potential (5.4), just as if, instead of (5.2), we take the exact scattering wavefunction on a nucleus, where solution of the inverse problem should lead to the exact nuclear potential $v(r - r')$. However, if we see that (5.1) can be represented with a high precision in the form of a set (a triplet) of plane waves (incident, reflected, and refracted), then we will be able to find such a potential, differing from (5.4), that is significantly simpler than (5.4) and yields the same triplet of waves with the same amplitudes. Precisely this potential will be the sought optical potential u_0. All corrections that are to be added to the triplet of waves for obtaining the more precise scattering function (5.7) can then be found by applying perturbation theory, taking as a perturbation the difference between the potential (5.4) and u_0.

Solution of the inverse problem in the general form represents a complex procedure. We shall not resort to it. Instead, the potential will be chosen in each concrete case so as to describe the principal part of the wavefunction (5.1) correctly.

First we shall consider a disordered medium occupying the semi-space $z > 0$, then an infinite linear periodic chain of nuclei, followed by a crystalline plane, and, ultimately, a crystalline semi-space. (These problems were dealt with from a somewhat different standpoint by *Kagan and Afanas'ev (1966).)

The exact solution of the scattering problem is found by the method of multiple scattering of waves (MSW) (developed by *Albeverio *et al.* (1988); *Baryshevskii *et al.* (1965*a*,*b*); *Foldy (1945); *Lax (1951, 1952)), but this solution has only been found in the above-mentioned cases, until now.

We shall not restrict ourselves to considering only issues concerning UCN, since the methods applied here are of a more general character and transition to the case of UCN is quite elementary. From time to time certain questions will be considered concerning the imaginary part of the potential that are closely related to the loss coefficient.

We shall now write down the main equation of the MSW theory.

5.2 EQUATIONS OF THE MSW THEORY

Throughout this chapter we shall assume the atoms to be fixed and motionless and the neutrons to undergo only s scattering, so the wavefunction of a neutron of given energy in the presence of a nucleus at point r_i has the form

$$\Psi(r) = \Psi_0(r) - b\Psi_0(r_i)\eta(r - r_i), \quad \eta(r) = \exp(ikr)/r,$$

where $\Psi_0(r)$ is the wavefunction of the incident neutron. In the presence of many nuclei we have

$$\Psi(\mathbf{r}) = \Psi_0(\mathbf{r}) - \sum_i \Psi_i b_i \eta(\mathbf{r} - \mathbf{r}_i), \tag{5.14}$$

where the coefficients Ψ_i characterize the wave exciting the ith nucleus and are determined by the following set of linear equations:

$$\Psi_i = \Psi_0(\mathbf{r}_i) - \sum_{j \neq i} \Psi_j b_j \eta(\mathbf{r}_i - \mathbf{r}_j), \tag{5.15}$$

where the summation on the right-hand side is performed over all j, except $j = i$. The set of equations (5.14) and (5.15) is the principal one in the theory. It is applicable when it suffices to take into account only the s scattering of UCN and when the distances between atoms are significantly larger than the radius of the nucleus.

For comparison it is useful to give the expression for the scattering of photons on particles exhibiting an isotropic polarizability α. If the incident photon is described by a field $E_0(r)$, then in the dipole approximation we have the following, by analogy with (5.2):

$$E(r) = E_0(r) + \alpha[E_0(r_i)k^2 + (E_0(r_i)\nabla)\nabla]\eta(r - r_i).$$

The corresponding set of equations, similar to (5.14) and (5.15), and its solutions have been examined, for instance, by *Baryshevskii *et al.* (1965*b*). From multiple scattering theory it is also possible to find the refraction coefficient and the Lorentz–Lorenz formula for permittivity in the static case.

5.3 REFLECTION FROM AMORPHOUS SUBSTANCES

Let us assume all the nuclei to be identical, disordered, and independent of each other. Since all the coordinates of the nuclei are random, the coefficients Ψ_i, describing the effective field inside the material, will also be random, as will the function $\Psi(r)$ itself. A random function is characterized by mean values and by correlations for a given probability density function. We shall now find the ensemble average of $\Psi(r)$ assuming the probability for a nucleus to be inside the volume element d^3r_i at the point r_i to equal d^3r_i/V, where V is the total volume of the material. We shall adopt the notation $\Psi_i = \Psi(\mathbf{r}_i)$. To determine the average wavefunction $\Psi(\mathbf{r})$, it is necessary, on the right-hand sides of equalities (5.14) and (5.15), to multiply each term of the sum by d^3r_i/V and to perform integration over the total volume, substituting, in doing so, the coherent part $\bar{\Psi}(\mathbf{r})$ for $\Psi(\mathbf{r})$. As a result we obtain the integral equation

$$\bar{\Psi}(\mathbf{r}) = \Psi_0(\mathbf{r}) - N_0 b \int \eta(\mathbf{r}-\mathbf{r}')\bar{\Psi}(\mathbf{r}')d^3\mathbf{r}',$$

where $N_0 = N/V$ and the bar over the function indicates that it is averaged over all the sites occupied by the nuclei. Applying the operator $-(\Delta + k^2)$ to both parts of this equation we obtain

$$[-\Delta + 4\pi N_0 b\theta(r \in V) - k^2]\bar{\Psi}(\mathbf{r}) = 0, \tag{5.16}$$

where we have taken into account that

$$(\Delta + k^2)\eta(\mathbf{r}) = -4\pi\delta(\mathbf{r}), \tag{5.17}$$

and Ψ_0 is assumed to satisfy eqn (5.17) without the right-hand side, and the nuclei occupy all the infinite semi-space $z > 0$.

The solution of eqn (5.16) represents a triplet of waves

$$\bar{\Psi}(\mathbf{r}) = \exp(\mathrm{i}\mathbf{k}_{0\parallel}\mathbf{r}_{\parallel})\{\theta(z<0)[\exp(\mathrm{i}k_{0\perp}z) + r^+\exp(-\mathrm{i}k_{0\perp}z)] + \theta(z>0)t^+\exp(\mathrm{i}k'_{0\perp}z)\},$$

where $r_{\parallel} = (x, y)$, $k_{0\parallel} = (k_{0x}, k_{0y})$ are vectors parallel to the entrance surface, and $k'_{0\perp}$, $k_{0\perp}$ are components of the wavevector normal to the entrance surface inside and outside the material under consideration, respectively.

Expression (5.16) shows that we have indeed obtained the potential (5.5) and that the approximate wavefunction derived with this potential represents the average, or coherent, part of the exact function.

An interesting problem arises here. It is that the amplitude b, and thus the potential in eqn (5.16), are complex quantities. In accordance with the optical theorem the imaginary part of b is $b'' = k\sigma/4\pi$, where σ is the total cross-section of scattering and absorption. If a nucleus does not absorb neutrons, then in the case of an amorphous substance $b'' = kb^2 \neq 0$, anyhow. Correspondingly, the imaginary part of the potential also differs from zero. But this means that the system behaves as if absorption existed, even though the nuclei themselves do not absorb. For example, if the reflection of neutrons from an infinite amorphous semi-space is considered, the reflected neutron intensity will turn out to be lower than the incident intensity, even when the condition of total reflection is fulfilled. The question arises: what becomes of the missing neutrons? The form of the optical potential provides no answer to this question, since it only describes a part of the wavefunction. To obtain an answer to the above question we must consider that part of the wavefunction that does not enter into $\bar{\Psi}$. From the results of the next chapter it will be clear that the missing neutrons are reflected along non-specular directions. Naturally, the potential (5.5) cannot describe non-specular reflection.

5.4 SCATTERING ON AN INFINITE CRYSTALLINE LINE

Consider an infinite line coinciding with the z axis, along which atoms with the same neutron scattering amplitude b are fixed in perfect order with a period d. If the neutron wavelength λ is larger than d, then we can expect the discreteness of the line to be of no significance when scattering occurs on it, so the line can be represented, by analogy with (5.5), by the potential

$$u = 4\pi N_1 b\delta(\boldsymbol{\rho}), \tag{5.18}$$

where N is the linear density of the atoms, i.e. $N = 1/d$, if d is the distance between atoms, and $\boldsymbol{\rho}$ is the radius vector perpendicular to the line . The solution of the Schrödinger equation with such a potential, when $\Psi_0(r)$ has the form of a plane wave, is

$$\Psi(r) = \exp(\mathrm{i}\mathbf{k}_0\mathbf{r}) - \pi\mathrm{i}(b/d)\mathscr{H}_0^{(1)}(k_{0\perp}r_\perp)\exp(\mathrm{i}k_{0\parallel}r_\parallel), \tag{5.19}$$

where $\mathscr{H}_0^{(1)}(x)$ is the Hankel function corresponding to an outgoing cylindrical wave and $k_{0\perp} = (k_{0x}^2 + k_{0y}^2)^{1/2}$. Here $r_\parallel = z$ and $\mathbf{r}_\perp = \boldsymbol{\rho} = (x, y)$.

Now we shall find the wavefunction without turning to the concept of potential, i.e. we shall solve eqn (5.15) of MSW theory, then substitute the solution into (5.14), and compare the result with the wavefunction (5.19) obtained with the aid of the Schrödinger equation. If the wavefunctions turn out to be identical, we will, subsequently, be justified in applying the potential (5.18). If, on the contrary, the wavefunctions differ, this will provide a hint for the most appropriate choice of the potential.

The exact solution, deduced in MSW theory, is presented in Appendix 5.1. The exact wavefunction has the form

$$\Psi(\mathbf{r}) = \exp(\mathrm{i}\mathbf{k}_0\mathbf{r}) - \mathrm{i}\pi(b_1/d) \sum_{m=-\infty}^{\infty} \mathcal{H}_0^{(1)}(k_{m\perp}r_\perp)\exp(\mathrm{i}k_{m\parallel}r_\parallel). \tag{5.20}$$

Let us compare (5.19) and (5.20). First of all, the formula (5.20) differs from (5.19) in that, instead of a sole outgoing cylindrical wave, it contains several. Each of the additional (with respect to (5.19)) cylindrical waves with the wavevector $\mathbf{k}_m$ corresponds to diffraction in the mth order. The wavevector $\mathbf{k}_m$ differs from the vector $\mathbf{k}_0$ of the incident wave by the reciprocal lattice vector directed along the axis of the crystalline line,

$$\tau_m = 2\pi m/d. \tag{5.21}$$

Since the scattering is elastic, the components of the wavevector, $\mathbf{k}_m$, are

$$k_{m\parallel} = k_{0\parallel} + \tau_m, \quad k_{m\perp} = \sqrt{k_0^2 - k_{m\parallel}^2} = \sqrt{k_{0\perp}^2 - \tau_m^2 - 2k_{0\parallel}\tau_m}. \tag{5.22}$$

Formula (5.20) represents the wavefunction in the presence of several scattering channels. If, in the space, we restrict ourselves to considering only two dimensions perpendicular to the axis of the line and assume the third dimension to be non-observable, then formula (5.19) will only describe the elastic scattering channel, while formula (5.20) describes both elastic and 'inelastic' scattering channels.

When $k_0^2 \ll \tau_1^2$, i.e. if the neutron energy is small, then all the waves undergoing diffraction are damped exponentially as the distance from the line increases, and at large distances the wavefunction is identical in form with (5.19). If only two dimensions are dealt with, then the above circumstance may be understood to indicate that the neutron energy is insufficient for transferring the line into an 'excited' state, even though 'excited' states do manifest themselves in a virtual manner through exponentially damped waves and their presence influences the scattering amplitude b_1.

The amplitude b_1 in (5.20) differs from b in (5.19), as it should, for otherwise the condition of unitarity would not be satisfied. To demonstrate the above, we expand a plane wave into cylindrical waves:

$$\exp(\mathrm{i}\mathbf{k}\mathbf{r}) = \exp(\mathrm{i}k_{0\parallel}r_\parallel) \sum_{n=-\infty}^{\infty} \mathrm{i}^n \exp(\mathrm{i}n\varphi)J_n(k_{0\perp}r_\perp). \tag{5.23}$$

In accordance with (5.19), only the cylindrical s wave is scattered, i.e. the wave $n = 0$. Since $J_0(x) = [\mathcal{H}_0^{(1)}(x) + \mathcal{H}_0^{(2)}(x)]/2$, where $\mathcal{H}_0^{(2)}(x)$ describes the incoming wave and $\mathcal{H}_0^{(1)}(x)$ describes the outgoing one, the S matrix in the cylindrical s wave is, as follows from (5.19) and (5.23):

$$S_s = 1 - 2\pi\mathrm{i}b/d. \tag{5.24}$$

In the absence of absorption and multichannel scattering (i.e. diffraction) the scattering matrix must be unitary, i.e. $|S_s| = 1$. Since in this case the S_s matrix is a number, this number has to be representable in the form $(1 - i\alpha)/(1 + i\alpha)$, where α is a real parameter. However, the amplitude b in the absence of absorption equals (5.10), i.e. $b = b_0/(1 + ikb_0)$ with b_0 real. If it is substituted into (5.24), then the resultant S matrix will be non-unitary. For it to become unitary, the amplitude b in (5.24) must be replaced with an amplitude of the form given by the expression:

$$\tilde{b}_1 = b_0(1 + i\pi b_0/d)^{-1} \tag{5.25}$$

with b_0 real.

Expression (5.25) was obtained on the basis of the most general argument without reference to the exact solution for the wavefunction (5.20). On the other hand, the exact solution, as shown in Appendix 5.1, involves a somewhat different quantity b_1:

$$b_1 = b\left(1 - ikb - (b/d)\sum_{m=-\infty}^{\infty}\ln(-k_{m\perp}^2/\tau_m^2)\right)^{-1}, \quad \tau_0 = 1/d. \tag{5.26}$$

Let us transform this expression to render its comparsion with (5.25) easier. The right-hand part is readily expressed through b_0 if $(1 - ikb)$ is factored out in the denominator and advantage is taken of the fact that $b/(1 - ikb) = b_0$. Further, we note that the arguments of certain logarithms occurring in (5.26) are negative. This is so in those cases when $k_{m\perp}^2 > 0$, i.e. when $k_{m\perp}$ is real. Setting $\ln(-x) = -i\pi + \ln x$, we obtain that in the presence of n diffracted waves and a single scattered one expression (5.26) reduces to the form

$$\begin{aligned} b_1 &= b_0[1 + i(n+1)\pi b_0/d - \gamma b_0/d]^{-1} \\ &= \tilde{b}_1(1 + in\pi\tilde{b}_1/d - \gamma\tilde{b}_1/d)^{-1}, \end{aligned} \tag{5.27}$$

where γ, as follows from Appendix 5.1, is given by

$$\gamma = \sum_{m=-\infty}^{\infty}\ln(|k_{m\perp}^2|/\tau_m^2) = \ln|\cos(kd) - \cos(k_{0\parallel}d)|. \tag{5.28}$$

That (5.27) differs from (5.25) is due to the fact that when there are n diffracted waves, the S_s matrix is no longer just a number, and the unitary condition for it is equivalent to the condition

$$|1 - 2\pi i b_1/d|^2 + n|2\pi b_1/d|^2 = 1.$$

We note an interesting peculiarity of expressions (5.27) and (5.28). Namely, from these expressions it follows that for those values of the incident wavevector k_0, for which among the diffracted waves there appears a wave with $k_{m\perp} = 0$, the amplitude b_1 becomes zero, i.e. the incident wave stops to undergo scattering on a crystalline line. From (5.20) it follows that in this case there remains only one diffracted wave, which propagates strictly along the line and extinguishes the incident wave at those points on the line where the scatterers are situated. Now let us estimate how close to zero the quantity $km_{m\perp}$ has to be for the effect to manifest itself. Clearly, for this to happen the condition $b_1 \ll b_0$ must be fulfilled. Then, from (5.27),

$$-\gamma b_0/d \gg 1, \quad \text{or} \quad -\ln(|k_{m\perp}^2|/\tau_m^2) \gg d/b_0,$$

i.e. $|k_{m\perp}^2| \ll \tau_m^2 \exp(-d/b_0)$. Correspondingly, from (5.22) it follows that $(k_\parallel - \tau_m)/\tau_m \ll \exp(-d/b_0)$, which is an astronomically small quantity, since $d/b_0 \propto 10^4$. Even if we were to assume the necessary degree of monochromaticity to have been achieved in the incident beam and the beam to have been collimated appropriately, for the angular divergence of the diffracted beam to remain as small, the length of the line on which the experiment should have to be performed would have to be significantly greater than the dimensions of the entire Universe.

Scattering on a crystalline line has been considered by *Kagan and Afanas'ev (1966) with the purpose of studying change in resonance position and width, when the scatterers are situated periodically along the line. The result obtained follows directly from the formulae presented. Namely, if we take as the scattering amplitude b the quantity

$$b = \frac{1}{2k}\frac{\Gamma_0}{E - E_0 + \mathrm{i}\Gamma/2}, \tag{5.29}$$

describing resonance scattering on a sole nucleus, and substitute it into (5.26), expression (5.27) assumes the form

$$b_1 = \frac{1}{2k}\frac{\Gamma_0}{E - E_0 - \Delta E + \mathrm{i}(\Gamma + \Delta\Gamma)/2},$$

i.e. the resonance is shifted by $\Delta E = \gamma\Gamma_0/2kd$, while the width alters by $\Delta\Gamma = -\Gamma_0 + \pi(n+1)\Gamma_0/kd$.

The quantity $\tilde{\Gamma}_0 = \pi\Gamma_0/kd$ represents the scattering width through any of the channels, simply elastic scattering on a homogeneous wire, or diffraction. The sum of the above widths, equal to $\Delta\Gamma + \Gamma_0$, substitutes the width of purely elastic scattering on an individual atom, Γ_0, since in the case of a crystalline line the scattering on an individual atom is absent. The width $\tilde{\Gamma}_0$ is constant at small k, but the ratio $\tilde{\Gamma}_0/\Gamma_0$ increases as k decreases, which indicates that the role played by coherence becomes more important as the wavelength increases. On the other hand, as the wavelength decreases the role attributed to coherence should fall, and the sum of the elastic scattering and diffraction widths should converge towards Γ_0. This actually happens, since for large k the number of diffraction channels, n, can be estimated by the quantity $n = kd/\pi$. Thus, as k increases the correction $\Delta\Gamma$ tends towards zero.

The form of the amplitude b undergoing a change does not mean that the probabilities of inelastic processes also change. But when the nuclei are ordered, the inelastic processes do also become coherent. If, upon capturing a neutron, a nucleus emits a particle of energy p^2/M, the wavefunction $\Psi_{in}(\mathbf{r})$ of this particle, in accordance with eqns (5.14) of MSW theory, must be written in the form

$$\text{a) } \Psi_{in}(\mathbf{r}) = -b^{in}\sum \Psi_i \eta_p(\mathbf{r}-\mathbf{r}_i), \quad \text{b) } \eta_p(\mathbf{r}) = \exp(ipr)/r, \tag{5.30}$$

where b^{in} is the inelastic scattering amplitude, which is of the same form as in (5.29), but instead of Γ_0 the numerator contains $(\Gamma_0\Gamma_{in})^{1/2}$, i.e. it involves the width of the inelastic process, Γ_{in}. Substituting Ψ_i from the solution of eqns (5.15) (see Appendix 5.1), we obtain the function of the daughter particle (we neglect here the recoil of the nucleus)

$$\text{a) } \Psi_{in}(\mathbf{r}) = -i\pi(b_1^{in}/d)\sum \mathscr{H}_0^{(1)}(p_{n\perp}r_\perp)\exp(i\mathbf{k}_{n\parallel}\mathbf{r}_\parallel), \quad \text{b) } p_{n\perp} = \sqrt{p^2 - k_{n\parallel}^2},$$

where

$$b_1^{in} = \frac{1}{2k}\frac{\sqrt{\Gamma_0\Gamma_{in}}}{E - E_0 - \Delta E + i(\Gamma + \Delta\Gamma)/2}. \tag{5.31}$$

Thus, the width of an inelastic process remains unchanged, but on the whole the probability of an inelastic process does change and the probability of capture from an elastic channel changes as a result of interference. The probability of capture, in this case, is altered not because the width Γ_0 changes, but mainly due to the shift ΔE appearing in the denominator.

Let us return to our main issue, however. Energies $k^2 \ll \tau_1^2$ are characteristic of UCN, so their scattering actually leads to the origination of a single scattered wave, i.e. at large distances from the wire the wavefunction for the UCN must be of the form (5.19), but the amplitude b_1 (eqn (5.27)) with $n = 0$ should be substituted for b. This expression differs from (5.25) by the presence of the summand $-\gamma b_0/d$ in the denominator. If d represents the interatomic distance, then this term in the denominator is readily estimated by the quantity $-(b_0/d)\ln(u_0 d^2) \approx 10^{-3}$, where for u_0 the usual expression $u_0 = 4\pi N_0 b$ is adopted. Thus, it may be asserted that the scattering of UCN on a crystalline wire can be described by the potential (5.18) with b replaced by the amplitude $\tilde{b}_1$ (eqn (5.25)) decreased somewhat (by 0.1%) as compared with the scattering on an individual atom of amplitude b_0. Taking advantage of the asymptotic expression for the Hankel function we readily obtain the scattering cross-section

$$\sigma = \frac{1}{k_{0\perp}}\left(\frac{2\pi b_\perp}{d}\right)^2. \tag{5.32}$$

This cross-section has the dimension of length, so it might be more correct to call it the transverse scattering length. When n diffracted waves are present the formation cross-section of each of them has the same value (5.32), so the total cross-section turns out to be $n + 1$ times greater. In the case of UCN the amplitudes of virtual (i.e. exponentially damped) diffracted waves decreases rapidly as the distance from the line increases, and already at a distance of the order of magnitude of d from it their sum does not exceed 0.01% of the amplitude of the scattered wave.

5.5 SCATTERING ON AN INFINITE CRYSTALLINE PLANE

We shall now consider an infinite plane perpendicular to the z axis, in which atoms are fixed in perfect order with identical periods along the x and y axes, all the atoms having the same neutron scattering amplitude b. If the neutron wavelength λ is greater than d, we can expect the discreteness of the plane to have no significance for the neutron scattering. The plane can then, by analogy with (5.5), be represented by the potential

$$u = 4\pi N_2 b_2 \delta(z), \tag{5.33}$$

where N_2 is the two-dimensional density, i.e. $N_2 = 1/d^2$, if d is the distance between adjacent atoms and the lattice on the plane is considered to be square. The solution of the Schrödinger equation with the potential (5.33) is

$$\begin{aligned} &\text{a) } \Psi(\mathbf{r}) = \exp(\mathrm{i}\mathbf{k}_0\mathbf{r}) - 2\pi\mathrm{i}(\tilde{b}_2/d)(1/k_{0\perp}d)\exp(\mathrm{i}\mathbf{k}_{0\parallel}\mathbf{r}_\parallel + \mathrm{i}k_{0\perp}|z|), \\ &\text{b) } \tilde{b}_2 = b/[1 + 2\pi\mathrm{i}(1/k_{0\perp}d)b/d]. \end{aligned} \tag{5.34}$$

The subscripts $\parallel$ and $\perp$ indicate vector components parallel and perpendicular to the plane, respectively.

We shall now find the wavefunction, without using the notion of potential, and compare the result with the wavefunction (5.33) obtained with the aid of the Schrödinger equation. If the wavefunctions turn out to be identical, this will mean that we are justified in utilizing the potential (5.34) further. If, on the contrary, the wavefunctions differ, this will provide a hint for a more correct choice of the potential.

The exact solution within the MSW theory is presented in Appendix 5.2. Unlike (5.34), the exact wavefunction contains, besides the scattered wave, diffracted waves also:

$$\begin{aligned} &\text{a) } \Psi(\mathbf{r}) = \exp(\mathrm{i}\mathbf{k}_0\mathbf{r}) - 2\pi\mathrm{i}(b_2/d)\sum_n (1/k_{n\perp}d)\exp(\mathrm{i}\mathbf{k}_{n\parallel}\mathbf{r}_\parallel + \mathrm{i}k_{n\perp}z), \\ &\text{b) } b_2 = b\Big/\Big(1 - \mathrm{i}kb + 2\pi\mathrm{i}\sum_n (1/k_{n\perp}d)b/d\Big). \end{aligned} \tag{5.35}$$

where

$$\text{a) } \mathbf{k}_{n\parallel} = \mathbf{k}_{0\parallel} + \boldsymbol{\tau}_n, \quad \text{b) } \boldsymbol{\tau}_n = 2\pi(m_n, n_n)/d, \quad \text{c) } k_{n\perp} = \sqrt{k^2 - k_{n\parallel}^2}, \tag{5.36}$$

where $\boldsymbol{\tau}_n$ is the reciprocal lattice vector which is parallel to the (x, y) plane.

Formulae (5.35) represent the wavefunction when several scattering channels exist. Formulae (5.34) only describe scattering in the 'elastic' channel, i.e.

scattering without alteration of the magnitude of the component of the wavevector normal to the plane, $k_\perp$.

When $k_0 < 2\pi/d$, all the diffracted waves die away exponentially as the distance from the plane increases. At great distances from it the wavefunctions (5.34a) and (5.35a) no longer differ in form from each other.

The amplitudes (5.34b) and (5.35b) differ from each other and the main difference is related to the requirement of unitarity. In the absence of diffraction, the amplitude of the transmitted wave is $1 - 2\pi i(\tilde{b}_2/d)(1/k_{0\perp}d)$ and the amplitude of the reflected wave is $-2\pi i(\tilde{b}_2/d)(1/k_{0\perp}d)$. In the absence of absorption the total intensity of scattered waves should be equal to the incident intensity. Therefore, the unitarity condition must be satisfied:

$$|1 - 2\pi i(\tilde{b}_2/d)(1/k_{0\perp}d)|^2 + |-2\pi i(\tilde{b}_2/d)(1/k_{0\perp}d)|^2 = 1.$$

However, when (5.34b) is substituted, this condition is not satisfied. It holds only if a real amplitude is substituted for b in (5.34b). It is not difficult to guess that b_0 should be taken as the real amplitude, since the isotropic scattering channel is suppressed in the scattering from the plane, and, subsequently, the corresponding elastic cross-section involved in the imaginary part of the amplitude must be dropped. Indeed, if $(1 - ikb)$ is factored out in the denominator of the right-hand side of (5.35b) and $b/(1 - ikb) = b_0$ is taken into account, then the expression for the exact amplitude, (5.35b), reduces to the form

$$b_2 = b_0 \Big/ \left(1 + 2\pi i \sum_n (1/k_{n\perp}d) b_0/d\right). \tag{5.37}$$

For sufficiently large k_0 diffraction also occurs, besides 'elastic' scattering, i.e. certain $k_{n\perp}$ (eqn (5.36c)) turn out to be real and the unitarity condition acquires the form

$$\begin{aligned} k_{0\perp}|1 - 2\pi i(b_2/d)(1/k_{0\perp}d)|^2 + 2\sum_n{}' k_{n\perp}|-2\pi i(b_2/d)(1/k_{n\perp}d)|^2 \\ - k_{0\perp}|-2\pi i(b_2/d)(1/k_{0\perp}d)|^2 = k_{0\perp}, \end{aligned} \tag{5.38}$$

where the prime on the sum indicates that summation is to be performed only over those τ for which $k_{n\perp}$ is a real quantity. The factor 2 in front of the sum takes into account that the diffracted waves travel in either direction away from the plane. The sum over τ also involves the term with $\tau = 0$. Since this term must have a coefficient equal to 1, a third term is introduced for compensation into (5.38). We can readily verify that condition (5.38) is satisfied by the amplitude

$$b_2 = b_0 \Big/ \left(1 + 2\pi i \sum_n{}' (1/k_{n\perp}d) b_0/d\right).$$

This amplitude differs from (5.37), but the difference is readily taken into account if the following quantity is substituted for b_0:

$$b_{02} = b_0 \bigg/ \left(1 + 2\pi \sum_n{}''(1/|k_{n\perp}|d)b_0/d\right), \tag{5.39}$$

where summation is performed only over those τ for which $k_{n\perp}^2 < 0$.

Note that there is an interesting peculiarity of expression (5.37). Namely, from this expression it follows that the same phenomenon occurs in the case of a crystalline plane as in the case of crystalline line: the amplitude b_2 is zero for those k_0 for which one of the diffracted waves has $k_{n\perp} = 0$, and of all the diffracted waves only one remains which travels strictly along the plane. It suppresses the incident wave at the lattice sites, as a result of which the plane becomes completely transparent.

Now let us estimate how close to zero the quantity $k_{n\perp}$ must be for the effect to manifest itself. Clearly, for this to happen the condition $b_2 \ll b_0$ must hold. Then, from (5.35b),

$$k_{n\perp} \ll 2\pi b_0/d^2.$$

Let the plane be of finite dimensions L. To observe the indicated effect it is necessary that the uncertainty $\Delta k = 2\pi/L$, due to finite dimensions of the plane, be significantly smaller than $k_{n\perp}$, i.e. that $L \gg d^2/b_0$. The above requirements are quite modest and readily fulfilled.

We shall now consider resonance scattering to see how the resonance position and width change when the scatterers are situated periodically in the plane (*Kagan and Afanas'ev 1966). As the scattering amplitude b we shall take the quantity (5.29) and substitute it into (5.35), whereupon expression (5.35b) assumes the form

$$b_2 = \frac{1}{2k}\frac{\Gamma_0}{E - E_0 - \Delta E + \mathrm{i}(\Gamma + \Delta\Gamma)/2}, \tag{5.40}$$

where the shift and broadening of the resonance, ΔE and $\Delta\Gamma$, respectively, are the following:

$$\text{a) } \Delta E = -\Gamma_0 \sum_n{}'' \pi/k|k_{n\perp}|d^2, \quad \text{b) } \Delta\Gamma = -\Gamma_0 + \Gamma_0 \sum_n{}' 2\pi/kk_{n\perp}d^2.$$

The quantity $\Gamma_{n0} = 2\pi\Gamma_0/kk_{n\perp}d^2$ represents the scattering width into the diffractive channel τ_n. The sum of the widths equal to $\Delta\Gamma + \Gamma_0$ substitutes for the purely elastic scattering width on an individual atom Γ_0, since such scattering is absent in the case of a plane. The width $\tilde{\Gamma}_{n0}$ increases as $k_{n\perp}$ decreases, which indicates an enhancement of the role played by coherence as the threshold is approached of diffraction with a transfer to the neutron of the reciprocal lattice vector τ. As the wavelength decreases the role of coherence should decline, as we know, and the total scattering and diffraction width should tend towards Γ_0. In the case under consideration this does not happen, since, whatever the wavelength, the wavevector $\mathbf{k}_0$ can be directed so as to make one of the $k_{n\perp}$ turn out to be close to zero, and then coherence will manifest itself fully. The absence of the expected transition $\Delta\Gamma \to 0$ is due to the assumption that the atoms are absolutely motionless and perfectly ordered in the crystalline plane.

The wavefunction of an inelastic process is computed with the aid of eqn (5.30), which, upon substitution of the coefficients Ψ_l for the plane, leads to the expression

$$\Psi_{\text{in}}(\mathbf{r}) = -2\pi i(b_2^{\text{in}}/d)\sum_n (1/p_{n\perp}d)\exp(i\mathbf{k}_{n\parallel}\mathbf{r}_\parallel + ip_{n\perp}|z|), \quad p_{n\perp} = \sqrt{p^2 - k_{n\parallel}^2},$$

where b_2^{in} is given by (5.40) and $(\Gamma_0\Gamma_{\text{in}})^{1/2}$ must be substituted for Γ_0 in the numerator. Thus, everything that was said about a crystalline wire remains valid here too.

However, let us return to our main task. For the energies characteristic of UCN, $k^2 \ll \tau_1^2$, τ_1 being the minimal reciprocal lattice vector, all the diffracted waves turn out to be virtual, i.e. they die away exponentially as the distance from the plane increases and there remains only the 'elastic' scattering leading to the wave triplet (5.34a). This is described well by the Schrödinger equation with the potential (5.33), in which the amplitude from expression (5.35) is substituted for b_2, while, instead of b_0, the amplitude b_{02} is taken from (5.39). This amplitude differs from b_0 by the quantity $\Delta b = b_0(2\pi b_0/d)$, which amounts to about 0.01% of b_0. Thus, the interaction potential (5.33) of a crystalline plane with UCN can be considered well founded.

We shall now deal with three-dimensional crystals.

5.6 SCATTERING OF NEUTRONS ON PERFECT THREE-DIMENSIONAL CRYSTALS

5.6.1 General equations

First, let us consider an infinite crystal consisting of identical crystalline planes parallel to each other, which we shall, for the present, consider totally arbitrary. We shall split the crystal up, mentally, with infinitesimal vacuum gaps to separate periods (Fig. 5.1), the partitions being in between the planes of atoms. Without loss of generality, we shall consider the potentials of the atoms to be symmetric with respect to the plane.

We choose a certain vacuum gap. We denote the wave travelling in it towards the planes lying to the right of the chosen gap by $\vec{\Psi}_0$, and the wave going to the left by $\overleftarrow{\Psi}_0$. Note that both waves are infinite-dimensional vectors, since they contain components corresponding to all possible directions of diffraction from one plane, even those that lead to waves dying away exponentially as the distance from the plane increases.

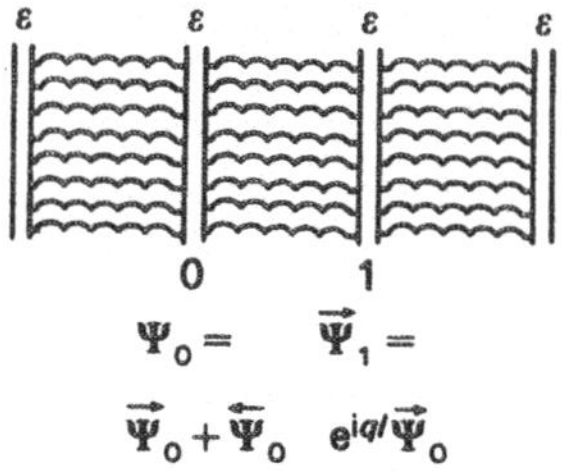

Fig. 5.1 Imaginary vacuum gaps of infinitesimal width are introduced in between the crystalline planes. Two adjacent gaps numbered 0 and 1 are chosen and the wave-functions in these gaps are indicated, consisting of waves travelling both to the right and to the left.

The wave $\overleftarrow{\Psi}_0$ is related to $\vec{\Psi}_0$ by the obvious relation

$$\overleftarrow{\Psi}_0 = R\vec{\Psi}_0, \tag{5.41}$$

where R is the reflection amplitude from a semi-infinite crystal, which naturally represents an infinite-dimensional matrix.

Now let us consider the next vacuum gap to the right. In it there exist similar waves $\vec{\Psi}_1$ and $\overleftarrow{\Psi}_1$ which, owing to the periodicity of the crystal, are related to the preceding waves by the relationship

$$\vec{\Psi}_1 = \exp(iql)\vec{\Psi}_0 \tag{5.42}$$

where l is the period width and q is the Bloch wavevector, which also represents an infinite-dimensional matrix.

Now consider the period placed between the two indicated adjacent gaps. The wave $\vec{\Psi}_0$ being diffracted on it is reflected with amplitude r and transmitted with amplitude t to the next vacuum gap. The amplitudes r and t are infinite-dimensional matrices, and we shall assume them to be known. For the wave $\vec{\Psi}_1$ we can write the self-consistent expression

$$\vec{\Psi}_1 = t\vec{\Psi}_0 + r\overleftarrow{\Psi}_1, \tag{5.43}$$

which signifies that $\vec{\Psi}_1$ represents the sum of the wave arriving from the preceding gap and the wave $\overleftarrow{\Psi}_1$ of the same gap that is reflected from one period to the left.

Substituting $\overleftarrow{\Psi}_1 = R\vec{\Psi}_1$ and (5.42) into (5.43) we obtain the relation

$$\exp(iql) = (1 - rR)^{-1}t. \tag{5.44}$$

Now consider the wave $\overleftarrow{\Psi}_0$; it can be represented in the form of the sum

$$\overleftarrow{\Psi}_0 = r\vec{\Psi}_0 + t\overleftarrow{\Psi}_1, \tag{5.45}$$

in which the interpretation of each summand is quite evident. Substituting

(5.41) and (5.42) into (5.45) we obtain one more relation:

$$R = r + tR\exp(iql). \tag{5.46}$$

Substitution of (5.44) into (5.46) leads to the following equation for R:

$$R = r + tR(1 - rR)^{-1}t, \tag{5.47}$$

which can also be written in the form

$$Rt^{-1}(1 - rR) = tR + rt^{-1}(1 - rR). \tag{5.48}$$

Substituting the solution of eqn (5.48) into (5.44) we find the Bloch phase factor. For a complete solution of the problem of a particle moving in an infinite crystal it is still necessary to find the vector $\vec{\Psi}_0$, which in this case represents the eigenfunction of the Bloch vector q.

The scattering problem on a semi-infinite crystal has, then, also been solved, since R and $\exp(iql)$ have already been found. The only difference from an infinite crystal is that as $\vec{\Psi}_0$ we can take an arbitrary vector, for example a vector with a sole component corresponding to the given incident plane wave.

It is also extremely easy to determine the diffraction on a crystal of finite thickness, i.e. on a crystal consisting of a finite number of infinitely extended crystal planes. Indeed, consider a crystal consisting of N such planes. Let us again return to the infinite crystal and note that it is also periodic with a period $L = Nl$. Now we introduce two vacuum slits at a distance L from each other and reason in the same manner as before, but replace the amplitudes r and t by the scattering and transmission amplitudes R_N and T_N, respectively, of N periods, which are also infinite-dimensional matrices. Expressions (5.44) and (5.46) will now acquire the form

$$\text{a) } \exp(iqL) = (1 - R_N R)^{-1} T_N, \quad \text{b) } R = R_N + T_N R\exp(iqL). \tag{5.49}$$

Expression (5.49a) is conveniently represented in the form

$$\exp(iqL) = T_N + R_N R\exp(iqL). \tag{5.50}$$

Resolving the set of equations (5.49b) and (5.50) with respect to R_N and T_N we obtain

$$T_N = (1 - R^2)\exp(iqL)[1 - R\exp(iqL)R\exp(iqL)]^{-1}, \tag{5.51}$$

$$R_N = [R - \exp(iqL)R\exp(iqL)]\,[1 - R\exp(iqL)R\exp(iqL)]^{-1}. \tag{5.52}$$

Substituting R and q here from (5.48) and (5.44) we express R_N and T_N through the given parameters r and t. Formula (5.51) contains all the results relevant to the *pendellösung* fringes of the intensity at the exit surface of the crystal in the case of Laue diffraction.

Thus, the key expressions are (5.47), (5.44), (5.51), and (5.52), which hold for the scattering of arbitrary waves on an arbitrary crystal in the case of arbitrary interaction of waves with the atoms of a crystal. An atom can be understood here to be a whole crystal cell. The condition, adopted earlier, that the interaction potential of the radiation with the atoms be symmetric relative to the crystal plane is no longer necessary. In the absence of symmetry the amplitudes r for reflection to the left and to the right of the plane differ in their phases, which are totally inherited by the amplitudes R and R_N, while the transmission amplitudes remain identical for transmission in both directions.

Our task now is to demonstrate how the above expressions are applied to practical computations. For this purpose we shall limit our consideration to the diffraction of spinless neutrons on a single-atom tetragonal crystal with its tetragonal axis perpendicular to the surface.

5.6.2 Solution for the practical case of a tetragonal crystal, the tetragonal axis of which is perpendicular to the surface

In the case of scattering on a crystal plane at the point z', we shall write the neutron wavefunction (5.35a) in the form

$$\psi(\mathbf{r}) = \exp(i\mathbf{k}_0\mathbf{r}) \sum_n K_n \exp(i\mathbf{k}_{n\|}\mathbf{r}_\|)\exp(ik_{n\perp}|z - z'| + ik_{0\perp}z'), \qquad (5.53)$$

where the following notation is introduced:

$$\text{a) } K_n = -2\pi i b_2/k_{n\perp} \equiv \bar{K}_n/(1-\bar{\kappa}), \quad \text{b) } \bar{K}_n = -2\pi i b_0/k_{n\perp}, \quad \text{c) } \bar{\kappa} = \sum_n \bar{K}_n. \qquad (5.54)$$

The sum (5.54c) is of a divergent form, but as already noted it actually represents quite a finite function of k_0.

From the form of the wavefunction (5.53) it follows that it can be represented as an infinite-dimensional vector expanded in the basis $\varphi_n = \exp(i\mathbf{k}_{n\|}\mathbf{r}_\|)$. The vector ψ of the wave reflected from the plane can be represented in the form $\psi_r = r\psi_0$, where ψ_0 corresponds to the incident wave and thus to the vector directed along the sole base unit vector φ_0. The components of the vectors ψ_r and ψ_0 are related to each other by the elements of the reflection matrix r. In a similar manner, the wave ψ_t transmitted by the plane has components related to the vector component ψ_0 by the elements of the transmission matrix t.The elements of the matrices r and t of period 1 with the crystal plane in the middle are, respectively:

$$\text{a) } r_{fn} = -2\pi i b_2 e_f e_n/k_{f\perp}, \quad \text{b) } t_{fn} = e_f(\delta_{fn} - 2\pi i b_2/k_{f\perp})e_n, \qquad (5.55)$$

where $e_f = \exp(ik_{f\perp}l/2)$ and δ_{fn} is the Kronecker symbol equal to 1 when $n = f$, and to zero when $n \neq f$.

Now we write the matrices (5.55) in the form of a product of matrices:

$$\text{a) } r = \mathbf{EKNE}, \quad \text{b) } t = \mathbf{E}(\mathbf{I} + \mathbf{KN})\mathbf{E} \tag{5.56}$$

where $\mathbf{I}$ is the unit matrix, the matrices $\mathbf{E}$ and $\mathbf{K}$ are diagonal, and the matrix elements of $\mathbf{E}$, $\mathbf{K}$, and $\mathbf{N}$ are

$$\text{a) } \mathbf{E}_{fi} = e_f \delta_{fi}, \quad \text{b) } [illegible] = K_f \delta_{fi}, \quad \text{c) } K_f = -2\pi i b_2 / k_{f\perp}, \quad \text{d) } \mathbf{N}_{fi} = 1. \tag{5.57}$$

From the condition of detailed balance and from comparison with (5.56a) it follows that the matrix sought, R, can be found in the form

$$\text{a) } R = \mathbf{EKRE}, \quad \text{b) } \mathbf{R}_{fi} = \mathbf{R}_{if}. \tag{5.58}$$

Now we substitute (5.56) and (5.58) into (5.48). Upon performing simple transformations (see Appendix 5.3) we obtain the expressions for the matrix elements of the reflection matrix R from the crystal semi-space:

$$\begin{gathered}\text{a) } R_{fi} = \mu s_f e_f \bar{K}_f s_i e_i / (1 - e_f^2 e_i^2), \quad \text{b) } \mu = 1 \Big/ \Big(1 + \sum_i s_i \bar{K}_i\Big), \\ \text{c) } \bar{K}_f = -2\pi i b_0 / k_{f\perp} = -i u_0 l / 2k_{f\perp}, \quad \text{d) } u_0 = 4\pi N_0 b_0, \\ \text{e) } N_0 = 1/d^2 l,\end{gathered} \tag{5.59}$$

where s_f satisfy the equations

$$s_f = 1 + \sum_i s_i \bar{K}_i + e_f^2 s_f \sum_i s_i \bar{K}_i e_i^2 / (1 - e_f^2 e_i^2). \tag{5.60}$$

Substituting (5.59) into (5.44) we obtain (see Appendix 5.4)

$$\text{a) } \exp(iql) = \mathbf{E}(\mathbf{I} + \bar{\mathbf{K}}\mathbf{NS})\mathbf{E}, \quad \text{b) } \mathbf{S}_{fi} = s_f \delta_{fi}. \tag{5.61}$$

Now we shall find the eigenfunctions Ψ_λ and the eigenvalues λ of the operator $\exp(iql)$. From (5.61a) and from the properties of the matrix $\mathbf{N}$ it follows that

$$\exp(iql)\Psi_\lambda = \mathbf{E}^2 \Psi_\lambda + \nu \mathbf{E}\bar{\mathbf{K}} n = \lambda \Psi_\lambda, \tag{5.62}$$

where n is a vector with all coordinates equal to unity and the constant ν is given by

$$\nu = \langle n | \mathbf{SE}\Psi \rangle = \sum_i s_i e_i \Psi_i. \tag{5.63}$$

Here the angle brackets indicate the scalar product of vectors. From eqn (5.62)

$$\Psi = \nu(\lambda \mathbf{I} - \mathbf{E}^2)^{-1} \mathbf{E}\bar{\mathbf{K}} n. \tag{5.64}$$

Substituting (5.64) into (5.63) we obtain the final equation for λ:

$$\langle n | (\lambda \mathbf{I} - \mathbf{E}^2)^{-1} \mathbf{E}\bar{\mathbf{K}} | n \rangle = 1. \tag{5.65}$$

Assuming Ψ to be normalized to unity we rewrite (5.64) and (5.65) in terms of the matrix components:

$$\text{a) } \psi_{i\lambda} = ve_i\bar{K}_i/(\lambda - e_i^2), \quad \text{b) } v^{-1} = \sum e_i^2 \bar{K}_i^2/(\lambda - e_i^2)^2, \tag{5.66}$$

$$\sum_i s_i e_i^2 \bar{K}_i/(\lambda - e_i^2) = 1. \tag{5.67}$$

5.6.3 Investigation of various processes

1 Total reflection of ultracold neutrons

Let the wavevector k be so small that all the diffracted waves die away exponentially. In this case all the summands in the sums (5.59) and (5.60) are small, with the exception of the term with the denominator which contains e_0^4. Setting all the s_i equal to unity for $i \neq 0$ and expanding the denominator $1 - e_0^4$ in powers of $k_{0\perp}^l$ we find, to the lowest approximation in powers of l, the following equation for $s_0 (x = s_0)$:

$$x = 1 + u_0 x^2/4k_{0\perp}^2,$$

the solution of which is

$$\text{a) } x = 2k_{0\perp}(k_{0\perp} - k'_{0\perp})/u_0, \quad \text{b) } k'_{0\perp} = \sqrt{k_{0\perp}^2 - u_0}, \tag{5.68}$$

where $k'_{0\perp}$ plays the role of the wavevector inside the substance with the potential u_0. Substituting (5.68a) into (5.59a) we obtain, upon performing elementary transformations, the well-known expression for the reflection amplitude of cold neutrons:

$$R = R_{00} = (k_{0\perp} - k'_{0\perp})/(k_{0\perp} + k'_{0\perp}), \tag{5.69}$$

from which we can see that, when $k_{0\perp}^2 < u_0$, total coherent reflection is observed. The relation $k^2 < u_0$ is characteristic of UCN, and therefore they undergo total reflection at all angles of incidence. Thus, in the three-dimensional case, we can also consider the interaction potential u_0 to be well founded.

Substituting (5.68a) into (5.67) we obtain

$$\text{a) } \lambda = \exp(iq_\lambda l) = e_0^2(1 + x\bar{K}_0), \quad \text{b) } q_\lambda = k_{0\perp} + \ln[1 + i(k'_{0\perp} - k_{0\perp})l]/il.$$

To the lowest order in l, also, q assumes the well-known expression for $k'_{0\perp}$ corresponding to the normal component of the wavevector inside the substance with the potential u_0.

Having found that the reflection of cold neutrons from a semi-infinite crystal is equivalent to their being reflected from a potential step of height u_0, we can take advantage of formulae (5.51) and (5.52) and find the transmission and reflection from a rectangular potential barrier of height u_0 and width a

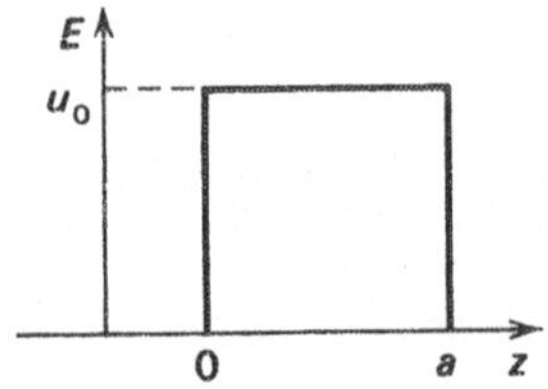

Fig. 5.2 Rectangular potential barrier of height u_0 and width a.

(Fig. 5.2). Since, in this case, all the quantities are scalars,

$$\begin{aligned} &\text{a)}\ T_a = \exp(ik_{0\perp} a)(1 - R^2)/[1 - R^2\exp(ik_{0\perp} a)], \\ &\text{b)}\ R_a = R[1 - \exp(2ik_{0\perp} a)]/[1 - R^2\exp(2ik_{0\perp} a)]. \end{aligned} \tag{5.70}$$

With the aid of these formulae and the recursive relations presented above it is possible to determine the reflection and transmission of a more complex step potential, for instance that shown in Fig. 5.3. To this end we introduce an infinitesimal vacuum slit in between two rectangular potentials and assign a subscript 1 to the amplitudes R and T for the first rectangle and a subscript 2 to the amplitudes for the second rectangle. Then the combined reflection amplitude for both rectangles will be

$$\text{a)}\ R_{12} = R_1 + R_2 T_1^2/(1 - R_1 R_2), \quad \text{b)}\ T_{12} = T_1 T_2/(1 - R_1 R_2). \tag{5.71}$$

Expressions (5.71) can be utilized to compute the reflection from a supermirror representing a periodic structure consisting of a pair of layers with two different potentials. Substituting $r = R_{12}$ and $t = T_{12}$ into eqn (5.47) we obtain a quadratic equation of the form

$$x^2 - 2ax + 1 = 0$$

for the reflection amplitude R from a semi-infinite supermirror, where the notation $x + R$, $a = (1 + r^2 - t^2)/2r$ is adopted. The solution of this equa-

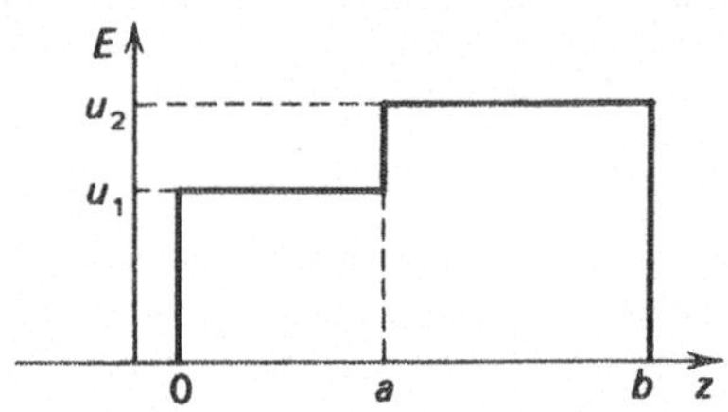

Fig. 5.3 Steplike rectangular potential. For computation of reflection from such a potential it is convenient to partition it mentally by an infinitely narrow gap with the zero potential at point a. The reflection from the total potential is then readily found, if the reflection from the potential in Fig. 5.2 is known.

tion may, for instance, be represented in a form similar to the Fresnel coefficient:

$$x_{1,2} = (\sqrt{a+1} \pm \sqrt{a-1})/(\sqrt{a+1} \mp \sqrt{a-1}).$$

Substituting here the quantity a we obtain

$$\text{a) } R = (\sqrt{+} - \sqrt{-})/(\sqrt{+} + \sqrt{-}), \quad \text{b) } \sqrt{\pm} = \sqrt{(r \pm 1)^2 - t^2}, \tag{5.72}$$

where the solution chosen for R turns out to be zero in the case of the fictitious potential ($r = 0$).

The formulae obtained make possible the description not only of UCN reflection, but also of all processes characteristic of dynamic diffraction theory. Now, since the technology for preparing supermirrors and diffraction gratings (Ioffe and Zabijakin 1988) can also be applied to obtain artificial crystals, it will be useful to demonstrate for practical examples how the above formulae permit the description of effects that are already known from the physics of thermal neutrons and that are explained within conventional dynamic diffraction theory. Let us, therefore, turn once again to expressions (5.59) and (5.60).

2. *Single-wave Bragg diffraction* (Fig. 5.4)

Let $k_{0\perp} l$ be close to π and let the other ones be far from it and yield only waves dying away exponentially. Then we can, again, retain a single term in the sums (5.59) and (5.60). The equation for $x \equiv s_0$ acquires the form

$$\text{a) } x = 1 - u_0 x^2/4k_{0\perp}\delta, \quad \text{b) } \delta = k_B - k_{0\perp}, \quad \text{c) } k_B = \pi/l, \tag{5.73}$$

where k_B is the so-called Bragg vector. Substituting the solution of eqn (5.73)

$$x = 2\sqrt{k_{0\perp}\delta}(\sqrt{k_{0\perp}\delta + u_0} - \sqrt{k_{0\perp}\delta})/u_0 \tag{5.74}$$

into (5.59) and approximating $k_{0\perp}$ by $(k_{0\perp} + k_B)/2$ we obtain the following expression for the Bragg reflection amplitude:

$$R = R_{00} = (\sqrt{k_B^2 - k_{0\perp}^2} - \sqrt{k_B^2 - k_{0\perp}^2 + 2u_0})/(\sqrt{k_B^2 - k_{0\perp}^2} + \sqrt{k_B^2 - k_{0\perp}^2 + 2u_0}),$$

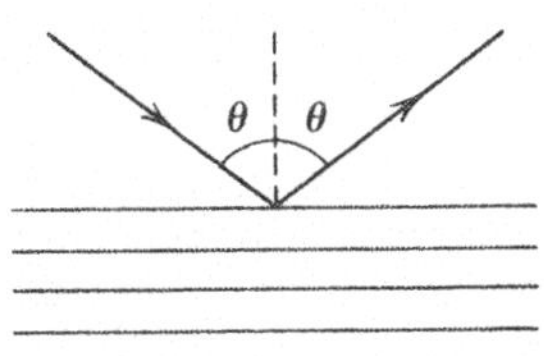

Fig. 5.4 Bragg reflection; here specular reflection occurs from the crystal plane.

from which we can see that total reflection, or so-called Bragg reflection, occurs within the interval $|k_B^2 - k_{0\perp}^2| \leqslant 2|u_0|$. This interval is known in dynamical diffraction theory as the Darwin table.

Substituting (5.74) into (5.60a) we obtain the following expression for the Bloch vector:

$$q_\lambda = k_{0\perp} + \ln[1 + i(\delta - \sqrt{\delta/k_{0\perp}}\sqrt{k_{0\perp}\delta + u_0})l]/il$$
$$\simeq k_B - \sqrt{(k_B^2 - k_{0\perp}^2)(k_B^2 - k_{0\perp}^2 + 2u_0)}/2k_{0\perp},$$

which can also be written approximately in the form

$$q = \sqrt{k_B^2 - \sqrt{(k_B^2 - k_{0\perp}^2)(k_B^2 - k_{0\perp}^2 + 2u_0)}}. \tag{5.75}$$

Expression (5.75) holds for $k_{0\perp} < k_B$ and, as $k_{0\perp}$ departs from k_B, it approaches $k_{0\perp}$. In the opposite case of $k_{0\perp} > k_B$ another equality holds:

$$q = \sqrt{k_B^2 + \sqrt{(k_{0\perp}^2 - k_B^2)(k_{0\perp}^2 - k_\beta^2 - 2u_0)}},$$

which also approaches $k_{0\perp}$ when $k_{0\perp}$ departs from k_B.

3. Double-wave Laue diffraction (Fig. 5.5)

Assume that only two travelling waves with $k_{0\perp}$ and $k_{1\perp}$ can be present in the crystal, while the other diffracted waves die away exponentially. For simplicity we shall consider the quantity $k_{i\perp}l$ to be far from π for both travelling waves, and then all the s_i can be set approximately equal to unity. The Bloch vector is more conveniently computed to the lowest approximation in l. The matrix elements of vector q are

$$q_{fi} = k_{f\perp}\delta_{fi} - (u_0/2k_{f\perp})s_i.$$

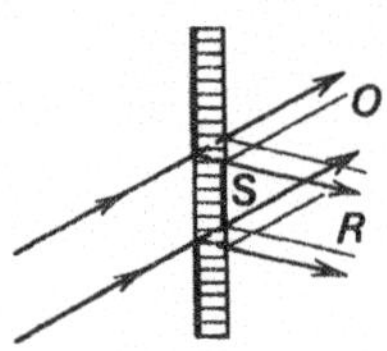

a

Fig. 5.5 The crystal planes on which the Bragg reflection occurs are perpendicular to the entrance surface. Inside the crystal two pairs of waves are formed with slightly differing wavevectors. Each pair contains both the incident wave and the wave reflected from the crystal planes. Upon leaving the crystal the two incident waves as well as the two reflected waves belonging to the two pairs interfere with each other, which leads to so-called 'pendulum' beats of intensity (pendellösung).

Setting $s_i = 1$ we find the eigenfunctions of the matrix q_{fi}:

$$\det\begin{bmatrix} k_0 - 1/2k_0 - q & -1/2k_0 \\ -1/2k_1 & k_1 - 1/k_1 - q \end{bmatrix} = 0,$$

where the k_i stand for the $k_{i\perp}$. The above equation can be rewritten in the form

$$(k_1 - 1/2k_1 - q)(k_0 - 1/2k_0 - q) - 1/4k_0k_1 = 0,$$

whence

$$q = [(k_0 + k_1)/2](1 - u_0/2k_0k_1) \pm \sqrt{(k_0 - k_1)^2(1 + u_0/2k_0k_1)^2/4 + u_0^2/4k_0k_1}\,.$$

Now we introduce $p = (k_0 + k_1)/2$. Then $k_i = p \pm \Delta/2p$ and $q = p - u_0/2p \pm (\Delta^2 + u_0^2)^{1/2}/2p$. In the vicinity of the exact condition of diffraction, when $\Delta \simeq u$, this expression can be rewritten in the form

$$\text{a) } q \simeq \sqrt{p^2 - u_0(1 \pm \sqrt{1 + y^2})}, \quad \text{b) } y = \Delta/u_0.$$

Now consider a crystal of finite width L. In computing the transmission of the crystal for the Laue geometry, in the expression for T_N R can be set equal to zero. Indeed, since all the s_i are close to unity, as noted above, and the denominators in (5.59) are not small, all the matrix elements of the reflection matrix R are proportional to $u_0 l/k_{0\perp}$ and are, therefore, small. As a result we have

$$T_N = \exp(iqL), \tag{5.76}$$

and the wavefunction at the exit is computed in accordance with the expression

$$\Psi \simeq \exp(iqL)\psi_0 \tag{5.77}$$

Since the vector Ψ propagating within the crystal only possesses two components, to compute the transmission we can apply spinor algebra. Expressing the matrix q through the Pauli matrices $\boldsymbol{\sigma}$ we have

$$\exp(iqL) = \exp(iq_0 L + i\mathbf{q}\boldsymbol{\sigma}L), \tag{5.78}$$

where q_0 and q represent four numerical parameters. Any vector Ψ_0 may be decomposed into the eigenvectors of the matrix (5.78) as follows:

$$\psi_0 = [(1 + \mathbf{b}\boldsymbol{\sigma})/2 + (1 - \mathbf{b}\boldsymbol{\sigma})/2]\psi_0, \tag{5.79}$$

where $\mathbf{b} = \mathbf{q}/|\mathbf{q}|$ is a unit vector. Substituting (5.79) into (5.77) we obtain

$$\Psi = \exp[i(q_0 + |\mathbf{q}|)L]\,[(1 + \mathbf{b}\boldsymbol{\sigma})/2]\psi_0 + \exp[i(q_0 - |\mathbf{q}|)L]\,[(1 - \mathbf{b}\boldsymbol{\sigma})/2]\psi_0. \tag{5.80}$$

Hence it follows that the amplitudes of the direct and diffracted waves, Ψ_0

and Ψ_d, are, respectively:

$$\text{a)}\ \Psi_0 = \exp(iq_0 L)[\cos(|\mathbf{q}|L) + ib_z \sin(|\mathbf{q}|L)],$$

$$\text{b)}\ \Psi_d = i(b_x + ib_y)\exp(iq_0 L)\sin(|\mathbf{q}|L).$$

From the form of the matrix q it follows that $|\mathbf{q}| = u_0(1 + y^2)^{1/2}/2p$, and, besides,

$$\text{a)}\ b_z^2 = y^2/(1 + y^2), \quad \text{b)}\ b_x^2 + b_y^2 = 1/(1 + y^2), \quad \text{c)}\ y = \Delta/u_0.$$

Thus, the intensity of the diffracted wave is

$$I_d = \sin^2(Lu_0\sqrt{1 + y^2}/2p)/(1 + y^2),$$

while the intensity of the transmitted wave is

$$I_0 = 1 - \sin^2(lu_0\sqrt{1 + y^2}/2p)/(1 + y^2),$$

which is in good agreement with conventional ideas.

4. Double-wave Bragg diffraction

Now let us consider double-wave Bragg diffraction, for which in conventional theory it is necessary to take advantage of the four-wave approximation. Here we shall only deal with the simplest symmetric case, when

$$\text{a)}\ k_{0\perp} = k_{1\perp} = k_B - \delta, \qquad \text{b)}\ k_B = \pi/l.$$

The equations for $x \equiv s_0$ and $y \equiv s_1$ have the form

$$\text{a)}\ x = 1 + x^2 u_0/2k_B\delta, \qquad \text{b)}\ y = x.$$

Their solution is

$$x = y = [k_B\delta - \sqrt{k_B\delta(k_B\delta - 2u_0)}]/u_0.$$

Substituting this into (5.59) we obtain the reflection amplitudes

$$R_{00} = R_{10} = (\sqrt{k_B\delta} - \sqrt{k_B\delta - 2u_0})/2(\sqrt{k_B\delta} + \sqrt{k_B\delta - 2u_0}).$$

When $k_B\delta < 2u_0$ total reflection occurs and half of the reflected intensity is directed backwards, or at an angle different from the specular one.

Note that in this section a novel formulation of the dynamical theory of diffraction was presented, which possesses a number of advantages compared with conventional theory. Thus, it is possible here to view in a new way one of the most interesting issues of the Laue diffraction, namely the direction of a neutron flux in a crystal. We shall not deal further with this point.

5.6.4 The case of small $k_{n\perp}$ in a three-dimensional crystal

In a three-dimensional crystal the same effect may be expected as in the case of a crystal line and a crystal plane in that in the presence of a diffraction wave with $k_{n\perp} = 0$ all the reflections, except the nth one, must vanish. Let us check this assumption.

For simplicity we shall consider all the $k_{f\perp}l$ to be far from πn. Then all the s_f can be considered close to 1. On the other hand the equation for s_n can be written in the form

$$x = 1 + xK + x^2K^2/l^2u_0, \qquad K = \bar{K}_n,$$

and its solution is

$$x \simeq (-K \pm \sqrt{K^2 - 4K^2/l^2u_0})l^2u_0/2K^2 \simeq \pm 2il\sqrt{u_0}/2K$$
$$= \pm 2k_{n\perp}/\sqrt{u_0}.$$

The reflection amplitude in this channel is, correspondingly,

$$R_{ni} = \pm i\mu\sqrt{u_0}/(1 - e_i^2),$$

and the reflection amplitudes to other channels are suppressed. Thus, in the case of a three-dimensional crystal the effect is absent.

Now let us see how the amplitudes of resonance scattering transform. We note that, in accordance with expression (5.59), the reflection amplitudes R_{fi} between any channels i and f contain the factor μb_0, which can be written as

$$b_3 = b_0 \Big/ \left(1 - 2\pi i(b_0/d)\sum{}'(s_i/k_{i\perp}d) - 2\pi(b_0/d)\sum{}''(s_i/|k_{i\perp}|d)\right). \tag{5.80}$$

This same amplitude will also enter into the equations for s_i, if they are written in the form

$$\text{a)}\ s_f = 1 + s_f\sum s_i\tilde{K}_i/(1 - e_f^2e_i^2), \quad \text{b)}\ \tilde{K}_i = -2\pi i b_3/k_{i\perp}.$$

It is surprising that the imaginary part of the amplitude b_3 differs in sign from the amplitudes b_1 and b_2. In addition, the virtual diffraction channels seem to lead to an enhancement of the amplitude b_0 and to an enhancement of the reflection between the channels, R_{fi}. This circumstance does not, moreover, lead to a violation of unitarity, since, as R increases, the fraction of neutrons transmitted by the crystal decreases and the balance with the incident flux is conserved.

It is now clear what happens when the Breit–Wigner formula is substituted into (5.80). The resulting expression will also be of the Breit–Wigner form, only with a shift in the resonance energy, ΔE, and a broadening $\Delta\Gamma$ equal, respectively, to the following:

$$\text{a)}\ \Delta E = \pi\Gamma_0\sum{}''(s_i/k|k_{i\perp}|d^2),$$
$$\text{b)}\ \Delta\Gamma = -\Gamma_0[1 + 2\pi\sum{}'(s_i/kk_{i\perp}d^2)]. \tag{5.81}$$

These formulae contain the factors s_i in the sums and, therefore, such sharp jumps do not occur as in the crystal line and crystal plane cases.

The new Breit–Wigner formula determines the elastic rescattering of a neutron. For determining the wavefunction of the daughter particle produced as a result of the capture of a neutron it is necessary to multiply the inelastic scattering amplitude b^{in} for a single nucleus by the neutron wavefunction at the given nucleus and sum over all the nuclei. We shall not do this here, since in this case, to take the crystal structure fully into account, we must also take into account rescattering of the daughter particle on the nuclei, and this would detract us too far from our main issue.

So, to conclude! The foundation of the potential u_0 for UCN can be considered complete. In the course of theoretical studies on this subject many interesting results have been achieved that pertain to diffraction problems and are related not only to UCN, but also to neutrons of higher energies and even to problems outside the scope of neutron physics. This supports the thesis, formulated in the Preface, that the physics of UCN makes it possible to view various phenomena in a new way and thus to reveal in these phenomena new facets which were previously outside the scope of the studies. We can now pass to other questions, but before completing this chapter it would be worthwhile to touch upon an interesting relationship between the phases of the reflection amplitude R and the transmission amplitude T of an arbitrary one-dimensional real potential, since it is derived from extremely lucid and illustrative arguments which, again, were evoked by the properties of UCN, thus demonstrating their expediency from the standpoint of knowledge.

5.7 RELATIONSHIP BETWEEN THE PHASES OF THE REFLECTION AND TRANSMISSION AMPLITUDES FOR AN ARBITRARY POTENTIAL

In the one-dimensional case, when both R and q are just numbers and, therefore, commute, expressions (5.51) and (5.52) clearly reveal that the phases of R_N and T_N differ by $\pi/2$. Now we shall show that this is not a chance circumstance, but must occur for any non-absorbing potential. Let us consider the plan of a thought experiment (Fig. 5.6). The neutron beam is divided by a semi-transparent plate into two parts which, upon undergoing total reflection from two additional identical mirrors, are brought together and retransmitted through another semi-transparent plate identical to the first one. We shall consider the interaction potential of the semi-transparent plate with the neutrons to be equal to the arbitrary potential of interest. The neutron intensity registered by the detectors $D_{1,2}$ must be equal to the incident beam intensity, which we shall consider equal to unity. With the aid of peripheral devices it is possible to create any difference of phases η between the neutrons following path I and those following path II. The intensity registered by detector D_1 is

$$J_1 = |rt + te^{i\eta}r|^2, \tag{5.82}$$

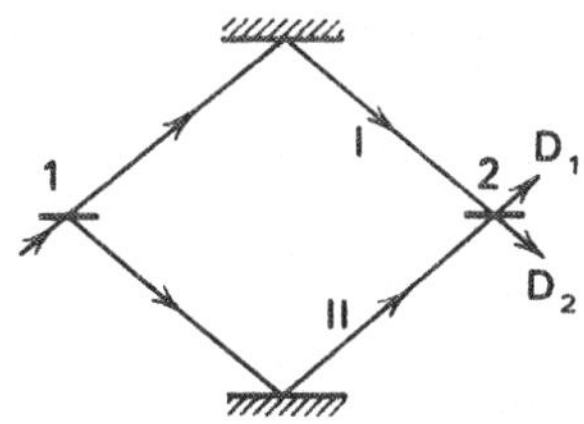

Fig. 5.6 Mental interference experiment to determine the relationship between the phases of the reflected and transmitted waves for an arbitrary potential utilized here as the partitioning plates 1 and 2.

and the one registered by detector D_2 is

$$J_2 = |r^2 + e^{i\eta}t^2|^2. \tag{5.83}$$

The total intensity $J_1 + J_2$ must equal unity, so

$$2|r|^2|t|^2(1 + \cos\eta) + |r^2 + t^2\exp(i\eta)|^2 = 1. \tag{5.84}$$

We let $r = |r|\exp(i\rho)$, $t = |t|\exp(i\tau)$, and take into account that $|r|^2 + |t|^2 = 1$. Substituting there into (5.84) we obtain

$$1 + 2|r|^2|t|^2[\cos\eta + \cos(2\tau - 2\rho + \eta)] = 1. \tag{5.85}$$

Since the phase η can vary arbitrarily, and $|rt|$ is generally not zero, then for (5.85) to be satisfied for any η the following is necessary:

$$2\tau - 2\rho = (2n + 1)\pi, \quad \tau = \rho + (n + \tfrac{1}{2})\pi, \tag{5.86}$$

which was to be proved.

APPENDICES

Appendix 5.1 Neutron scattering on a crystalline line

$$\text{a)}\ \Psi(\mathbf{r}) = \Psi_0(\mathbf{r}) - b\sum_i \psi_i\eta(\mathbf{r} - \mathbf{r}_i), \quad \text{b)}\ \psi_i = \Psi_0(\mathbf{r}_i) - b\sum_{j\neq i}\psi_j\eta(\mathbf{r}_i - \mathbf{r}_j). \tag{1}$$

$$\text{a)}\ \Psi_0(\mathbf{r}) = \exp(i\mathbf{k}_0\mathbf{r}), \quad \text{b)}\ \psi_i = \psi(\mathbf{r}_i), \quad \text{c)}\ \eta(\mathbf{r}) = \exp(ikr)/r. \tag{2}$$

$$\text{a)}\ k = |\mathbf{k}_0|, \quad \text{b)}\ \mathbf{r}_n = (0, 0, nd), \quad \text{c)}\ kd \to k, \quad \text{d)}\ \mathbf{r}/d \to \mathbf{r}, \quad \text{e)}\ b/d \to b. \tag{3}$$

$$\text{a)}\ \psi_n = C\exp(ik_{0\parallel}z_n), \quad \text{b)}\ C = 1 - bCS, \quad \text{c)}\ S = \sum_{n\neq 0}\eta(z_n)\exp(ik_{0\parallel}n). \tag{4}$$

$$\text{a)}\ S = \sum_{n\neq 0}\exp(ik|n| + ik_{0\parallel}n)/|n|, \quad \text{b)}\ \frac{1}{|n|} = \int_0^\infty ds\exp(-s|n|). \tag{5}$$

$$S = 2 \sum_{n=1}^{\infty} \int_0^{\infty} ds \exp[(ik - s)n] \cos(k_{0\|} n). \tag{6}$$

$$\sum_1^{\infty} \exp(-n\alpha) = \exp(-\alpha)/[1 - \exp(-\alpha)]. \tag{7}$$

$$\text{a) } S = -ik - \ln(f), \quad \text{b) } f = 4 \sin[(k_{0\|} + k)/2] \sin[(k_{0\|} - k)/2]. \tag{8}$$

$$\text{a) } \sin z = z \prod_{n=1}^{\infty} (1 - z^2/\pi^2 n^2), \quad \text{b) } f = -k_{0\perp}^2 \prod_{n \neq 0} (-k_{n\perp}^2/\tau_n^2)^2. \tag{9}$$

$$\text{a) } k_{n\perp}^2 = k^2 - k_{n\|}^2, \quad \text{b) } k_{n\|} = k_{0\|} + \tau_n, \quad \text{c) } \tau_n = 2\pi n. \tag{10}$$

$$\text{a) } k_{n\perp}^2 > 0, \quad \text{b) } \ln(-k_{n\perp}^2) = -i\pi + \ln(k_{n\perp}^2),$$

$$\text{c) } S = -ik + i\pi(m + 1) - \ln|f|. \tag{11}$$

$$\exp(ikr)/r = (1/2\pi^2) \int \exp(i\mathbf{pr}) d^3p/(p^2 - k^2 - i\varepsilon). \tag{12}$$

$$\text{a) } \sum_{n=N}^{M} g(n) = \sum_{m=-\infty}^{\infty} \int_N^M dx f(x) \exp(i\tau_m x), \quad \text{b) } \sum_n \exp(iqn) = 2\pi \sum_m \delta(q - \tau_m). \tag{13}$$

$$\sum_n \eta(\mathbf{r} - \mathbf{r}_n) \exp(ik_\| n) = 2 \sum_m \exp(ik_{m\|} z) \int_0^{\infty} J_0(p_\perp r_\perp) p_\perp dp_\perp/(p_\perp^2 - k_{m\perp}^2 - i\varepsilon). \tag{14}$$

$$\int x J_0(x) dx/(x^2 - a^2 - i\varepsilon) = \oint dx \mathcal{H}_0^{(1)}(x)/2(x - a - i\varepsilon) = i\pi \mathcal{H}_0^{(1)}(a). \tag{15}$$

$$\text{a) } \Psi(\mathbf{r}) = \exp(i\mathbf{k}_0\mathbf{r}) - i\pi b C \sum_m \exp(ik_{m\|} z) \mathcal{H}_0^{(1)}(k_{m\perp} r_\perp), \quad \text{b) } C = b/(1 + bS). \tag{16}$$

$$\text{a) } b_1 = bC, \quad \text{b) } b_1 = b/\{1 - ikb + i\pi(m + 1)b - b\ln[2|\cos(k) - \cos(k_{0\|})|]\}. \tag{17}$$

$$\text{a) } x \ll 1, \quad \text{b) } \mathcal{H}_0^{(1,2)}(x) \simeq (\pm 2i/\pi)\ln(x), \quad \text{c) } J_0(x) \simeq 1 - x^2/4. \tag{18}$$

$$\text{c) } x \gg 1, \quad \text{d) } \mathcal{H}^{(1,2)}(x) \simeq \sqrt{\pm 2/\pi i x} \exp(\pm ix). \tag{19}$$

$$\exp(i\mathbf{k}_0\mathbf{r}) = \exp(ik_{0\|} z) \sum i^n \exp(in\varphi) J_n(k_{0\perp} r_\perp). \tag{20}$$

$$\text{a) } j_0^{\text{in}} = k_{0\perp} 2\pi r_\perp |\mathcal{H}^{(2)}(k_{0\perp} r_\perp)/2|^2 = 1, \quad \text{b) } (1 - 2\pi i b_1)^2 + m(2\pi b_1)^2 = 1. \tag{21}$$

The MSW equations are written out in (1). The utilized notation is given in (2). The quantity k represents the length (3a) of the wavevector of the incident plane wave. The coordinates of the scattering nuclei are indicated in (3b). All the lengths will be given in terms of the period d and we shall also use the dimensionless quantities (3c–e). The solution of eqns (1b) will be sought in the form (4a). We substitute (4a) into (1b). Equation (1b) reduces to the form (4b), where S stands for the sum (4c). Substituting (2c) into (4c) we obtain (5a). Utilizing the obvious equality (5b) we reduce (5a) to the form (6). Performing in (6) summation in accordance with the rule (7) and then integrating we obtain (8). We take advantage of the expansion (9a). Then (8b) will be represented in the form (9b), where the notation (10) is introduced. When (11a) holds, we utilize relation (11b). Then (8a) can be written in the form (11c). We make use of the Fourier expansion of the spherical wave (13) and the rule for reducing the sum to the integral (13a). Substituting for $g(n)$ the exponential $\exp(iqn)$ into (13a) we obtain (13b). Substituting (4a) into (1a) we obtain the sum (14). Using expansion (12) and performing the integration over $p_{\|}$ and the angle we perform the remaining integration according to the rule (15). As a result (1a) reduces to the form (16a), where the solution (16b) of eqn (4b) is also substituted. We introduce the notation (17a). Substituting (16b) and (11c) here we obtain (17b). Now we take advantage of the asymptotic formulae (18) and (19) and of the expansion of a plane wave, eqn (20). We then compute the converging and diverging fluxes. Thus, the computation of the converging flux is given in (21a). In the absence of absorption the sum of the divergent fluxes should equal the converging flux. This requirement leads to the condition of unitarity, eqn (21b).

Appendix 5.2 Neutron scattering on a crystalline plane

$$\psi_n = \exp(\mathrm{i}\mathbf{k}_0\mathbf{r}_n) - b\sum_{n'}\theta(n' \neq n)\psi_{n'}\eta(r_n - r_{n'}), \quad \text{b)}\ \psi_n = C\exp(\mathrm{i}\mathbf{k}_{0\|}\mathbf{r}_n). \tag{1}$$

$$\text{a)}\ \mathbf{r}_i = (m_i, n_i, 0), \quad \text{b)}\ r_i = \sqrt{m_i^2 + n_i^2}, \quad \text{c)}\ \mathbf{k}_{0\|} = (k_{0x}, k_{0y}, 0). \tag{2}$$

$$\text{a)}\ C = 1 - bCS, \quad \text{b)}\ S = \sum_n \theta(n \neq 0)s_n, \quad \text{c)}\ s_n = \eta(\mathbf{r}_n)\exp(\mathrm{i}\mathbf{k}_{0\|}\mathbf{r}_n). \tag{3}$$

$$\text{a)}\ \eta(\mathbf{r}) \to \eta(\mathbf{r}) - \eta_\Lambda(\mathbf{r}), \quad \text{b)}\ \eta_\Lambda(r) = \exp(-\Lambda r)/r, \quad \text{c)}\ S = \lim_{\Lambda\to\infty}(S_\Lambda). \tag{4}$$

$$\text{a)}\ S_\Lambda = \sum_n (s_n - s_{n\Lambda}) - ik - \Lambda, \quad \text{b)}\ s_{n\Lambda} = \eta_\Lambda(r_n)\exp(\mathrm{i}\mathbf{k}_{0\|}\mathbf{r}_n). \tag{5}$$

$$\text{a)}\ \sum_{m,n} f(m,n) = \sum_n \int f(\mathbf{r}_\|)\mathrm{d}^2r_\|\exp(\mathrm{i}\boldsymbol{\tau}_n\mathbf{r}_\|), \quad \text{b)}\ \boldsymbol{\tau}_n = 2\pi(M_n, N_n, 0). \tag{6}$$

$$\text{a)}\ \int \eta(\mathbf{r})\exp(\mathrm{i}\mathbf{k}_\|\mathbf{r}_\|)\mathrm{d}^2r_\| = 2\pi\mathrm{i}/k_\perp, \quad \text{b)}\ k_\perp = \sqrt{k^2 - k_\|^2}. \tag{7}$$

$$S_\Lambda = 2\pi\sum_n\left(1/\sqrt{-k_{n\perp}^2} - 1/\sqrt{\Lambda_{n\perp}^2}\right) - ik - \Lambda. \tag{8}$$

$$\text{a)}\ k_{n\perp}^2 = k^2 - k_{n\|}^2, \quad \text{b)}\ \mathbf{k}_{n\|} = \mathbf{k}_{0\|} + \boldsymbol{\tau}_n, \quad \text{c)}\ \Lambda_{n\perp}^2 = \Lambda^2 + k_{n\|}^2. \tag{9}$$

$$f(\mathbf{k}, \mu, 1/\Lambda) = (2\pi)^{2\mu}\sum_n \theta(k_{n\perp}^2 < 0)[(-k_{n\perp}^2)^{-\mu} - (\Lambda_{n\perp}^2)^{-\mu}] - \Lambda^{2(1-\mu)}. \tag{10}$$

$$\text{a) } \mu > 1, \quad \text{b) } g(\mathbf{k}, \mu) = (2\pi)^{2\mu} \sum_n \theta(k_{n\perp}^2 < 0)(-k_{n\perp}^2)^{-\mu} = f(\mathbf{k}, \mu, 0). \tag{11}$$

$$\text{a) } S = -ik + 2\pi i \sum_n \theta(k_{n\perp}^2 > 0) k_{n\perp}^{-1} + g(\mathbf{k}, \tfrac{1}{2}), \quad \text{b) } C = 1/(1 + bS). \tag{12}$$

$$\text{a) } \Psi(\mathbf{r}) = \Psi_0(\mathbf{r}) - b \sum_n \psi_n \eta(\mathbf{r} - \mathbf{r}_n) \exp(i\mathbf{k}_{0\parallel}\mathbf{r}_n), \quad \text{b) } \Psi_0(\mathbf{r}) = \exp(i\mathbf{k}_0\mathbf{r}). \tag{13}$$

$$\Psi(\mathbf{r}) = \exp(i\mathbf{k}_0\mathbf{r}) - 2\pi i b_2 \sum_n k_{n\perp}^{-1} \exp(i\mathbf{k}_{n\parallel}\mathbf{r}_\parallel + ik_{n\perp}|z|). \tag{14}$$

$$\text{a) } b_2 = bC, \quad \text{b) } b_2 = b/(1 - ikb + ib\alpha + b\gamma) = b_0/(1 + ib_0\alpha + b_0\gamma). \tag{15}$$

$$\text{a) } \alpha = 2\pi \sum k_{n\perp}^{-1} \theta(k_{n\perp} > 0), \quad \text{b) } \gamma = g(\mathbf{k}, \tfrac{1}{2}), \quad \text{c) } b_0 = b/(1 - ikb). \tag{16}$$

$$\text{a) } k \ll 1, \quad \text{b) } g(\mathbf{k}, \mu) = \sum (\mathbf{k}\bar{\nabla}_x)^n g(\mathbf{x}, \mu)_{x=0}. \tag{17}$$

$$\text{a) } g(0, \mu) = \sum_n \theta(n \neq 0)(M_n^2 + N_n^2)^{-\mu}, \quad \text{b) } g(0, \tfrac{1}{2}) = -1. \tag{18}$$

We represent the solution of the set of MSW equations (1a) in the form (1b), where the notation (2) is adopted, and the θ function equals 1 if the equality indicated in its argument is fulfilled, and zero otherwise. Substituting (1b) into (1a) we obtain eqn (3a), where the notation (3b, c) is utilized. To extend the sum (3b) to all n we substitute (4a, b) into (3c). Then (3b) will be written in the form of the limit (4c), where the notation (5) is introduced. We use the rule (6) for summation of an arbitrary convergent series. Performing integration, as indicated in (7), we reduce the sum (5a) to the form (8), where the notation (9) is adopted. For calculating the series (8) we consider the function (10) which coincides with (8) when $\mu = \frac{1}{2}$. When $\mu > 1$, the limit of function (10), as $\Lambda \to \infty$, can be found to equal (11b). The value (4c) of the series (3b) can be found by the analytic continuation of the function (11b) to the point $\mu = \frac{1}{2}$. As a result we obtain (12a). Substituting (12a) into the solution (12b) of eqn (3a) we find all the coefficients (1a). Substituting them into the wavefunction (13), and once again taking advantage of the rule (6) and of the integral (7), we reduce it to the form (14) where the notation (15, 16) is introduced and b_0, as in (16c), is the scattering length of an individual atom. At low neutron energies, when (17a) holds, the function (11b) can be expanded into the series (17b). The first term of the series is represented in the form (18a) and, as found by *Glaser (1973), it equals (18b).

Appendix 5.3 Diffraction on a semi-infinite crystal

$$W(\mathbf{k} \to \mathbf{k}') k_\perp d^3k d^3k' = W(\mathbf{k}' \to \mathbf{k}) k'_\perp d^3k' d^3k. \tag{1}$$

$$W(\mathbf{k} \to \mathbf{k}') = \sum_\tau (k'_\perp/k_\perp) |R(\mathbf{k} \to \mathbf{k}')|^2 \delta(k'_\parallel - k_\parallel - \tau) \delta(k^2 - k'^2) 2k'_\perp. \tag{2}$$

a) $W(\mathbf{k} \to \mathbf{k}') = k'_{\perp} S(\mathbf{k}, \mathbf{k}')$, b) $R_{fi} = \tilde{R}_{fi}/k_{f\perp}$, c) $\tilde{R}_{fi} = \tilde{R}_{if}$. (3)

a) $k_{i\perp}^2 = k^2 - (\mathbf{k}_{0\parallel} - \boldsymbol{\tau}_i)^2$, b) $\boldsymbol{\tau}_i = (\tau_{xi}, \tau_{yi}, 0)$. (4)

a) $(1 - rR)^{-1} t = \exp(iql)$, b) $R = r + tR\exp(iql)$. (5)

a) $R = r + tR(1 - rR)^{-1} t$, b) $r = \mathbf{EKNE}$,

c) $t = \mathbf{E}(1 + \mathbf{KN})\mathbf{E}$, d) $R = \mathbf{EKRE}$. (6)

a) $\mathbf{E}_{fi} = e_f \delta_{fi}$, b) $e_i = \exp(ik_{i\perp}/2)$, c) $\mathbf{K}_{fi} = \delta_{fi} K_f$. (7)

a) $K_f = -2\pi i b_2/k_{f\perp}$, b) $\mathbf{N}_{fi} = 1$, c) $\mathbf{R}_{fi} = \mathbf{R}_{if}$. (8)

a) $\mathbf{NAN} = A\mathbf{N}$, b) $A = \sum_{f,i} \mathbf{A}_{fi}$, c) $\mathbf{AN} = \mathbf{A}_l \mathbf{N}$, d) $\mathbf{NA} = \mathbf{NA}_r$. (9)

a) $(\mathbf{A}_l)_{fi} = \delta_{fi} \sum_n \mathbf{A}_{fn}$, b) $(\mathbf{A}_r)_{fi} = \delta_{fi} \sum_n \mathbf{A}_{ni}$. (10)

a) $Rt^{-1}(1 - rR) = rt^{-1}(1 - rR) + tR$, b) $t^{-1} = \mathbf{E}^{-1}(1 + \mathbf{KN})^{-1}\mathbf{E}^{-1}$. (11)

a) $(1 + \mathbf{KN})^{-1} = 1 - \beta\mathbf{KN}$,

b) $(1 + \mathbf{KN})(1 - \beta\mathbf{KN}) = 1 + (1 - \beta - \beta\kappa)\mathbf{KN}$. (12)

a) $\kappa = \sum_i K_i = -2\pi i b_1 \sum_i k_{i\perp}^{-1}$, b) $1 - \beta\kappa = \beta$, c) $\beta = 1/(1 + \kappa)$. (13)

a) $t^{-1} r = \beta \mathbf{E}^{-1}\mathbf{KNE}$, b) $rt^{-1} = \beta \mathbf{EKNE}^{-1}$, c) $rt^{-1} r = \beta\kappa \mathbf{EKNE}$. (14)

$\mathbf{R} - \beta \mathbf{D}_{\mathrm{L}} \mathbf{N} - \beta \mathbf{D}_{\mathrm{L}} \mathbf{N} \mathbf{D}_{\mathrm{R}} \mathbf{E}^2 = \beta \mathbf{N} + \beta \mathbf{N} \mathbf{D}_{\mathrm{R}} \mathbf{E}^2 + \mathbf{E}^2 \mathbf{R} \mathbf{E}^2$. (15)

a) $\mathbf{D}_{\mathrm{L,R}} = D_f^{\mathrm{L,R}} \delta_{fi}$, b) $D_f^{\mathrm{L}} = \sum_f \mathbf{R}_{fi} K_i$, c) $D_i^{\mathrm{R}} = \sum_f \mathbf{R}_{fi} K_f e_f^2$. (16)

a) $\mathbf{R} - \mathbf{E}^2 \mathbf{R} \mathbf{E}^2 = \beta(\mathbf{D}_{\mathrm{L}} + 1)\mathbf{N}(\mathbf{D}_{\mathrm{R}} \mathbf{E}^2 + 1)$, b) $\mathbf{R}_{fi} = \beta s_f v_i/(1 - e_f^2 e_i^2)$. (17)

a) $s_f = 1 + \beta s_f \sum_i v_i K_i/(1 - e_f^2 e_i^2)$, b) $v_i = 1 + \beta v_i e_i^2 \sum_f s_f K_f e_f^2/(1 - e_f^2 e_i^2)$. (18)

a) $v_i = 1 + \beta v_i \sum_f s_f K_f/(1 - e_f e_i) - \beta v_i \sum_f s_f K_f$, b) $\kappa = \beta \sum_f s_f K_f$. (19)

a) $v_i(1 + \kappa) = 1 + \beta v_i \sum_f s_f K_f/(1 - e_f e_i)$, b) $v_i = s_i \mu$, c) $\mu = 1/(1 + \kappa)$. (20)

a) $R_{fi} = \beta\mu e_f K_f s_f s_i e_i/(1 - e_f^2 e_i^2)$, b) $s_f = 1 + \beta\mu s_f \sum_i s_i K_i/(1 - e_f^2 e_i^2)$. (21)

a) $b_2 \beta = b_0$, b) $\bar{K}_f = -2\pi i b_0/k_{f\perp}$, c) $\kappa = \sum s_i \bar{K}_i$. (22)

$$R_{fi} = \mu e_f \bar{K}_f s_f s_i e_i/(1 - e_f^2 e_i^2). \tag{23}$$

$$s_f = 1 + \sum_i s_i \bar{K}_i + s_f e_f^2 \sum_f s_i \bar{K}_i e_i^2/(1 - e_f^2 e_i^2). \tag{24}$$

The probability of elastic scattering, $W(\mathbf{k} \to \mathbf{k}')$, involving a change of the direction of the wavevector $\mathbf{k}$ of the incident neutron to that of the reflected one, $\mathbf{k}'$, must satisfy the condition of detailed balance, eqn (1). In the case of diffraction with a reflection amplitude R the probability W is written in the form (2), where the vector components assigned the subscript $\perp$ are perpendicular and the components assigned the subscript $\parallel$ are parallel to the entrance surface of the reflecting body, and summation is performed over all the vectors τ of the reciprocal lattice for which $k'_\perp$ is a real quantity. From (1) it follows that the probability W must be representable in the form (3a), where S is a symmetric function relative to its arguments. For expression (2) to satisfy condition (3a), the amplitude R must be representable in the form (3b, c), where the notations (4) are adopted. We substitute (5a) into (5b) and obtain eqn (6a) for the matrix of reflection amplitudes, R. The matrices of reflection and transmission amplitudes of a single period of length l in the middle of which there is a crystal plane can, in accordance with Appendix 5.2, be written as (6b) and (6c), respectively. We write the sought amplitude R in the form (6d), similar to (6b), where the matrix elements of the matrices $\mathbf{E}$, $\mathbf{K}$, and $\mathbf{N}$ are indicated in (7) and (8a, b), and $\mathbf{R}$ is the symmetric matrix (8c). In multiplying by an arbitrary matrix $\mathbf{A}$ we take advantage of the properties (9, 10) of the matrix $\mathbf{N}$. We rewrite eqn (6a) in the form (11a). For computing the matrix (11b) we make use of relation (12a), which upon substitution into (12b) with (9a, b) leads to (13a) and to eqn (13b) for the parameter β, the solution of which is indicated in (13c). Substituting (12a) into (11b) and making use of expression (6b) and the properties (9, 10) of the matrix $\mathbf{N}$, we readily calculate the products (14). Substituting (14) into (11a) we reduce (11a) to the form (15) containing the diagonal matrices $\mathbf{D}$ defined in (16). Equation (15) is rewritten in the form (17a). Expression (17b) follows for the matrix elements of $\mathbf{R}$ with the unknown parameters s and v, for which eqns (18) are obtained by substitution of (17b) into (16b, c). The set of equations (18b) can be rewritten in the form (19a). We introduce the notation (19b) and rewrite (19a) in the form (20a). Comparison of (20a) with (18a) reveals that the solution of (20a) has the form (20b), where the notation (20c) is utilized. Substituting (20b) into (17b) and (18a) we obtain (21). We now introduce the notation (22a, b). With its aid expressions (19b) and (21) are rewritten as (22–24), respectively.

Appendix 5.4 Determination of the Bloch wavevector q

$$\exp(iql) = (1 - rR)t^{-1}. \tag{1}$$

$$\text{a) } r = \mathbf{EKNE}, \quad \text{b) } t = \mathbf{E}(1 + \mathbf{KN})\mathbf{E}, \quad \text{c) } R = \mathbf{EKRE}. \tag{2}$$

$$\text{a) } rR = \mathbf{EKND_R E}, \quad \text{b) } rRrR = \text{R} rR, \quad \text{c) } (1 - rR)^{-1} = 1 + \rho r R. \tag{3}$$

$$\text{a) } (1 - rR)(1 + \rho rR) = 1, \quad \text{b) } 1 + \text{R}\rho = \rho, \quad \text{c) } \rho = 1/(1 - \text{R}). \tag{4}$$

$$\text{a) } \text{R} = \sum_{f,i} e_f R_{fi} K_i e_i, \quad \text{b) } R_{fi} = \beta\mu e_f K_f s_f s_i e_i/(1 - e_f^2 e_i^2). \tag{5}$$

$$\text{a) } s_f = 1 + \beta\mu s_f \sum_i s_i K_i/(1 - e_f^2 e_i^2), \quad \text{b) } \beta = 1/(1 + \kappa), \quad \text{c) } \mu = 1/(1 + \text{K}). \tag{6}$$

$$\text{a) } \text{K} = \sum_i K_i = -2\pi i b_2 \sum_i k_{i\perp}^{-1}, \quad \text{b) } \kappa = \beta \sum s_i K_i. \tag{7}$$

$$\text{R} = \beta\mu \sum K_f s_f \sum K_i s_i/(1 - e_f^2 e_i^2) - \kappa^2\mu/\beta = \sum K_f(s_f - 1) - \kappa^2\mu/\beta$$
$$= \kappa\mu/\beta - \kappa = (\text{K} - \kappa)\mu. \tag{8}$$

$$\text{a) } \rho = \beta/\mu, \quad \text{b) } \exp(iql) = \mathbf{E}[1 + \rho\mathbf{KN}(\mathbf{D_R E}^2 + 1)]\mathbf{E} = \mathbf{E}(1 + \bar{\mathbf{K}}\mathbf{NS})\mathbf{E}. \tag{9}$$

$$\text{a) } (\mathbf{D_R E}^2 + 1)_{fi} = \delta_{fi} v_i = \mu\delta_{fi} s_i, \quad \text{b) } \mathbf{S}_{fi} = s_f \delta_{fi}, \quad \text{c) } \bar{K}_f = \beta K_f. \tag{10}$$

$$\text{a) } \bar{K}_i = -iu_0 l/2k_{i\perp}, \quad \text{b) } l \to 0, \quad \text{c) } q_{fi} = k_{f\perp}\delta_{fi} - u_0 s_i/2k_{f\perp}. \tag{11}$$

$$s_f = 1 + us_f \sum s_i/2k_{i\perp}(k_{f\perp} + k_{i\perp}), \quad \text{c) } R_{fi} = s_f s_i u/2k_{f\perp}(k_{f\perp} + k_{i\perp}). \tag{12}$$

We substitute matrix (2) into eqn (1). Taking advantage of the property of matrix **N** we perform preliminary calculations of the product (3a, b) and the reciprocal matrix (3c). The equation for the parameter ρ is obtained from the condition (4a) from which (4b, c) follow. The quantity R is determined in (3b) and is equal to (5a). We substitute (5b) into (5a) and make use of eqns (6a) with the parameters indicated in (6b, c) and (7). Upon performing transformations (8) we obtain (8a). Substituting (2b), (3c), and (9a) into (1) we obtain (9b), where we use the value of the diagonal matrix (10a), obtained in Appendix 5.3 and introduced in the notation (10b, c). Expression (10c) can be rewritten in the form (11a), where $u_0 = 4\pi b_0/ld^2$ is a potential characteristic of UCN. To the first order in 1, which is equivalent to the limit (11b), (11c) follows from (9b). In the limit (11b) expression (5a) and equations (6a) transform into (13), respectively.

6

Storage of UCN in material traps

The main characteristic property of UCN is that they can be kept in a hermetically sealed vessel for a long time. The storage time τ is limited by the lifetime of the neutron itself, τ_β, and by the time τ_l due to losses occurring in collisions of the UCN with the walls of the trap. The time τ_l is determined by the cross-sections of absorption, σ_a, and inelastic scattering, σ_{ie}, on the nuclei of the material of the walls and by the geometric dimensions of the trap l. The smaller the loss cross-section $\sigma_l = \sigma_a + \sigma_{ie}$ and the larger the value of l, the longer is the time τ_l. If τ_l is significantly greater than τ_β, then the storage time τ is determined mainly by the lifetime of the neutron itself, $\tau_\beta \approx 1000$ s. Studies reveal, however, that in a weakly absorbing and sufficiently large trap the time τ is significantly smaller than the expected value. This means that the actual value of τ_l is essentially smaller than the theoretical value, sometimes even by two orders of magnitude.

In principle, such a discrepancy may be explained by the surface of the trap walls being insufficiently clean. There may be present on the surface of the walls either strongly absorbing impurities or impurities exhibiting large inelastic scattering cross-sections. Strongly absorbing impurities such as ^{3}He, ^{6}Li, Cd, Gd, and others are rarely encountered in nature, so their presence in traps seems quite improbable. Chlorine is somewhat more widespread naturally, but for it to lead to such a significant decrease of the time τ_l, its amount on the walls of the trap would have to be extremely large. Thus, the small time τ_l is most likely due to the presence on the walls of impurities with a large inelastic scattering cross-section. Hence it follows that the walls of a trap contain a large amount of hydrogen.

Hydrogen is ubiquitous, and in principle there is nothing strange in all surfaces being contaminated by hydrogen-containing substances, even though these substances must be quite abundant. If they are in the form of organic films, they can be removed by heating the vessel up to 200–400°C. Heating, however, does not increase τ sufficiently (Egorov *et al.* 1974; I.M. Frank 1979; Groshev *et al.* 1971; Steyerl and Trusted 1974). On the other hand, even if the hydrogen does dissolve in the surface layer of the wall and is not removed by heating, its cross-section σ_{ie} must nevertheless depend on the temperature, and therefore τ should increase when the trap is cooled; but this effect was not observed. Thus, the hydrogen contamination hypothesis was discarded. A dead-end situation had arisen.

To overcome these difficulties, an experiment (Strelkov and Hetzelt 1978) was performed in which the trap for UCN was surrounded by thermal neutron counters. It turned out that during the storage of the UCN the thermal neutron counters registered neutrons, precisely the quantity vanishing from the trap. It was therefore concluded that the small τ_l was due to the large cross-section σ_{ie}.

The experiment (Strelkov and Hetzelt 1978) concluded the first stage in the development of the physics of UCN. The second stage began when the hydrogen contamination hypothesis again came into being. If we consider the hydrogen to be responsible for the small τ_l, then, first, its presence on the wall of the trap must be verified in some other way; secondly, it must be understood why its cross-section σ_{ie} is independent of or weakly dependent on the temperature; and, thirdly, a method must be found of getting rid of the hydrogen.

In this chapter methods will be described for deriving the loss coefficient from experiments on UCN storage, and we shall discuss attempts to explain the results of these experiments both at the first stage, when the presence of hydrogen was repudiated, and at the second stage, when hydrogen was considered the principal cause of the excess loss coefficient. The search for the causes of the excess loss coefficient not related to hydrogen might be considered senseless and no place should be given to it in this book, but the above conclusion must be considered too premature, since, as demonstrated at the end of this chapter, the contamination of the walls of a trap by hydrogen is still not verified. Moreover, there are also arguments against such contamination.

6.1 THE STORAGE CURVE AND THE LOSS COEFFICIENT

Experiments on UCN storage are as follows (see drawing in Fig. 6.1(a)). A vessel is filled with neutrons and sealed. The neutrons are then kept in the sealed vessel for a certain exposure time t_{exp}, after which the vessel is opened and the neutrons remaining in it are counted by a detector yielding $N(t_{exp})$. Further, the dependence $N(t_{exp})$ is plotted, which is known as the storage curve, and is similar in the first approximation (Fig. 6.1(b)) to the dependence of radioactive decay:

$$\text{a)}\ N(t_{exp}) = N(0)\exp(-t_{exp}/\tau), \quad \text{b)}\ 1/\tau = 1/\tau_\beta + 1/\tau_l \tag{6.1}$$

where $N(0)$ is the number of accumulated neutrons.

A trap represents, for instance, a horizontal copper pipe $\sim$10 cm in diameter and 1–3 m long, inside which a vacuum of $\sim 10^{-3}$–10^{-4} mm Hg is maintained. In the first experiments the number $N(0)$ amounted to about 7 neutrons within a volume of $\sim$25 litres (Groshev *et al.* 1971), and it was further successfully extended to 36 cm^{-3} (Steyerl and Malik 1986).

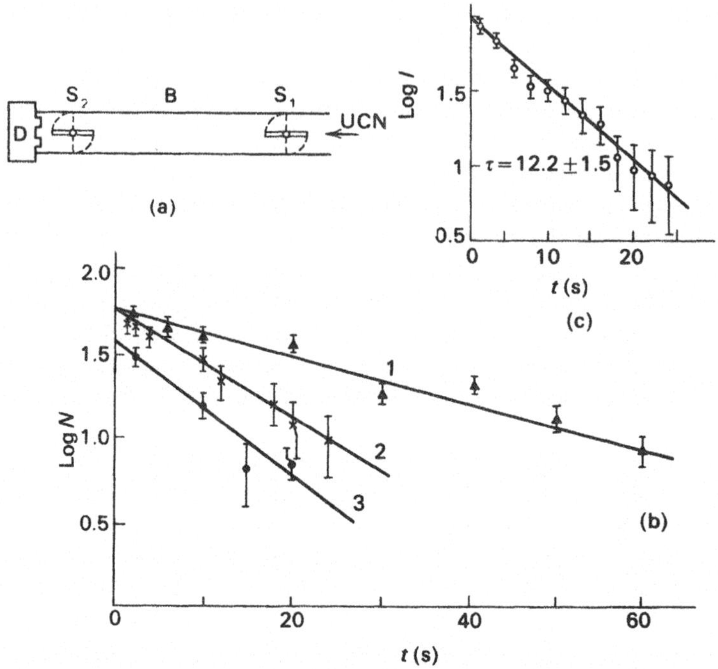

Fig. 6.1 Measurement of UCN storage time in sealed vessels (Groshev *et al.* 1971). (a) Experimental layout: B, vessel (in this case a cylindrical pipe 14 cm in diameter and 174 cm long); $s_{1,2}$, input and output shutters. (b) Storage curves: 1, chemically polished copper pipe, $\tau = 33$ s; 2, untreated copper foil, $\tau = 14$ s; 3, pyrographite, $\tau = 14$ s. (c) Emptying curve.

All the remaining neutrons are registered by the detector only if they reach it simultaneously. But a certain time is required to register the neutrons remaining in the trap. When the trap is opened and the detector starts counting the remaining neutrons, if a time analyser is switched on, the time dependence of the count of the detector will be obtained, which is known as the emptying curve (Fig. 6.1(c)) and from which it is possible to determine the characteristic emptying time τ_{emp}. The total count of the detector represents the integral of the emptying curve, and, if integration is extended to ∞, the total count N_d will be

$$N_d = N(0)[\tau/(\tau + \tau_{emp})]\exp(-t_{exp}/\tau). \tag{6.2}$$

Thus, if expression (6.1) holds, then the storage curve retains an exponential form when the finite time τ_{emp} is taken into account. In practice, however, the storage curve differs from an exponential because the time τ_1 depends on the neutron angular distribution and velocity v (Fig. 6.2). Let us determine

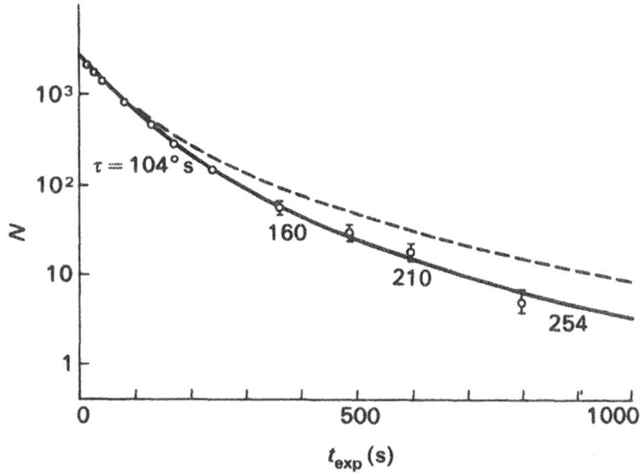

Fig. 6.2 Storage curve for a copper vessel cleaned by an argon glow discharge (Golub *et al.* 1981). The figures are the storage times corresponding to the slopes of the curve at the points indicated by arrows. The slopes decrease as the exposure time increases, since the faster neutrons exhibit a larger loss coefficient and are lost more rapidly. As a result, the mean neutron velocity in the trap decreases. The coefficient η found from this curve is three times greater than the theoretical value. The dashed line demonstrates the storage curve, from which the lifetime of the neutron itself, τ_β, is excluded.

the dependence of τ_l on v, assuming the angular distribution to be isotropic. We represent τ_l in the form

$$\tau_l = t_f/\bar{\mu}, \tag{6.3}$$

where $\bar{\mu}$ is the loss coefficient averaged over all angles of incidence and $t_f = \bar{l}/v$ is the mean time of free flight between two collisions of a neutron with the wall. Here $\bar{l}$ is the mean free path $\bar{l} = 4V/S$, where V is the volume of the trap and S is the surface of its walls. The time t_f, and consequently τ_l, depends on the geometric dimensions of the trap and increases as its dimensions increase. However, the time τ_l can only be increased up to certain limits, taking advantage of the geometric dimensions of the trap, since gravity will manifest itself when the dimensions become large. The time t_f cannot exceed $2v/g$, where g is the acceleration of free fall. Thus, gravity diminishes the storage time τ, so the loss coefficient deduced from experiments without taking gravity into account turns out to be too high.

The dependence of the loss coefficient averaged over the angles of incidence on the velocity v is determined by the expression (Appendix 6.1)

$$\bar{\mu} = (2/y^2)\,(\arcsin y - y\sqrt{1 - y^2})\eta, \quad y = v/v_{\text{lim}}. \tag{6.4}$$

From expressions (6.3) and (6.4) it follows that the storage curve may exhibit different forms in different experiments even when the traps have the same geometry, if the accumulated neutron spectra in them differ. The storage curve is purely exponential and depends only on the properties of the material of the trap—strictly speaking, in the single case when neutrons of extremely low energies are stored on an infinite horizontal plane with a perfect specular surface. In this case averaging over the angles is not required, and μ can be approximated by the expression $\mu = 2\eta v_\perp / v_{\text{lim}}$, where $v_\perp$ is the vertical component of the velocity. Since $t_f = 2v_\perp/g$ and $\tau_l = 2v_\perp/g\mu$, the velocities in the numerator and the denominator cancel out, and as a result $\tau_l = v_{\text{lim}}/\eta g$ is obtained. For $v_{\text{lim}} = 5\ \text{m s}^{-1}$ and $\eta = 5 \times 10^{-4}$ we obtain $\tau_l \approx 1000$ s. Thus, by decreasing the neutron velocity we can obtain relatively high storage times.

The real physical parameter, however, is not τ_l but the reduced loss coefficient η, which is not so easily extracted from experimental data. To this end it is necessary to know the neutron spectrum in the trap, as well as the dependence on the velocity v not only of $\tau_l(v)$ but also of $\tau_{\text{emp}}(v)$. The time $\tau_{\text{emp}}(v)$ is found from simple gas kinetic arguments: if the density of neutrons of velocity v is $n(v) = N(v)/V$, then the flux incident upon the exit opening of area s_{emp} is

$$n(v)vs_{\text{emp}}/4 = N(v)vs_{\text{emp}}/4V = N(v)\tau_{\text{emp}}. \tag{6.5}$$

Thus, $\tau_{\text{emp}}(v) = 4V/s_{\text{emp}}v$. If the exit opening is covered by a window partially reflecting neutrons with a reflection coefficient $|R|^2$, the emptying time increases:

$$\tau_{\text{emp}}(v) = 4V/vs_{\text{emp}}(1 - |R|^2). \tag{6.6}$$

Since this expression follows from (6.5), it holds only when the neutrons are isotropically distributed, i.e. in the case of sufficiently diffuse reflection from the walls when the area s_{emp} is small or when $|R|^2 \approx 1$. Under other conditions the outflow violates isotropy, formula (6.5) is no longer valid, and the emptying time may turn out to be larger than (6.6). Moreover, if the reflection from the walls exhibits a high degree of specularity, the emptying curve may turn out to be totally non-exponential. Monte Carlo calculations (Antonov *et al.* 1982*b*) confirm the above reasoning.

Now let us assume the reflection of UCN from the walls to be sufficiently diffuse and the emptying curve to be exponential and characterized by a known parameter $\tau_{\text{emp}}(v)$. In this case the reduced loss coefficient η is determined by fitting the following expression to the measured storage curve:

$$N_d(t_{\text{exp}}) = \int N_d(t_{\text{exp}}, v)\text{d}v = \int \text{d}v\rho(v)\varepsilon(v)\{\tau(v)/[\tau(v) + \tau_{\text{emp}}(v)]\}$$
$$\times \exp[-t_{\text{exp}}/\tau(v)][\![1 - \exp\{-t_{\text{cnt}}[1/\tau(v) + 1/\tau_{\text{emp}}(v)]\}]\!], \tag{6.7}$$

where $\rho(v)$ is the spectrum, $\int \rho(v)\text{d}v = N(0)$ is the number of neutrons accumulated in the trap, $\varepsilon(v)$ is the registration efficiency for those neutrons that have overcome the potential barrier of the detector, and the last factor of the integrand takes into account the finite counting time of the detector, t_{cnt}, after the opening of the trap.

Unfortunately, the neutron spectrum in the trap is also not always known. In certain cases the limits of the spectrum can only be indicated approximately: $v_1 < v < v_2$. In such cases the spectrum is sometimes assumed to have the form of an interval of the Maxwellian distribution

$$\rho(v) = N(0)\theta(v_1 < v < v_2)3v^2/(v_2^3 - v_1^3). \tag{6.8}$$

It must be pointed out that formula (6.7) is practically never used in dealing with storage curves; at any rate no such indication exists in the literature. Usually, it is assumed that $t_{cnt} = \infty$ and τ_{emp} is constant (see Appendix 6.1). Moreover, the entire storage curve is often described by the simple exponential formula (6.2), in which τ_l is represented in the form $\bar{l}/\overline{v\bar{\mu}}(v)$, where $v\bar{\mu}$ is averaged over the spectrum, thus yielding the coefficient η. The coefficient η in Table 9.23 was obtained precisely in this manner. If necessary, gravity is also taken into account. In Appendix 6.2 the procedure is demonstrated for obtaining $\overline{v\bar{\mu}}(v)$ in the cases of a monochromatic and a Maxwellian spectrum respectively. Utilizing an exponential expression, instead of (6.7), is justified only in the case of small exposure times, when the exponential in the integrand can be expanded in powers of t_{cnt}, limiting the expansion to the linear term. Furthermore, τ_{emp} must be significantly smaller than the storage time τ.

The reduced loss coefficient is related to the loss cross-section, since

$$\eta = u_0''/u_0' = b''/b', \qquad b'' = k\sigma_l(k)/4\pi. \tag{6.9}$$

The relation (6.9) for the imaginary part of the scattering amplitude b'' is not obvious, generally speaking. Indeed, for a rigidly fixed nucleus it follows from the optical theorem that $b'' = k\sigma_{tot}/4\pi$, where $\sigma_{tot} = \sigma_{el} + \sigma_a$. The elastic scattering cross-section must be excluded from η, since elastic scattering does not lead to losses of neutrons. The inelastic scattering cross-section σ_{ie} does not enter into the expression for the imaginary part b'' in the optical theorem. However, as shown in Appendices 6.4–6.8, taking into account nuclear vibrations leads to the necessity of including σ_{ie} in the imaginary part b''. Upon determining η from experiments we can find the loss cross-section $\sigma_l = 4\pi\eta b/k$. For comparison with known cross-sections (* Hughes and Schwartz 1958) it is convenient to reduce the quantity σ_l to the standard thermal point $k_T = 2\pi/\lambda_T$, where $\lambda_T = 1.08$ Å (* Gurevich and Tarasov 1965). At this point $\sigma_l(k_T) = 2.16b\eta \times 10^{-8}\,cm^2$.

6.2 THE FIRST STAGE IN THE INVESTIGATION OF UCN STORAGE

For nearly 10 years experiments on UCN storage in traps made of various materials (copper, nickel, beryllium, stainless steel, pyrographite, teflon, glass, etc.) yielded values of the loss coefficient η within the range $3\text{–}10 \times 10^{-4}$; the respective loss cross-sections reduced to the thermal point exceeded the tabulated quantities (* Hughes and Schwartz 1958) by 4–8 barn. Thus, for example, the tabulated value for copper $\sigma_l \approx \sigma_a = 2.2$ barn, and therefore the loss coefficient η should be 1.45×10^{-4} (for copper $b = 7.9 \times 10^{-13}$ cm), but the value of η derived from experimental data was three times higher, corresponding to the excess $\sigma_l \approx 4$ barn. In beryllium the main contribution to σ_l is given by the quantity $\sigma_{ie} \approx 0.1$ barn, and the excess cross-section also amounts to ~ 4 barn.

It seems unlikely for the excess loss cross-section to be due to approximately the same contamination of the wall surfaces of all the traps by strongly absorbing elements, since they are not so abundant in nature. Thus, for instance, in a surface layer 100 Å thick 20 at.% of chlorine ($\sigma_a \approx 15$ barn) is required (Ignatovich and Stepanov 1974), while its abundance relative, say, to hydrogen amounts to 2×10^{-7} (* Kay and Laby 1962). In a layer of the same thickness about 0.2 at.% are required of cadmium ($\sigma_a = 2500$ barn), while its abundance with respect to hydrogen is $\sim 2 \times 10^{-11}$.

It seems more reasonable to assume that the excess loss cross-section is caused by inelastic scattering on hydrogenous contaminations. But if the excess cross-section is inelastic, then a temperature dependence of the storage time should be observed. Indeed, the inelastic one-phonon scattering cross-section may be represented in the form (Appendix 6.6)

$$\sigma_{ie}(k) = \frac{\sigma_{el}}{M}\frac{p}{k}, \tag{6.10}$$

where M is the ratio of the mass of the nucleus to the neutron mass and p plays the part of a characteristic momentum of the scattered neutron. Relation (6.10) is valid for coherent, σ_{coh}, and incoherent, σ_{inc}, scattering, while p is (for simplicity we take a cubic crystal), respectively, the following:

$$p_{coh} = \sum_{\tau} \int (d^3q/2\pi)\delta(\omega_q - (\mathbf{q} + \tau)^2)n(\omega_q/T)\exp[-\omega_q B(T)], \tag{6.11}$$

$$p_{inc} = \int d\omega \sqrt{\omega} g(\omega) n(\omega/T)\exp[-\omega B(T)]. \tag{6.12}$$

In (6.11) and (6.12) the last factor in the integrand is the Debye–Waller factor, τ is the reciprocal lattice vector, ω_q is the phonon energy ($\omega_q = 2mcq/h$ for small q where c is the velocity of sound), and integration over q is performed in (6.11) over the first Brillouin zone. (Note that at $\tau = 0$ the integral in (6.11) becomes equal to zero, since $2mc/h > q$, always.) The quantity $n(x) = 1/(\exp x - 1)$ is the phonon density and $g(\omega)$ is the phonon spectrum.

The cross-sections σ_{ie} are usually not large owing to the mass of the nucleus occurring in the denominator (an exception is hydrogen, $M = 1$). In the Debye approximation, at high temperatures $p \approx T/\theta^{1/2}$, where θ is the

Debye temperature. Incidentally, a linear dependence of p on T at high temperatures is obtained for other frequency spectra also. When $T \to 0$ the spectrum can always be approximated by the Debye spectrum, and $p_{inc} \sim T^{7/2}$, while p_{coh} may decrease exponentially: $p_{coh} \sim \omega_0^{1/2} \exp(-\omega_0/T)$, where ω_0 is a certain characteristic frequency. Thus, if the excess cross-section is inelastic, then the reduced loss coefficient η should change with the temperature, at least linearly.

In the experiment by Groshev *et al.* (1971) the copper bottle was heated to 250°C, while the graphite bottle was cooled to 100 K, but in both cases no change of the storage time was observed. In the experiment by Egorov *et al.* (1974) a nickel bottle was submitted to prolonged annealing at 150°C, upon which measurements were performed at room temperature and at 200°C. Again no change of the storage time was observed. Finally, Luschikov *et al.* (1978) measured the storage curve for a glass vessel at room temperature without and after annealing at 500°C, at high temperatures (500°C), and at liquid nitrogen temperatures (80 K). Annealing led to an increase in the storage time by a factor of 2. In a vessel 8 cm in diameter and 2.5 m long the value $\tau = 160$ s was achieved after annealing. This corresponds to $\eta \approx 10^{-4}$ and to an excess loss cross-section of 2.5 barn, which is approximately 10 times higher than the loss cross-section in the glass itself, due in the main to inelastic scattering. After annealing the storage time no longer changed as the temperature varied. Thus, it seems that excess losses cannot be attributed to inelastic scattering.

We shall now discuss the hypotheses put forward to explain the data obtained and to verify them.

6.3 EXPLANATIONS FOR THE EXCESS UCN LOSSES

6.3.1 Leaks

Clearly, in all experiments the UCN could escape through leaks that might exist near the entrance and exit shutters. It is quite improbable for the leaks related to shutters of different constructions to turn out to be nearly identical, but naturally they cannot be excluded on the basis of such arguments. Let us estimate the influence of a leak. The flux of UCN through a slit of area S_{sl} is $(N(v)v/4V)S_{sl}$ and the emptying time through this leak is, correspondingly, $\tau_{sl} = 4V/S_{sl}v$. We shall consider that neutrons in the leak do not return, otherwise it would be necessary to introduce the reflection coefficient, although it can be included in the effective area of the leak. Using the time τ_{sl} the effective storage time τ_{eff} is determined through the time τ, introduced earlier, with the aid of the relation

$$1/\tau_{eff} = 1/\tau + 1/\tau_{sl} = 1/\tau_\beta + 1/\tau_l + 1/\tau_{sl}. \qquad (6.13)$$

Let the storage time τ be mainly determined by the emptying time of the UCN from the leak. Then the area s_{sl} must be given by

$$s_{sl} = 4V/v\tau. \tag{6.14}$$

If the trap is a cylindrical pipe of diameter D and length L, covered by a shutter at the end, then the width of the slit of the leak must be

$$d_{sl} = DL/v\tau.$$

Correspondingly, in the experiments by Egorov *et al.* (1974) and Luschikov *et al.* (1978) the leak should have been ~0.5 mm wide. In Egorov *et al.* the leaks were controlled by a detector, which revealed no leakage of UCN past the shutters. This result, however, cannot indicate unambiguously that the existence of the leak can be neglected, since a neutron in the leak may be absorbed in it without even hitting the detector.

A special experiment was performed by Kosvintsev *et al.* (1976*b*) to reveal the influence of controlled leaks on the storage curve. The trap for the UCN was a cylinder 64 cm in diameter and 105 cm long. The UCN storage time in it amounted to about 50 s. If this time is determined by the leak, then the area of the leak must be 40 cm^2. When the diameter of the shutters is ~ 20 cm, such a leak must be ~ 0.5 cm wide. In fact, the width of the leak did not exceed 0.5 mm, while its being increased under control by a factor of 10 diminished the storage time only by about 10%.

Thus, it could be concluded that in the given case the leak was not the cause of UCN losses in the vessel. Note, however, that this does not point to a negligible influence of a leak in all cases. As the storage time increases and the velocity of accumulated neutrons decreases, the leak starts to play an ever more significant part. With the aid of (6.13) it is readily shown that, as $v \to 0$, a leak leads to an effective enhancement of the parameter η by a factor of $1 + (3/4\eta)(s_{sl}/S)(v_{lim}/v)$ (where S is the total area of the lateral walls of the trap) which is greater, the smaller the velocity v.

6.3.2 Acoustics (Gerasimov *et al.* 1973)

If the UCN are heated by acoustic, instead of thermal, vibrations, the excess loss coefficient will naturally be independent of the temperature. It might seem, on the basis of general arguments, that the heating involving acoustic oscillations can be indefinitely large. But let us see what actually can happen here.

An elastic wave $\boldsymbol{\xi}(\mathbf{r}, t) = \boldsymbol{\xi}_0(iq\mathbf{r} - i\omega_s t)$ propagating in a medium leads, first, to a modulation of the particle number density $N_0 \to N_0/(1 + \nabla\boldsymbol{\xi}(\mathbf{r}, t))$, and, secondly, to a distortion of the surface. Therefore, the interaction potential of a neutron with matter occupying the infinite semi-space $z > 0$ can be written in the form

$$u = u_0\theta[z + \xi_\perp(\mathbf{r}_\parallel, t) > 0]/[1 + \nabla\boldsymbol{\xi}(\mathbf{r}, t)]$$

$$\approx u_0\theta(z > 0) - u_0\theta(z > 0)\nabla\boldsymbol{\xi}(\mathbf{r}, t) + u_0\xi_\perp(\mathbf{r}_\parallel, t)\delta(z). \qquad (6.15)$$

Here the potential has been expanded in ξ. The first term is the unperturbed interaction potential of a neutron with the medium, the second is the perturbation caused by modulation of the density, and the third represents the perturbation caused by distortion of the surface.

It is reasonable to utilize this expansion for the potential for frequencies of the acoustic wave $\omega_s > \omega_{\text{lim}} = E_{\text{lim}}/\hbar \approx 10^8\ \text{s}^{-1}$. In the case of lower frequencies ω_s it is more reasonable to consider the wall as impinging upon the neutron as a whole with a velocity $v_s = \omega_s\xi_\perp$ and to consider the normal component of a reflected neutron's velocity to alter by $2v_s$ when reflection occurs. The resulting change in energy of the neutron may turn out to be greater than when a single quantum of energy $h\omega_s$ is absorbed. Thus, UCN heating by acoustic oscillations must be treated differently depending on whether ω_s is greater or less than ω_{lim}.

Let $\omega_s > \omega_{\text{lim}}$. In this case we can take advantage of expansion (6.15) and solve the problem of inelastic scattering for a neutron impinging upon the surface by applying perturbation theory. We shall consider the two perturbating terms in the potential (6.15) separately. The first term represents a volume effect and gives a correction to the loss coefficient owing to scattering on the sound, μ_{sV}, while the second is a surface term, and we shall denote the corresponding correction by μ_{sS}.

From Appendix 6.5 it follows that

$$\mu_{sV} = (k_{0\perp}/k_\perp)u_0(\mathbf{q}\xi_0)^2/[(k'_\perp - q_\perp)^2 + (k''_{0\perp})^2], \qquad (6.16)$$

where

$$k_\perp = \sqrt{k_0^2 + \omega_s - (k_{0\parallel} + \mathbf{q}_\parallel)^2}, \quad k'_\perp = \sqrt{k_\perp^2 - u_0}, \quad k''_{0\perp} = \sqrt{u_0 - k_{0\perp}^2}.$$

Since $\omega_s = q2mc/\hbar \gg q^2$, we have $k_\perp \approx \omega_s^{1/2} \gg q$ and expression (6.16) leads to the simpler one $\mu_{sV} \approx (\omega_{\text{lim}}/\omega_s)^{3/2}(q\xi_0)^2$. We shall express the oscillation amplitude ξ_0 through the density of acoustic energy $\mathscr{E}_s = \rho\omega_s^2\xi_0^2$. Taking advantage of the relation $\omega_s/q = c$ we obtain for μ_{sV} the estimate

$$\mu_{sV} = (\omega_{\text{lim}}/\omega_s)^{3/2}(\mathscr{E}_s/\rho c^2).$$

Setting $\rho \approx 5\ \text{g cm}^{-3}$ and $c \approx 10^5\ \text{cm s}^{-1}$ we obtain $\rho c^2 \approx 5\ \text{kJ cm}^{-3}$. If the worst case of $\omega_s = \omega_{\text{lim}}$ is chosen, then to obtain $\mu_{sV} \approx 10^{-4}$ it is necessary to have a density of acoustic energy $\mathscr{E}_s = 0.5\ \text{J cm}^{-3}$, or an intensity $I_s = c\mathscr{E} \approx 5 \times 10^4\ \text{W cm}^{-2}$. In acoustics the intensity of sound is usually measured in decibels, $10\lg(I/I_0)$, where the unit for comparison is $I_0 = 10^{-16}\ \text{W cm}^{-2}$. Thus, the required energy is 200 decibels. The existence of such a high density of acoustic energy inside a material seems quite improbable for a frequency of $10^8\ \text{s}^{-1}$.

The following expression for μ_{ss} follows from Appendix 6.5:

$$\mu_{ss} = \frac{k_{0\perp}}{k_{\perp}} u_0 \xi_{0\perp}^2 \approx \left(\frac{\omega_{lim}}{\omega_s}\right)^{1/2} \frac{\xi_0^2}{\lambda_{lim}^2},$$

where the same approximation as before is utilized. Since $\lambda_{lim} \approx 10^{-6}$ cm, it is possible to obtain $\mu_{ss} \approx 10^{-4}$ for $\omega_s = \omega_{lim}$ only when $\xi_0 \approx 10^{-8}$. The intensity of the corresponding sound amounts to 5×10^{-2} W cm^{-2}, or ~ 140 decibels. If the sound is considered to be confined within the material and to leak through its surface with a probability equal to the ratio of impedances from both sides of the surface, and if in the case of a metal this ratio is assumed to amount to 10^{-7}, then the material within which the sound is confined with an intensity of 140 decibels must represent a source of sound with an intensity equal to 70 decibels. Such an intensity corresponds to the volume of sound of a masculine voice at a distance of 3 m (Kay and Laby 1962, p. 74). Clearly we are dealing here with frequencies that are far beyond the range of audibility.

Now let us consider the second version, $\omega_s < \omega_{lim}$. In this case each collision of a neutron with the wall leads to the normal component of its velocity changing by $\sim 2\omega_s \xi_1$, and since the neutron velocity in the velocity space undergoes a random walk when a neutron impinges upon the wall and the oscillation phase turns out to be undefined, after n collisions with the wall the square velocity is incremented by $2n\omega_s^2 \xi_0^2$. The number of collisions that a neutron can undergo before its velocity exceeds the limiting velocity is $n_{lim} = v_{lim}^2/2\omega_s^2 \xi_0^2$. On the other hand, the number of collisions can be expressed through the equivalent loss coefficient $n_{lim} = 1/\mu_{eff}$. Assuming $\mu_{eff} \approx 10^{-4}$ and $v_{lim} \approx 6$ m s^{-1}, we find that $\omega_s \xi_0 \approx 4$ cm s^{-1}. The intensity of the corresponding acoustic energy $I = \rho\omega_s^2 \xi_0^2 c \approx 1$ W cm^{-2}, or 160 decibels. If this energy is stored within the walls of the trap and 10^{-7} of it is emitted into the surrounding space, then in the vicinity of the trap there should exist a considerable acoustic background.

No direct measurements of the vibrations of the walls of the trap were performed; however, indirect data testify against the neutrons being heated by vibrations of the walls. An enhancement of the UCN extraction time from a vessel was observed by Steyerl and Trusted (1974) as the exposure time was increased; this signified a softening of the spectrum in the trap, instead of it becoming harder. Subsequently, an experiment was performed with very slow neutrons (Kosvintsev *et al.* 1979), in which UCN were stored in a copper vessel, the top of which was covered by a polyethylene piston. The storage curve did not alter when the polyethylene piston was replaced by a copper one. Hence it was concluded that even if heating by vibrations of the wall does occur, it increases the UCN energy by not more than 7×10^{-11} eV in each collision with the wall. That heating is not caused by acoustic oscillations of the walls of a trap is confirmed by the fact that, when the trap is totally decoupled from the pumps that are the main source of acoustic vibrations, no change is observed in the storage curve.

6.3.3 Quasi-elastic scattering

It could be possible that all the processes considered elastic, when dealing with the scattering of thermal neutrons, are actually quasi-elastic, but with the change in energy being so small that it can always be neglected compared with the energy of a thermal neutron. In the case of UCN this change in energy may turn out to be significant.

Both coherent and incoherent quasi-elastic processes exist. The incoherent processes are related to the slow migration of individual atoms or with the fluctuations of their spins (Blokhintsev and Plakida 1976). But the incoherent quasi-elastic scattering cross-section is approximately equal to the incoherent elastic scattering cross-section, $\sigma_{\text{el}}^{\text{inc}}$, which does not increase as the neutron energy decreases, contrary to σ_{ie} and σ_{a}. Therefore, it amounts to a small fraction of σ_{l} in the UCN energy region and cannot lead to a significant enhancement of η.

Coherent quasi-elastic processes are related to the fluctuation of temperature, or entropy. The corresponding scattering cross-section amounts to the fraction $\gamma = c_{\text{p}}/c_{\text{V}} - 1$ of the one-phonon inelastic scattering cross-section, where c_{p} and c_{V} are the heat capacities for constant pressure and volume, respectively. In a solid-state body the coefficient $\gamma \approx 10^{-5}$, and so the corresponding addition here also leads only to a negligible enhancement of the loss coefficient η.

Finally, the question arises of whether coherent reflection from the wall is not itself actually quasi-elastic. In principle this is possible, but the change in energy in the quasi-elastic scattering must be extremely small, otherwise the inelasticity would be noticeable in the total reflection of thermal neutrons. If, on the other hand, the change in energy is small, then after several collisions the neutron becomes, with a certain probability, over-limiting and escapes from the trap. But these processes are totally similar to heating by acoustic vibrations of the wall, and they lead to the same change in the spectrum of the confined neutrons.

6.3.4 Peculiarities of inelastic scattering in total reflection

Although the total reflection was itself taken into account in determining the contribution of inelastic scattering to the loss coefficient, not all the features of the process were accounted for, however. It will be shown that these peculiarities do not lead to a significant change in the contribution of inelastic scattering, but they must nevertheless be considered briefly and qualitatively. The peculiarities at issue are the following: the Debye–Waller factor, conservation of the normal momentum component, the difference between phonon and plane waves, and inelastic scattering on clusters.

The Debye–Waller factor (see Appendix 6.8) characterizes the mean square displacement of atoms due to their oscillations,

$$\tilde{w} = \exp[-\boldsymbol{\kappa}^2 B(T)],$$

where $\boldsymbol{\kappa} = \kappa_0 - \kappa$ is the momentum transfer and

$$B(T) = \int \mathrm{cth}(\omega/2T)g(\omega)\mathrm{d}\omega/6M\omega;$$

$\tilde{\omega}$ is known to enter into the scattering amplitude as a factor, and therefore, at first glance, it lowers the amplitude and subsequently the potential barrier also. Usually, in determining the height of the potential barrier advantage is taken of the refractive index $n^2 = 1 - 4\pi N_0 b/k^2$, i.e. the change in phase of the neutron wave within the material is calculated in the case of forward scattering. But in forward scattering $k = 0$ and hence $\tilde{w} = 1$. Now consider the quantity $\tilde{w}$ from the point of view of reflection. This is more natural, since the height of a potential barrier is determined by the maximum energy E_{lim} at which total reflection occurs. In reflection, however, the momentum transfer is $2k_\perp$, i.e. the potential barrier may be lower. Actually, the momentum transfer must be taken inside the medium, where in the case of total reflection it equals $2ik''_\perp$. Therefore $k^2 < 0$ and, correspondingly, $\tilde{w} > 1$. Thus, instead of lowering the height of the potential barrier, the Debye–Waller factor acts as if to raise the potential barrier. At any rate, however, $E_{\mathrm{lim}} = u_0$. When $k_\perp^2 > u_0$, the Debye–Waller factor $w = \exp[4(u_0 - k_\perp^2)B(T)]$ becomes smaller than unity. A rise of the potential barrier at $k_\perp^2 < u_0$ simply indicates (see Appendix 6.8) a change in phase of the reflected wave, i.e. the phase φ of the reflected wave will be, instead of $\varphi_0 = 2\arccos(k_\perp/u_0^{1/2})$, the following:

$$\varphi = 2\arccos(k_\perp/\sqrt{u_0 \tilde{w}}).$$

The change in phase is totally caused by the smearing of the boundary of the potential barrier due to oscillations of the nuclei. Further, in the course of investigating the role of roughnesses, it will be shown that, generally, when a smearing of width l exists, the reflection amplitude can be computed as if the potential remained steplike, but were of a height

$$u = u_0 \tilde{w}_l = u_0 \exp[(u_0 - k_\perp^2)l^2]. \tag{6.17}$$

Note that the factor $\tilde{w}_l$ leads simultaneously to an enhancement of both the real and imaginary parts of the potential, and therefore we might expect the reduced loss coefficient $\eta = u''_0/u'_0$ not to change. However, if there exists an imaginary part of the potential, it must also be taken into account in the exponent of the exponential (6.17). This leads to the multiplication of u''_0 by $1 + u'_0 l^2$. Hence we can see that the smearing, on the whole, leads to an enhancement of the loss coefficient; but a smearing caused by thermal oscillations increases it by not more than 0.01%, which is negligible.

6.3.5 Non-conservation of the normal component of the momentum

The matrix element of the plane phonon wave between the initial and finite neutron states, $\Psi_0(\mathbf{r})$ and $\Psi(\mathbf{r})$, respectively, is proportional to the quantity

$$\int \Psi^*(\mathbf{r}) e^{i\mathbf{q}\mathbf{r}} \Psi_0(\mathbf{r}) dz \sim 1/[(k'_\perp - q_\perp) - ik''_{0\perp}], \tag{6.18}$$

instead of the δ function, as when the momentum is strictly conserved. In expression (6.18) $k'_\perp$ indicates the normal component inside the medium of the wavevector of the scattered neutron. The square of the matrix element (6.18) is a Lorentzian with its centre at $k'_\perp = q_\perp$ and of width $k''_{0\perp}$. The absence of rigorous restrictions on the normal component of the momentum may lead to those inelastic processes being permitted which are otherwise forbidden. In the general case the inelastic coherent scattering cross-section can be represented in the form

$$\sigma_{\text{ie}}^{\text{coh}} = \int A(\mathbf{k}, \mathbf{q}) d^3k d^3q \delta(\mathbf{k}_\parallel - \mathbf{q}_\parallel - \mathbf{k}_{0\parallel}) \\ \times \delta(k^2 - \omega_q - k_0^2)/[(k'_\perp - q_\perp)^2 + (k''_{0\perp})^2]. \tag{6.19}$$

where only the factors related to conservation laws are shown explicitly. Integration over d^3q discards three δ functions, and if, instead of the Lorentzian, a δ function were also present, the cross-section (6.19) would in certain cases be zero. Owing to the Lorentzian it always differs from zero. For instance, the term with $\tau = 0$ occurring in (6.11) becomes different from zero. This circumstance leads to part of the loss coefficient μ_{sv}, eqn (6.16), differing from zero as a result of scattering on the sound.

When the strict conservation laws do not forbid inelastic scattering, substitution of the Lorentzian for the δ function only leads to the appearance of a small correction to the cross-section of the order of magnitude of $k''_{0\perp}/Q$, where Q represents a scale that characterizes the range within which the coefficient $A(\mathbf{k}, \mathbf{q})$ in expression (6.19) changes. The Debye momentum $\theta^{1/2}$ or $T^{1/2}$ may serve as such a scale.

In principle, however, we can imagine a process in which the coefficient $A(\mathbf{k}, \mathbf{q})$ in (6.19) is proportional to the factor $(k'_\perp - q_\perp - k_{0\perp})$. In this case the inelastic cross-section will be zero if the law of momentum conservation is rigorously obeyed, otherwise it will differ from zero. Such processes would remain unnoticed in the scattering of thermal neutrons, but might play an important role in the storage of UCN.

Even smaller restrictions are imposed on the normal component of the momentum in the scattering of neutrons on surface ripples. Now the function $\delta(z)\exp(iq_\parallel r_\parallel)$ must be substituted for the plane phonon wave in the matrix element (6.18). The corresponding contribution to the loss coefficient of the wave of amplitude ξ_0 and with the wavevector q was considered earlier for acoustic oscillations. In the case of phonons it is necessary to set $\xi_0 = 1/(M\omega_q)^{1/2}$, to take into account the thermal factor $n(\omega_q/T)$ and to perform averaging over all the wavevectors q:

$$\mu_S = \int \frac{k_{0\perp}}{k_\perp} \frac{u_0}{M\omega_q} n\left(\frac{\omega_q}{T}\right) d^3q \frac{1}{N_0(2\pi)^3}. \tag{6.20}$$

From (6.20) it can readily be seen that $\mu_S \approx k_0 b/M \approx 10^{-7}$, i.e. the correction to the loss coefficient is negligible.

The phonon wavefunction in the semi-space $z > 0$ can be represented in the form $\xi_0\varphi(z)\exp(iq_\parallel r_\parallel - i\omega_q t)$, where

$$\varphi(z) = e_{in}\exp(-iq_\perp z) + R_1 e_1 \exp(iq_{1\perp} z) + R_2 e_2 \exp(iq_{2\perp} z),$$

i.e. there exists a wave with polarization e_{in} incident on the interface and two reflected waves. (More reflected waves than one may exist, the phonon being a vector particle (*Landau and Lifshits 1965; *Ezawa 1971).) Reflected waves differ in polarization and have different velocities, so the normal components of their wavevectors also differ.

An incident phonon wave $\exp(-iq_\perp z)$ mainly scatters a neutron inelastically backward into the vacuum, while the reflected phonon waves scatter them forward, towards the medium. Because the normal component of the momentum is not rigorously conserved, the reflected phonon wave may also contribute to backward scattering into the vacuum, but this contribution is $\sim k''_{0\perp}/q_\perp$ times smaller than that of the incident phonon wave. The whole packet of phonon waves participates in the scattering coherently, and here the surface ripple, as well as the density modulation, must be taken into account. Therefore, the neutron scattering amplitudes on the incident and scattered phonon waves, as well as the scattering amplitude on the surface ripple caused by the phonon waves, add up and lead to various interference terms:

$$\mu = \int A(q) | \{(\boldsymbol{\kappa e}_{in})/[k''_{0\perp} + i(k'_\perp + q_\perp)] + (\boldsymbol{\kappa e}_1)R_1/[k''_{0\perp} + i(k'_\perp - q_{1\perp})]$$
$$+ (\boldsymbol{\kappa e}_2)R_2/[k''_{0\perp} + i(k'_\perp - q_{2\perp})] + (\boldsymbol{\kappa}\varphi(0))\}|^2 d^3q.$$

(The factors of no interest to us are denoted by $A(q)$.) The largest contribution to this integral, however, is given by the resonance terms, for example the first one, when $q_\perp = -k'_\perp$. Although the range of integration over $q_\perp$, in which this term yields the main contribution, is small ($\Delta q_\perp \approx k''_{0\perp}$), its value, nevertheless, is $q/k''_{0\perp} \approx 10^2$ times greater than all the remaining summands. Thus, the corrections to μ that are due to the distortion of phonon waves in the vicinity of the surface are negligible.

At the interface more waves of a special type are formed, such as Rayleigh waves which die away exponentially inside the medium:

$$\xi(\mathbf{r}, t) = \sqrt{2q_\perp} C\xi_0 \mathbf{e}_\mathrm{R} \exp(-q_\perp z + \mathrm{i}\mathbf{q}_\parallel \mathbf{r}_\parallel - \mathrm{i}\omega_q t).$$

Here the amplitude of the wave $\xi_0 = 1/(M\omega_q)^{1/2}$ and the factor $(2q_\perp)^{1/2}$ from the normalizing factor $(2q_\perp)^{1/2}C$ are factored out, so the quantity $C \approx 1$ is dimensionless. The damping factor $q_\perp$ equals $q_\parallel[1 - (c_\mathrm{R}/c)^2]^{1/2}$, where c may be either c_l or c_t, i.e. the propagation velocity of the longitudinal or transverse waves, respectively. The quantity $q_\perp$ is large in the case of thermal Rayleigh waves, i.e. such a wave dies away, as it travels from the surface, at a distance equal to several atomic layers. For this reason the contribution of Rayleigh waves to the loss coefficient is $k''_{0\perp}/q_\perp \approx k_0/q_\mathrm{D} \approx 10^{-2}$ times smaller than the contribution of ordinary phonons.

The scattering of UCN on Rayleigh waves can be represented as the diffraction of thermal neutrons on a static diffraction grating, if one passes to a reference frame moving together with the Rayleigh wave with a velocity c_R (Frank 1975). Such an approach provides a clear result and permits estimation of the contribution of Rayleigh waves to μ by the quantity $u_0\xi_0^2 k_{0\perp}/\omega_q^{1/2} \approx 10^{-6}$ (Appendix 6.5).

6.3.6 Clusters

All the mechanisms considered only contributed small corrections to the probability of inelastic scattering, but in principle there exists a mechanism which might enhance this probability significantly. It is related to the possibility of the formation of surface clusters. Since the surface layer may be quite loose, it can be imagined to consist of clusters of n atoms each. If n is not too large, i.e. the dimensions of the clusters are smaller than the UCN wavelength, then the scattering on each cluster can be considered in the s-wave approximation. The coherent scattering amplitude on a cluster will then be nb, and the elastic scattering cross-section will be n^2 times greater than the elastic scattering cross-section on a single nucleus. It may seem that this circumstance should in itself lead to an enhancement of inelastic scattering, if we take into account the possibility of clusters undergoing oscillations. Nevertheless, no enhancement exists in the one-phonon approximation, and this is readily verified. Indeed, the probability of inelastic scattering is proportional to the number of clusters per unit volume, N_0/n, and inversely proportional to the cluster mass nM, where M is the ratio between the mass of a nucleus and the neutron mass. Therefore, the factor n^2 is compensated. We can, however, imagine the clusters oscillating with a low frequency, i.e. many quanta are simultaneously transferred to the neutron in scattering.

This means that scattering on a cluster can be treated as scattering on a free particle of mass nM. But in this case the scattering cross-section is proportional to the square root of the mass, instead of the mass itself. The total probability of inelastic scattering at a temperature of the walls equal to T can be estimated by the quantity $(nM)^{1/2}bT^{1/2}/M$ that is $(nM)^{1/2}$ times larger than in the case of scattering on ordinary phonons. To increase the loss coefficient by a factor of 100, when the mass $M = 10$, it is necessary to set $n = 1000$. The temperature dependence in this case turns out to be $\sim T^{1/2}$, but the momentum of the heated neutron is $n^{1/2}$ times smaller, i.e. the energy is only slightly higher than the limiting energy. We can also imagine the dimension of a cluster to be proportional to $n \sim 1/T^{1/2}$, in which case the inelastic scattering probability on clusters becomes totally independent of the temperature.

The cluster model is not very attractive and it seems difficult to accept, but it may turn out to be useful in another field—in studies of the interaction of UCN with gases. In gases the formation of cluster macromolecules may take place. Such aggregates are, in principle, observable with the aid of UCN. Their quantity and dimensions may be estimated depending on temperature.

6.3.7 The influence of inhomogeneities in a material on the loss coefficient

Here we shall not consider the quite obvious case when inhomogeneities in the material are made up of an impurity with a large loss cross-section σ_{imp}, because if the atomic concentration of the impurity averaged over the surface layer of width λ_{lim} is c, then the quantity $c\sigma_{imp}$ must simply be added to the total loss cross-section σ_l of the medium. A more interesting case is provided by inhomogeneities distorting the potential of the medium, but which do not introduce any losses of their own. Alteration of the form of the potential may lead to an enhancement of the loss coefficient μ, while the reduced loss coefficient η remains intact. Correspondingly, the fraction of inelastic scattering and capture in the loss coefficient also remains unchanged. In addition, if the inhomogeneities on the surfaces of all substances are approximately the same, the storage time must vary from substance to substance to the same extent as the coefficient η. In practice, however, the spreading out of the storage time turns out to be significantly smaller. These arguments indicate that inhomogeneities are unlikely to play a significant part in causing the excess loss coefficient. We shall deal with them nevertheless, because they are of interest in themselves. First we shall discuss roughnesses, then pores, and finally films.

Roughnesses

Roughnesses play a double role. First, they lead to scattering, i.e. to non-specular reflection of UCN, and, secondly, they enhance the loss coefficient.

(Note that roughnesses may only enhance the loss coefficient, but never lead to its decrease (Stepanov 1975)). Questions related to the reflection of UCN from a rough surface have been considered by Ignatovich (1973), Stepanov and Shelagin (1974), Steyerl (1972*c*), and others.

Now imagine the profile of a surface being described by a random function $z = \zeta(\mathbf{r}_\parallel)$, with a known correlation function (we shall consider it to be Gaussian):

$$\langle \zeta(\mathbf{r}_\parallel)\zeta(\mathbf{r}'_\parallel)\rangle = \zeta_0^2 K(\mathbf{r}_\parallel - \mathbf{r}'_\parallel), \quad K(\mathbf{r}) = \exp(-\mathbf{r}^2/2l^2). \tag{6.21}$$

It is natural to set the mean value $\langle \zeta(\mathbf{r}_\parallel)\rangle$ equal to zero. The parameters ζ_0 and l represent the root-mean-square height and the correlation length of the roughnesses, respectively.

To find the general solution for the problem of reflection of UCN from an arbitrary rough surface is quite a difficult task. Therefore we shall only consider certain limit cases: first, when the height ζ_0 is small compared with the depth to which the neutron penetrates the medium, $1/k''_{0\perp}$; and, secondly, when the height ζ_0 is arbitrary, but then the correlation length l is negligible. Another limit case exists when l is significantly greater than the UCN wavelength, but in this case the role played by the roughnesses reduces to a general enhancement of the area of the reflecting surface, each section of which being considered plane but inclined at a certain angle to the average interface.

Small roughnesses may be taken into account using perturbation theory. We expand the neutron interaction potential with the medium $u = u_0\theta(z + \zeta(\mathbf{r}_\parallel) > 0)$ in powers of $\zeta(\mathbf{r}_\parallel)$ and only consider the linear and quadratic terms:

$$u_0\theta(z + \zeta(\mathbf{r}_\parallel) > 0) = u_0\theta(z > 0) + \zeta(\mathbf{r}_\parallel)u_0\delta(z) + u_0\zeta^2(\mathbf{r}_\parallel)\delta'(z)/2. \tag{6.22}$$

The linear term in $\zeta(\mathbf{r}_\parallel)$ perturbation occurring in (6.22) leads to non-specular reflection, the indicatrix of which (see Appendix 6.10) is of the form

$$W_{\text{ref}}(\Omega_0, \Omega) = 4kk_{0\perp}k_\perp^2\zeta_0^2 K(\kappa_\parallel) = (2/\pi)\zeta_0^2 l^2 kk_{0\perp}k_\perp^2 \exp(-\kappa_\parallel^2 l^2/2),$$
$$K(\kappa_\parallel) = \int K(\mathbf{r}_\parallel)\exp(i\kappa_\parallel\mathbf{r}_\parallel)d^2\mathbf{r}_\parallel/(2\pi)^2, \tag{6.23}$$

where

$$k_{0\perp} = k_0\cos\theta_0, \qquad k_\perp = k_0\cos\theta,$$
$$k_\parallel^2 = k_0^2(\sin^2\theta_0 + \sin^2\theta + 2\sin\theta_0\sin\theta\cos\varphi).$$

Thermal neutrons incident at the angle of total reflection are scattered by the roughnesses both outwards and into the reflecting medium. The scattering indicatrix for the major part of angles can be written in the form

$$W_{\text{scat}}(\Omega_0, \Omega) = kk_{0\perp}u_0\zeta_0^2 K(\kappa_\parallel). \tag{6.24}$$

Expression (6.24) seems to violate the condition of detailed balance, in accordance with which function (6.24) should be of the form $F(\Omega_0, \Omega)\cos\theta$ (see Chapter 4), where the function $F(\Omega_0, \Omega)$ is symmetric relative to its arguments. Actually, however, no violation exists. The symmetry turns out to be hidden here, just as in the expression

$$|\exp[i(\varphi_1 + \varphi_2)]| = \exp(-2\text{Im}\,\varphi_1),$$

if the imaginary part of the phase φ_2 is zero.

Roughnesses alter the value of the loss coefficient (see Appendix 6.10). It is as if their influence could be decomposed into two components. First of all, they smear the interface, as a result of which the sharp jump in the step of the potential is replaced by a smearing of width ζ_0. The smearing increases the time the neutron stays near the wall, and so the loss coefficient increases by a factor of $Q = 1 + u_0\zeta_0^2$. But the depth of the smearing can be identified with ζ_0 only in the case of frequent roughnesses with correlation length $l = 0$. When $l \neq 0$, the smearing depth is γ times smaller than ζ_0, where

$$\gamma = 1 - \int K(k_\parallel)\mathrm{d}^2k_\parallel k''_{0\perp}/k''_\perp.$$

Obviously, when the correlation is total ($K(r_\parallel) = 1$) there are practically no roughnesses, and the smearing vanishes. Correspondingly, γ will be zero and Q will be unity. Secondly, roughnesses lead to part of the neutrons undergoing non-specular reflection. The loss coefficient depends on the angle between the neutron velocity and the normal to the surface. Losses occur partly when the neutron enters the substance and partly when it leaves the medium. In the case of specular reflection these losses are identical. If the wavevector of a neutron impinging upon a surface is k_0 and the probability for a neutron to be reflected with the wavevector k is α, then the loss coefficient alters by the quantity $\Delta\mu = \alpha[\mu(k_\perp) - \mu(k_{0\perp})]/2$. The probability α is determined by the reflection indicatrix (6.23). Summarizing the above, we arrive at the conclusion that the loss coefficient in reflection from a rough surface is described by the expression

$$\mu(k_{0\perp}) = \mu_0(k_{0\perp})\{1 + u_0\zeta_0^2[1 - \int K(\kappa_\parallel)\mathrm{d}^2k_\parallel k''_{0\perp}/k''_\perp]\}$$
$$+ \int 2kk_{0\perp}k_\perp^2\zeta_0^2 K(\kappa_\parallel)[\mu_0(k_\perp) - \mu_0(k_{0\perp})]\mathrm{d}\Omega, \qquad (6.25)$$

where μ_0 stands for the loss coefficient in the case of reflection from a perfect plane boundary. Evidently, when $\zeta_0 \to 0$ as $l \to \infty$ the loss coefficient (6.25) transforms into $\mu_0(k_{0\perp})$.

As $l \to 0$ we obtain

$$\mu(k_{0\perp}) = \mu_0(k_{0\perp})(1 + u_0\zeta_0^2), \qquad (6.26)$$

i.e. enhancement of the losses is only due to the smearing of the interface, since it increases the depth to which the neutron penetrates the medium. Note that for small ζ_0 the increase of μ is proportional to ζ_0^2.

We shall now turn to high, but frequent ($l \to 0$), roughnesses when the surface can be pictured as covered with thick fur. This case is equivalent to a smearing of the surface. In other words, instead of a sharp step, the neutron interaction potential with the medium is described by a smooth function gradually increasing from zero to u_0. The problem of reflection from such a smeared potential can be solved analytically (*Landau and Lifshits 1963) if the potential is considered to be described by the expression

$$u(z) = \frac{u_0}{1 + \exp(-z/h)},$$

where h serves as the smearing parameter.

The loss coefficient in reflection from such a potential ($u_0'' \neq 0$) is computed in Appendix 6.11. The expression for it, averaged over the angles of incidence, can be reduced to the form

$$\mu = \mu_0[1 + f(h\sqrt{u_0})], \quad f(x) = \begin{cases} \pi^2 x^2/3, & x \to 0, \\ 2x, & x \to \infty. \end{cases} \tag{6.27}$$

Thus, when the parameter h is sufficiently large the loss coefficient may achieve arbitrarily high values. Such enhancement is readily explained. In approaching the surface a neutron gradually decelerates and the time it stays inside the substance is longer the more gentle the increase of the potential, i.e. the smaller is the acceleration of the neutron. The existence of a furry surface, however, does not seem very real. To increase the loss coefficient by a factor of 10 the length of the fur must be ~ 1000 Å. Pictures obtained with the aid of an electron microscope reveal, for instance, that the roughnesses on an electropolished copper surface are ~ 100 Å in height and that their cross-sectional dimensions amount to ~ 300 Å, while the distances between roughnesses are ~ 1000 Å. These roughnesses can only enhance the loss coefficient by a factor of 2 or 3.

For small h it is easy to calculate the phase of the reflected wave. Its value is as if the potential u_0 were $\exp[(u_0 - k_{0\perp}^2)\pi^2 h^2/3]$ times higher. The quantity $\pi^2 h^2/3$ plays the part of the root-mean-square height ζ_0^2. This follows directly from a comparison of formulae (6.26) and (6.27) for small h. In this case ζ_0 also represents the depth of the potential smearing. Thus, the smearing ζ_0 acts as if it increases the height of the potential $\exp[\zeta_0^2(u_0 - k_{0\perp}^2)]$ times. A similar effect was encountered in the discussion of the Debye–Waller factor.

Pores

Pores are most interesting from a theoretical point of view. Here we encounter the beautiful phenomenon of resonance capture, which has counterparts both in nuclear physics and in elementary particle physics. Bound states of energy $E_r > 10^{-7}$ eV may exist in a spherical pore of sufficiently large radius, $r > 100$ Å. If the energy of the incident neutron is equal to the resonance energy, then the neutron may, with a high probability, penetrate into the pore by the tunnel effect, leading to an enhancement of the loss coefficient.

Resonance states have a finite width Γ, which is the sum of the width Γ_l due to losses in the walls and the width Γ_t describing the probability of tunnelling out from the pore. The width Γ_l equals h/τ_l, where τ_l is the storage time inside the pore, $\tau_l = r/v\mu \approx 10^{-5}$ s, from which it follows that $\Gamma_l \approx 10^{-10}$ eV, i.e. $\Gamma_l/E_r \approx 10^{-3}$. The width Γ_t is equal to h/τ_{ext}, where $\tau_{ext} = r/vw$ is the time required for tunnelling out of the pore. In the denominator of the expression for this time, instead of the loss coefficient μ there occurs the probability of tunnelling out from the pore that is proportional to $\exp(-2k''z_0)$, where z_0 is the distance of the pore from the surface and $k'' = (u_0 - k^2)^{1/2}$. In the presence of a pore the loss coefficient can be represented by the conventional Breit–Wigner formula (see Appendix 6.12)

$$\Delta\mu = \Gamma_l\Gamma_t/[(E - E_r)^2 + (\Gamma_l + \Gamma_t)^2]. \tag{6.28}$$

Hence it follows that if $\Gamma_t \gg \Gamma_l$, then at the resonance $\Delta\mu = \Gamma_l/\Gamma_t \approx \eta\exp(2k''z_0)$, i.e. it is large when the pore is far from the surface. This result is readily explained, since in penetrating a far-away pore the resonance neutron has to pass through a thick layer of substance greater than the usual penetration depth. Naturally, the farther the pore from the surface, the smaller is the range of energies within which the resonance enhancement of the loss coefficient manifests itself, i.e. the smaller the probability for the neutron to happen to be just in the required range. Moreover, even at the exact resonance, if the pore is too far from the surface when $\Gamma_t \ll \Gamma_l$, the resonance correction to the loss coefficient starts to decrease exponentially as z_0 increases.

If the loss coefficient (6.28) is averaged over the angles of incidence, the spectrum, and the dimensions and distribution of the pores, the quantity $\Delta\mu = -12V_p\mu_0\ln\mu_0$ is obtained, where V_p is the fraction of the volume occupied by the pores. When $V_p \approx 2$–3% and $\mu_0 \approx 10^{-4}$, $\Delta\mu$ may turn out to be two–three times larger than μ_0.

Films

If on the surface of the reflecting substance with a potential u_0 a film of width h is deposited with a potential u_1, then the loss coefficient can be quite easily calculated for any h and u_1, since the reflection from such a system is found analytically. Applying the method developed in Chapter 5 we obtain for the amplitude of the reflected wave

$$R = [r_{01}^{+} + r_{12}^{+}\exp(2ik_1h)]/[1 + r_{01}^{+}r_{12}^{+}\exp(2ik_1h)],$$

where $r_{01} = (k_0 - k_1)/(k_0 + k_1)$, $r_{12} = (k_1 - k_2)/(k_1 + k_2)$, and $k_0 \equiv k_{0\perp}$, $k_1 \equiv k_{1\perp} = (k_0^2 - u_1)^{1/2}$, $k_2 = (k_0^2 - u_0)^{1/2}$ are, respectively, the normal components of the wavevector in vacuum, in the film, and in the reflecting substance. We shall consider the condition of total reflection to be satisfied, i.e. $k_2 = ik_2'' = i(u_0 - k_0^2)^{1/2}$. In this case R can be represented in the form $\exp(-2i\psi)$, where

$$\psi = -\operatorname{arctg}[(k_1/k_0)\operatorname{tg}\varphi], \quad \varphi = k_1 h - \operatorname{arctg}(k_2''/k_1).$$

The loss coefficient is defined in the conventional way:

$$\mu = 1 - |R|^2 = 1 - \exp(-4\psi''),$$

where ψ'' is the imaginary part of the phase ψ. It is readily found by expanding ψ into a series in powers of the imaginary part of the potentials u_1'' and u_0'': $\psi'' = u_1'' \partial\psi/\partial u_1 + u_0'' \partial\psi/\partial u_0$. Calculations reveal that when $k_0^2 > u_1$, $u_1'' = 0$ and $\mu = K\mu_0$, where

$$K = k_1^2 u_0/(u_0 - u_1)(k_0^2 \cos^2\varphi + k_1^2 \sin^2\varphi).$$

The gain coefficient K exhibits the following peculiarities. When $k_0^2 \to u_1$, the quantities $K = u_0/(u_0 - u_1) > 1$, i.e. a non-absorbing film with a potential $u_1 > 0$, but $u_1 < u_0$ enhances the absorption. Further, as the phase φ increases, the coefficient K oscillates and may be either larger or smaller than unity. It may be that the anomalous spectra observed by Kosvintsev *et al.* (1980*d*) can be explained by the above effect.

If the potential of the film is negative, $u_1 = -|u_1|$, then when $k_0 = 0$ the coefficient $K = u_0/(u_0 + |u_1|)\sin^2\varphi$. If φ is close to zero, then the loss coefficient may increase indefinitely as $k_0 \to 0$. The condition $\varphi = 0$ at $k_0 = 0$ is equivalent to the appearance of a bound level in the film. It determines the thickness of the film, h, corresponding to the appearance of the consecutive level. The first level appears when the thickness of the film satisfies the condition $[|u_1|/(u_0 + |u_1|)]^{1/2} = \cos(h|u_1|^{1/2})$. The thickness corresponding to a water film is 400 Å. When the films are thicker, the formation of a bound state is possible, in which motion is normal to the film with negative energy.

An incident neutron travels along the normal with positive energy of motion and can happen to occupy a bound state only as a result of scattering on the roughnesses or of inelastic interaction with the wall. When such processes exist, the loss coefficient can be written in a form similar to expression (6.28), obtained for a pore, but with negative resonance energy:

$$\mu = \Gamma_1\Gamma_2/[(E + |E_r|)^2 + (\Gamma_1 + \Gamma_2)^2],$$

where Γ_1 characterizes the lifetime of an ultracold neutron in a bound state before it is absorbed in the wall, and Γ_2 is the probability of occupying a bound state, and hence also the probability of leaving it as a result of the same scattering processes on the roughnesses or of inelastic interaction. In principle, such conditions can be imagined when the resulting loss coefficient is significantly greater than μ_0.

Substances with a negative potential are usually good absorbers. The general expression for the additional loss coefficient due to the imaginary part of the potential u_1'' is readily obtained. In the case of thin films the additional loss coefficient is $\Delta\mu = 4k_0 h(u_1''/u_0')$. We leave it to readers to derive the general expression.

6.3.8 Fundamental causes of excess losses

In principle, we cannot exclude the possibility of such manifestations of the laws of quantum mechanics within the UCN energy range that are not observable at higher energies. For this reason certain hypotheses have been considered that concern the essence of the neutron and that are within the scope of fundamental physics. The small storage time τ led to the idea that, owing to some special properties, the neutron overcomes the potential barrier created by the wall with less difficulty than it should following conventional quantum mechanics. Some of the hypotheses put forward in this direction are still not sufficiently elaborate, and therefore we shall not touch upon them. We shall only consider the hypotheses of wave packets (Ignatovich 1978*c*) and of hidden states (Ignatovich and Luschikov 1981).

Packets

The penetration of a potential barrier by a neutron may be pictured as something like a long train overcoming a hill. This is the known classical problem of whether a body of mass M can overcome a hill of height H if its kinetic energy is less than MgH. The answer turns out to be positive, but then the body must be sufficiently elongated for the potential energy of the raised part to be less than the total energy of the body (*Vlasov and Lupandin 1979).

In accordance with present-day concepts the wavefunction of a particle can always be represented in the form of a set of plane waves,

$$\Psi(\mathbf{r}, t) = \int f(\mathbf{k})\mathrm{d}^3 k \exp(\mathrm{i}\mathbf{kr} - \mathrm{i}k^2 t/2), \tag{6.29}$$

each of which satisfies the free motion Schrödinger equation, while all together they form a certain localized state. Among the plane waves of a packet there are those for which $k_\perp^2 > u_0$. As a result of these waves the packet may overcome the potential barrier as a whole, i.e. in the absence of real absorption the loss coefficient in the case of a neutron impinging upon the wall can be represented as the probability for the wave packet to penetrate the wall:

$$\mu = \int |f(k)|^2 |t^+(k_\perp)|^2 k_\perp' \mathrm{d}^3 k \theta(k_\perp^2 > u_0) / \int |f(k)|^2 k_\perp \mathrm{d}^3 k.$$

In accordance with conventional quantum mechanics the coefficients $f(k)$ are determined by the initial conditions, i.e. they depend on the story of preparing the packet. Thus, for instance, upon being reflected from the barrier a packet splits up. Waves of high energy leave through the wall, while those of low energy are reflected. It is to these waves that UCN correspond. We can, however, imagine a free neutron to be always described by one and the same wave packet, which cannot change in scattering. Such a point of view does not really contradict conventional quantum mechanics, since in spirit it corresponds to the known idea of packet reduction. According to this idea, if an absorbing nucleus is found in the path of a neutron described by a plane wave, the neutron will either be absorbed or not, but it will not represent that part of a plane wave with a hole of area equal to the absorption cross-section σ_a.

Proceeding on the basis of these notions we shall attempt, without diverging from conventional quantum mechanics, to construct a wavefunction of the form

$$\Psi(\mathbf{r}, t) = f(\mathbf{r} - \mathbf{k}t), \tag{6.30}$$

which is, to a certain extent, localized and does not spread out. It turns out that the free non-stationary Schrödinger equation is satisfied by the function

$$\Psi(\mathbf{r}, t) = cj_0(s|\mathbf{r} - \mathbf{p}t|)\exp(\mathrm{i}\mathbf{pr} - \mathrm{i}\omega t), \tag{6.31}$$

where c is a constant, $j_0(x) = \sin x/x$ is the spherical Bessel function, and $\omega = (p^2 + s^2)/2$. This function differs in form from (6.30) only by a phase factor, the presence of which is quite admissible since it does not hinder localization of the particle. The function (6.31) can be constructed as follows. First we find the stationary spherically symmetric solution $\Psi(\mathbf{r})\exp(-\mathrm{i}s^2 t/2)$ of the Schrödinger equation. The following will serve as this solution: $\Psi(\mathbf{r}) = j_0(sr)$. Then we pass to the reference frame moving with velocity p. As a result of appropriate transformations we obtain (6.31). Applying the derivatives of (6.31) with respect to time and coordinates it is possible to construct other localized, although not spherically symmetric, solutions.

The obtained packet does not spread out and it is not normalizable, i.e. the integral $\int|\Psi(\mathbf{r}, t)|^2 \mathrm{d}^3 r$ taken over the entire space diverges. However, for solving various scattering problems normalizability is not required, and the packet (6.31) is just as admissible as a plane wave. The coefficients $f(\mathbf{k})$ in (6.29) for the packet (6.31) are $f(\mathbf{k}) = \delta[(\mathbf{k} - \mathbf{p})^2 - s^2]$, i.e. all the wave-vectors lie on a sphere of radius s circumscribing the point p. Anomalous penetration through the barrier starts with such values of p for which $p + s > k_{\mathrm{lim}}$. The parameter s characterizes the dimension of the packet, and, as follows from *Shull (1969) and Scheckenhofer and Steyerl (1977), the dimensions of a packet are not smaller than 10^{-4} cm, so $s < 10^4\ \mathrm{cm}^{-1}$, i.e. $s < 10^{-2} k_{\mathrm{lim}}$. Thus, anomalous penetration can be explained only for neutrons of energies very close to the limiting energy.

In principle, wave packets of another sort can also be constructed:

$$\Psi(\mathbf{r}, t) = c\exp(\mathrm{i}\mathbf{pr} - \mathrm{i}\omega t)\exp(-s|\mathbf{r} - \mathbf{p}t|)/|\mathbf{r} - \mathbf{p}t|, \tag{6.32}$$

where $\omega = (p^2 - s^2)/2$. This packet is normalizable but does not satisfy the Schrödinger equation at all points. An exception is the point $r = pt$. In other words, for the function (6.32) we must write either the inhomogeneous equation

$$(i\partial_t + \Delta/2)\Psi(\mathbf{r}, t) = c\delta(\mathbf{r} - \mathbf{p}t),$$

or the non-linear equation

$$\left[i\partial_z + \frac{\Delta}{2} + \frac{2\pi}{c}\delta\left(\Delta\frac{1}{|\Psi(\mathbf{r}, t)|}\right)\right]\Psi(\mathbf{r}, t) = 0.$$

With the aid of this function the excess loss coefficient can be described approximately. For $s \ll k_{\text{lim}}$ we have

$$\mu = (16s/\pi)\int k^3 k' dk\theta(k > k_{\text{lim}})/(k^2 - p_\perp^2)^2(k + k')^2,$$

$$k' = \sqrt{k^2 - u_0},$$

and to explain the excess loss coefficient it is sufficient to adopt $s \approx 4 \times 10^{-4} p$.

One of the consequences of this hypothesis is the deep penetration of the neutron inside the reflecting wall. It was checked by Novopoltsev *et al.* (1988). In the experiment, the distribution of radioactive copper in the copper wall was measured after the copper had been exposed to the UCN flux using the activation method. Thin layers of copper were evaporated on float glass and after exposure dissolved in sulphuric acid. The pills of copper sulphide were prepared from solution and the activity of Cu^{64} was measured. The results of the experiment are in agreement with usual quantum mechanics. No anomalous penetration inside the matter was observed.

Hidden states

The notion that there might be hidden states of elementary particles is discussed in the book by *Gel'fer *et al.* (1975). Applied to UCN this idea is as follows (Ignatovich and Luschikov 1981). Let the neutron have two states, $|n\rangle$ and $|\tilde{n}\rangle$, differing from each other by the value of a certain quantum number (let us call it hiddenness), and let the neutron always be produced in the state $|n\rangle$, owing to the conservation of hiddenness. But in the Universe there exist fields slightly violating the law of hiddenness conservation, so the eigenstates of the neutron in vacuum are $|1\rangle = (1 - \alpha^2)^{1/2}|n\rangle + \alpha|\tilde{n}\rangle$ and $|2\rangle = \alpha|n\rangle - (1 - \alpha^2)^{1/2}|\tilde{n}\rangle$, where the parameter α characterizes the degree of hiddenness violation. If the energies of the states $|1\rangle$ and $|2\rangle$ differ, the wavefunction of the neutron becomes non-stationary:

$$\Psi(t) = \sqrt{1 - \alpha^2}(\sqrt{1 - \alpha^2}|n\rangle + \alpha|\tilde{n}\rangle)e^{-iE_1 t} + \alpha(\alpha|n\rangle - \sqrt{1 - \alpha^2}|\tilde{n}\rangle)e^{-iE_2 t}. \tag{6.33}$$

The probability of finding the neutron in the state $|\tilde{n}\rangle$ is $4\alpha^2(1-\alpha^2) \times \sin^2(\Delta Et/2)$, where $\Delta E = E_1 - E_2$. Now imagine that a neutron in the state $|\tilde{n}\rangle$ interacts with matter in a totally different manner than in the state $|n\rangle$: for instance, let its potential have the opposite sign. Then the component $|\tilde{n}\rangle$ will pass easily through the wall. The state $|\tilde{n}\rangle$ will again be regenerated in the reflected neutron, and in the subsequent collision the neutron will again be able to go through the wall with a probability $\sim\alpha^2$.

We can make the assumption that $|n\rangle$ and $|\tilde{n}\rangle$ have opposite magnetic moments $\boldsymbol{\mu}$, as in the case of the neutron and the antineutron. In this case the coefficient α in (6.33) decreases as the magnetic field increases and it might seem that, as the uniform magnetic field increases in the trap from 0 up to 100 G, the UCN storage time should increase. Direct measurements, however, contradict this hypothesis (Kosvintsev *et al.* 1982*b*).

6.3.9 Measurement of the dependence of the loss coefficient μ on the UCN velocity v

The various theoretical principles considered in the preceding subsections can be verified by measuring the dependence $\mu(v)$. If the surface of the walls is perfectly smooth, this dependence should be of the form (6.4). At small velocities v, i.e. at $y = v/v_{\text{lim}} \ll 1$, the function $\mu(v)$ depends linearly on the velocity, $\mu(v) \approx \frac{4}{3}\eta y$. As y approaches unity the dependence $\mu(v)$ becomes steeper. Thus, for instance, when $y = 1$ we have $\mu = \pi\eta$.

The experimental layout for measuring $\mu(v)$ is shown in Fig. 6.3. The trap (Groshev *et al.* 1976) for the storage of UCN was placed in the bar of the crank of a gravitational spectrometer. When the crank was rotated, the trap was displaced vertically. If a neutron had an energy E at the bottom, then at

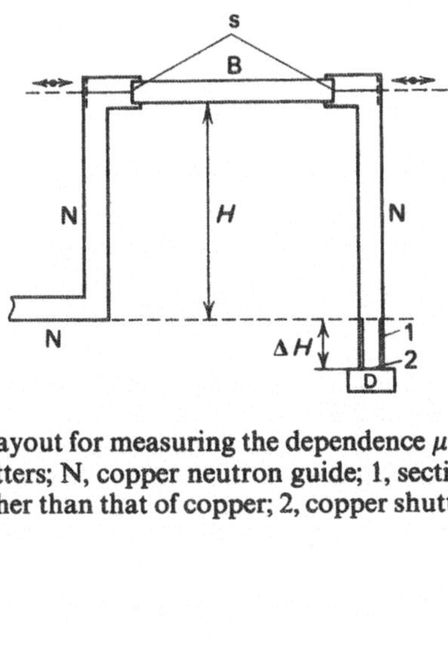

Fig. 6.3 Experimental layout for measuring the dependence $\mu(v/v_{\text{lim}})$. B, copper vessel for storing UCN; s, shutters; N, copper neutron guide; 1, section of the neutron guide with limiting energy higher than that of copper; 2, copper shutter (Groshev *et al.* 1976).

the top its energy was $E - mgH$. After exposure in the vessel the neutrons were extracted on to a detector that was lowered with respect to the input horizontal neutron guide by a height ΔH and covered with a copper foil (both the trap and the neutron guide are also made of copper). The detector could only register neutrons with energies higher than the E_{lim} of copper. Inside the input neutron guide the energy of these neutrons turns out to be in the spectrum range $E_{\text{lim}} - mg\Delta H < E < E_{\text{lim}}$. In the trap raised to the height H this part of the spectrum transforms into the band $E_{\text{lim}} - mg(H + \Delta H) < E < E_{\text{lim}} - mgH$. By altering H the whole range of energies can be scanned, and by changing ΔH it is possible to vary the width of the interval of the spectrum under study. In this case this interval amounted to ~ 25 neV (one point was even obtained with the interval 10 neV) and was sufficiently narrow for the neutrons to be considered monochromatic and the storage curve to be exponential. The results of the experiment are presented in Fig. 6.4. The dependence obtained $\mu(v)$ is in reasonable agreement with expression (6.4) and differs from the theoretical one only by the reduced coefficient η, which is three times greater. The sensitivity of the experiment was naturally insufficient to reveal certain subtleties related, for example, to the smearing of the surface (the corresponding theoretical curves are shown in the same figure). If the curve $\mu(v)$ were to exhibit any deviations from the theoretical curve (6.4), even qualitative ones, this could suggest in which direction one should search for the causes of the anomaly in τ, but this did not happen.

Somewhat earlier (Steyerl and Trusted 1974) another experiment was performed to determine the dependence $\mu(v)$. Here the UCN were stored in a

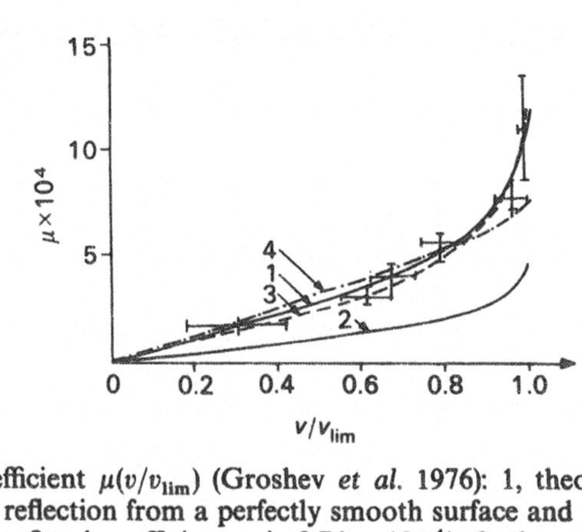

Fig. 6.4 Loss coefficient $\mu(v/v_{\text{lim}})$ (Groshev *et al.* 1976): 1, theoretical dependence corresponding to reflection from a perfectly smooth surface and fitted to the experimental points (the fitted coefficient η is 3.74×10^{-4}); 2, the same curve but with theoretical value $\eta = 1.45 \times 10^{-4}$; 3, loss coefficient in reflection from a potential with smeared boundary (the smearing width is 110 Å); 4, theoretical loss coefficient in the presence of a water film 55 Å thick on the surface of the walls.

trap of pyrographite, while the detector was covered with shutters with various E_{lim}. As E_{lim} increases, the spectral interval of the neutrons studied becomes narrower. From measurements of the dependence of the storage time on the height E_{lim} it is possible to derive the dependence $\mu(v)$. From the measurement results and their analysis it can be concluded that $\mu(v)$ increases more steeply than the function (6.4), as $v \to v_{\mathrm{lim}}$. This experiment, however, was less straightforward than that by Groshev *et al.* (1976), and its mathematical analysis depends on several assumptions.

6.4 THE SECOND STAGE IN STUDIES OF UCN STORAGE

6.4.1 Discovery of the heating of UCN

The first stage in the investigation of UCN storage led to a contradiction. To overcome this situation an experiment (Strelkov and Hetzelt 1978) was performed. This experiment was to provide an unambiguous answer to the question concerning the whereabouts of the neutrons vanishing from the trap: are they absorbed, heated, or do they penetrate the barrier in an anomalous way? The layout of the experiment is shown in Fig. 6.5. The trap in which the UCN are stored is surrounded by peripheral gas counters containing ^3He. Neutrons passing through the wall of the trap should be registered by the counters; if, on the other hand, the neutrons are absorbed in the walls, then the peripheral counters will not count them. By varying the pressure in the counters it is possible to estimate the energy of the outgoing neutrons.

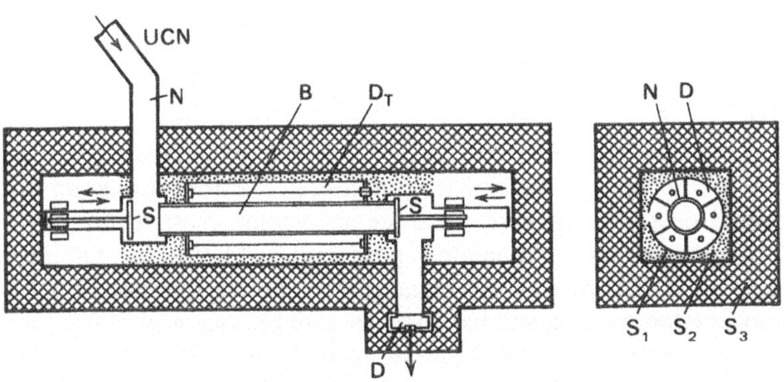

Fig. 6.5 Experimental layout for observing UCN heating (Strelkov and Hetzelt 1978). D_T, counters of thermal neutrons (CTN); S_1, shielding from thermal neutrons (Cd, 1.5 mm); S_2, shielding from fast neutrons (B_4C); S_3, polyethylene filled with boron. In the vicinity of the reactor CM-2 such a shielding led to suppression of the CTN background down to a single pulse per second.

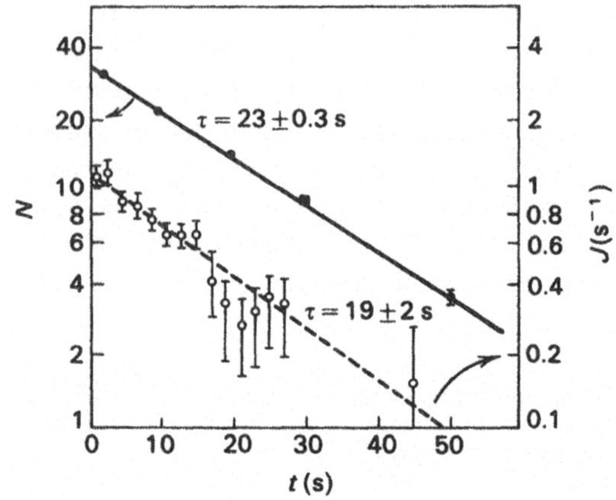

Fig. 6.6 Results of experiment to observe UCN heating. The solid curve is the storage curve (the left ordinate); the dashed curve is the CTN counting rate depending on time after the input valve is closed (right ordinate). The figures indicate the storage times found from the curves (Strelkov and Hetzelt 1978).

In Fig. 6.6 the storage curve is shown for UCN in a copper electropolished vessel as well as the counting rate of the peripheral counters depending on the exposure time of UCN in the vessel. These two curves should naturally be identical in form, since the count of the peripheral detectors is proportional to the number of neutrons present at a given moment in the trap. The integral of the second curve over time from zero to t_{exp} equals the number of UCN that have left the trap during the time t_{exp}. Analysis revealed that 75% of the neutrons that vanish from the trap pass through its walls, and the energy of these neutrons is close to the thermal energy (Stoica and Strelkov 1978*b*; Stoica *et al.* 1978) of ~ 10 meV. Hence it can be concluded that the excess loss coefficient is related to the heating of UCN in a single collision with the wall. The gradual heating of UCN by acoustic oscillations could be excluded, since in this case the neutrons acquiring an energy slightly higher than the limiting energy would penetrate the material of the trap (copper, in this case) and would be absorbed there, since the absorption cross-section for copper significantly exceeds σ_{ie}. Anomalous penetration of UCN through the barrier was also excluded, as was the enhancement of the loss coefficient due to inhomogeneities on the surface such as roughnesses and pores, since the inhomogeneities do not alter the relationship between the inelastic scattering and absorption, which for copper is $\sigma_{ie}/\sigma_a \approx 0.1$.

The only reasonable explanation for the results of the experiment (Strelkov and Hetzelt 1978) could be the heating of UCN on hydrogen present in the surface regions of the copper. Thus, the hypothesis affirming the existence of hydrogen was once more revived. Let us estimate the amount of hydrogen on the surface necessary to obtain for copper an excess loss coefficient $\Delta\eta \approx 3 \times 10^{-4}$ twice as large as the theoretical value $\eta_{th} \approx 1.4 \times 10^{-4}$. Let the atomic concentration of hydrogen in the surface layer of width λ_{lim} be c and the scattering cross-section on a hydrogen atom be $\sigma_{H,ie}(k_T)$ at the thermal point k_T corresponding to the neutron velocity $v_T \approx 2200\ \mathrm{m\,s^{-1}}$. Then $\Delta\eta = ck_T\sigma_{H,ie}(k_T)/4\pi b$ and the total amount of hydrogen per unit surface area $N_H = cN_0\lambda_{lim}$ is given by

$$N_H = \frac{4\pi N_0 b \bar{\lambda}_{lim} \Delta\eta}{k_T \sigma_{H,ie}(k_T)} = \frac{v_{lim}}{v_T} \frac{\Delta\eta}{\sigma_{H,ie}(k_T)}.$$

We do not know the exact cross-section $\sigma_{H,ie}$ for hydrogen situated near the surface. Nevertheless we shall consider it to be firmly bound in the lattice and the cross-section to be determined in the usual way, i.e. it can be estimated by the elastic incoherent scattering cross-section $\sigma_{el} \approx 80$ barn. Then for $\Delta\eta \approx 3 \times 10^{-4}$ we have, for estimating the necessary amount of hydrogen in any arbitrary substance, the simple formula $N_H = (v_{lim}/6) \times 10^{16}\ \mathrm{cm^{-2}}$. Thus, to obtain such a value $\Delta\eta$ in copper, approximately $1 \times 10^{16}\ \mathrm{cm^{-2}}$ is required.

6.4.2 Straightforward observation of hydrogen on the surface of a substance

Approximately at the same time a technique was devised for investigating surface regions of substances by means of nuclear reactions (Lanford and Golub 1977). To measure the contents of hydrogen the nuclear reaction $p(^{15}N, ^{12}C)\alpha\gamma$ is utilized, which in the laboratory reference system exhibits a resonance at the energy of ^{15}N equal to $E_r = 6.385$ MeV. If a beam of ^{15}N ions directed towards the surface has a higher energy, then, upon entering the substance, it first decelerates down to the energy E_r and at the appropriate depth undergoes interaction with the hydrogen, the amount of which is determined from the intensity of γ-quanta produced with the characteristic energy of 4.43 MeV. Figure 6.7 demonstrates that in a surface layer 100 Å thick (approximately the resolution of the technique) there is indeed enough hydrogen to explain the losses of UCN. Thus, the hydrogen hypothesis acquired a new confirmation.

6.4.3 The theoretical explanation of the weak temperature dependence of the time τ

A paradoxical situation has arisen. On the one hand, the excess loss coefficient was shown to be due to inelastic scattering, most likely on the hydrogen, sufficient amounts of which are present in the surface regions of a substance, while on the other hand previous experiments could not detect any temperature dependence of the loss coefficient, which should doubtless have mani-

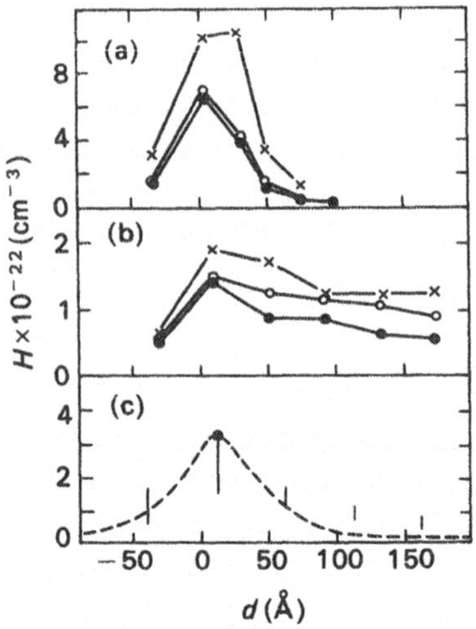

Fig. 6.7 Distribution of hydrogen in a material depending on the distance from the surface: (a) copper mechanically polished, then electropolished and cleaned with acetone and alcohol; (b) pyrolytic graphite degassed for a week at a vacuum of 10^{-4} mm Hg, whereupon it was exposed in the atmosphere for 6 months; (c) glass etched with hydrofluoric acid. Upon measuring the concentration of hydrogen at room temperature (×), the samples were baked at 190°C for 18 h (○) and, finally, the distribution of hydrogen was measured after additional baking at 265°C for 13 h (●). Measurement in the glass was performed after baking at 210°C for 2.5 days. The dashed line corresponds to the minimal amount of hydrogen which was in agreement with experimental data. The difficulty here is that the surface of the glass becomes charged when bombarded by ions (Lanford and Golub 1977).

fested itself in inelastic scattering. It can be concluded that the hydrogen in the surface regions exhibits an inelastic scattering cross-section with an anomalously weak temperature dependence. Hydrogen firmly bound in a lattice scatters like an oscillator, and at high temperatures σ_{ie} varies linearly with the temperature T, while at low temperature its dependence on T is not weaker than $T^{7/2}$. Theoretically, however, we can also consider models that lead to an even weaker dependence of σ_{ie} on T.

Scattering on a free hydrogen atom provides the weaker temperature dependence $\sigma_{ie} \sim T^{1/2}$. A free gas should readily escape when heated, but for some reason hydrogen stays in place. This may be explained by the fact that the hydrogen is firmly bound with a substance in the direction normal to its surface, while it can move quite freely along the surface itself. This is why the model of a two-dimensional gas (Ignatovich and Satarov 1977) came into being, which gives a cross-section $\sigma_{ie} \sim T^{1/2}$, just like a free gas. These conclusions were confirmed by Blokhintsev and Plakida (1977). We note, however, that the scattering cross-section on a free atom is four times smaller than on a bound atom, and therefore to obtain the same excess reduced loss coefficient $\Delta\eta \approx 3 \times 10^{-4}$ four times more hydrogen is required in the medium.

An even weaker temperature dependence can be obtained if the hydrogen in the medium is considered to be a two-level system with a distance between the levels of $\Delta E \approx 20$ K. In this case inelastic scattering on hydrogen is related to the transition of the hydrogen from the upper level to the lower one and to a momentum transfer of $p \approx (\Delta E)^{1/2}$ to the neutron. The cross-section of this process depends on the population of the upper level, which does not change with the temperature until T reaches 20 K. The cross-section then decreases exponentially as the temperature decreases further. The two-level model can be improved somewhat. If the quantity ΔE changes from point to point in the substance, then the rapid exponential decrease of the cross-section with temperature becomes smoother (Plakida 1981).

We note, however, that the smaller the value of ΔE, the smaller is the momentum transfer to the neutron and the smaller is the inelastic scattering cross-section, so consequently the greater is the amount of hydrogen required to explain the excess $\Delta\eta$. A similar problem arises if the scattering on hydrogen is considered to be mainly quasi-elastic.

6.4.4 Search for the temperature dependence of the coefficient η

While theoreticians were busy trying to reduce the dependence of $\sigma_{\mathrm{H,ie}}$ on temperature, experimental physicists were repeatedly attempting to find this dependence. Finally, it began to be observed.

Kosvintsev *et al.* (1979) performed experiments on the storage of 20 neV UCN in an aluminium vessel. The vacuum in the vessel was maintained by utilizing oil diffusion pumps with a nitrogen trap. (We draw attention to the way vacuum is maintained, since, generally speaking, it determines how clean the walls of the vessel are.) The vessel underwent annealing for 2 h at 250°C. At 300 K the coefficient η is $(5.9 \pm 1.2) \times 10^{-4}$ before the annealing and $(3.0 \pm 0.6) \times 10^{-4}$ after the annealing (the theoretical value of η is 0.23×10^{-4}). The temperature dependence of the coefficient η after annealing (Fig. 6.8) was quite noticeable and did not contradict the idea that the excess loss coefficient $\Delta\eta$ is totally due to the heating of UCN on hydrogen atoms.

A search was performed (Kosvintsev *et al.* 1980*c*) for the temperature dependence of μ in a copper vessel shut off from the main neutron guide by an aluminium window which hindered the flow of any impurities from the neutron guide into the vessel. The temperature of the vessel could vary between 80 and 800 K. In one case it was even made to reach 20 K. In Fig. 6.9 the dependence of the loss coefficient μ on temperature is shown for three cases: an annealed copper vessel, a non-annealed copper vessel, and a vessel with beryllium deposited on its walls. Curve 1 corresponds to the cooling of a

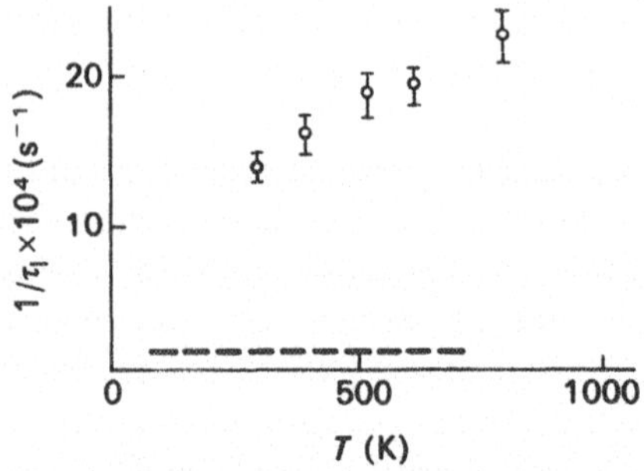

Fig. 6.8 Experimentally measured dependence of $1/\tau_1$ on temperature in an aluminium vessel that was first baked for 2 h at a temperature of 250°C; ○ experimental points. The dashed line corresponds to losses caused only by τ_β (Kosvintsev *et al.* 1979).

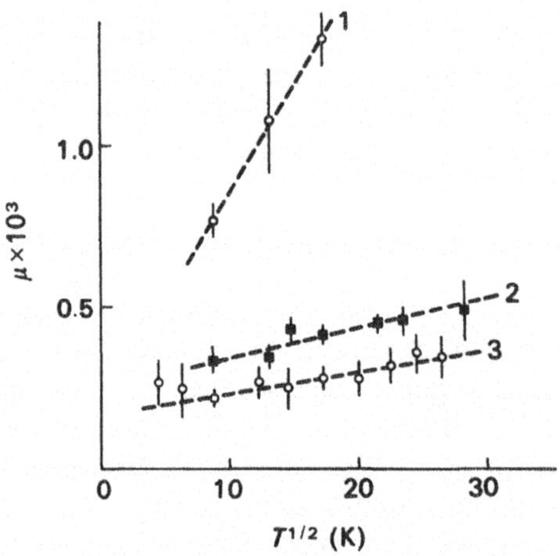

Fig. 6.9 Loss coefficient μ depending on temperature (Kosvintsev *et al.* 1980*c*): 1, electropolished copper trap before outgassing; 2, the same trap, but degassed; 3, beryllium trap obtained by vacuum deposition of beryllium on the walls of the copper trap. The quality of the vacuum deposition was not controlled, but subsequently (in 1.5 years) it turned out that the beryllium partly crumbled away from the walls.

non-degassed vessel. The strong temperature dependence is explained by the large amount of hydrogen on the walls. As a result of degassing for 6 h at a temperature of 800 K the loss coefficient decreased and its temperature dependence (curve 2) became significantly weaker. The minimal loss coefficient exceeded the theoretical value by a factor of 2. Such an excess is not in itself too great, since it may be explained by the surface being rough, i.e. by the actual surface area being twice the apparent surface. Curve 3 was obtained after beryllium was deposited on the walls of the vessel. The excess loss coefficient is about the same as in the case of the annealed copper vessel. Unfortunately, the quality of the deposited beryllium was not under control, and subsequently (in a year), when the vessel was opened, it turned out that the beryllium partly peeled off and did not cover the walls of the vessel completely, so the behaviour of curve 3 testifying in favour of a weak temperature dependence started to give rise to doubts.

6.4.5 Attempts to remove the hydrogen

If the excess loss coefficient is assumed to be due to the presence of hydrogen we can pose the problem of removing it to achieve a real storage time in the trap identical with the theoretical value. To achieve this objective we can use several methods: annealing at high temperature, annealing in oxygen, depositing a new substance on the surface, bombardment of the surface with ions to clean it, and deuteration.

Attempts at deuteration undertaken by Alfimenkov *et al.* (1980) and I.M. Frank (1979), by treating the surface in heavy water vapour at high temperature or with alkali, NaOD, and acids, D_2SO_4 and DNO_3, were unsuccessful. The treatment led to no change in the loss coefficient. High-temperature annealing in the case of an aluminium trap (Kosvintsev *et al.* 1979), for example, was also not entirely successful. Great success was achieved in the annealing of an aluminium vessel in an oxygen atmosphere (Kosvintsev *et al.* 1980*a*) for 1 h at a temperature of 700 K and a pressure of 0.1 mm Hg. The loss coefficient η was reduced in this case to 0.5×10^{-4}. However, the resulting state of the surface was unstable, and in 3–4 hours the coefficient η increased again.

Kosvintsev *et al.* (1978*b*) attempted to remove the hydrogen by depositing a substance on the walls. The experimental layout is shown in Fig. 6.10. The neutrons were made to fill up a raised vessel which contained sprayers of metals under its top. The energy of the accumulated neutrons was about 20 neV so they could not reach the sprayers. The storage time was measured for deposits on the walls of aluminium, zinc, copper, lead, bismuth, beryllium. The vacuum was maintained at a level of 10^{-5} mm Hg, but nevertheless the loss coefficient exhibited an excess of 3×10^{-4} for all the substances. Storage was also carried out in a dynamic mode, when the deposition was performed during storage. Aluminium was deposited under a vacuum of 10^{-6} mm Hg at rates of 1 Å s^{-1} and 30 Å s^{-1}. The storage time did not change within 25%.

Kosvintsev *et al.* (1978*b*) and Alfimenkov *et al.* (1980) tried unsuccessfully to remove the excess $\Delta\eta$ by vacuum depositing heavy water and CO_2 on to the walls at liquid nitrogen temperature. On the other hand, Kosvintsev *et al.* (1982*a*) showed that it was possible to lower η to a level of 5×10^{-5} by deposition of D_2O, which is only twice the theoretical value of η for a perfect surface.

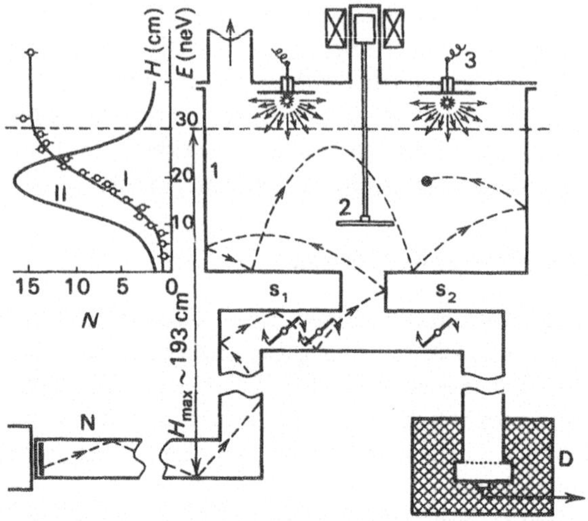

Fig. 6.10 Experimental layout for studying UCN storage in a vessel with a vacuum-deposited surface: 1, storage vessel; 2, input valve also playing the part of the exit valve; 3, metal sprayer. The inset demonstrates the measured integral spectrum of accumulated neutrons (I) and the fitted Gaussian differential spectrum (II). The dashed curve represents a possible neutron trajectory in the vessel (Kosvintsev *et al.* 1987*b*).

A significant decrease in the loss coefficient was achieved by ion cleaning the surface with a glow discharge in an atmosphere of argon and deuterium (Mampe *et al.* 1981). Thus, for example, it became possible to reduce the value of η to 5×10^{-5} in aluminium (Kosvintsev *et al.* 1980*a*) and to 3×10^{-5} in quartz. It is noteworthy that these results were obtained at a pressure of 10^{-3} mm Hg maintained by oil pumping with the aid of diffusion pumps. In Fig. 6.11 storage curves are shown for a quartz vessel, the surface of which was submitted to various treatments.

To find out how vacuum and oil pumping influence the coefficient η experiments were performed with a stainless steel vessel under the above conditions and in a vacuum of 10^{-7} mm Hg, as well as in the case of oil-less pumping. In both cases the glow discharge turned out, to the same extent, to provide for the possibility of making the coefficient η approach the theoretical value.

Mampe *et al.* (1981) observed certain effects that are worth mentioning. First of all, vacuum deposition of beryllium on the walls led to a high value of η (1×10^{-4}). It became impossible to decrease this value with the aid of a glow discharge. This result is in agreement with the results of Kosvintsev *et al.* (1978*b*). Further, it was noted that a discharge in oxygen, which should seemingly lead to the formation of H_2O molecules and the removal of

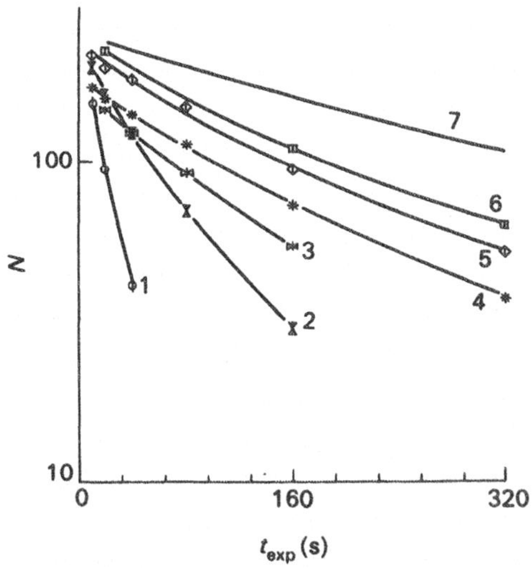

Fig. 6.11 Storage curves for a quartz vessel, the surface of which was submitted to various treatments: 1, cleaning in alcohol; 2, cleaning by ultrasound with RBS 25 as a solvent and rinsing in distilled water; 3, cleaning in a glow discharge, then exposing to the atmosphere for 2 days, after which procedure 2 is repeated; 4, cleaning in a discharge for 30 min in an atmosphere of (80% Ar, 20% D_2); 5, cleaning in the same discharge but in an atmosphere of D_2O ($\tau = 280$ s); 6, cleaning with the same discharge as for curve 3, but with an admixture of deuterated methyl alcohol vapour ($\tau = 290$ s); 7, computed storage curve for a perfect surface of the trap walls, when the accumulated neutrons exhibit a Maxwellian spectrum. The coefficient η was assumed to be equal to 10^{-5} for quartz and 10^{-4} ($\tau = 375$ s) for the end shutters made of stainless steel. The width of the slits in the vicinity of the two valves was assumed to be equal to 10^{-4} (Mampe *et al.* 1981).

hydrogen from the trap walls, actually led to worse results than a discharge in argon or deuterium. The best results were obtained with discharges in deuterium and in vapours of D_2O and deuterated methyl alcohol. A surface bombarded with deuterium ions turns out to be insensitive to the influence of the atmosphere and even of hydrogen.

6.4.6 Control of the surface submitted to various cleaning procedures

La Marche *et al.* (1981) controlled the amount of hydrogen with the aid of ^{15}N ions after the surface was treated in various ways.

Deuteration was performed in two different ways: by boiling the samples in D_2O for 15 min or by etching them in DF acid for 45 min and subsequently rinsing them in D_2O. The samples were also kept in heavy water for a month, after which they were placed wet in a vacuum chamber, and the remaining amount of hydrogen in them was measured at a vacuum of 10^{-6} mm Hg.

Experiments were performed with copper, graphite, and glass. It turned out that after all these procedures the amount of hydrogen present in the samples decreased by less than a factor of 2, as compared with the non-treated samples, and was $(2–5) \times 10^{16}\ cm^{-2}$.

The next experiment concerned the vacuum deposition of a pure substance on to a surface. When lead was vacuum deposited at a rate of 7 monolayers/s the hydrogen content on the surface dropped to $5 \times 10^{15}\ cm^{-2}$, but in 1 hour after the vacuum deposition was stopped it increased to the stable quantity of $8 \times 10^{15}\ cm^2$. The hydrogen content in aluminium fell to $0.2 \times 10^{15}\ cm^{-2}$ when the deposition rate was 5 monolayers/s, but after the spraying was stopped it increased in 1 hour, in a vacuum of $(1–3) \times 10^{-6}$ mm Hg, to $(2–3) \times 10^{15}\ cm^{-2}$. These data somewhat contradict the results of dynamic vacuum deposition (Kosvintsev *et al.* 1978*b*).

The cleaning of a surface by ion bombardment was studied by Mampe and Bugeat (1980) and La Marche *et al.* (1981). Mampe and Bugeat bombarded the surface with ^{11}B ions, which have a resonance interaction cross-section with hydrogen at an energy of 1.8 MeV in the laboratory reference frame. The corresponding reaction $p(^{11}B, 2\alpha)\alpha$ permits the determination of the hydrogen content from the intensity of the α-particles within the energy range from 2 to 4 MeV. The ion bombardment and the control were, in this case, performed simultaneously. It was shown that non-treated copper and stainless steel surfaces contain $\sim(2–3) \times 10^{16}\ cm^{-2}$ of hydrogen, and that the bombardment makes the amount of hydrogen drop to $2 \times 10^{15}\ cm^{-2}$, while after the bombardment is stopped it increases to $6 \times 10^{15}\ cm^{-2}$ after several hours in a vacuum of $10^{-6}–10^{-8}$ mm Hg.

Similar results were obtained by La Marche *et al.* in cleaning stainless steel with a glow discharge in argon. A discharge in an atmosphere of oxygen made it possible to obtain nickel, copper, and lead surfaces with a stable hydrogen content of $(2–4) \times 10^{15}\ cm^{-2}$.

Annealing at a temperature of 1000°C (stainless steel, SiO_2, Ni, W, Ti, Ta, Zr, Fe) led to the concentration of hydrogen being lowered to a level of $\sim 10^{14}\ cm^{-2}$, after which, in a vacuum of 10^{-6} mm Hg, it increased in several hours to $\sim(4–5) \times 10^{15}\ cm^{-2}$. However, if the vacuum is maintained at the level of 5×10^{-9} mm Hg, then by baking a nickel film at only 500°C it is possible to obtain in several minutes a surface that contains only $\sim 5 \times 10^{14}\ cm^{-2}$ after 24 h (Bugeat and Ligeon 1979).

In Japan, Kawabata *et al.* (1988) also measured the concentration of hydrogen on the surface of different metals and tried to remove it. The elastic recoil detection analysis (ERDA) method was used to profile the hydrogen with the help of the beam of 50 MeV argon ions. The results of measurements for two types of samples were reported: vacuum evaporated nickel layers and beryllium plates. Evaporation of nickel layers of thickness 600–2000 Å was performed in different vacuum conditions. After evaporation, all samples were exposed to atmospheric gas. It was found that the concentration of hydrogen does not depend on vacuum if the vacuum is better than 10^{-6} Torr. This shows, that contamination by hydrogen occurs during exposure to air and not during evaporation for vacuum conditions better than 10^{-6} Torr. In nickel samples of thickness 2000 Å evaporated on aluminium substrate, the density of hydrogen on the surface was found to be 1.7×10^{16} atoms cm^{-2}, and that on the interface between nickel and aluminium was found to be 1.2×10^{16} atoms cm^{-2}. For the copper substrate the density on the interface was found to be 2.9×10^{16} atoms cm^{-2}.

Measurements of beryllium were performed with commercial plates washed by ultrasonic cleaning in acetone. The hydrogen concentration on the surface was $(2\text{–}4) \times 10^{16}$ atoms cm^{-2}.

The attempt to replace hydrogen by deuterium was only partly successful. Only 25 per cent of the hydrogen could be replaced. The heating by argon ions to nearly 100°C did not diminish the number of hydrogen atoms on the surface of samples.

6.4.7 Measurement of partial loss coefficients

Since the hydrogen on a surface is mainly responsible for inelastic scattering, σ_{ie}, its contribution to the loss coefficient η can be determined from the amount of heated neutrons. We shall represent the total reduced loss coefficient in the form $\eta = \eta_a + \eta_{ie}$. The inelastic scattering cross-section of the main substance is usually negligible in comparison with the σ_{ie} of hydrogen and therefore $\eta_{ie} \approx \eta_H$, i.e. the partial loss coefficient related to heating is nearly totally determined by the hydrogen. If we succeed in noticing a change in the partial coefficients η_a and η_{ie} as a result of absorption and inelastic scattering corresponding to a change in the amount of hydrogen present on the walls, then we will be able to find out whether the excess loss coefficient is really fully determined by the presence of hydrogen on the walls.

The flux of heated neutrons is

$$J_{up} = K\eta_{ie},$$

where K is a certain proportionality factor depending on the UCN spectrum. If η_a is considered constant, which is not always true, then

$$\eta = \eta_a + J_{up}/K,$$

i.e. the coefficient η is a linear function of J_{up}. This function can be measured in an experiment, the layout of which is shown in Fig. 6.12 (Kosvintsev *et al.* 1983). A stack of plates of known area is placed in a vessel, in which the UCN are stored, and the common loss coefficient is measured. If the loss coefficient of the vessel without plates is known, then we may determine the coefficient η related only to the plates. The flux J_{up} from the plates is also measured, with

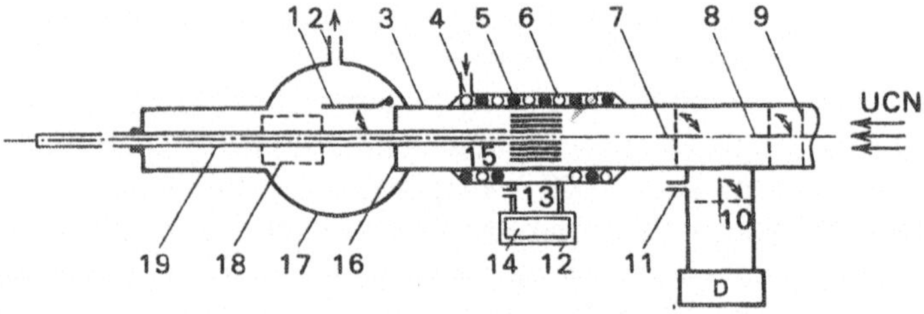

Fig. 6.12 Experimental layout for measuring UCN heating processes in collisions with a surface: 1 and 10, shutters; 2 and 11, outlets leading to the pumps; 3, vessel for UCN storage; 4, pipe for coolant; 5, heating filament; 6, outer casing; 7 and 8, shutters; 9, membrane; 12, cadmium jacket; 13, gas bulb; 14, counter of heated neutrons; 15, stack of plates; 16, window; 17, gate valve chamber; 18, radiator in position of removal; 19, stem (Kosvintsev *et al.* 1983).

the aid of a thermal neutron counter, in the given UCN flux. These two experiments determine a single point on the (η, J_{up}) plane. The amount of hydrogen present on the plates depends on the way their surfaces are treated. Changing the treatment changes η and, consequently, J_{up}. By repeating this process several times we can find a set of points in the (η, J_{up}) plane, fit them with a straight line, and by extrapolating the line to the point $J_{up} = 0$ find η_a. Comparison of the resultant η_a with the theoretical value reveals whether η_a contains an excess part not related to η_{ie} and, consequently, not related to the presence of hydrogen.

In Fig. 6.13 the results are presented of an experiment (Kosvintsev *et al.* 1983) with copper plates (the stack contained 80 plates 8 μm wide and each $56 \times 60\ \mathrm{cm}^2$ in area) and beryllium plates (three coiled plates 0.8 mm wide and each $440\ \mathrm{cm}^2$ in area). The extrapolated loss coefficient for copper was $\eta_a = (1.51 \pm 0.3) \times 10^{-4}$ $(\eta_{th} = 1.45 \times 10^{-4})$, and for beryllium it was $\eta_a = (-0.25 \pm 0.40) \times 10^{-4}$ $(\eta_{th} = 5 \times 10^{-6})$. The result of this experiment does not contradict the assumption that the excess loss coefficient is totally due to hydrogen.

The velocity of the heated neutrons was also determined in the experiment (Kosvintsev *et al.* 1983). To this end a chamber filled with ^{3}He under variable pressure was placed in between the stack of plates and the thermal neutron detector. When the pressure p was changed, the counting rate of the detector also changed, since the transmission of the chamber is proportional to $\exp(-N_0\sigma_T v_T pd/v_{eff}p_0)$, where v_T is the velocity of the thermal neutron ($v_T \approx 2200\ \mathrm{m\,s^{-1}}$), σ_T is the absorption cross-section of a neutron by a ^{3}He nucleus when the neutron velocity is v_T ($\sigma_T = 5340$ barn), d is the width of the gas chamber, N_0 is the number of atoms per unit volume at normal pressure p_0 (the Loschmidt number), and v_{eff} is the effective velocity of a heated neutron, which was just the quantity to be determined by Kosvintsev *et al.*

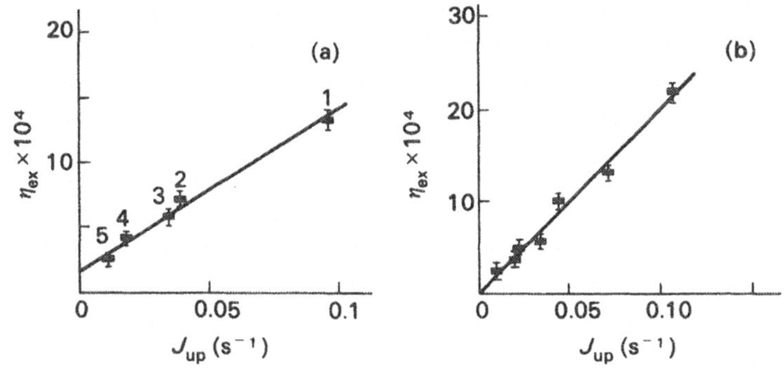

Fig. 6.13 Dependence of experimental coefficient η on the flux of heated neutrons, J_{up}. (a) For copper: 1, upon electropolishing oil vapour is deposited on the surface of the plates; 2, electropolishing is repeated; 3, chemical etching of the plates is performed in H_3PO_4; 4, after annealing at 550 K for 2 h; 5, after additional annealing at 700 K for 3 h. (b) For beryllium: the points from right to left are obtained after the following treatments of the beryllium plates: 1, etching, cleaning, and annealing for 0.5 h at 700 K; 2, annealing for 0.5 h at 850 K; 3, annealing for 1.5 h at 850 K; 4, annealing for 0.5 h at 900 K; 5, annealing for 0.7 h at 900 K; 6, annealing for 1 h at 950 K; 7, annealing for 0.5 h at 1000 K (Kosvintsev *et al.* 1983).

(1983). By measuring the dependence of the change in the counting rate of the thermal neutron detector on the pressure in the gas chamber, we find v_{eff}. It turned out that in well-annealed copper* $v_{eff} \approx 1.5v_T$, in beryllium $v_{eff} \approx v_T$, and the dependence of the velocity of heated neutrons on the temperature of the copper and of the beryllium is approximately $v_{eff} \sim T^{1/2}$. Hence it follows that the flux of heated neutrons should also change, as the temperature changes, proportionally to $T^{1/2}$. The measured dependence $J_{up}(T)$ (Fig. 6.14) can actually be described by the curve $J_{up} \approx T^{1/2}$.

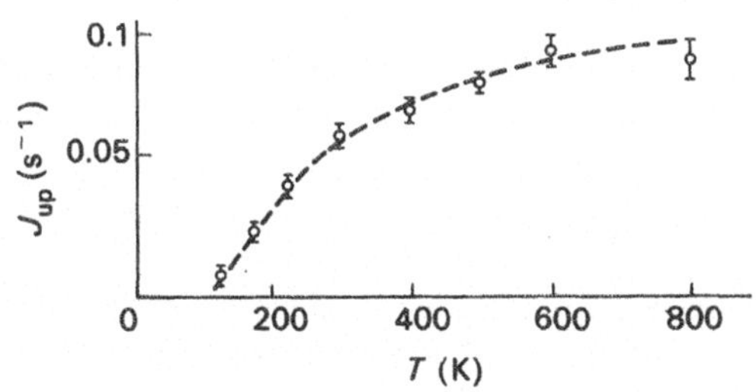

Fig. 6.14 Dependence of the flux J_{up} of heated neutrons on the temperature T of degassed beryllium plates (Kosvintsev *et al.* 1983).

---

* Some other experiments (Kalchev and Strelkov 1989) yielded $v_{eff} \approx 0.8v_T$.

Yet another experiment was performed (besides Kosvintsev *et al.* 1983) to determine the partial coefficients $\eta_{a,ie}$ (Ignatovich *et al.* 1982). In this experiment the neutrons passed through the chamber in a continuous flow, and the chamber was surrounded by thermal neutron counters (TNCs). The ^{4}He gas could be fed into the interior of the chamber. When the pressure of helium, p, changed, both the counting rate of the UCN detector, J_d, and the counting rate of the TNC, J_{up}, also changed. Heating of the UCN in the chamber takes place as a result of inelastic collisions not only with the walls, but also on the atoms of the gas. The heating rate on ^{4}He atoms is

$$1/\tau_{He} = p/(p\tau)_{He}.$$

The constant $(p\tau)_{He}$ equals 343 mm Hg. Assuming $1/\tau_l = 1/\tau_a + 1/\tau_{ie}$:

$$J_d = \varepsilon_d T_0/\tau_{emp}[2/\tau_{emp} + 1/\tau_\beta + 1/\tau_{ie} + p/(p\tau)_{He}], \qquad (6.34)$$

$$J_{up} = \varepsilon_{up} J_0[1/\tau_{ie} + p/(p\tau)_{He}]/[2/\tau_{emp} + 1/\tau_\beta + 1/\tau_{ie} + p/(p\tau)_{He}], \qquad (6.35)$$

where the detector efficiencies, ε_d and ε_{up}, are introduced. From expressions (6.34) and (6.35) it follows that $1/J_d$ and J_{up}/J_d are linear functions of the pressure p.

Having measured these dependences, we can determine (if the efficiencies ε are known) the quantities J_0, τ_{emp}, τ_a, and τ_{ie}, from which it is possible to find η_a and η_{ie} if the UCN spectrum is known. Experimental results (Fig. 6.15) reveal that $\eta_{ie}/\eta_a = 0.1$ after the copper vessel has been annealed at 500°C. Without annealing $\eta_{ie}/\eta_a = 0.3$. Thus, in this experiment hydrogen did not manifest itself in any way. The measured η_{ie} is quite adequately described by inelastic scattering of the copper atoms themselves. Moreover, when the temperature of the chamber walls changed, the flux J_{up} varied proportionally to T.

6.5 CONCLUDING REMARKS

A great deal of experimental and theoretical material has been accumulated on the storage of UCN, and to be fair we should discuss all that has been done. The exposition above may have given the impression that no problems exist in UCN storage, but hydrogen contamination exists and appropriate methods of cleaning surfaces are required to achieve the predicted UCN storage times. Moreover, in some cases the predicted limit has almost been achieved. It must be noted, however, that as the results of certain experiments are contradictory, the analysis of experimental data involves quantities that are also uncertain. Thus, for example, in Mampe *et al.* (1981) the spectrum of accumulated neutrons was not known, and in determining η the area adopted for the leaks was quite arbitrary. The temperature dependence of the loss coefficient was insufficiently well investigated.

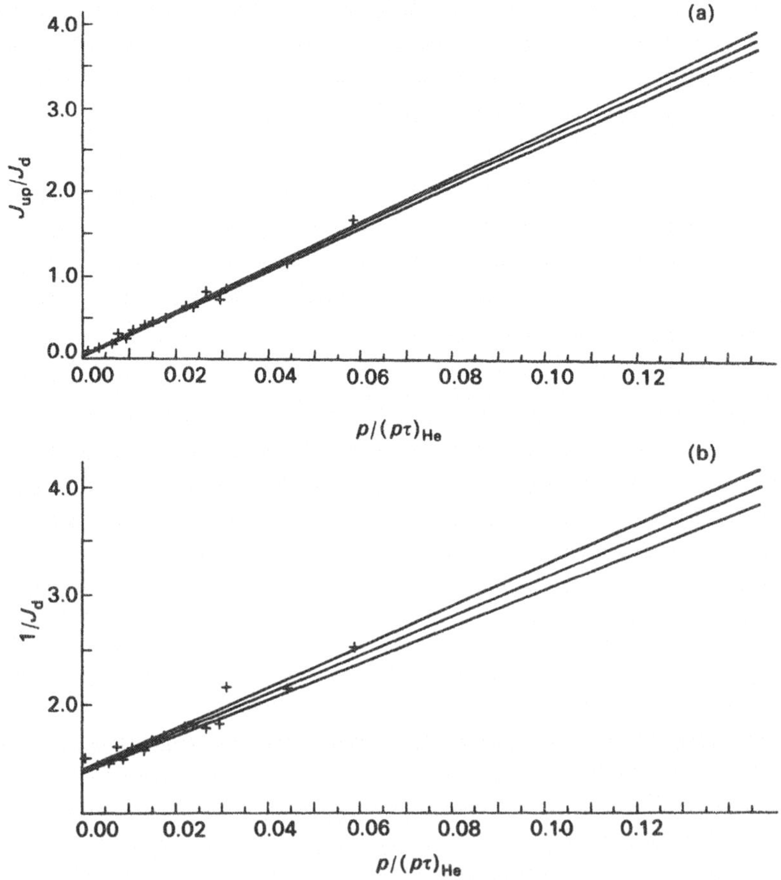

Fig. 6.15 Dependence on $p/(p\tau)_{He}$, in the case of a copper electropolished vessel, of (a) J_{up}/J_d and (b) $1/J_d$ (Ignatovich *et al.* 1982).

At present it seems that the most complete answer to all questions can be obtained with the aid of an experiment such as that by Ignatovich *et al.* (1982) together with direct measurements of storage times, emptying times, spectra, and various detector efficiencies. These experiments must be performed at various wall temperatures, under various cleaning conditions, and with substances exhibiting small cross-sections σ_a. It would be best of all to carry out these experiments at the Leningrad Institute of Nuclear Physics UCN source with the highest available intensity. Thus all questions could be answered and the way paved for applying the method of UCN storage to studies of numerous interesting physical phenomena related to the properties of surfaces.

Experiments involving powders (Golikov *et al.* 1987, 1988*b*) also represent interesting studies that may clarify the nature of the UCN excess interaction with the surface of a substance. These experiments, although differing strongly from storage experiments naturally, do possess certain advantages compared with the latter. Powders exhibit a peculiarity in that the loss coefficient in a single interaction of a neutron with the surface of a substance turns out to be strongly enhanced, owing to multiple scattering in the course of diffusion. The powder then occupies a small volume in space, and therefore it is more suitable for measuring the temperature dependence of the loss coefficient, for example, or for studying the change in the loss coefficient as a result of annealing under various vacuum conditions.

The above remarks were to conclude this chapter on UCN storage in hermetically sealed vessels, but a fairly recent publication (Ageron *et al.* 1985) appeared just in time, since an analysis of this work may arouse readers' interest in the problem. The following conclusions were formulated: in a beryllium trap the excess loss coefficient is due to the presence on the beryllium surface of hydrogen amounting to 10 at.%. We shall now see that these results actually lead to a totally opposite conclusion: the excess coefficient $\Delta\eta$ cannot be explained by the presence of hydrogen, otherwise it must be inferred that the hydrogen cross-section σ_{ie} is extremely small and depends very weakly on the temperature. The point is that in the analysis of the results, which are unquestionable, there is an elementary mistake which led to the erroneous conclusion.

The experiment was as follows. A 1000 Å layer of beryllium was vacuum deposited on the walls of a stainless steel vessel (a pipe 6.7 cm in diameter and 300 cm long). The vessel was separated from the neutron guide by two aluminium windows. The pumping was performed using a turbo-molecular pump through a nitrogen trap and a cryostat, cooled by helium. The storage time was measured at the temperatures 6.5, 77.0, 200.0, and 300.0 K. The results were independent of whether the vessel was pumped out during the UCN storage or was disconnected from the cryopump. The storage times for the temperatures were 250, 225, 170, and 110 s, respectively.

Now we shall make use of the relation $1/\tau = 1/\tau_\beta + 1/\tau_{sl} + 1/\tau_l$. Here $1/\tau_\beta \approx 1.1 \times 10^{-3}\ \text{s}^{-1}$. Ageron *et al.* estimate the width of the leak between the vessel and the shutter that covers it not to exceed 10^{-3} cm and, therefore, $1/\tau_{sl} \approx 0.3 \times 10^{-3}$. As a result

$$1/\tau_l = 1/\tau - 1.4 \times 10^{-3}.$$

At 6.5 K we obtain $1/\tau_l = 2.6 \times 10^{-3}\ \text{s}^{-1}$. To derive η from the value of τ_l Ageron *et al.* applied the expression

$$1/\tau_l = (\bar{v}/2r)\eta \approx 80\eta,$$

where r is the radius of the vessel and $\bar{v}$ is the average UCN velocity, which

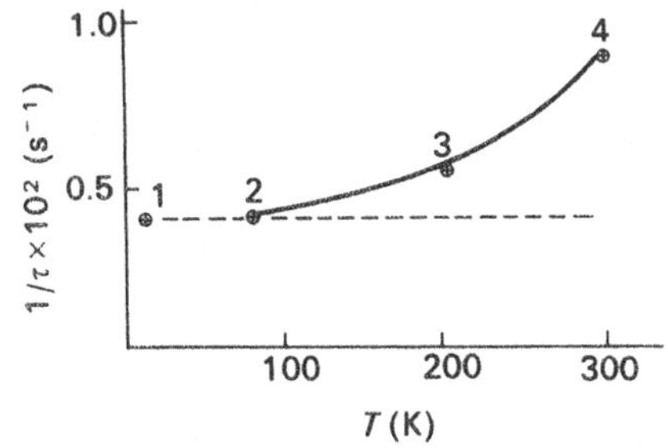

Fig. 6.16 The quantity $1/\tau$ for a stainless steel vessel with beryllium vacuum-deposited on to its walls at various temperatures: 1, 6.5 K; 2, 77 K; 3, 200 K; 4, 300 K (Ageron *et al.* 1985).

was estimated to be 540 cm s^{-1}. From this expression it follows that $\eta = 1/80\tau_l$. At a temperature of 6.5 K all the losses are related only to absorption, and therefore $\eta = \eta_a$. If the atomic concentration c of hydrogen is small (as estimated by Ageron *et al.* (1985), $c = 0.1$), then

$$\eta_a = (c\sigma_{aH} + \sigma_{aBe})/2\lambda b_{Be} = (c\sigma_{aH}/\sigma_{aBe} + 1)\sigma_{aBe}/2\lambda b_{Be}. \tag{6.36}$$

For beryllium we have $b_{Be} = 0.77 \times 10^{-12}$ cm, $\sigma_{aBe} = 0.005$ barn for $\lambda = 1.08$ Å, and therefore the last factor in (6.36) is 3×10^{-7}. To obtain the experimental value of η_a the first factor in (6.36) must be ~ 80. Since the ratio of the absorption cross-sections for hydrogen and beryllium is ~ 40, it follows that $c = 2$ instead of 0.1 as indicated in Ageron *et al.* (1985). The mistake was that the total loss cross-section for beryllium at 300 K, σ_l, was erroneously substituted into the last factor in (6.36) for σ_a, which led to 5×10^{-6} instead of 3×10^{-7}. Thus, there should be two times more hydrogen on the beryllium, or, to be more precise, in a 200 Å surface layer, than atoms of the beryllium itself. But since the hydrogen inelastic cross-section is significantly greater than σ_{ie} for beryllium, then at a temperature of 300 K the quantity η should be 20 times larger and the time τ_l 20 times smaller than the observed values, otherwise σ_{ie} for hydrogen must be considered to be small and to depend weakly on the temperature (Fig. 6.16).

It now seems opportune to end our discussion, since it can be stated definitely that the problem of UCN storage still awaits a satisfactory solution.

APPENDICES

Appendix 6.1 General relations between τ_l and η without gravity

a) $N(t) = \int N(v, t)\mathrm{d}v$, b) $N(v, t) = 4\pi V n(v, t)$. (1)

$\mathrm{d}N(v, t)/\mathrm{d}t = -\int v_\perp n(v, t)\mu(v_\perp)\mathrm{d}\Omega \mathrm{d}S = -N(v, t)v\bar{\mu}(v)S/4V.$ (2)

$$\text{a) } v_\perp = v\cos\vartheta_n, \quad \text{b) } \mu(v_\perp) = 2\eta x/\sqrt{1-x^2}, \quad \text{c) } x = v_\perp/v_{\text{lim}}. \tag{3}$$

$$\text{a) } \bar{\mu}(v) = \int \mu(v_\perp)\cos\vartheta_n \mathrm{d}\Omega_n/\pi = 4\eta \int_0^y x^2\mathrm{d}x/y^2\sqrt{1-x^2}, \quad \text{b) } y = v/v_{\text{lim}}. \tag{4}$$

$$\bar{\mu}(v) = (2\eta/y^2)\,(\arcsin y - y\sqrt{1-y^2}) = \eta\begin{cases}\pi, & y \to 1\\ 4y/3, & y \to 0.\end{cases} \tag{5}$$

$$\text{a) } N(v,t) = N(v,0)\exp[-t/\tau_1(v)], \quad \text{b) } \tau_1(v) = 4V/Sv\bar{\mu}(v). \tag{6}$$

$$N(t) = \int N(v,0)\exp[-t/\tau_1(v)]\mathrm{d}v = N(0)\langle\exp[-t/\tau_1(v)]\rangle. \tag{7}$$

$$\text{a) } N(0) = \int N(v,0)\mathrm{d}v, \quad \text{b) } \langle f(v)\rangle = \int f(v)N(v,0)\mathrm{d}v/N(0). \tag{8}$$

$$\text{a) } \langle\exp[-t/\tau_1(v)]\rangle \simeq \exp(-t/\tau_1), \quad \text{b) } 1/\tau_1 = \langle 1/\tau_1(v)\rangle = \chi/\tau_1(v_{\text{lim}}). \tag{9}$$

$$\text{a) } 1/\tau(v_{\text{lim}}) = \pi\eta v_{\text{lim}}S/4V, \quad \text{b) } \chi = \int N(v,0)v\bar{\mu}(v)\mathrm{d}v/\pi\eta v_{\text{lim}}. \tag{10}$$

$$\text{a) } N(v,0) = N(0)3v^2/v_{\text{lim}}^3, \quad \text{b) } \chi = 12\int_0^1 y\mathrm{d}y\int_0^y x^2\mathrm{d}x/\pi\sqrt{1-x^2}. \tag{11}$$

$$\text{a) } \chi = 6\int_0^1 x^2\sqrt{1-x^2}\mathrm{d}x/\pi = 3\int\sqrt{z(1-z)}\mathrm{d}z/\pi = 3/8, \quad \text{b) } \tau_1 = (8/3)\tau_1(v_{\text{lim}}). \tag{12}$$

In the case of an isotropic distribution the number of neutrons in the trap at time t is represented by the integral (1a), where $N(v, t)$ is the spectral density which is related to the volume spectral density $n(v, t)$ by expression (1b), and V is the volume of the trap. The decrease in the number of UCN in the trap is determined by eqn (2) in which S is the surface area of the walls of the trap, $v_\perp$ is the velocity component normal to the surface of the wall, eqn (3a), and μ is the loss coefficient represented in (3b, c). The mean loss coefficient introduced in (2) is defined in (4). Performing integration in (4) we obtain (5), where the asymptotic behaviour is indicated as $v \to 0$ and $v \to v_{\text{lim}}$. The solution (6a) of eqn (2) contains the spectral storage time (6b). Substituting (6a) into (1a) we obtain (7), where $N(0)$ is the total number of neutrons accumulated in the trap, (8a), and the angle brackets indicate averaging, (8b). We adopt the approximate equality (9a) which holds best when t is small. The quantity τ_1 is defined in (9b). It can be expressed through two parameters: (10a) and (10b). We shall adopt the spectrum of accumulated UCN in the form (11a). Substituting (11a) into (10b) we obtain (11b). By changing the order of integration we obtain (12a). Substituting (12a) into (9b) we find (12b).

Appendix 6.2 Calculation of τ_1 for a vertical cylindrical trap with gravity taken into account

$$\text{a) } n(z,v) = \rho(v^2 + 2gz), \quad \text{b) } \rho(v^2) = n(0,v). \tag{1}$$

$$\text{a) } n(z,v)v_z\mathrm{d}^3v = \rho(v'^2)\theta(v'^2 > 2gz)v'_z\mathrm{d}^3v', \quad \text{b) } v_z'^2 = v_z^2 + 2gz. \tag{2}$$

$$\text{a) } v_z \mathrm{d}v_z = v'_z \mathrm{d}v'_z, \quad \text{b) } v_{x,y} = v'_{x,y}, \quad \text{c) } v_z \mathrm{d}^3 v = v'_z \mathrm{d}^3 v'. \tag{3}$$

$$\text{a) } N(t) = \int n(z, v, t)\mathrm{d}^3 r \mathrm{d}^3 v, \quad \text{b) } N'(t) = -\int v_\perp \mu(v_\perp) n(z, v, t)\mathrm{d}S\mathrm{d}^3 v. \tag{4}$$

$$\text{a) } N'(t) = -\pi \int v^3 \bar{\mu}(v) n(z, v, t)\mathrm{d}S\mathrm{d}v, \quad \text{b) } \bar{\mu}(v) = (2\eta/y^2)\kappa(y). \tag{5}$$

$$\text{a) } y = v/v_{\text{lim}}, \quad \text{b) } \kappa(y) = \arcsin y - y\sqrt{1-y^2} = \begin{cases} \pi/2, & y \to 1 \\ 2y^3/3, & y \to 0 \end{cases} \tag{6}$$

$$\text{a) } \rho(v^2) = c(t)\delta(v^2 - v_{\text{b}}^2), \quad \text{b) } n(z, v, t) = c(t)\delta(v^2 - v_{\text{b}}^2 + 2gz). \tag{7}$$

$$\text{a) } 2gh > v_{\text{b}}^2, \quad \text{b) } N(t) = (2\pi v_{\text{b}}^3 S_{\text{b}}/3g)c(t), \quad \text{c) } c = (3g/2\pi S_{\text{b}} v_{\text{b}}^3)N. \tag{8}$$

$$N'(t) = -\pi\eta v_{\text{lim}}^2 \left(\kappa(y_{\text{b}})S_{\text{b}} + 2\pi r \int_0^\infty \kappa(y_z)\mathrm{d}z \right) c(t)$$

$$= -N(t)[1/\tau_{\text{b}} + 1/\tau_{\text{w}}]. \tag{9}$$

$$\text{a) } 1/\tau_{\text{b,w}} = \chi_{\text{b,w}}/\tau_{\text{lim}}, \quad \text{b) } \tau_{\text{lim}} = v_{\text{lim}}/g\eta, \quad \text{c) } \chi_{\text{b}} = 3\kappa(y_{\text{b}})/2y_{\text{b}}^3. \tag{10}$$

$$\text{a) } y_z^2 = y_{\text{b}}^2 + z/h_{\text{lim}}, \quad \text{b) } h_{\text{lim}} = v_{\text{lim}}^2/2g, \quad \text{c) } \chi_{\text{w}} = 3\pi r h_{\text{lim}} \kappa_1(y_{\text{b}})/S_{\text{b}} y_{\text{b}}^3. \tag{11}$$

$$\kappa_1(y) = \int_0^\infty \kappa(y_z)\mathrm{d}z/h_{\text{lim}} = 2\int_0^y \kappa(x)x\mathrm{d}x = (y^2 - \tfrac{3}{4})\kappa(y) + (y^3/2)\sqrt{1-y^2}. \tag{12}$$

$$\text{a) } 2gh < v_{\text{b}}^2, \quad \text{b) } N(t) = [2\pi S_{\text{b}}(v_{\text{b}}^3 - v_{\text{c}}^3)/3g]c(t), \tag{13}$$

$$\text{a) } N'(t) = -N(t)/\tau_1,$$

$$\text{b) } 1/\tau_1 = 1/\tau_{\text{b}} + 1/\tau_{\text{w}} + 1/\tau_{\text{c}} = (\chi_{\text{b}} + \chi_{\text{w}} + \chi_{\text{c}})/\tau_{\text{lim}}. \tag{14}$$

$$\text{a) } \chi_{\text{b}} = 3\kappa(y_{\text{b}})/2(y_{\text{b}}^3 - y_{\text{c}}^3), \quad \text{b) } \chi_{\text{c}} = 3\kappa(y_{\text{c}})/2(y_{\text{b}}^3 - y_{\text{c}}^3). \tag{15}$$

$$\text{a) } \chi_{\text{w}} = 3\pi r h_{\text{lim}}[\kappa_1(y_{\text{b}}) - \kappa_1(y_{\text{c}})]/S_{\text{b}}(y_{\text{b}}^3 - y_{\text{c}}^3), \quad \text{b) } y_{\text{c}}^2 = y_{\text{b}}^2 - h/h_{\text{lim}}. \tag{16}$$

$$\text{a) } \rho(v^2) = c(t)\theta(v_1^2 < v^2 < v_2^2), \quad \text{b) } n(z, v, t) = c(t)\theta(v_1^2 < v^2 + 2gz < v_2^2). \tag{17}$$

$$\text{a) } \theta(u^2 < v^2 < w^2) = 2\int_u^w \delta(v^2 - x^2)x\mathrm{d}x, \quad \text{b) } v_1^2 < 2gh < v_2^2. \tag{18}$$

$$\text{a) } N(t) = [4\pi(v_2^5 - v_{2\text{h}}^5 - v_1^5)S_{\text{b}}/15g]c(t), \quad \text{b) } v_{2\text{h}}^2 = v_2^2 - 2gh. \tag{19}$$

$$\text{a) } \chi_{\text{b}} = 15[\kappa_1(y_2) - \kappa_1(y_1)]/4(y_2^5 - y_{2\text{h}}^5 - y_1^5), \quad \text{b) } y_q = v_q/v_{\text{lim}}. \tag{20}$$

$$\chi_{\text{w}} = 15\pi r h_{\text{lim}}[\kappa_2(y_2) - \kappa_2(y_{2\text{h}}) - \kappa_2(y_1)]/2S_{\text{b}}(y_2^5 - y_{2\text{h}}^5 - y_1^5). \tag{21}$$

$$\text{a) } \chi_{\text{c}} = 15\kappa_1(y_{2\text{h}})/4(y_2^5 - y_{2\text{h}}^5 - y_1^5), \quad \text{b) } \kappa_2(y) = 2\int_0^y \kappa_1(x)x\mathrm{d}x. \tag{22}$$

$$\kappa_2(y) = (y^2 - 5/12)\kappa_1(y) - y^2[(6y^2 - 5)\kappa(y) + 2y^3\sqrt{1-y^2}]/12. \tag{23}$$

$$\text{a) } \kappa_1(y) = \begin{cases} \pi/8, & y \to 1 \\ 4y^3/15, & y \to 0, \end{cases} \quad \text{b) } \kappa_2(y) = \begin{cases} \pi/32, & y \to 1 \\ 8y^7/105, & y \to 0. \end{cases} \tag{24}$$

Now we shall assume the trap to have perfectly specular non-absorbing walls. We shall prove the dependence of the volume spectral density on the height z to be (1a), i.e. it is determined by the density (1b) at the bottom of the trap, which we shall further consider isotropic. Expression (1a) follows from the condition of flux conservation (2a) and the law of energy conservation (2b). From (2b) follows (3a). Owing to the walls being specular we have (3b). From (3a) and (3b) follows (3c), from which follows (1a). From (1a) it follows that when the distribution at the bottom, (1b), is isotropic, the distribution will also be isotropic everywhere above the bottom. This means that the requirement of perfect specularity in the reflection may be dropped, since no reflection law can violate isotropy. We shall neglect the anisotropy that arises from absorption. The total number of particles and the rate at which it decreases with time are given by expressions (4a) and (4b), respectively. Integration over the angles in (4b) reduces it to the form (5b) where the average loss coefficient (5b) is introduced together with the notation (6).

Now we consider the monoenergetic spectrum (7a). Substituting it into (1a) we obtain (7b). If the height of the trap h is large, (8a), then upon substitution of (7b) into (4a) we obtain (8b) from which follows (8c), where S_b is the area of the bottom of the trap and v_b is the total velocity of the UCN at the bottom of the trap. Substituting (7b) into (4b) we obtain (9) where the storage times are introduced, which are related to losses occurring in collisions with the bottom (τ_b) and with the lateral walls (τ_w). We represent them in the form (10a) where the parameter (10b) is introduced. From comparison with (9) we obtain (10c) with the notations (11a, b) and (11c), involving the function (12). The integral in (12) is readily integrated by parts and by a change of variables, $x = \cos\varphi$. When the height of the vessel, (13a), is small, substitution of (7b) into (4a) leads to (13b). We represent (4b) in the form (14a); then (14b) holds for the total storage time due to losses, where the quantities assigned the subscript c are related to losses occurring in the ceiling of the trap. Comparison with (4b) leads to (15) and (16).

We shall now assume the spectrum at the bottom to be of the form (17a). Then (1a) will take the form (17b). We shall make use of the representation (18a) and assume the height of the trap to satisfy condition (18b). Substituting (17b) into (4) we obtain (19)–(23). The asymptotic behaviour of functions (12) and (23) is indicated in (24). The substitution (18a) is applicable to any spectrum.

Appendix 6.3 The Green function for the Schrödinger equation with a potential step

$$(i\partial_t + \Delta - u_0\theta(z > 0))G(\mathbf{r}, \mathbf{r}', t - t') = \delta(\mathbf{r} - \mathbf{r}')\delta(t - t'). \tag{1}$$

$$G(\mathbf{r}, \mathbf{r}', t) = \int G(k_\perp, z, z')\exp[i\mathbf{k}_\parallel(\mathbf{r} - \mathbf{r}')_\parallel - iEt]\,d^2k_\parallel dE/(2\pi)^3. \tag{2}$$

$$\text{a) } [\Delta_z + k_\perp^2 - u_0\theta(z>0)]G(k_\perp, z, z') = \delta(z-z'), \quad \text{b) } k_\perp^2 = E - k_\parallel^2. \tag{3}$$

$$G(k_\perp, z, z') = [\theta(z>z')\mathcal{H}_+(k_\perp, z)\mathcal{H}_-(k_\perp, z') + \theta(z<z')\mathcal{H}_+(k_\perp, z')\mathcal{H}_-(k_\perp, z)]/w. \tag{4}$$

$$\text{a) } \mathcal{H}_+(k, z) = \theta(z<0)[\exp(ikz) + r_+(k)\exp(-ikz)] + \theta(z>0)t_+(k)\exp(ik'z), \quad \text{b) } k'_\perp = \sqrt{k_\perp^2 - u_0}. \tag{5}$$

$$\mathcal{H}_-(k, z) = \theta(z>0)[\exp(-ik'z) + r_-(k)\exp(ik'z)] + \theta(z<0)t_-(k)\exp(-ikz). \tag{6}$$

$$w = \mathcal{H}'_+(k_\perp, z)\mathcal{H}_-(k_\perp, z) - \mathcal{H}'_-(k_\perp, z)\mathcal{H}_+(k_\perp, z) = 2ik_\perp t_-(k_\perp) = 2ik'_\perp t_+(k_\perp). \tag{7}$$

$$G(k_\perp, z, z')_{z\to\pm\infty} = \exp(\pm ik_{\pm\perp} z)\mathcal{H}_\mp(k_\perp, z')/2ik_{\pm\perp}. \tag{8}$$

$$\text{a) } k_{+\perp} = k'_\perp, \quad \text{b) } k_{-\perp} = k_\perp, \quad \text{c) } r_- = -r_+, \quad \text{d) } t_- = 2k'_\perp/(k'_\perp + k_\perp). \tag{9}$$

$$G(\mathbf{r}, \mathbf{r}', t) = \frac{\int \mathcal{H}_\mp(k_\perp, z')\exp[\pm ik_{\pm\perp}z + i\mathbf{k}_\parallel(\mathbf{r}-\mathbf{r}')_\parallel - iEt]d^2k_\parallel dE}{2ik_{\pm\perp}(2\pi)^3}. \tag{10}$$

$$\text{a) } k_\perp^2 \gg u_0, \quad \text{b) } k_{\pm\perp} \simeq k_\perp, \quad \text{c) } \mathcal{H}_\mp(k_\perp, z') \simeq \exp(i\mathbf{k}_\perp\mathbf{z}), \quad \text{d) } \mathbf{k}_\perp = \pm k_\perp. \tag{11}$$

$$G(\mathbf{r}, \mathbf{r}', t)_{z\to\pm\infty} \simeq \int \exp[i\mathbf{k}(\mathbf{r}-\mathbf{r}') - iEt]d^2k_\parallel dE/2ik_\perp(2\pi)^3. \tag{12}$$

$$\mathbf{k} = (\mathbf{k}_\parallel, \pm k_\perp) = (\mathbf{k}_\parallel, \mathbf{k}_\perp). \tag{13}$$

The solution of eqn (1) is represented in the form (2), where the one-dimensional Green function satisfies eqn (3) and is represented in the form of the combination (4) of independent eigenfunctions (5) and (6) of the one-dimensional Schrödinger equation. The Wronskian of these solutions is given in (7). At a large distance from the reflecting boundary expression (4a) may be represented in the more simple form (8), where the notations (9) are utilized. Substituting (8) into (2) we obtain (10). In case (11a) holds, the approximate equalities (11b–d) also hold, and expression (10) reduces to the form (12), in which the notation (13) is used.

Appendix 6.4 Scattering to the first order of perturbation theory with the total reflection phenomena taken into account

$$[i\partial_t + \Delta - u_0\theta(z>0) - \delta u(\mathbf{r}, t)]\Psi(\mathbf{r}, t) = 0. \tag{1}$$

$$\Psi(\mathbf{r}, t) = \Psi_0(\mathbf{r}, t) + G\cdot\delta u\Psi_0 + \dots. \tag{2}$$

$$G\cdot f = \int G(\mathbf{r}, \mathbf{r}', t-t')f(\mathbf{r}', t')d^3r'dt'. \tag{3}$$

$$\delta\Psi(\mathbf{r}, t) = G\cdot\delta u\Psi_0 = \int\delta\Psi(\mathbf{k})\exp(i\mathbf{k}\mathbf{r} - iEt)d^2k_\parallel dE/2i|\mathbf{k}_\perp|(2\pi)^3. \tag{4}$$

$$\delta\Psi(\mathbf{k}_\pm) = \int \delta u(\mathbf{r}, t)\mathscr{H}_+(k_{0\perp}, z)\mathscr{H}_\mp(k_\perp, z)\exp(\mathrm{i}\kappa_\| \mathbf{r}_\| - \mathrm{i}\omega t)\mathrm{d}^3 r \mathrm{d}t. \tag{5}$$

$$\text{a) } \mathbf{k} = (\mathbf{k}_\|, \mathbf{k}_\perp), \quad \text{b) } \mathbf{k}_{\pm\perp} = \pm k_{\pm\perp}, \quad \text{c) } \mathbf{k}_\pm = (\mathbf{k}_\|, \pm k_{\pm\perp}). \tag{6}$$

$$\text{a) } \kappa_\| = \mathbf{k}_{0\|} - \mathbf{k}_\|, \quad \text{b) } \omega = E_0 - E, \quad \text{c) } E_0 = k_0^2, \quad \text{d) } E = k^2. \tag{7}$$

$$n_\pm = \int [\delta\Psi^*(\mathbf{r}, t)\overleftrightarrow{\nabla}_\perp \delta\Psi(\mathbf{r}, t)]\mathrm{d}^2 r_\| \mathrm{d}t/2\mathrm{i}$$
$$= \int |\delta\Psi(\mathbf{k}_\pm)|^2 \mathrm{d}^2 k_\| \mathrm{d}E/4k_{\pm\perp}(2\pi)^3. \tag{8}$$

$$\text{a) } n_\pm = T_\mathrm{t} J_\pm, \quad \text{b) } J_\pm = \sigma_\pm \varphi_0, \quad \text{c) } \varphi_0 = k_{0\perp}, \quad \text{d) } \sigma_\pm = S_\mathrm{s} W_\pm. \tag{9}$$

$$W_\pm = n_\pm / k_{0\perp} S_\mathrm{s} T_\mathrm{t} = \int |\delta\Psi(\mathbf{k}_\pm)|^2 \mathrm{d}^2 k_\| \mathrm{d}E/4k_{0\perp} k_{\pm\perp}(2\pi)^3 S_\mathrm{s} T_\mathrm{t}. \tag{10}$$

$$\text{a) } \delta u(\mathbf{r}, t) = \theta(z > 0)\delta u(\mathbf{r}, t), \quad \text{b) } \delta\Psi(\mathbf{k}_\pm) = t_+(k_{0\perp})\delta u_\pm(\kappa, \omega). \tag{11}$$

$$\delta u_+(\kappa, \omega) = \delta u(\kappa_-, \omega) + r_-(k_\perp)\delta u(\kappa_+, \omega). \tag{12}$$

$$\text{a) } \delta u_-(\kappa, \omega) = t_+(k_\perp)\delta u(\kappa_-, \omega), \quad \text{b) } \kappa_\pm = (\kappa_\|, \mathrm{i}k_{0\perp}'' \pm k_\perp'). \tag{13}$$

$$\delta u(\kappa, \omega) = \int \delta u(\mathbf{r}, t)\exp(\mathrm{i}\kappa \mathbf{r} - \mathrm{i}\omega t)\mathrm{d}^3 r \mathrm{d}t. \tag{14}$$

$$W_\pm = |t_+(k_{0\perp})|^2 \int |\delta u_\pm(\kappa, \omega)|^2 \mathrm{d}^2 k_\| \mathrm{d}E/4k_{0\perp} k_{\pm\perp}(2\pi)^3 S_\mathrm{s} T_\mathrm{t}. \tag{15}$$

$$\text{a) } |t_+(k_{0\perp})|^2 = 4k_{0\perp}^2/u_0, \quad \text{b) } \mathrm{d}E = 2k_{\pm\perp}\mathrm{d}k_{\pm\perp}, \quad \text{c) } \mathrm{d}^3 k_\pm = \mathrm{d}^2 k_\| \mathrm{d}k_{\pm\perp}. \tag{16}$$

$$W_\pm = 2k_{0\perp}\int |\delta u_\pm(\kappa, \omega)|^2 \mathrm{d}^3 k_\pm / u_0 (2\pi)^3 S_\mathrm{s} T_\mathrm{t}. \tag{17}$$

The solution of eqn (1) to first order in δu is (2), where the notation (3) is utilized. We write the correction to the non-perturbed function in the form (4). Substituting into (2) the Green function from Appendix 6.3 we obtain (5), where the notations (6) and (7) are applied. The total number of particles that are reflected or enter the medium, n_- and n_+, respectively, is determined by expression (8) and also by (9a), where T_t is the total observation time and $J_\pm$ is the particle flux within and outside the medium, determined in (9b). It is written in terms of the initial flux density (9c) and the cross-section (9d), where S_s is the total area of the reflecting surface, while W is the probability of scattering which, as a result of (8) and (9), is represented in the form (10). If δu differs from zero only inside the medium, expression (11a), then expression (5) can be represented in the form (11b), where use is made of the notations (12)–(14). Substituting (11b) into (10) we obtain (15). With the aid of (16) we reduce (15) to the form (17).

Appendix 6.5 Inelastic scattering to the first order of perturbation theory

$$\text{a) } \delta u(\mathbf{r}, t) = u_0 \nabla\xi\theta(z > 0), \quad \text{b) } \xi = \xi_0 \mathbf{e}\exp(\mathrm{i}\mathbf{q}\mathbf{r} - \mathrm{i}\omega_q t), \quad \text{c) } \mathbf{e} = \mathbf{q}/q. \tag{1}$$

$$\text{a) } \omega_q \gg u_0, \quad \text{b) } W = k_{0\perp}\int |\delta u(\kappa, \omega)|^2 \mathrm{d}^2 k_\| \mathrm{d}E/k_\perp u_0 (2\pi)^3 S_\mathrm{s} T_\mathrm{t}. \tag{2}$$

$$\text{a) } \delta u(\kappa, \omega) = \int \delta u(\mathbf{r}, t)\exp(\mathrm{i}\kappa \mathbf{r} - \mathrm{i}\omega t)\mathrm{d}^3 r \mathrm{d}t, \quad \text{b) } \kappa = (\kappa_\|, \mathrm{i}k_{0\perp}'' - k_\perp). \tag{3}$$

$$\text{a) } \boldsymbol{\kappa}_\parallel = \mathbf{k}_{0\parallel} - \mathbf{k}_\parallel, \quad \text{b) } \omega = E_0 - E, \quad \text{c) } k_\perp = \sqrt{E - k_\parallel^2},$$

$$\text{d) } k''_{0\perp} = \sqrt{u_0 - k_{0\perp}^2}. \tag{4}$$

$$\delta u(\boldsymbol{\kappa}, \omega) = (2\pi)^3 u_0 \mathbf{q}\boldsymbol{\xi}\delta(\mathbf{k}_\parallel - \mathbf{q}_\parallel)\delta(E - \omega_q)/[k''_{0\perp} + \mathrm{i}(k_\perp - q_\perp)]. \tag{5}$$

$$|\delta u(\boldsymbol{\kappa}, \omega)|^2 = (2\pi)^3 S_s T_t (u_0 \mathbf{q}\boldsymbol{\xi})^2 \delta(\mathbf{k}_\parallel - \mathbf{q}_\parallel)\delta(E - \omega_q)/[k''^2_{0\perp} + (k_\perp - q_\perp)^2]. \tag{6}$$

$$\text{a) } W = (k_{0\perp}/k_\perp) q^2 (\xi_0^2/l_0^2)/[k''^2_{0\perp} + (k_\perp - q_\perp)^2], \quad \text{b) } l_0^2 = 1/u_0. \tag{7}$$

$$\text{a) } \mathrm{d}P = n(\omega_q/T) g(\omega_q/\theta_\mathrm{D}) \mathrm{d}^3 q/3N_0(2\pi)^3, \quad \text{b) } n(x) = 1/[\exp(x) - 1]. \tag{8}$$

$$\text{a) } g(x) = \exp(-x), \quad \text{b) } \int g(\omega_q/\theta_\mathrm{D}) \mathrm{d}^3 q/(2\pi)^3 = 3N_0, \quad \text{c) } \xi_0^2 = 1/M\omega_q. \tag{9}$$

$$\text{a) } k_\perp^2 \simeq \omega_q - q_\parallel^2, \quad \text{b) } 1/k_\perp[k''^2_{0\perp} + (k_\perp - q_\perp)^2] \simeq (2\pi/k''_{0\perp})\delta(\omega_q - q^2). \tag{10}$$

$$\text{a) } \bar{W} = \int W \mathrm{d}P = 2(k_{0\perp}/k''_{0\perp})\eta, \quad \text{b) } \eta = k_\mathrm{a}\sigma(k_\mathrm{a})/4\pi b. \tag{11}$$

$$\text{a) } \sigma(k_\mathrm{a}) = 2(b^2/3Mk_\mathrm{a}) \int n(\omega_q/T) g(\omega_q/\theta_\mathrm{D}) \mathrm{d}^3 q \delta(\omega_q - q^2), \quad \text{b) } \omega_q = cq. \tag{12}$$

$$\text{a) } \sigma(k_\mathrm{a}) = 2(4\pi b^2/3Mk_\mathrm{a}) n(c^2/T) \exp(c^2/\theta_\mathrm{D}) c/k_\mathrm{a}, \quad \text{b) } c = 2m[c]/\hbar. \tag{13}$$

$$\text{a) } \delta u(\mathbf{r}, t) = u_0 \xi_\perp \delta(z), \quad \text{b) } \delta u(\boldsymbol{\kappa}, \omega) = (2\pi)^3 u_0 \zeta_{0\perp} \delta(\mathbf{k}_\parallel - \mathbf{q}_\parallel)\delta(\omega + \omega_q)$$

$$\text{c) } |\delta u(\boldsymbol{\kappa}, \omega)|^2 = (2\pi)^3 S_s T_t u_0^2 \xi_{0\perp}^2 \delta(\mathbf{k}_\parallel - \mathbf{q}_\parallel)\delta(E - \omega_q). \tag{14}$$

$$\text{a) } W = (k_{0\perp}/k_\perp)(\xi_{0\perp}^2/l_0^2), \quad \text{b) } l_0^2 = 1/u_0, \quad \text{c) } k_\perp = \sqrt{\omega_q - q_\parallel^2}. \tag{15}$$

$$\text{a) } \xi = \zeta_0 \mathbf{e} \exp(\mathrm{i}\mathbf{q}_\parallel \mathbf{r}_\parallel - q''_\perp z - \mathrm{i}\omega_q t), \quad \text{b) } q''_\perp = 1/l_\mathrm{R} = q_\parallel \sqrt{1 - c_\mathrm{R}/c_\mathrm{l}}. \tag{16}$$

$$\text{a) } \omega_q = c_\mathrm{R} q_\parallel, \quad \text{b) } \mathbf{e} = \mathbf{q}/|q|, \quad \text{c) } W \simeq k_{0\perp}\xi_0^2(q_\parallel^2 + q''^2_\perp)/k_\perp l_0^2(q''^2_\perp + k_\perp^2). \tag{17}$$

$$\mathrm{d}P = n(\omega_q/T) g(\omega_q/\theta_\mathrm{D}) \mathrm{d}^2 q_\parallel/2N_0 l_\mathrm{R}(2\pi)^2. \tag{18}$$

$$\text{a) } c_\mathrm{R}^2 \gg \theta_\mathrm{D}, \quad \text{b) } k_\perp = \sqrt{c_\mathrm{R} q_\parallel}, \quad \text{c) } \bar{W} = \int W \mathrm{d}P \simeq f(T/\theta_\mathrm{D}) k_{0\perp} b T^{5/2}/M c_\mathrm{R}^5. \tag{19}$$

$$f(\alpha) = \int x^{3/2} \exp(-\alpha x) \mathrm{d}x/(\exp x - 1) = \Gamma(5/2)\zeta(5/2, T/\theta_\mathrm{D} + 1). \tag{20}$$

Consider the perturbation (1a) in the presence of waves of elastic deformation, (1b), where ζ_0 is the oscillation amplitude and $\mathbf{e}$ is the polarization vector (1c). If the frequency of oscillations is large, expression (2a), then the probability (2b) of inelastic scattering into and out of the medium is approximately the same and is expressed through the Fourier transform (3a) of the perturbation (1a). The momentum and energy transfers are presented in (3b) and (4). When (2a) holds, the quantities $k_{0\parallel}$ and E_0 in (4a) can be neglected. In this case expression (3a) reduces to the form (5), from which follows (6). Substituting (6) into (2b) we obtain (7a), where the notation in (7b) is introduced. If the waves (1b) are phonons, then their number in the element $\mathrm{d}^3 q$ in thermal equilibrium equals (8a), where (8b) characterizes the number of phonons of given energy. We assume the frequency spectrum in the form (9a), and we shall determine the Debye temperature θ_D from condition (9b). The phonon amplitude is determined from (9c). Owing to (4c) and (5) the approximation (10a) holds, while owing to $k''_{0\perp}$ being small the approximation (10b) holds. We substitute (10b) into (7a) and integrate with (8a). We obtain (11a, b), where k_a is an arbitrary wavevector and the cross-section $\sigma(k_a)$ is determined by expression (12a). We assume (12b) to be a dispersion relation; then integration in (12a) leads to (13a), where use is made of the notation (13b) in which $[c]$ is the velocity of a phonon, exhibiting the conventional dimensionality of velocity.

Now we shall consider scattering on surface waves with a very small depth of penetration, for which the perturbation can be represented in the form (14a). Substitution of (14a) into (3a) yields (14b) from which follows (14c). Substituting (14c) into (2b) we obtain (15a), where use of the notations (15b) and (15c) is made.

To estimate the contribution of Rayleigh waves to the inelastic scattering of UCN, they can be represented by the expression (16a). As the damping depth within the medium, l_R, we shall assume the maximum value determined from (16b), where c_R is the velocity of the Rayleigh waves following from the dispersion relation (17a) and c_l is the velocity of the longitudinal sound. As the polarization vector we shall assume the quantity (17b); then, substituting (16a) into (1a), (3a), and (2b), we obtain (17c). The number of Rayleigh waves in the element $d^2q_\parallel$ is (18). The relation (19a) is characteristic of Rayleigh waves and therefore the approximation (19b) can be utilized. Integrating (17c) with (18) and neglecting the difference between c_R and c_l we obtain (19c), where the function (20) is introduced and expressed through the Riemann ζ function.

Appendix 6.6 The heating of UCN at reflection from the wall

$$\delta u(\mathbf{r}, t) = 4\pi \sum_l b_l \delta(\mathbf{r} - \mathbf{r}_l - \boldsymbol{\xi}(\mathbf{r}_l, t)). \tag{1}$$

$$\delta u(\boldsymbol{\kappa}, \omega) = 4\pi \sum_l b_l \chi(\mathbf{r}_l, \omega) \exp(i\boldsymbol{\kappa}\mathbf{r}_l). \tag{2}$$

$$\text{a)}\ \chi(\mathbf{r}_l, \omega) = \int \chi(\mathbf{r}_l, t) \exp(-i\omega t) dt, \quad \text{b)}\ \chi(\mathbf{r}_l, t) = \exp[i\boldsymbol{\kappa}\boldsymbol{\xi}(\mathbf{r}_l, t)]. \tag{3}$$

$$W = 2k_{0\perp} \int \langle |\delta u(\boldsymbol{\kappa}, \omega)|^2 \rangle d^3k / u_0 (2\pi)^3 S_s T_t. \tag{4}$$

$$\langle |\delta u(\boldsymbol{\kappa}, \omega)|^2 \rangle = \delta_{coh} + \delta_{inc}. \tag{5}$$

$$\text{a)}\ \delta_{coh} = (4\pi b)^2 T_t \sum_{l'l} I(\mathbf{r}_{l'l}, \omega) \exp(i\mathbf{k}\mathbf{r}_l - i\mathbf{k}^*\mathbf{r}_l), \quad \text{b)}\ \mathbf{r}_{lm} = \mathbf{r}_l - \mathbf{r}_m. \tag{6}$$

$$\delta_{inc} = (4\pi)^2 b_{inc}^2 T_t \sum_l S_{inc}(\omega) \exp(-2k''_{0\perp} z_l). \tag{7}$$

$$\text{a) } I(\mathbf{r}, \omega) = \int G_{\text{coh}}(\mathbf{r}, t)\exp(\mathrm{i}\omega t)\mathrm{d}t, \quad \text{b) } S_{\text{inc}}(\omega) = I(0, \omega). \tag{8}$$

$$G_{\text{coh}}(\mathbf{r}_l, t) = G(\mathbf{r}_l, t) = \langle \chi^*(\mathbf{r}_l, t)\chi(0, 0)\rangle. \tag{9}$$

$$\delta_{\text{inc}} = (4\pi)^2 b_{\text{inc}}^2 N_0 T_t S_s S_{\text{inc}}(\omega)/2k''_{0\perp}. \tag{10}$$

$$\sum_{l,l'} f(z_l - z_{l'})\theta(z_l, z_{l'} > 0)\exp(-\alpha z_l - \alpha z_{l'}) = \sum_l f(z_l)\exp(-\alpha|z_l|)$$

$$\times \sum_{l'} \theta(z_{l'} > 0)\exp(-2\alpha z_{l'}). \tag{11}$$

$$\delta_{\text{coh}} = (4\pi b N_0)^2 T_t S_s \sum_\tau S_{\text{coh}}(\boldsymbol{\kappa} + \boldsymbol{\tau}, \omega)/2k''_{0\perp}. \tag{12}$$

$$S_{\text{coh}}(\boldsymbol{\kappa}, \omega) = \int I(\mathbf{r}, \omega)\exp[-\mathrm{i}\mathrm{Re}(\boldsymbol{\kappa})\mathbf{r} - \mathrm{Im}(\kappa_\perp)z]\mathrm{d}^3 r. \tag{13}$$

$$W_{\text{inc}} = 2(k_{0\perp}/k''_{0\perp})b_{\text{inc}}^2 \int S_{\text{inc}}(\omega)\mathrm{d}^3 k/b(2\pi)^2 = 2(k_{0\perp}/k''_{0\perp})\eta_{\text{inc}}. \tag{14}$$

$$\text{a) } \eta_{\text{inc}} = k_a \sigma_{\text{inc}}(k_a)/4\pi b, \quad \text{b) } \sigma_{\text{inc}}(k_a) = b_{\text{inc}}^2 \int S_{\text{inc}}(\omega)\mathrm{d}^3 k/\pi k_a. \tag{15}$$

$$W_{\text{coh}} = 2(k_{0\perp}/k''_{0\perp})bN_0 \sum_\tau S_{\text{coh}}(\boldsymbol{\kappa} + \boldsymbol{\tau}, \omega)\mathrm{d}^3 k/(2\pi)^2 = 2(k_{0\perp}/k''_{0\perp})\eta_{\text{coh}}. \tag{16}$$

$$\text{b) } \eta_{\text{coh}} = k_a \sigma_{\text{coh}}(k_a)/4\pi b, \quad \text{b) } \sigma_{\text{coh}} = b^2 N_0 \sum_\tau S_{\text{coh}}(\boldsymbol{\kappa} + \boldsymbol{\tau}, \omega)\mathrm{d}^3 k/\pi k_a. \tag{17}$$

$$\xi(\mathbf{r}, t) = \sum_{qj} \xi_{qj}\mathbf{e}_{qj}[\hat{a}_{qj}\exp(\mathrm{i}\mathbf{q}\mathbf{r} - \mathrm{i}\omega_{qj}t) + \hat{a}_{qj}^+\exp(-\mathrm{i}\mathbf{q}\mathbf{r} + \mathrm{i}\omega_{qj}t)]. \tag{18}$$

$$\text{a) } \xi_{qj} = 1/\sqrt{M\omega_{qj}}, \quad \text{b) } \chi(\mathbf{r}, t) = \prod_{qj} \exp(\alpha_{qj}\hat{\alpha}_{qj} + \beta_{qj}\hat{\alpha}_{qj}^+). \tag{19}$$

$$\alpha_{qj} = \mathrm{i}\xi_{qj}(\boldsymbol{\kappa}\mathbf{e}_{qj})\exp(\mathrm{i}\mathbf{q}\mathbf{r} - \mathrm{i}\omega_{qj}t), \quad \beta_{qj} = \mathrm{i}\xi_{qj}(\boldsymbol{\kappa}\mathbf{e}_{qj})\exp(-\mathrm{i}\mathbf{q}\mathbf{r} + \mathrm{i}\omega_{qj}t). \tag{20}$$

$$\text{a) } \sum_{qj} f_j(\mathbf{q}) = \sum_j \int f_j(\mathbf{q})\mathrm{d}^3 q/3N_0(2\pi)^3, \quad \text{b) } \sum_{qj} f_j(\omega) = \sum_j \int f_j(\omega)g(\omega)\mathrm{d}\omega/3. \tag{21}$$

$$G(\mathbf{r}, t) = \exp[-\mathrm{Re}(\boldsymbol{\kappa}^2)B(T) + |\boldsymbol{\kappa}|^2 C(\mathbf{r}, t, T)]. \tag{22}$$

$$\text{a) } B(T) = \int g(\omega)\mathrm{cth}(\omega/2T)\mathrm{d}\omega/3M\omega, \quad \text{b) } \mathrm{Re}(\boldsymbol{\kappa}^2) = (\mathrm{Re}\,\boldsymbol{\kappa})^2 - (\mathrm{Im}\,\boldsymbol{\kappa})^2. \tag{23}$$

$$C(\mathbf{r}, t, T) = \int \mathrm{ch}(\omega_q/2T + \mathrm{i}\mathbf{q}\mathbf{r} - \mathrm{i}\omega_q t)\mathrm{d}^3 q/3N_0 M\omega_q \mathrm{sh}(\omega_q/2T)(2\pi)^3. \tag{24}$$

$$C(0, t, T) = \int g(\omega)\mathrm{ch}(\omega/2T - \mathrm{i}\omega t)\mathrm{d}\omega/3N_0 \mathrm{sh}(\omega/2T)M\omega. \tag{25}$$

$$\text{a) } G(\mathbf{r}, t) \simeq |\boldsymbol{\kappa}|^2 C(\mathbf{r}, t, T)\exp[-\mathrm{Re}(\boldsymbol{\kappa}^2)B(T)], \quad \text{b) } \boldsymbol{\kappa}^2 \simeq \mathbf{k}^2 = E. \tag{26}$$

$$\sigma_{\text{coh}}^{\text{lp}}(k_a) = (\sigma_{\text{coh}}^{\text{el}}/3Mk_a)\sum_\tau \int g(\omega_q/\theta_D)n(\omega_q/T)\delta((\mathbf{q} + \boldsymbol{\tau})^2 - \omega_q)\mathrm{d}^3 q/2\pi. \tag{27}$$

$$\sigma_{\text{inc}}^{\text{lp}}(k_a) = (\sigma_{\text{inc}}^{\text{el}}/3Mk_a)\int g(E/\theta_D)n(E/T)\sqrt{E}\mathrm{d}E. \tag{28}$$

$$\text{a) } \sigma_{\text{coh}}^{\text{el}} = 4\pi b^2, \quad \text{b) } \sigma_{\text{inc}}^{\text{el}} = 4\pi b_{\text{inc}}^2. \tag{29}$$

The Fourier transform of the perturbation (1) is (2), where the notation (3) is used. The probability of scattering on high-frequency oscillations reduces to the form (4). We introduce the notation (5). Isotopic and spin averaging leads to the expressions (6) and (7), in which the notations (8) and (9) are used. Summation in (7) leads to (10), while summation in (6), performed according to (11), leads to (12)–(13). Substituting (10) into (4) we obtain (14), where the notation (15) is introduced and k_a represents an arbitrary small momentum. Substituting (12) into (4) we obtain (16), where the notations (17) are introduced. The displacement of an atom at point **r** is represented in the form of the phonon expansion (18). The phonon amplitude is indicated in (19a). Substituting (18) into (3b) we obtain (19b) with the notations indicated in (20). We substitute (19b) into (9) and perform thermodynamic averaging as indicated in Appendix 6.7. Since phonons with differing frequencies ω, different wavevectors q, and belonging to differing modes j are independent of each other, the averaging can be performed for them separately. As a result (9) transforms into a product of exponentials. Having then performed summation of the exponents of the exponentials in accordance with the rules (21a) and (21b) we obtain (22), where the notations (23) and (24) are applied. Expression (22) is utilized for calculating the coherent scattering cross-section. For the incoherent cross-section it is necessary in (22) to substitute (25) for (24). The approximation (26) of expression (22) can be utilized to compute one-phonon cross-sections. Substituting (26) into (8), (17b), and (15) we obtain the one-phonon scattering cross-sections (17) and (28), where the notation (29) is utilized.

Appendix 6.7 Computation of thermodynamic averages

$$\text{a) } \hat{F}(x) = \exp[x(\alpha\hat{a} + \beta\hat{a}^+)], \quad \text{b) } [\hat{a}, \hat{a}^+]_- = 1, \quad \text{c) } [\partial/\partial p, p]_- = 1. \tag{1}$$

$$\hat{F}(x) = \hat{F}_N(x) = \exp_N[f_N(x) + x(\alpha\hat{a} + \beta\hat{a}^+)] = \exp_A[f_A(x) + x(\alpha\hat{a} + \beta\hat{a}^+)]. \tag{2}$$

$$\exp_N(\alpha\hat{a} + \beta\hat{a}^+) \equiv \hat{F}_{1N} = \exp(\beta\hat{a}^+)\exp(\alpha\hat{a}). \tag{3}$$

$$\exp_A(\alpha\hat{a} + \beta\hat{a}^+) \equiv \hat{F}_{1A} = \exp(\alpha\hat{a})\exp(\beta\hat{a}^+). \tag{4}$$

$$d\hat{F}(x)/dx = (\alpha\hat{a} + \beta\hat{a}^+)F(x) = (\alpha\hat{a} + \beta\hat{a}^+)\hat{F}_N(x) = [(\alpha a + \beta a^+ + \alpha\partial/\partial a^+)F_N\hat{]}_N. \tag{5}$$

$$\text{a) } d\hat{F}(x)/dx = d\hat{F}_N(x)/dx = \{[f'_N(x) + \alpha\hat{a} + \beta\hat{a}^+]\hat{F}_N(x)\}_N, \quad \text{b) } \hat{F}(0) = \hat{F}_N(0) = 1. \tag{6}$$

$$\text{a) } \partial F_N/\partial a^+ = x\beta F_N, \quad \text{b) } f'_N(x) = x\alpha\beta, \quad \text{c) } f_N(0) = 0, \quad \text{d) } f_N = x^2\alpha\beta/2. \tag{7}$$

$$\hat{F}_1(\alpha, \beta) = \hat{F}(1) = \exp(\alpha\hat{a} + \beta\hat{a}^+) = \exp(\alpha\beta/2)\exp(\beta\hat{a}^+)\exp(\alpha\hat{a}). \tag{8}$$

$$\hat{F}_{1A} = \{\exp[\alpha(a + \partial/\partial a^+)]\exp(\beta a^+)\}^{\wedge}_{N} = \exp(\alpha\beta)\exp(\beta\hat{a}^+)\exp(\alpha\hat{a}). \quad (9)$$

$$\text{a) } \langle \hat{F}_{1N}\rangle = (1 - z)\sum_n \langle n|\hat{F}_{1N}|\rangle z^n, \quad \text{b) } z = \exp(-\omega/T). \quad (10)$$

$$\text{a) } |n\rangle = (\hat{a}^+)^n|0\rangle/\sqrt{n!}, \quad \text{b) } \exp(\alpha\hat{a})|n\rangle = (\hat{a}^+ + \alpha)^n|0\rangle/\sqrt{n!}. \quad (11)$$

$$\text{a) } \hat{a}|0\rangle = 0, \quad \text{b) } \langle n|\hat{F}_{1N}|n\rangle = \langle 0|(\hat{a} + \beta)^n(\hat{a}^+ + \alpha)^n|0\rangle/n!. \quad (12)$$

$$\text{a) } \langle \hat{F}_{1N}\rangle = (1 - z)\langle 0|\hat{Q}_A(z)|0\rangle,$$

$$\text{b) } \hat{Q}_A(z) = \exp_A[z(\hat{a} + \beta)(\hat{a}^+ + \alpha)]. \quad (13)$$

$$\hat{Q}_A(z) = \hat{Q}_N(z) = \exp_N[\varphi(z) + \psi(z)(\hat{a} + \beta)(\hat{a}^+ + \alpha)]. \quad (14)$$

$$\partial\hat{Q}_A(z)/\partial z = (\hat{a} + \beta)\hat{Q}_A(z)(\hat{a}^+ + \alpha)$$
$$= [(a + \beta + \partial/\partial a^+)(a^+ + \alpha + \partial/\partial a)Q_N(z)]^{\wedge}_{N}. \quad (15)$$

$$\partial\hat{Q}_A(z)/\partial z = [\![\{1 + \psi(z) + [1 + \psi(z)]^2(a^+ + \alpha)(a + \beta)\}Q_N(z)]\!]^{\wedge}_{N}. \quad (16)$$

$$\partial\hat{Q}_N(z)/\partial z = \{[\varphi'(z) + \psi'(z)(a^+ + \alpha)(a + \beta)]Q_N(z)\}^{\wedge}_{N}. \quad (17)$$

$$\text{a) } \varphi'(z) = 1 + \psi(z), \quad \text{b) } \psi'(z) = [1 + \psi(z)]^2, \quad \text{c) } Q_A(0) = 1. \quad (18)$$

$$\text{a) } \varphi(0) = \psi(0) = 0, \quad \text{b) } \psi(z) = z/(1 - z), \quad \text{c) } \varphi(z) = -\ln(1 - z). \quad (19)$$

$$\text{a) } \langle 0|\hat{Q}_N(z)|0\rangle = \exp[\varphi(z) + \alpha\beta\psi(z)], \quad \text{b) } \langle \hat{F}_{1N}\rangle = \exp[\alpha\beta z/(1 - z)]. \quad (20)$$

$$\langle \hat{F}_1(\alpha, \beta)\rangle = \exp[\alpha\beta/2 + \alpha\beta z/(1 - z)] = \exp[\alpha\beta(1 + z)/2(1 - z)]. \quad (21)$$

$$\langle F(\alpha, \beta)F(\gamma, \delta)\rangle = \exp[(\alpha\beta + \gamma\delta)/2 + \alpha\delta + (\alpha + \gamma)(\beta + \delta)z/(1 - z)]$$
$$= \exp[(\alpha\beta + \gamma\delta)(1 - z)/2(1 + z) + (\alpha\delta + z\gamma\beta)/(1 - z)]. \quad (22)$$

Consider the operator (1a) with the commutation relations (1b). We note that the property (1b) is also exhibited by the commutator (1c) of the numerical quantity p. By permutation of $\hat{a}$ and $\hat{a}^+$ operator (1a) may be written in various ways and, for example, be reduced to the normal and antinormal forms. We shall assume the normal and antinormal forms to be given by expression (2a) with the unknown functions $f_{N,A}(x)$ and notations (3) and (4). Let us find function $f_N(x)$ in (2). To this end we differentiate (1a) and (2) with respect to x. We obtain (5) and (6a), respectively. Now we reduce both results to the normal form. To do this it is convenient in (5) to pass from operators to numerical functions, to substitute (1c) for (1b), and to perform all operations as indicated in (7a). Then, having arranged a and a^+ in the normal order, we transform them once again into operators, which is indicated by the circumflex over the last square bracket. Comparison of (5) and (6) with each other leads to eqn (7b). Owing to the evident equality (6b) we have the boundary condition (7c). The solution of (7b, c) is (7d). Substituting (7d) into

(2) we obtain (8) for $x = 1$. Making use of the substitution of (1c) for (1b) we can readily obtain the normal form (9) of operator (4). The thermodynamic average of operator (3) is represented in the form (10a), where the notation (10b) and (11a) is utilized. We shall now take advantage of the obvious relation (11b) following from (1b) and (12a), which leads to (12b). Substituting (12b) into (10a) we obtain (13a) with the notation (13b). Now we shall reduce (13b) to the normal form. We assume it to be given by expression (14) with the unknown functions φ and ψ. Proceeding as before we calculate the derivative (15) of operator (13b), then pass to numerical quantity a, substitute (1c) for the commutator (1b), perform the corresponding operations (16), and obtain the normal form of the derivative. Then we differentiate (14) and obtain (17). Compare (16) and (17). We obtain eqns (18a, b) which, because of (18c), are supplemented by the boundary conditions (19a). The solution of these equations is given in (19b, c). As a result of (12a) relation (20a) holds for (14). Substituting (19b, c) and (20a) into (13a) we obtain (20b). From (3), (8), and (20b) follow (21) and (22).

Appendix 6.8 The Debye–Waller factor in the case of total reflection

$$\text{a) } u(\mathbf{r}, t) = 4\pi b \sum_l \delta(\mathbf{r} - \mathbf{r}_l - \boldsymbol{\xi}(\mathbf{r}_l, t)), \quad \text{b) } \langle u \rangle = u_0 \exp[4k''^2_{0\perp} B(T)]. \tag{1}$$

$$\text{a) } \langle \exp(\mathrm{i}\mathbf{q}\boldsymbol{\xi}) \rangle = \exp[-q^2 B(T)/2], \quad \text{b) } q^2 = -4k''^2_{0\perp}. \tag{2}$$

$$\langle u \rangle = (b/2\pi^2) \int \mathrm{d}^3 p \sum \exp[\mathrm{i}\mathbf{p}(\mathbf{r} - \mathbf{r}_l)] \langle \exp[\mathrm{i}\mathbf{p}\boldsymbol{\xi}(\mathbf{r}_l, t)] \rangle = 4\pi b \sum f(\mathbf{r} - \mathbf{r}_l). \tag{3}$$

$$\text{a) } f(\mathbf{r}) = [2\pi B(T)]^{-3/2} \exp[-r^2/2B(T)], \quad \text{b) } \sum f(l) = \int f(\mathbf{r}) N_0 \mathrm{d}^3 r. \tag{4}$$

$$\text{a) } \langle u \rangle = u_0 \Phi(z/z_{\mathrm{T}}), \quad \text{b) } z_{\mathrm{T}} = \sqrt{2B(T)},$$

$$\text{c) } \Phi(x) = \int_{-\infty}^{x} \exp(-y^2) \mathrm{d}y / \sqrt{\pi}. \tag{5}$$

$$\text{a) } \delta u = \langle u \rangle - u_0 \theta(z > z_{\mathrm{T}}) \simeq u_0 \theta(-z_{\mathrm{T}} < z < z_{\mathrm{T}})(z/z_{\mathrm{T}} + 1)/2,$$

$$\text{b) } \delta \Psi = G_0 \circ \delta u \Psi_0. \tag{6}$$

$$\text{a) } G_0(\mathbf{r}) = -\exp(\mathrm{i}kr)/4\pi r, \quad \text{b)} (\Delta + k^2) G_0(\mathbf{r}) = \delta(\mathbf{r}). \tag{7}$$

$$\text{a) } G_0(\mathbf{r}) = \int \exp(\mathrm{i}\mathbf{p}_\parallel \mathbf{r}_\parallel + \mathrm{i}p_\perp |z|) \mathrm{d}^2 p / (2\pi)^2 2\mathrm{i}p_\perp, \quad \text{b) } p_\perp = \sqrt{k^2 - p_\parallel^2}. \tag{8}$$

$$\text{a) } \delta \Psi = r_1 \Psi_0, \text{ b) } r_1 = u_0 z_{\mathrm{T}} / 2\mathrm{i}k_{0\perp}, \text{ c) } R = r_1 + t_1^2 r_+ / (1 - r_1 r_+). \tag{9}$$

$$\text{a) } t_1 = 1 + r_1, \text{ b) } R = r_+ + t_+^2 r_1, \text{ c) } t_+ = 1 + r_+, \text{ d) } t_+^2 = 4k_{0\perp}^2 r_+ / u_0. \tag{10}$$

$$\text{a) } R \simeq r_+(1 - 2\mathrm{i}k_{0\perp} z_{\mathrm{T}}) = \exp[-\mathrm{i}(\varphi + \delta\varphi)], \quad \text{b) } \varphi = \arccos(k_{0\perp}/\sqrt{u_0}). \tag{11}$$

$$\text{a) } \varphi + \delta\varphi = \arccos(k_{0\perp}/\sqrt{u}), \quad \text{b) } u = u_0[1 + 4k''_{0\perp} \sqrt{2B(T)}]. \tag{12}$$

Consider the total potential (1a). Its being averaged over thermal oscillations, yielding ξ, may be expected to lead to expression (1b), since the square momentum transfer within a medium, occurring in the Debye–Waller factor (2a), is equal to (2b). Note that thermal oscillations apparently enhance the interaction potential in total reflection. We shall show that this is indeed so. Writing the δ function in (1a) in the form of a Fourier transform and taking advantage of the results of Appendices 6.6 and 6.7, from which follows (2a), we obtain the averaged potential in the form (3), where the notation (4a) is adopted. In the case of total reflection discreteness is not essential, and therefore the sum in (3) can be transformed into an integral, as shown in (4b), resulting in (3) being transformed into the form (5a), where the notation (5b, c) is adopted. Thus, we see that the resultant potential differs from a rectangular step not in height, as indicated in (1b), but by a smeared edge, the smearing depth of which is determined by the uncertainty in the position of the atom relative to the equilibrium position, i.e. by the oscillation amplitude (5b) in this case. We shall show that the reflection phase from the potential (5a) is the same as from a potential with a sharp edge but larger height. For simplicity we replace the smeared edge by the linear expression (6a). Consider scattering on the layer (6a), assuming it to be narrow and neglecting the remaining step. This scattering is described by the scattered wave (6b), where Ψ_0 is the incident plane wave and G_0 is the Green function (7a) satisfying equation (7b). We represent (7a) in the form (8), and (6b) in the form (9a), where the reflection amplitude from the layer r_1 is introduced. Substituting (6a) and (8a) into (6b) we obtain (9b). Now we shall take into account the remaining step. By the method of recursive relations developed in Chapter 5 we obtain the combined reflection amplitude from the layer and the step in the form (9c). Expanding (9c) and taking into account (10a), we obtain to the first order of r_1 the expression (10b), where the notation (10c) is adopted. Since equality (10d) holds, (10b) reduces to the form (11a) with the non-perturbed phase (11b) and the perturbation represented in the form of a phase correction. Assuming equality (12a) to be valid, we find (12b), which confirms qualitatively the equivalence of a potential with a smeared edge and a potential with a sharp edge but larger height. The obtained linear increase is only due to the utilized approximation, (6a). A more complete account of the smearing (5a) leads to (1b).

Appendix 6.9 Stationary perturbations. Reflection of UCN from a rough surface

$$\delta\Psi(\mathbf{r}, t) = \exp(-\mathrm{i}Et)\delta\Psi(\mathbf{r}). \tag{1}$$

$$\delta\Psi_{\pm}(\mathbf{r}) = \int \delta\psi_{\pm}(\mathbf{k})\exp(\mathrm{i}\mathbf{k}_{\pm}\mathbf{r})\mathrm{d}^2 k_{\parallel}/2\mathrm{i}k_{\pm\perp}(2\pi)^2. \tag{2}$$

$$\delta\psi_{\pm}(\mathbf{k}) = \int \delta u(\mathbf{r})\mathscr{H}_{+}(k_{0\perp}, z)\mathscr{H}_{\mp}(k_{\perp}, z)\exp(\mathrm{i}\kappa_{\parallel}\mathbf{r}_{\parallel})\mathrm{d}^3 r. \tag{3}$$

$$\text{a) } k_\perp^2 \gg u_0, \quad \text{b) } \delta\psi_\pm(\mathbf{k}) \simeq t_+(k_{0\perp})\delta u(\boldsymbol{\kappa}_\pm), \quad \text{c) } \boldsymbol{\kappa}_\pm = (\boldsymbol{\kappa}_\parallel, ik''_{0\perp} \mp k_\perp). \tag{4}$$

$$\text{a) } \delta u(\boldsymbol{\kappa}_\pm) = \int \exp(i\boldsymbol{\kappa}_\pm \mathbf{r})\delta u(\mathbf{r})d^3r, \quad \text{b) } \boldsymbol{\kappa}_\parallel = \mathbf{k}_{0\parallel} - \mathbf{k}_\parallel. \tag{5}$$

$$J_\pm = \int k_\perp |\delta\psi_\pm(\mathbf{k})|^2 d^2k_\parallel/4k_\perp^2(2\pi)^2 = k_{0\perp}^2 \int |\delta u(\boldsymbol{\kappa}_\pm)|^2 d^2k_\parallel/u_0 k_\perp(2\pi)^2. \tag{6}$$

$$\text{a) } \sigma_\pm = J_\pm/k_{0\perp}, \quad \text{b) } d\sigma_\pm(k) = (k_{0\perp}/k_\perp)|\delta u(\boldsymbol{\kappa}_\pm)|^2 d^2k_\parallel/u_0(2\pi)^2. \tag{7}$$

$$\text{a) } k_\perp^2 < u_0, \quad \text{b) } \delta\psi_-(\mathbf{k}) = t_+(k_{0\perp})t_+(k_\perp)\delta u(\boldsymbol{\kappa}), \quad \text{c) } \boldsymbol{\kappa} = (\boldsymbol{\kappa}_\parallel, ik''_{0\perp} + ik''_\perp). \tag{8}$$

$$\text{a) } dJ_-(\mathbf{k}) = 4k_{0\perp}^2 k_\perp |\delta u(\boldsymbol{\kappa})|^2 d^2k_\parallel/u_0^2(2\pi)^2, \quad \text{b) } d^2\mathbf{k}_\parallel = kk_\perp d\Omega. \tag{9}$$

$$\text{a) } d\sigma_-(\boldsymbol{\Omega}) = 4kk_{0\perp}k_\perp^2|\delta u(\boldsymbol{\kappa})|^2 d\Omega/u_0^2(2\pi)^2, \quad \text{b) } dW_-(\boldsymbol{\Omega}) = d\sigma_-(\boldsymbol{\Omega})/S_s. \tag{10}$$

$$\text{a) } \delta u = 4\pi b_N \delta(\mathbf{r} - \mathbf{r}_N), \quad \text{b) } \delta u(\boldsymbol{\kappa}) = 4\pi b_N \exp(i\boldsymbol{\kappa}\mathbf{r}_N). \tag{11}$$

$$\text{a) } k_\perp^2 < u_0, \quad \text{b) } |\delta u(\boldsymbol{\kappa})|^2 = (4\pi b_N)^2 \exp[-2(k''_{0\perp} + k''_\perp)z_N]. \tag{12}$$

$$d\sigma_-(\boldsymbol{\Omega})/d\Omega = 16b_N^2(kk_{0\perp}k_\perp^2/u_0^2)\exp[-2(k''_{0\perp} + k''_\perp)z_N]. \tag{13}$$

$$\text{a) } k_\perp^2 > u_0, \quad \text{b) } d\sigma(\boldsymbol{\Omega})/d\Omega = 4b_N^2(kk_{0\perp}/u_0)\exp[-2k''_{0\perp}z_N]. \tag{14}$$

$$\text{a) } \delta u(\mathbf{r}) = u_0\zeta(\mathbf{r}_\parallel)\delta(z), \quad \text{b) } \langle\zeta\rangle = 0, \quad \text{c) } \langle\zeta(\mathbf{r})\zeta(\mathbf{r}')\rangle = \zeta_0^2 K(\mathbf{r}_\parallel - \mathbf{r}'_\parallel). \tag{15}$$

$$\text{a) } \langle|\delta u(\boldsymbol{\kappa})|^2\rangle = u_0\zeta_0^2 S_s K(\boldsymbol{\kappa}_\parallel), \quad \text{b) } K(\mathbf{r}_\parallel) = \exp(-r_\parallel^2/2l^2). \tag{16}$$

$$\text{a) } K(\boldsymbol{\kappa}) = \int K(\mathbf{r})\exp(i\boldsymbol{\kappa}\mathbf{r})d^2r = 2\pi l^2 \exp(-\kappa^2 l^2/2), \quad \text{b) } k_\perp^2 < u_0. \tag{17}$$

$$W_{\text{ref}}(\boldsymbol{\Omega}_0 \to \boldsymbol{\Omega}) = dW_-(\boldsymbol{\Omega})/d\Omega = (2/\pi)kk_{0\perp}k_\perp^2\zeta_0^2 l^2 \exp(-\kappa_\parallel^2 l^2/2). \tag{18}$$

$$\text{a) } k_\perp^2 > u_0, \quad \text{b) } W_{\text{scat}}(\boldsymbol{\Omega}_0 \to \boldsymbol{\Omega}) = (1/2\pi)kk_{0\perp}u_0\zeta_0^2 l^2 \exp(-\kappa_\parallel^2 l^2/2). \tag{19}$$

$$\text{a) } \int f(\mathbf{n})d\Omega_n = 1, \quad \text{b) } \int f(\mathbf{n})2d^3n\delta(n^2 - 1)\int\delta(\mathbf{k} - \mathbf{k}_0 + 2\mathbf{n}(\mathbf{n}\mathbf{k}_0))d^3k = 1. \tag{20}$$

$$\text{a) } (\mathbf{n}\mathbf{k}_0) = \kappa/2, \quad \text{b) } \boldsymbol{\kappa} = \mathbf{k}_0 - \mathbf{k}, \quad \text{c) } \int f(\boldsymbol{\kappa}/\kappa)d\Omega_\kappa = 1. \tag{21}$$

$$W_{\text{ref}}(\boldsymbol{\Omega}_0 \to \boldsymbol{\Omega}) = f(\boldsymbol{\kappa}/\kappa). \tag{22}$$

In the case of a stationary perturbation the scattering is elastic and the correction to the wavefunction (1) is represented in the form (2) with the Fourier coefficients (3) computed in accordance with Appendix 6.4. If (4a) is satisfied after scattering, then (3) reduces to the form (4b) with the notations (4c) and (5). In this case the total flux of particles from the interface is (6). The scattering cross-section is determined with the aid of (7a) and, in particular, it follows from (6) that the differential cross-section is (7b). In the case (8a), (8b) follows from (3) with the notation (8c). The differential flux of neutrons reflected from the surface in the element $d^2k_\parallel$, found with the aid of (6), equals (9a). Using the transformation (9b) and the relation (7a) we obtain the differential cross-section (10a), with the aid of which the differential reflection probability (10b) is determined, where S_s is the total area of the reflecting surface. Now we shall consider the perturbation (11a) corresponding to the impurity atom with an amplitude b_N situated at point r_N within the reflecting medium. Substituting (11a) into (5a) we obtain (11b). In the case (12a) we obtain (12b). Substituting (12b) into (10a) we obtain (13). In the case (14a) one must make use of expressions (4) and then the resultant scattering cross-section in both directions from the interface acquires the form (14b). Now consider the perturbation (15a) corresponding to small roughnesses on the surface $z = 0$. We assume the roughnesses to represent a random process characterized by the mean quantities (15b, c). Substituting (15a) into (5a) we obtain (16a). We assume the correlation function to be of the form (16b). Its Fourier transform equals (17a). In the case (17b) we utilize expressions (10), (16a), and (17a), and as a result we obtain expression (18) for the reflection indicatrix in the region of total reflection angles. Outside this region (19a) is valid and it is necessary to make use of expressions (7b), (10b), (16a), and (17a). As a result we obtain expression (19b) for the scattering indicatrix in both directions from the interface.

For large and smooth roughnesses it is possible to introduce the angular distribution for the normals $f(n)$. We shall take advantage of the normalization condition (20a). We transform it to the form (20b), supplementing it with a second integral characterizing specular reflection of UCN at each point of the surface. Using relations (21a, b) and changing the order of integration in (20b) we obtain, as a result, (21c), from which expression (22) follows for the reflection indicatrix.

Appendix 6.10 The loss coefficient in reflection from a surface with small roughnesses

$$\delta u(\mathbf{r}) = u_0[\theta(z > \zeta(\mathbf{r}_\parallel)) - \theta(z > 0)] \simeq \delta_1 u + \delta_2 u. \tag{1}$$

$$\text{a) } \delta_1 u = u_0 \zeta(\mathbf{r}_\parallel)\delta(z), \quad \text{b) } \delta_2 u = u_0 \zeta^2(\mathbf{r}_\parallel)\delta'(z)/2. \tag{2}$$

$$\begin{aligned}\Psi &= \Psi_0 + G_0 \circ \delta u \Psi_0 + G_0 \circ \delta u G_0 \circ \delta u \Psi_0 \\ &= \Psi_0 + G_0 \circ (\delta_1 u + \delta_2 u)\Psi_0 + G_0 \circ \delta_1 u G_0 \circ \delta_1 u \Psi_0.\end{aligned} \tag{3}$$

$$G_0(\mathbf{r}) = -\exp(ikr)/4\pi r = -\int \exp(i\mathbf{p}\mathbf{r})\mathrm{d}^3 p/(2\pi)^3(p^2 - k^2 - i\varepsilon). \tag{4}$$

$$\text{a) } G_0(\mathbf{r}) = \int \exp(i\mathbf{p}_\parallel \mathbf{r}_\parallel + ip_\perp |z|)\mathrm{d}^2 p/(2\pi)^2 2ip_\perp, \quad \text{b) } p_\perp = \sqrt{k^2 - p_\parallel^2}. \tag{5}$$

$$\text{a) } (\Delta + k^2)G_0(\mathbf{r}) = \delta(\mathbf{r}), \quad \text{b) } \delta_1 \Psi = G_0 \circ \delta_1 u \Psi_0 = \hat{r}_1 \Psi_0,$$

$$\text{c) } \hat{r}_1 = G_0 \circ \delta_1 u. \tag{6}$$

$$\text{a) } r_{21} = \langle | G_0 \circ \delta_2 u | \rangle_s, \quad \text{b) } r_{22} = \langle | G_0 \circ \delta_1 u G_0 \circ \delta_1 u | \rangle_s = \langle | \hat{r}_1 \hat{r}_1 | \rangle_s. \tag{7}$$

$$\text{a) } \hat{r}_1 \cdot \Psi_0 = \int \exp(i\mathbf{k}\mathbf{r}) r_1(\boldsymbol{\kappa}_\parallel)\mathrm{d}^2 k_\parallel, \quad \text{b) } r_1(\boldsymbol{\kappa}_\parallel) = u_0 \zeta(\boldsymbol{\kappa}_\parallel)/2ik_\perp(2\pi)^2. \tag{8}$$

$$\text{a) } \boldsymbol{\kappa}_{\|} = \mathbf{k}_{0\|} - \mathbf{k}_{\|}, \quad \text{b) } k_{\perp} = \sqrt{k^2 - k_{\|}^2},$$

$$\text{c) } \zeta(\boldsymbol{\kappa}) = \int \zeta(\mathbf{r}_{\|}) \exp(i\boldsymbol{\kappa}\mathbf{r}_{\|}) d^2 r_{\|}. \tag{9}$$

$$\text{a) } r_{21}^{\pm} = \mp(u_0 \zeta_0^2/2), \quad \text{b) } r_{21}^{-} = -r_{21}^{+} = r_{21}, \quad \text{c) } r_p^{\pm} = \hat{r}_1 + r_{21}^{\pm} + r_{22}. \tag{10}$$

$$\text{a) } t_p = 1 + \hat{r}_1 + r_{22}, \quad \text{b) } R = r_p^{+} + t_p r_+ (1 - r_p^{-} r_+)^{-1} t_p. \tag{11}$$

$$R \simeq r_p^{+} + (1 + \hat{r}_1 + r_{22}) r_+ (1 + r_p^{-} r_+ + r_p^{-} r_+ r_p^{-} r_+ + \ldots)(1 + \hat{r}_1 + r_{22}). \tag{12}$$

$$\text{a) } \delta_1 \hat{R} = \hat{r}_1 + \hat{r}_1 r_+ + r_+ \hat{r}_1 + r_+ \hat{r}_1 r_+ = (1 + r_+)\hat{r}_1(1 + r_+),$$

$$\text{b) } t_+ = 1 + r_+, \tag{13}$$

$$\text{a) } \delta_1 \hat{R} = t_+ \hat{r}_1 t_+,$$

$$\text{b) } \delta_{22} R = t_+ r_{22} t_+ + \langle t_+ \hat{r}_1 r_+ \hat{r}_1 t_+ \rangle_s = t_+ \langle \hat{r}_1 t_+ \hat{r}_1 \rangle_s t_+. \tag{14}$$

$$\text{a) } \delta_{21} R = r_{21}^{+} + r_+ r_{21}^{-} r_+ = r_{21}(1 - r_+^2),$$

$$\text{b) } 1 - r_+^2 = t_+ t_- = 4ik_{0\perp} k_{0\perp}'' r_+ / u_0. \tag{15}$$

$$\text{a) } [t_+(k_\perp)]^2 = (4k_\perp^2/u_0) r_+(k_\perp), \quad \text{b) } R_s = r_+ + \delta_{22} R + \delta_{21} R. \tag{16}$$

$$R_s = r_+(k_{0\perp})\{1 - \zeta_0^2[2ik_{0\perp}k_{0\perp}'' + k_{0\perp}\int K(\boldsymbol{\kappa}_{\|}) d^2 k_{\|}(k_\perp - ik_\perp'')/2\pi^2]\}. \tag{17}$$

$$\text{a) } R_s = r_+(k_{0\perp})(1 - \zeta_0^2 \alpha' + i\zeta_0^2 \alpha''), \quad \text{b) } |R_s|^2 = (1 - \mu_0)(1 - 2\zeta_0^2 \alpha'). \tag{18}$$

$$\alpha' = u_0''(k_{0\perp}/k_{0\perp}'')[1 - \int K(\boldsymbol{\kappa}_{\|}) d^2 k_{\|} k_{0\perp}''/k_\perp''(2\pi)^2]$$
$$+ k_{0\perp} k \int K(\boldsymbol{\kappa}_{\|}) k_\perp^2 d\Omega/2\pi^2. \tag{19}$$

$$\text{a) } \mu = 1 - |R_s|^2 - \int k_\perp \langle |\delta_1 \hat{R}|^2 \rangle / k_{0\perp}, \quad \text{b) } \mu_0(k_\perp) = 2(u_0''/u_0')(k_\perp/k_\perp''). \tag{20}$$

$$\int k_\perp \langle |\delta_1 \hat{R}|^2 \rangle / k_{0\perp} = \int |t_+(k_\perp)|^2 |t_+(k_{0\perp})|^2 \zeta_0^2 K(\boldsymbol{\kappa}_{\|}) d^2 \kappa_{\|} / 4k_\perp k_{0\perp}(2\pi)^2. \tag{21}$$

$$\text{a) } |t_+(k_\perp)|^2 = (4k_\perp^2/u_0')[1 - \mu_0(k_\perp)/2], \quad \text{b) } d^2 k_{\|} = kk_\perp d\Omega. \tag{22}$$

$$\int k_\perp \langle |\delta_1 \hat{R}|^2 \rangle / k_{0\perp} = \int W_{\text{ref}}(\boldsymbol{\Omega}_0 \to \boldsymbol{\Omega})[1 - \mu_0(k_\perp)/2 - \mu_0(k_{0\perp})/2] d\Omega. \tag{23}$$

$$\mu(k_{0\perp}) = \mu_0(k_{0\perp})[1 + u_0 \zeta_0^2(1 - \gamma)]$$
$$+ \int W_{\text{ref}}(\boldsymbol{\Omega}_0 \to \boldsymbol{\Omega}) d\Omega [\mu_0(k_\perp) - \mu_0(k_{0\perp})]/2. \tag{24}$$

$$\text{a) } \gamma = \int_{-\infty}^{\infty} K(\boldsymbol{\kappa}_{\|}) d^2 k_{\|} k_{0\perp}''/k_\perp''(2\pi)^2,$$

$$\text{b) } W_{\text{ref}}(\boldsymbol{\Omega}_0 \to \boldsymbol{\Omega}) = \zeta_0^2 k_{0\perp} k k_\perp^2 K(\boldsymbol{\kappa}_{\|})/\pi^2. \tag{25}$$

Let us expand the perturbation (1) in powers of ζ. Taking into account quadratic terms only we represent (1) in the form of a sum of perturbations, (2). The wavefunction is written to the second order in ζ in.the form (3), where Ψ_0 is the incident plane wave and G is the Green function (4) which can be represented in the form (5) and satisfies eqn (6a). Thus, the perturbation is considered separately from all the remaining part of the potential. To the first order in ζ the correction to the wavefunction equals (6b). It is convenient to introduce the reflection operator (6c). It suffices to consider the quadratic corrections only for the specular direction. In accordance with expression (3) it is possible to determine the two corrections (7) to the amplitude of specular reflection, where the angle brackets indicate averaging of the roughnesses over the ensemble, performed as in Appendix 6.9. The result of the action of operator (6a) on the incident wave is given in (8a). The reflection matrix element is represented by the expression (8b), where notation (9) is adopted. The perturbation (2b) is non-symmetric, so the correction (7a) exhibits differing signs, (10a), when reflection occurs from the left and (10b) from the right (signs + and −, respectively). The total reflection amplitude to the left and to the right due to the total potential (1) is (10c). The potential (2b) gives no contribution to the transmission amplitude, so the total transmission amplitude for potential (1) is the same in both directions and equals (11a). Now we shall take into account the contribution of the remaining part of the wall. To this end we shall make use of the method of recursive relations and consider the perturbation (1) to be separated from the remaining wall by an infinitesimal slit. As shown in Chapter 5, the total reflection amplitude from the wall and from perturbation (1) is (11b). Expanding it in powers of r_p we obtain (12). To the first order in ζ the total amplitude of non-specular reflection is (13a). Using (13b) we obtain (14a). The quadratic corrections to the specular reflection amplitude are (14b) and (15a). We shall now utilize (15b) and (16a). We write out separately the specular reflection amplitude (16b). Substituting (15b), (15), and (16a) into (16b) we obtain expression (17) which involves the Fourier transform of the correlation function of the roughnesses, $K(x_{\|})$ and the average square height of the roughnesses ζ_0^2. We represent (17) in the form (18a). Then up to ζ_0^2 we obtain the coefficient of specular reflection (18b). The parameter α', as follows from a comparison of (17) and (18a), is equal to (19). The loss coefficient is determined by expression (20a), where integration in the last term is performed over all directions of non-specular reflection. The loss coefficient in reflection from a perfectly smooth wall is (20b). Using (14a) and averaging, as shown in Appendix 6.9, we represent the summand in the form (21), and then making use of relations (22) we rewrite it in the form (23). Substituting (18b), (19), and (23) into (20a) we obtain (24), where the notation (25) is used. We note that expression (25b) obtained here for the reflection indicatrix coincides with the expression found in Appendix 6.9.

Appendix 6.11 The loss coefficient in reflection from a surface with dense roughnesses

$$\text{a) } lu_0^2 \approx 0, \quad \text{b) } u = u_0/[1 + \exp(-z/h)], \quad \text{c) } z = [z]/h,$$
$$\text{d) } k = [k]h. \tag{1}$$

$$\text{a) } [d^2/dz^2 - u + k^2]\Psi(z) = 0, \quad \text{b) } u = u_0/(1 + e^{-z}), \quad \text{c) } u_0 = [u_0]h^2. \tag{2}$$

$$\text{a) } k^2 > u_0,$$
$$\text{b) } \Psi(z) = \mathscr{H}_+(k, z) = cF[i(k - k'), -i(k + k'), 1 - 2ik', -e^{-z}]. \tag{3}$$

$$\text{a) } k^2 < u_0, \quad \text{b) } \Psi(z) = \mathscr{H}_+(k, z) = cF(k'' + ik, k'' - ik, 1 + 2k'', -e^{-z}). \tag{4}$$

$$\text{a) } k' = \sqrt{k^2 - u_0}, \quad \text{b) } k'' = \sqrt{u_0 - k^2}, \quad \text{c) } k = K_\perp = K\cos\vartheta. \tag{5}$$

$$\text{a) } z \to -\infty, \quad \text{b) } \mathscr{H}_+(k, z) = \exp(ikz) + r_+(k)\exp(-ikz). \tag{6}$$

$$\text{a) } k^2 < u_0, \quad \text{b) } r_+(k) = \frac{\Gamma(2ik)\,\Gamma(k'' - ik)\,\Gamma(1 + k'' - ik)}{\Gamma(-2ik)\,\Gamma(k'' + ik)\,\Gamma(1 + k'' + ik)} = e^{-2i\varphi}. \tag{7}$$

$$\text{a) } \varphi = \varphi' - i\varphi'', \quad \text{b) } u_0'' \ll u_0', \quad \text{c) } \varphi'' = u_0''\,d\varphi/du_0', \quad \text{d) } \mu(k) = 4\varphi''. \tag{8}$$

$$\text{a) } d\ln\Gamma(x)/dx = -\gamma - 1/x + \sum_{n=1}^{\infty} 1/n(x + n), \quad \text{b) } \gamma = 0.577. \tag{9}$$

$$\text{a) } \mu(k) = \mu_0(k)\left(1 + 2\sum_{n=1}^{\infty} u_0/[(n + k'')^2 + k^2]\right), \quad \text{b) } \mu_0(k) = 2\eta k/k''. \tag{10}$$

$$\text{a) } \eta = u_0''/u_0',$$
$$\text{b) } \bar{\mu}(K) = \int_0^{\pi/2} \mu(k\cos\vartheta)2\cos\vartheta\, d\cos\vartheta = \bar{\mu}_0(K)[1 + f(u_0)]. \tag{11}$$

$$\text{a) } f(u) = \begin{cases} u_0\pi^2/3 = [u_0]h^2\pi^2/3, & h \to 0 \\ 2\sqrt{u_0} = 2h\sqrt{[u_0]}, & h \to \infty, \end{cases}$$
$$\text{b) } \bar{\mu}(K) = \bar{\mu}_0(K)(1 + [u_0]\zeta_0^2). \tag{12}$$

$$\text{a) } \zeta_0 = \pi h/\sqrt{3}, \quad \text{b) } \zeta_0 \to \infty, \quad \text{c) } \bar{\mu}(K) \to (2/\pi)\zeta_0\sqrt{[u_0]/3}\,\bar{\mu}_0(K). \tag{13}$$

$$\text{a) } k^2 > u_0, \quad \text{b) } |r_+(k)|^2 = \text{sh}^2(\pi h[k - k'])/\text{sh}^2(\pi h[k + k']). \tag{14}$$

We shall consider the roughnesses to be dense, if the correlation length l satisfies condition (1a). In this case the interaction of a neutron with the medium can be modelled, for instance, by the function (1b), which permits analytic solution of the Schrödinger equation. We shall pass to the dimensionless variables (1c, d), where the quantities in the square brackets have a natural dimensionality. The Schrödinger equation is written in the form (2a), where the potential equals (2b), the quantity u_0 having become dimensionless in accordance with (2c). The degrees of motion along the reflecting surface have been excluded from eqn (2a) and, therefore, k represents the normal component of the wavevector with the subscript $\perp$ dropped. If (3a) holds, then the solution of eqn (2a) is represented by the hypergeometric function (3b) (*Landau and Lifshits 1963), where c is a normalizing constant. If (4a) is valid, then (4b) represents the solution of (2a). The adopted notation is decoded in (5), where K is the total value of the wavevector of the incident neutron and ϑ is the angle of incidence. The constant c is chosen to provide for (6b) being valid in the asymptotic region (6a). In the case (7a) the reflection amplitude is (7b), where $\Gamma(x)$ is the Euler Γ function. As one can see from (7), when the quantity u_0 is real, the magnitude of the reflection amplitude is unity. If $u_0 = u_0' - iu_0''$, i.e. it contains an imaginary part, then (8a) occurs. When (8b) holds, the imaginary part of the phase is computed with the aid of (8c) and the loss coefficient equals (8d). The phase φ equals the sum of the logarithms of the Γ functions, the derivatives of which are indicated in (9a) (*Whittaker and Watson 1927). Performing (8c) and taking into account (9) we reduce (8d) to the form (10a), where the expressions (10b) and (11a) are utilized for the loss coefficient in the case of an ideal smooth surface. Now we average (10a) over the angles. We obtain expression (11b) which involves the function f exhibiting the asymptotic behaviour (12a). A similar averaging of the loss coefficient found in Appendix 6.10 yields (12b) in the limit (1a). Comparing (12b) with (12a) at small h we find that the potential (1b) represents small and dense roughnesses well, if h is related to ζ_0 by relation (13a). Then for high and dense roughnesses we obtain (13c) for large h in the limit (13b), upon substitution of (12a) into (11b).

When (14a) holds, the reflection coefficient from the potential (1b) equals (14b) (*Landau and Lifshits 1963).

Appendix 6.12 UCN scattering on a pore

$$\text{a) } \psi_{is} = Cj_0(k_i r) = C\sin(k_i r)/k_i r, \quad \text{b) } u_i = -iv'', \quad \text{c) } k_i = \sqrt{k^2 - u_i}. \tag{1}$$

$$\text{a) } \psi_{0s} = Dk_0(k''r) = D\exp(-k''r)/k''r, \quad \text{b) } u_0 = u_0' - iu_0'',$$
$$\text{c) } k'' = \sqrt{u_0 - k^2}. \tag{2}$$

$$\text{a) } \mathrm{d}\ln\psi_{is}/\mathrm{d}r|_{r=a} = \mathrm{d}\ln\psi_{0s}/\mathrm{d}r|_{r=a}, \quad \text{b) } \varphi(k) = 0, \quad \text{c) } \bar{E}_r = \bar{k}_r^2. \tag{3}$$

$$\text{a) } \varphi(k) \equiv \mathrm{ctg}k_i a + k''/k_i, \quad \text{b) } E_r = \bar{E}_r - i\Gamma_1/2, \quad \text{c) } v'', u_0'' \ll E_r,$$
$$\text{d) } \Gamma_1 \ll E_r. \tag{4}$$

$$\text{a) } k_r = \bar{k}_r + \Delta, \quad \text{b) } \delta\varphi(k_r) = -iu''\partial\varphi/\partial u + iv''\partial\varphi/\partial k_i^2 + \Delta\partial\varphi/\partial k = 0. \tag{5}$$

$$\text{a) } \Delta = -i\Gamma_1/4\bar{k}_r,$$
$$\text{b) } \Gamma_1 u_0'(1 + \bar{k}_r''a)/4\bar{k}_r^3\bar{k}_r'' = v''(au_0' + \bar{k}_r'')/2\bar{k}_r^3 + u_0''/2\bar{k}_r\bar{k}_r''. \tag{6}$$

$$\text{a) } \Gamma_1 = 2[\eta_0\bar{k}_r^2 + \eta_v\bar{k}_r''(au_0' + \bar{k}_r'')]/(1 + \bar{k}_r''a), \quad \text{b) } \eta_{0,v} = u_0''/u_0', v''/u_0'. \tag{7}$$

$$\text{a) } \psi_{0s}(r) = \hat{i}_0(k''r) + r_a\hat{k}_0(k''r), \quad \text{b) } \psi_{is}(r) = t_a\hat{j}_0(kr). \tag{8}$$

$$\text{a) } \hat{f}(qr) = f(qr)/f(qa), \quad \text{b) } i_0(x) = \mathrm{sh}\,x/x, \quad \text{c) } k''(\hat{i}_0' + r_a\hat{k}_0') = kt_a\hat{j}_0'. \tag{9}$$

$$\text{a) } 1+r_a=t_a, \quad \text{b) } r_a=(k''\hat{i}'_0 - k\hat{j}'_0)/(k\hat{j}'_0-k''\hat{k}'_0), \quad \text{c) } \hat{f}' = f'(a)/f(a). \tag{10}$$

$$\text{a) } r_a = A/(E - \bar{E}_r - \mathrm{i}\Gamma_1/2),$$
$$\text{b) } A \simeq 2\bar{k}_r^2 \bar{k}_r''^2[\operatorname{cth}(\bar{k}_r'' a) + 1]/u'(1 + \bar{k}_r'' a) > 0. \tag{11}$$

$$\text{a) } \psi_{0s}(r) = \psi_a(r) + r_a\psi_a(a)\hat{k}_0(k''r), \quad \text{b) } \psi_{is}(r) = t_a\psi_a(a)\hat{j}_0(kr). \tag{12}$$

$$\exp(-k''r)/r = \int \exp(-k''_\perp|z - z'| + \mathrm{i}\mathbf{k}_\parallel \mathbf{r}_\parallel)\mathrm{d}^2k_\parallel/2\pi k''_\perp. \tag{13}$$

$$\text{a) } \mathrm{d}r(k) = ar_a\mathrm{d}^2k/2\pi k''_\perp, \quad \text{b) } e_{0a}(k) = \exp(k''a - k''_\perp z'). \tag{14}$$

$$e_{a0}(k) = \int_a \exp(-k''_\perp z + \mathrm{i}\mathbf{k}_\parallel \mathbf{r}_\parallel) Y_{00}(\boldsymbol{\Omega})\mathrm{d}\Omega = \exp(-k''_\perp z')\operatorname{sh}(k''a)/k''a. \tag{15}$$

$$\psi_a = [1 - \int e_{a0}(k)r_-(k_\perp)e_{0a}(k)\mathrm{d}r(k)]^{-1}e_{a0}(k_0)t_+(k_{0\perp})\psi_0. \tag{16}$$

$$\text{a) } \psi_r = \int \exp(\mathrm{i}\mathbf{kr})t_-(k_\perp)e_{0a}(k)\mathrm{d}r_a(k)\psi_a, \quad \text{b) } \sigma_s = J_{ns\perp}/k_{0\perp}. \tag{17}$$

$$J_{ns\perp} = \int k_\perp |r_a t_-(k_\perp)e_{0a}(k)\psi_a|^2 a^2 \mathrm{d}^2k_\parallel/k''^2_\perp. \tag{18}$$

$$\text{a) } \mathrm{d}r_a(k) = Aa\mathrm{d}^2k_\parallel/2\pi k''_\perp(E - E_r - \mathrm{i}\Gamma_1/2), \quad \text{b) } \mathrm{d}^2k_\parallel = kk_\perp\mathrm{d}\Omega. \tag{19}$$

$$\mathrm{d}\sigma_s(\boldsymbol{\Omega}_0 \to \boldsymbol{\Omega}) = 4\pi\Gamma_s(\boldsymbol{\Omega}_0)\mathrm{d}\Gamma_s(\boldsymbol{\Omega})/k^2\{(E - E_r - \varepsilon)^2$$
$$+ [\Gamma_1 + \int \mathrm{d}\Gamma_s(\boldsymbol{\Omega})]^2/4\}. \tag{20}$$

$$\mathrm{d}\Gamma_s(\boldsymbol{\Omega}) = 2(kk_\perp^2/k''u')A\exp(-2k''_\perp z' + k''a)\operatorname{sh}(k''a)\mathrm{d}\Omega/\pi. \tag{21}$$

$$\varepsilon = A\int [u' - 2k_\perp^2 + 2\mathrm{i}k_\perp k''_\perp \theta(k_\parallel > k)]\mathrm{d}^2k_\parallel \exp(-2k''_\perp z'$$
$$+ k''a)\operatorname{sh}(k''a)/2\pi k''_\perp k''u'. \tag{22}$$

$$\text{a) } \Gamma_s(\boldsymbol{\Omega}_0) = 2(k^2k_{0\perp}/k''u')A\exp(-2k''_{0\perp}z' + k''a)\operatorname{sh}(k''a), \quad \text{b) } \Gamma = \int \mathrm{d}\Gamma. \tag{23}$$

$$\text{a) } \langle\Gamma\rangle = \int \Gamma(\boldsymbol{\Omega})\cos\vartheta_n \mathrm{d}\Omega/\pi = \int \mathrm{d}\Gamma,$$
$$\text{b) } \langle\sigma_s\rangle = 4\pi\Gamma_s^2/k_r^2[(E - E_r)^2 + \Gamma^2/4]. \tag{24}$$

$$\text{a) } E_r = \bar{E}_r + \varepsilon, \quad \text{b) } \Gamma = \Gamma_1 + \Gamma_s,$$
$$\text{c) } \langle\sigma_a\rangle = 4\pi\Gamma_s\Gamma_1/k_r^2[(E - E_r)^2 + \Gamma^2/4]. \tag{25}$$

1. A pore in an infinite medium

The wavefunction inside the pore (for simplicity we only take the s state into account) is (1a), and outside the pore it is (2a), where C and D are constants. We shall assume the potential inside the pore (it may be filled with a gas) to be (1b), and outside the pore to be (2b), from which (1c) and (2c) follow. From condition (3a) follows eqn (3b), where notation (4a) is adopted. The solution of the equation for $u''_0 = v'' = 0$ determines the bound level (3c). When $v'', u''_0 \neq 0$ we have (4b) instead of (3c). We shall assume (4c); then (4d) holds. We shall seek for the solution of eqn (3b) in the form (5a). Expanding (4a) in

v'', u_0'' and Δ we obtain eqn (5b) for Δ. We substitute (6a) into (5b) and reduce it to the form (6b), from which follows (7a), where the notation (7b) is utilized. Now we shall consider scattering of sub-limiting waves on a pore. Outside the pore the wave function is (8a), while inside the pore it is (8b), where the notation (9a, b) is used and r_a and t_a, respectively, represent the reflection and transmission amplitudes on the boundary of the pore. From the continuity conditions for the derivative and the function itself at $r = a$, eqns (9c) and (10a) follow, the solution of which is (10b), where the notation (10c) is adopted. At $E \simeq \bar{E}_r$ (10b) is represented in the form (11).

2. Scattering on a pore situated within the semi-space $z > 0$

The wavefunction outside the pore is (12a), while inside the pore it is (12b). For determining ψ_a we shall take into account rescattering between the surface $z = 0$ and the pore. We shall take advantage of the expansion (13). It follows from (13) that with the coefficient (14a) the pore generates a wave $\exp(ik_{\parallel} r_{\parallel} - k''|z' - z|)$, which at the boundary $z = 0$ has the amplitude (14b). Upon reflection from the boundary it returns to the pore and contributes to the spherical wave (15). In accordance with recursive relations (see Chapter 5) we have for ψ_a the relation (16), where ψ_0 is a plane wave incident from the semi-space $z < 0$, and the propagation functions equal (14b) and (15). The wavefunction describing scattering by the pore into the semi-space $z < 0$ is (17a). The scattering cross-section (17b) where $k_{0\perp}$ represents the normal component of the incident plane wave is determined by the flux of scattered waves, eqn (18). Substituting (11a) into (14a) we obtain (19a). Substituting (14b), (15), (16) and (19) into (18), and applying relation (19b) we obtain (20), where we introduce the differential width (21), the shift of the resonance level (22), and the width (23a), which determine the probability for the neutron to happen to be inside the pore. The integral of (21), as indicated in (23b), equals the total scattering width. Averaging (23a) over the angles in accordance with (24a) also leads to the total elastic scattering width. Integration of (20) over Ω and averaging over Ω_0 in accordance with (24a) yields (24b). If we introduce the notation (25a, b), then from (24b) it follows that the absorption cross-section by the pore can be written in the form (25b).

Appendix 6.13 Estimation of the correction to the loss coefficient due to the reflecting material being porous

$$\langle\!\langle \sigma_a \rangle\!\rangle = 4\pi\int(2E dE/u_0^2)\Gamma_s\Gamma_1/\bar{k}_r^2[(E - E_r)^2 + \Gamma^2/4]$$
$$\simeq 16\pi^2\Gamma_s\Gamma_1/u_0^2\Gamma. \quad (1)$$

$$\text{a) } \Gamma_s \gg \Gamma_1, \quad \text{b) } \langle\!\langle \sigma_a \rangle\!\rangle \simeq 16\pi^2\Gamma_1/u_0^2 = 32\pi^2\eta\bar{k}_r^2/u_0^2(1 + \bar{k}_r''a). \quad (2)$$

$$\text{a) } \Delta\mu = \int n dz' \langle\!\langle \sigma_a \rangle\!\rangle = 32\pi^2 n z'_{max}\eta\bar{k}_r^2/u_0^2(1 + \bar{k}_r''a), \quad \text{b) } \Gamma_s(z'_{max}) = \Gamma_1. \quad (3)$$

$$\Gamma_s(z) = 4\int [\bar{k}_r^3 \bar{k}_r'' \bar{k}_{r\perp}^2 / u'^2 \pi(1 + \bar{k}_r'' a)] \exp(-2\bar{k}_{r\perp}'' z + 2\bar{k}_r'' a) \mathrm{d}\Omega. \tag{4}$$

$$\text{a) } b = \bar{k}_r'' a, \quad \text{b) } \alpha = \bar{k}_r'' / \sqrt{u_0'}, \quad \text{c) } y = z/a, \quad \text{d) } \bar{k}_r^2 / u_0' \simeq b^2 \simeq \alpha^2 \simeq 1. \tag{5}$$

$$\Gamma_s(z) = [8\alpha \bar{k}_r^2 / (1 + b)] \exp(2\mathrm{b}) \int_\alpha^1 x \sqrt{1 - x^2} \exp(-2ybx/\alpha) \mathrm{d}x. \tag{6}$$

$$\text{a) } \Gamma_s(z) \simeq [4\alpha^2 \bar{k}_r^2 / (1 + b)] \exp[-2b(y - 1)] \alpha / by,$$

$$\text{b) } \Gamma_1 = 2\eta \bar{k}_r^2 / (1 + b). \tag{7}$$

$$\text{a) } z'_{\mathrm{max}} = a[1 - \ln(\eta/2)/2], \quad \text{b) } V_{\mathrm{p}} = (4\pi/3) a^3 n. \tag{8}$$

$$\text{a) } \Delta\mu \simeq 16\pi^2 n z'_{\mathrm{max}} \eta / u' \simeq -12\mu V_{\mathrm{p}} \ln \mu, \quad \text{b) } \mu = c\eta \simeq \pi\eta. \tag{9}$$

Let us average the loss cross-section over the incident neutron spectrum (1). We will find that, when (2a) is fulfilled, the losses are independent of the distance from the centre of the pore to the surface, and the result (1) can be approximated by expression (2b). The correction to the loss coefficient due to the pores in the case when they are homogeneously distributed in the substance equals (3a), where n is the number of pores per unit volume and z'_{max} is the distance from the centre of a pore to the surface, determined by eqn (3b), for which condition (2a) is violated. The dependence of the elastic scattering width of the pore on the distance z from the surface follows from (4). Now we introduce the parameters (5a–c). For simplicity we shall assume that all the pores are small and that relations (5d) hold for them. The integral (4) reduces to the form (6) and is approximated by expression (7a). The width determined by losses in a pore equals (7b). Upon substitution of (7) into eqn (3b) we find the solution (8a). We introduce the porosity (8b). Substituting (8a) into (3a) and taking advantage of the approximation (5d) we reduce (3a) to the form (9a), where the approximation (9b) is applied for the non-perturbed loss coefficient μ.

7

The interaction of UCN with a magnetic field

In this chapter the storage of neutrons in magnetic traps will be considered, as well as issues of UCN polarization and methods of analysing it, and also methods of UCN spin reversal. Magnetic traps have the advantage that no absorption or inelastic scattering exists in them within the range of thermal energies. There are, however, certain peculiarities that must be taken into account in estimating the UCN storage time in these traps. The magnetic field in a trap is extremely inhomogeneous and can confine neutrons within a restricted region of space only in the case of a certain neutron spin projection on to the field direction. When the spin direction changes, and in an inhomogeneous field it is sure to undergo changes, the magnetic 'walls' will stop being walls and a neutron will leave the trap with ease. It is noteworthy that a spin flip is most probable when the neutron is as far as possible from the 'magnetic wall', and not when it impinges upon the 'wall'.

Thus, one of the reasons of a possible UCN leakage from a magnetic trap is depolarization of the neutrons. A second reason may be the gradual heating of UCN occurring in the presence of small acoustic oscillations in the walls. These issues of the depolarization and gradual heating of UCN are dealt with in Section 7.1. In Section 7.2 all the experiments performed so far with magnetic traps are considered.

Section 7.3 is devoted to issues of UCN polarization and its analysis. The interaction potential of a neutron with a magnetized medium is $u_0 - \boldsymbol{\mu}\mathbf{B}$ (in the adopted notation the dimensionality of μ is $2m[\mu]/\hbar^2$). Hence it follows that when the energy of the neutrons is $u_0 - \mu B < E < u_0 + \mu B$, only those neutrons can penetrate a thin magnetized film, the polarization of which is opposite to the direction of the magnetic field $\mathbf{B}$. It might seem that all the neutrons passing through the film should be totally polarized; however, an analysis of their polarization performed with the aid of a similar magnetized film reveals that this is not so. The possible reasons are discussed in Section 7.3.

In analysing the polarization it is necessary to flip the neutron spin. This can be achieved by two methods, resonance or adiabatic, which will be dealt with in Section 7.4. The contents of this section are directly related to the search for the EDM of the neutron, to which the beginning of the next chapter is devoted.

7.1 THEORY OF UCN STORAGE IN MAGNETIC TRAPS

An inhomogeneous magnetic field $\mathbf{B}(r)$ acts on a neutron in two ways. First, it makes the neutron magnetic moment μ precess about the direction of the field $\mathbf{B}$, and, secondly, it alters the value and direction of the neutron velocity. Such influence exercised not only upon UCN but also on neutrons of arbitrary energies, is described by the single potential interaction energy $-\boldsymbol{\mu}\mathbf{B}(r)$. In the case of neutrons of thermal or higher energies the purely mechanical influence of the field on the motion of the neutrons may usually be neglected, but in the case of UCN this influence is decisive. It is precisely because of this influence that magnetic storage of UCN is possible.

UCN storage in ordinary solid-state vessels is possible because the walls of the vessels create a repulsive, i.e. positive, potential, so for magnetic storage it is necessary, for example, to surround a certain region of the space with a magnetic 'rampart' sufficiently high for the potential energy $-\boldsymbol{\mu}\mathbf{B}(r)$ to be positive and to exceed the neutron kinetic energy. Such a 'rampart' can be created using multipole systems. Imagine an infinitely long cylindrical region surrounded by infinitely long alternating poles. In the vicinity of the axis of this region the magnitude of the field will increase smoothly from the centre towards the periphery. Thus, for example, in the case of a six-pole network the increase of the field is proportional to r^2, where r is the distance from the axis. If one digresses from the motion along the axis of the six-pole magnet, then the problem of the behaviour of a neutron in the field of a six-pole magnet near its axis is equivalent to the problem of a two-dimensional symmetric harmonic oscillator, which has quite an obvious solution: inside a parabolic well there are discrete energy levels and the particle may be only in bound states corresponding to these levels.

For the potential energy of a neutron in a rising magnetic field to be positive it is necessary that the magnetic moment always be antiparallel to the field $\mathbf{B}(r)$. If, for some reason, the magnetic moment turns out to be parallel to the field, then the potential interaction energy of the neutron with the field will turn out to be negative, and the neutron will, instead of being repulsed by the magnetic 'rampart', be drawn into it. But is it possible for $-\boldsymbol{\mu}\mathbf{B}(r)$ to be positive always, i.e. for the moment μ always to follow the direction of $\mathbf{B}$? From most general arguments the answer to this question is in the negative. Indeed, if the magnetic moment of the neutron could always follow the field $\mathbf{B}(r)$, then the quantity $P = \mu B(r)/|\mu||B(r)|$ would assume two discrete values, $+1$ and -1, and these values would be good, i.e. conserved, quantum numbers. It is known, however, that the eigenvalues of

the operator P are conserved only when it commutes with the total Hamiltonian. The total Hamiltonian contains a summand corresponding to the kinetic energy, which is represented by a sum of derivatives with respect to the coordinates, and, since the operator P depends on the coordinates (the direction of the unit vector $\mathbf{B}/B$ depends on the coordinates) it does not commute with the Hamiltonian, and, consequently, the projection of $\boldsymbol{\mu}$ on to $\mathbf{B}$ is not conserved in time. This result may give rise to the pessimistic conclusion that magnetic storage of UCN is not altogether possible, but actually everything is much better. The point is that the neutrons do not have to be stored infinitely long. Moreover, they cannot be stored for a time longer than their own lifetime, so for effective storage the magnetic moment is only required not to deviate too much from the direction of $-\mathbf{B}$ during the neutron lifetime of 1000 s. It turns out that this goal is quite attainable.
required not to deviate too much from the direction of $-\mathbf{B}$ during the neutron lifetime of 1000 s. It turns out that this goal is quite attainable.

Having resigned ourselves to the fact that sooner or later the magnetic moment will reverse its direction with respect to the field $\mathbf{B}(r)$, we shall try to estimate the time it will take to do so. There may be different approaches to the solution of this problem. The author of the very first work, devoted to magnetic storage, V.V. Vladimirskii (1960), applied the semi-classical method, i.e. he neglected the influence of the field on the motion of the neutron and only took into account the precession of the magnetic moment about the direction of the field $\mathbf{B}(r)$. The neutron was then represented as a pointlike particle moving uniformly within a given region of space and in a straight line with a velocity v. The problem can then be simplified by using a reference frame related to the neutron, since in this system the neutron no longer moves in an inhomogeneous field, but the field changes with time in the vicinity of the motionless neutron. Here the field changes in both magnitude and direction.

We can now readily clarify under which conditions $\boldsymbol{\mu}$ will have time to follow the direction of $\mathbf{B}$ and under which conditions it will not. If the field is high, and it rotates with an angular velocity $\omega_1 = |\dot{B}|/B$ that is small compared with the angular velocity of the spin precession, $\omega_0 = \mu B/\hbar$, then μ follows the direction of the field adiabatically. In this case the so-called conditions of adiabaticity (see Appendix 7.1) are satisfied. When these conditions are fulfilled the projection of $\boldsymbol{\mu}$ on to $\mathbf{B}$ can be considered conserved. In a magnetic trap, however, the adiabaticity conditions are not always satisfied. If in the trap there exist regions where $\mathbf{B}$ is low and the particle may pass near these regions, at these sites the precession frequency ω_0 is also small and adiabaticity may be violated. Vladimirskii calculated the probability W of a spin flip of the neutron when it passes near such a region (Appendix 7.1) and obtained the quantity $w = \exp(-\pi\omega_0/\omega_1)$. Hence the conclusion follows that magnetic traps must be designed so as to avoid nodes, i.e. points at which the field turns to zero. In a six-pole magnet the field is zero at the axis, so it is desirable to create an additional homogeneous field parallel to the axis.

Matora (1972*b*) performed a calculation of UCN depolarization in a six-pole magnet in a semi-classical approximation. A classical trajectory was defined for the neutron and the probability for a spin flip to occur was calculated in a reference frame related to the neutron. In the case of a fixed polarization along the field $\mathbf{B}(r)$ a classical trajectory in a plane perpendicular to the axis of the six-pole magnet is described by the radius vector $\rho = (\rho_{ell}(\varphi))^{1/2}$, where $\rho_{ell}(\varphi)$ is the radius vector of the elliptic trajectory. The calculation revealed that even in the worst case the probability of spin flip did not exceed $10^{-4}\,s^{-1}$. Now when the neutron moves along a straight line passing through the centre of symmetry a spin flip never even occurs. This is explained by such a curve being the location of identical inclinations of the field $\mathbf{B}(r)$.

The problem of UCN storage in a magnetic bottle can be fully solved with the aid of quantum mechanics, taking into account the influence of the field $\mathbf{B}(r)$ on both the spin and the translational motion of the neutron. The quantum-mechanical problem turned out to be extremely unusual. If the projection of $\boldsymbol{\mu}$ on to $\mathbf{B}$ were not to change with time, then, as indicated at the very beginning of this section, the problem would be totally equivalent to that of a plane symmetric oscillator, and a neutron inside a trap would have stable discrete levels. Since the projection of $\boldsymbol{\mu}$ on to $\mathbf{B}$ is not conserved, however, these discrete levels are not stable and we have to deal with the decay of a non-stable system, similar, for example, to α-decay. But the essential difference from the α-decay is that here, while the production of an α-particle in α-decay is due to quantum-mechanical tunnelling through the barrier, the potential itself, which confines the neutron within the trap, changes (Ignatovich 1974*c*).

If we turn to the Schrödinger equation describing the behaviour of a neutron in a magnetic field $\mathbf{B}(r)$ (in Appendices 7.2–7.4 the general case of a $2(l+1)$-pole magnet is dealt with), then we will see that precisely the commutator of the projection of $\boldsymbol{\mu}$ on to $\mathbf{B}$ with the total Hamiltonian is responsible for the neutron leaving the trap. If this commutator is assumed to be a perturbation, then the time the neutron is confined within the trap can be found by applying perturbation theory. The perturbation, however, is not everywhere small. Moreover, it exhibits singularities at the nodal points, where the field $\mathbf{B}$ becomes equal to zero. However, even though the perturbation may be large, we can apply perturbation theory if the initial state changes insignificantly. If, on the other hand, the perturbation changes the initial state strongly, then it can be asserted that this given state decays rapidly, and such information will be sufficient for us.

By applying the procedure of perturbation theory we arrive at the following conclusion. A magnetic trap actually represents a potential well in which a neutron has discrete energy levels characterized by two quantum numbers: the radial n and the azimuthal m numbers. All these levels, with the exception of those for which $m = 0$, are unstable, but while in α-decay the least stable levels are the upper ones, here the least stable are the lower ones. (There is an interesting analogy with Rydberg atoms; see, for example, *Walther (1986).) The least stable levels among the upper levels are those with small m, corresponding to elongated orbits, while levels with $m \approx n \gg 1$, corresponding to nearly circular orbits, turn out to be extremely long lived. An exception is represented by the levels with $m = 0$, corresponding to trajectories which conventionally may be considered to be straight lined and passing through the centre. Such levels are absolutely stable. We have termed these trajectories conventionally straight-lined, because actually a neutron in the field of a multipole magnet does not travel along straight-lined trajectories. Trajectories with $m = 0$ can be thought of as extremely elongated ellipses, the region of $r = 0$ being totally inaccessible to the UCN.

The above conclusion holds for the field of any multipole magnet, independently of whether the number of homopolar poles is even or odd, while the number of not very stable orbits is high when the number of poles used is high. In six-pole traps 10 cm in diameter, creating fields of 10 kG, unstable levels decay in 0.1 s, while the stable levels live for more than 10^9 s if the lifetime of the neutron itself is not taken into account. This result does not contradict semi-classical calculations (Vladimirskii 1960).

Hitherto we have discussed magnetic traps of the same type as solid-state traps, i.e. those that represent a region of space surrounded by a magnetic 'rampart' repelling neutrons. However, traps of another sort are also feasible when the field draws the neutrons in. The magnetic field of a straight wire can serve as such a trap. For the neutrons to be confined in this case the potential energy $-\boldsymbol{\mu}\mathbf{B}$ and the total energy $\mathscr{E}$ must be negative. It is remarkable that neutrons can be kept indefinitely long in such a trap, since a spin flip leads to the neutron kinetic energy becoming negative. In practice this means that the spin flip is forbidden by the law of conservation of energy. But does this mean that we were wrong to assert that projection of the magnetic moment on the field direction cannot be conserved? The answer is no. We were right. In the case of magnetic storage in the field of a straight line current, the projection is not conserved either. So the neutron spin cannot be completely non-parallel to the field, but admixture of parallel projection is small, and it leads only to an increasing of bound levels of energy. It turns out that there exists an exact analytic solution (* Pron'ko and Stroganov 1977) for the quantum- mechanical problem of the behaviour of UCN in the magnetic field of a straight wire (see Appendix 7.5). The problem with traps of this type is how to fill them up with neutrons, since the energy of the neutrons arriving from the reactor is positive, but in the trap it should be negative. There are various solutions to this problem. For example, one may switch on the current rapidly

(Ignatovich and Pokotilovskii 1976). In this case the kinetic energy of the neutrons that happen to be near the wire has no time to change, while the drop in total energy equals the potential energy. The neutrons are then captured by the field. The neutrons will also be captured by the field if the current is switched on slowly. In this case the total energy decreases by a quantity equal to the potential energy averaged over the total volume within which the UCN were confined initially.

One more version of a trap combining both a magnetic field and the gravitational field is interesting. In this version the neutrons are stored in a plane, representing a Vladimirskii magnetic mirror (Vladimirskii 1960), i.e. a set of straight-line conductors, along which current flows in alternating directions. In such a system the field dies away exponentially as the distance from the plane increases. Therefore, in the absence of any additional homogeneous constant field the storage time of the neutron (if only its energy is not too low) in such a trap will be short, since upon being reflected from the magnetic cushion the neutron rises upwards and arrives in the region of low magnetic field, where its magnetic moment can no longer follow the field and therefore retains its direction. Then the neutron, having gone along a parabolic trajectory, again impinges upon the magnetic cushion, but now its magnetic moment may turn out to be antiparallel to the magnetic field, and, instead of reflecting the neutron, the field will draw it in. The resulting storage time may be estimated by the quantity $2v/g < 1$ s (Ignatovich and Terekhov 1976*b*). A weak additional homogeneous field, parallel to the conductors of the mirror, improves the situation significantly.

Such is the situation with the leakage of UCN from magnetic traps due to depolarization. Now we shall estimate the influence of oscillations of the magnetic field on the UCN storage time. Let the magnetic potential energy $u_m(r) = -\boldsymbol{\mu}\mathbf{B}(r)$ vary, as a result of fluctuations of currents, as $[1 + \varepsilon(t)]u_m(r)$, where $\varepsilon(t)$ is the relative fluctuation of the field. If a neutron happens to be at a point r, then, owing to the change in potential energy, its total energy also changes as

$$\Delta E = \Delta t \varepsilon'(t) u_m(r(t)), \tag{7.1}$$

where $r(t)$ is the radius vector of the neutron at time t. The total change in energy during time t is represented by the integral of (7.1).

We shall consider two cases. Let $\varepsilon(t)$ be a random function of time, such that its average value $\bar{\varepsilon}$ equals zero, and $\overline{\varepsilon(t)\varepsilon(t')} = \varepsilon_0^2 K(t-t')$, where $K(t)$ is the correlation function. Then the root-mean-square variation of energy during the time t can be represented in the form

$$\overline{(\Delta E)^2} = \iint_0^t \varepsilon_0^2 K''(t'-t'') u_m(r(t')) u_m(r(t''))\, dt'\, dt''. \tag{7.2}$$

Let* $K(t) = \exp(-t/t_0)$. Then the maximum value of the integral (7.2) may be estimated by the quantity† $\varepsilon_0^2 t E_{\text{lim}}^2/t_0$. Setting $\overline{(\Delta E)^2} = E_{\text{lim}}^2$ we find an estimate for the storage time, $\tau = t_0/\varepsilon_0^2$. If the field fluctuation amplitude amounts to 0.1%, then for $t_0 \approx 0.1$ s‡ we have $\tau \approx 10^5$ s. In the case of a periodic process with a frequency coinciding with one of the frequencies of the periodic orbital motion of the neutron, changes in the field affect the UCN storage time more significantly. Consider $\varepsilon(t) = \varepsilon_0 \sin(\omega t)$. Then

$$\Delta E = \pi \varepsilon_0 u_{\text{m}}(\omega) n,$$

where

$$u_{\text{m}}(\omega) = \int_0^{t_0} \cos \omega t u_{\text{m}}(r(t)) \omega \, \mathrm{d}t/\pi,$$

$t_0 = 2\pi/\omega$ is the period of motion, and $n = t/t_0$. For estimation purposes we may set $u_{\text{m}}(\omega) \approx E_{\text{lim}}$, i.e. the maximum energy attained by the neutrons confined in the trap. Assuming, further, that $\Delta E = E_{\text{lim}}$, we find an estimate for the storage time, $\tau = t_0/\pi\varepsilon_0$. For the same parameters as above, we obtain the value for the storage time, $\tau \approx 100$ s*.

7.2 EXPERIMENTS WITH A MAGNETIC TRAP

To confine neutrons with energies up to E_{Cu} magnetic fields of ≈ 30 kG are required as well as high currents and a powerful cooling system. Actually, without using superconductors, UCN storage has hitherto been achieved at fields of ≈ 2.5 kG. The maximum energy E_{lim} of the neutrons confined by such a field amounts to about 15 neV. The number of neutrons with $E < E_{\text{lim}}$ is small when E_{lim} is low, and, when the number of accumulated neutrons is small, it becomes extremely laborious to perform experiments to study UCN storage. One must also take into account that the magnetic fields fall off quite slowly, which leads to a decrease of the phase space of the trap and to an additional decrease in the number of neutrons being accumulated in the trap.

The first magnetic trap for UCN (Fig. 7.1) was designed in the Institute of Theoretical and Experimental Physics (Moscow) (Abov *et al.* 1977). It consists of a bowl with built-in coils in the lateral walls and the bottom. The

---

* Such a correlation function is not realistic, but it permits an upper estimate to be obtained for the quantity under study.

† If, as t_0, the time of the mean free path between the magnetic walls is assumed, then, as in Abov *et al.* (1983*b*), it follows that the increase in energy with time is proportional to the square root of the number of collisions, n: $\Delta E = (\langle(\Delta E)^2\rangle)^{1/2} \sim \varepsilon_0 n^{1/2}$, $n = t/t_0$.

‡ For instance, in the case of network fluctuations.

* In Abov *et al.* (1983*b*) UCN heating was also computed on high-frequency vibrations of the magnetic walls. The effect is insignificant, since the vibrations are averaged over the time t_0.

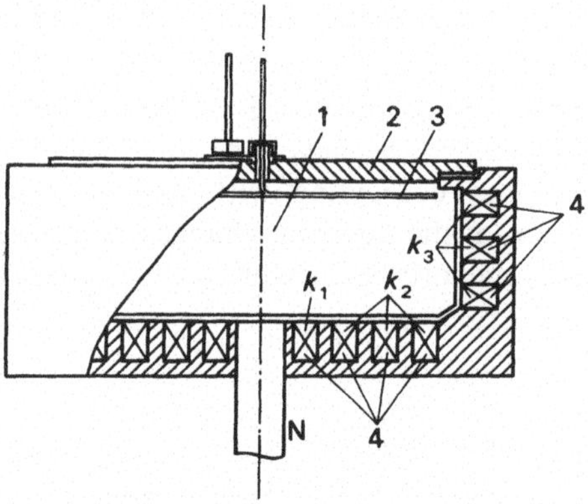

Fig. 7.1 A magnetic trap from the Institute of Theoretical and Experimental Physics (Moscow) (Abov *et al.* 1977): 1, vacuum chamber 34 cm high and 104 cm in diameter; 2, steel cover; 3, correcting coil; 4, the main coils; the current in coil k_1 is 1000 A, in coil k_2 600 A, and in coil k_3 450 A.

currents in adjacent coils flow in opposite directions. The stainless steel vacuum chamber has an entrance opening for the neutrons at the lower end and is situated inside the bowl. The top of the bowl containing the chamber is covered with a steel plate, beneath which there is a correcting coil (Fig. 7.2). Without the correcting coil the magnetic field inside the bowl exhibits three circular nodal lines. When the correcting coil is used, only several nodal points remain in the trap (Vasil'ev and Rojnin 1984).

The central coil in the bottom of the bowl represents a magnetic shutter. When the trap is closed, the current passing through this coil is ~1000 A.

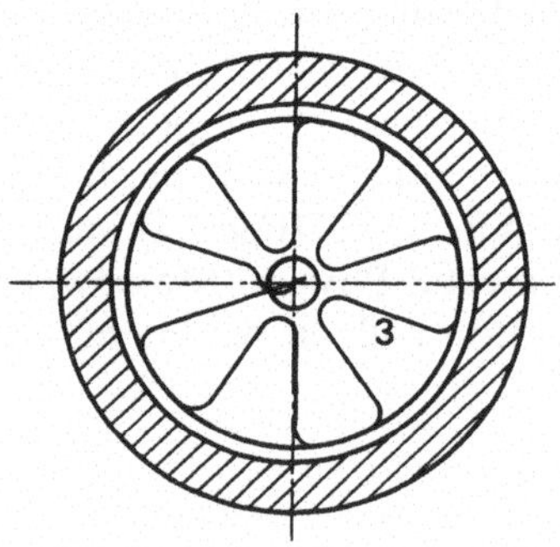

Fig. 7.2 Form of correcting coil of the trap shown in Fig. 7.1 (the notation is the same). The coil is made of a tubular busbar packed into the cover of the chamber and cooled with water. The current in the busbar is 1100 A. The magnetic field of the coil transforms the three nodal lines in the trap in Fig. 7.1 into nine nodal points (Vasil'ev and Rojnin 1984).

Approximately the same current (1100 A) is made to pass through the correcting coil, while the current in the other coils is 450–600 A. The maximum field of the magnetic wall amounts to 2.5 kG. This value determines $E_{\rm lim}$ precisely. For a flux of 86 s^{-1} in the input neutron guide (Abov *et al.* 1983*a*) it was only possible to accumulate about three neutrons, which were definitely confined by the magnetic field. The background counting rate of the UCN detectors was only 0.008 s^{-1}.

To reveal purely magnetic storage part of the bottom of the trap (20% of the total area) was covered with polyethylene, which served as an effective heater of the UCN and rapidly removed the high-energy UCN not held by the magnetic field. In Fig. 7.3 the measured storage curve is presented for a magnetic trap, which has a corresponding storage time of $\tau = 303 \pm 37$ s.

Before the experiment of Abov *et al.* (1983*a*) two other experiments had been performed. In one, Kosvintsev *et al.* (1976*a*) investigated combined wall–magnetic storage (Fig. 7.4). One of the walls of the copper trap was

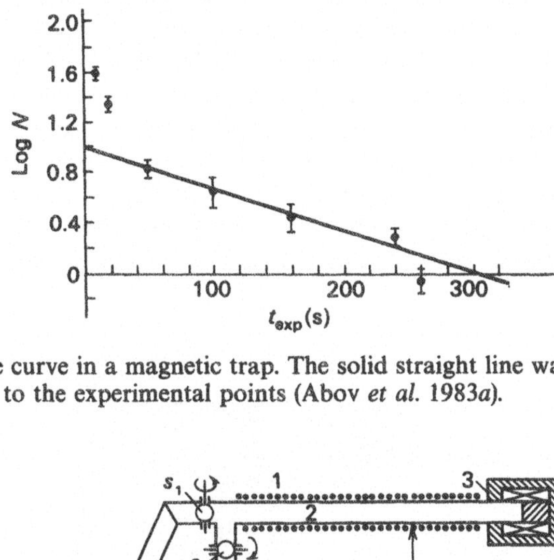

Fig. 7.3 Storage curve in a magnetic trap. The solid straight line was obtained by a least-squares fit to the experimental points (Abov *et al.* 1983*a*).

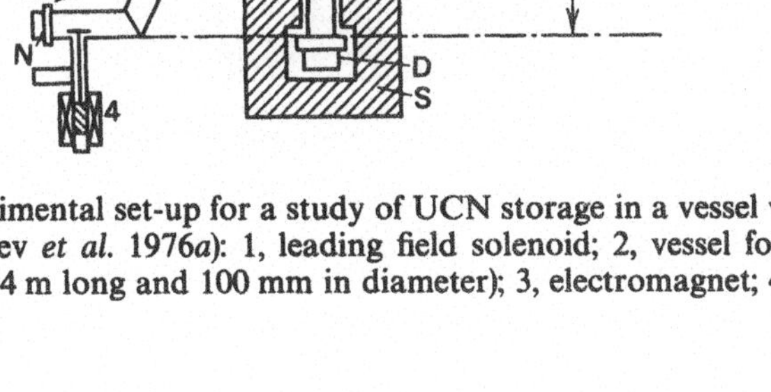

Fig. 7.4 Experimental set-up for a study of UCN storage in a vessel with a magnetic wall (Kosvintsev *et al.* 1976*a*): 1, leading field solenoid; 2, vessel for UCN storage (copper pipe 2.4 m long and 100 mm in diameter); 3, electromagnet; 4, pumping out.

covered with polyethylene and near it a magnetic field of 5–6 kG was created. In addition, a homogeneous field of 50 G could be created in the trap, which reduced the probability of UCN depolarization. The storage time in the copper trap without polyethylene and in the absence of the field amounted to 25 ± 2 s, and with polyethylene it was 10.5 ± 1.5 s; when the magnetic wall was switched on it increased to 15 ± 2 s, and then up to 22 ± 3 s when the additional homogeneous field was switched on.

In the other experiment, Kosvintsev *et al.* (1978*a*) described an arrangement similar to that of Abov *et al.* (1983*a*), but with a smaller trap and lower power. Here accumulation of 1.05 ± 0.15 neutrons was achieved and the measured storage time was 35 ± 10 s.

These experiments demonstrated that magnetic storage of UCN is quite feasible. However, for a detailed investigation of the behaviour of neutrons in a magnetic trap experiments have to be performed with UCN sources of significantly higher intensities.

The most successful and effective study of the magnetic storage of UCN was the experiment with a toroidal trap (Kügler *et al.* 1978, 1985) carried out in Grenoble. The first to put forward the idea of a toroidal trap was V.V. Vladimirskii (1960), who noted that neutrons of energies significantly higher than UCN energies can be confined in a toroidal trap (Fig. 7.5). Indeed, the velocity of the confined neutrons along the circular axis of the toroid is, in principle, not restricted in any way. It is only important that the velocity component normal to the walls of the toroid be smaller than the limiting velocity determined by the magnitude of the magnetic field. Having noted

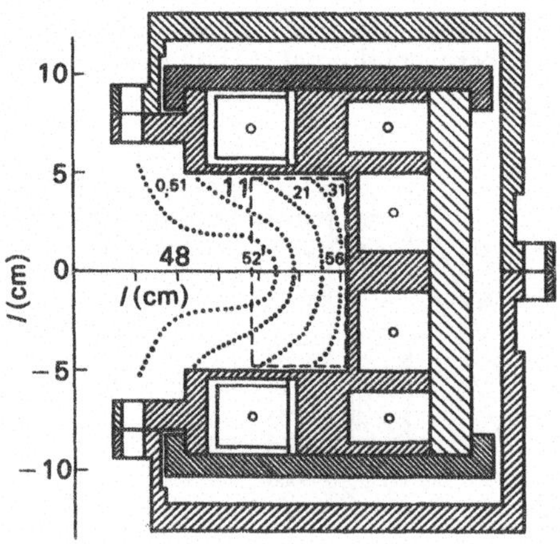

Fig. 7.5 Cross-section of toroidal magnetic trap. The shaded area indicates a copper alloy. The dotted lines show the levels of the magnetic field. The channels in the body of the magnet are filled with liquid helium (Kügler *et al.* 1978).

this fact, however, Vladimirskii expressed some doubt about the possibility of the practical realization of such a trap, since inevitable inhomogeneities of the magnetic field should lead either to diffuse reflection of a neutron from the magnetic wall or to a small change in the reflection angle, which would, as the number of revolutions of the neutron increased, gradually transform the energy of longitudinal motion into that of transverse motion. In turn, this resulted in the neutron being capable, in a subsequent collision with the magnetic wall, of overcoming the potential energy of the magnetic barrier.

Vladimirskii's idea was further developed by I.M. Matora (1972), who showed that a toroidal trap, which can be represented as an infinite six-pole magnet coiled up into a toroid, can undergo significant improvement. Actually, the three poles pointing towards the inside of the toroid can be totally eliminated. In this case, it is much easier to construct the trap and, more importantly, the nodal line which usually coincides with the circular axis of the toroid is eliminated. Elimination of the inner poles cannot worsen the conditions of confinement, since a neutron moving at small angles to the wall of the trap cannot approach the centre of the toroid. If we use the reference frame rotating with the neutron about the centre of the toroid, the above fact may be interpreted as a result of the action of a centrifugal repulsive force driving the neutron away from the centre.

The absence of the nodal line in toroidal traps with excluded inner poles represents their main advantage. Although, as shown by Matora (1972), the lifetime of a neutron on a major part of the orbits is large even in the presence of the nodal line, the presence of magnetic roughnesses may lead, with a certain probability, to the neutron skipping over to disadvantageous orbits and being depolarized in the vicinity of a nodal point. Generally, any asymmetry in the field (Golub 1978*a*) enhances the depolarization of a neutron. If a six-pole magnet is horizontal, then we must also take into account the gravitational field which leads to the neutron orbits 'sagging' and approaching the nodal point from the top. This leads to an additional decrease in the confinement time in a horizontal multipole system (Matora 1972). Evidently, the same effect would also occur in a simple toroidal trap, but since the nodal line is shifted towards the region inaccessible to the neutrons when the inner poles are removed, the misgivings concerning their depolarization can also be discarded.

The toroidal trap used in the Grenoble experiment differed from that of Matora (1972) in that it had only two poles removed, instead of three. The magnetic field was created by superconducting coils and two additional poles were introduced which produced a decapole component to diminish the influence of possible roughnesses in the magnetic field. The trap was constructed in Bonn, taken to Grenoble, and installed at the high-flux reactor. The superconducting coils created a field of 35 kG, the trap was 1.2 m in diameter, and it confined neutrons with velocities of 20 m s^{-1}. The supply of neutrons into the trap was performed with the aid of a bent glass neutron guide, the inner walls of which were covered with vacuum-deposited nickel. The neutron guide was supplied with a servomechanism which allowed its rapid removal, so as not to decrease the amount of neutrons released in the trap. Registration was carried out by a proportional counter with ^{3}He which could be inserted into and removed from the trap. From the results obtained it follows that the number of accumulated neutrons amounted to 100 and that, after an initial rapid drop with a lifetime of 100 s, the storage curve exhibits asymptotic behaviour with a damping factor of 918 $\pm$ 15 s (Fig. 7.6). The aim of the experiment was to determine the lifetime of the neutron itself. The storage time turned out to be not less than 918 s, in good agreement with the quantity τ_β accepted at present.

A six-pole spherical magnetic trap was also constructed in Bonn and transported to Grenoble. It consists of three superconducting coils positioned as shown in Fig. 7.7. These coils create a field within the trap close to that of a six-pole magnet. The trap was tested with neutral sodium atoms. The

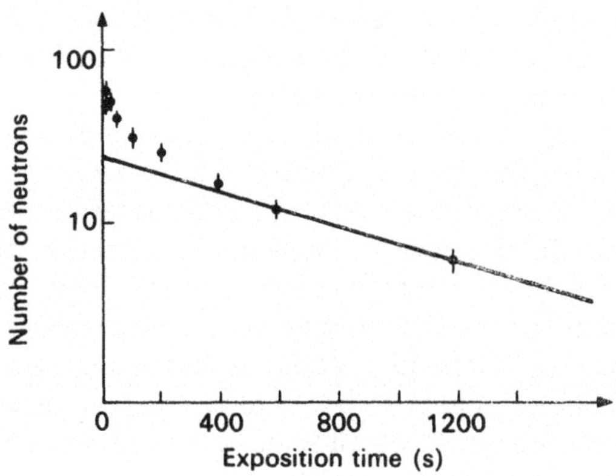

Fig. 7.6 Storage curve in a toroidal trap (Kügler *et al.* 1978).

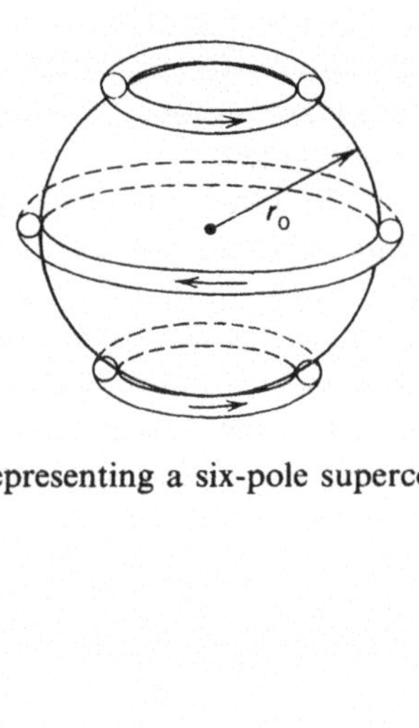

Fig. 7.7 A magnetic trap representing a six-pole superconducting sphere (Kügler *et al.* 1985).

experiment was as follows. A vaporizer with sodium was placed inside the trap. When the magnetic field was switched on, part of the evaporated atoms with low velocities were driven by the magnetic field from the walls and thus a sodium atmosphere was created, the existence of which could be observed by resonance fluorescence. The neutral atoms have a magnetic moment close to that of an electron, i.e. a thousand times greater than the magnetic moment of the neutron. This means that the potential barrier produced by the magnetic field is a thousand times higher for the atoms than for the neutrons. The superconducting coils created a field of 30 kG which was capable of confining sodium atoms with velocities up to $\sim 38\ \mathrm{m\,s^{-1}}$.

At a certain moment the vaporizer was rapidly removed from the trap and at the same time a sodium lamp was switched on, its light being directed towards the trap. A photomultiplier was placed perpendicularly to the light beam to register the bursts of resonance fluorescence emitted in a direction perpendicular to the incident light. The results were as folows. In the absence of the magnetic field only a background emission was seen. In the presence of the magnetic field an intense burst was observed, with an intensity dying away exponentially in ~ 3.2 s. The reason for such a short lifetime of the atoms in the trap may lie in the resonance absorption of light by sodium atoms leading to their depolarization. The experiment was performed once and not repeated, since the sodium atoms condensed, after evaporation, on the walls of the vacuum chamber and on the windows, thus rendering impossible the observation of storage during subsequent runs. The purpose of the six-pole spherical bottle is the storage of neutrons. Inside the 'sphere' there is superfluid helium. The helium is irradiated with thermal neutrons which, as a result of inelastic scattering, are partly converted into UCN. Calculations reveal that in a 1000 s exposure at the Grenoble reactor it will be possible to accumulate up to 1600 neutrons in the trap, whereupon the liquid helium will be removed and the neutrons kept in vacuum. At any rate, even in the presence of superfluid helium in the trap the storage time may be quite significant.

All the magnetic traps described above can be used in a similar way to storage neutral atoms (*Koelman *et al.* 1988, *Masuhara *et al.* 1988, and *Silvera and Walraven 1986). The magnetic moment of atoms is much higher than that of neutrons, so magnetic storage is more effective for them. The atomic density in a trap can be of $10^{14}\ \mathrm{cm^{-3}}$; much higher than that for neutrons, which, even in solid traps, cannot exceed $100\ \mathrm{cm^{-3}}$. But the threshold of registration for atoms is also higher: it is of $10^{8}\ \mathrm{cm^{-3}}$, whereas for neutrons it may be less than $10^{-4}\ \mathrm{cm^{-3}}$ (Abov *et al.* 1988). It is interesting to point out some terminology used in atomic physics. The atoms stored like neutrons in a space surrounded by a high field 'rampart' are called low field seekers, while atoms, stored like neutrons in the case of straight line current, are called high field seekers. The experiment on storage of sodium atoms, described above, is not very spectacular. For more successful experiments on the storage of polarized neutral atoms we refer the readers to the review by *Silvera and Walraven (1986).

7.3 UCN POLARIZATION AND ITS ANALYSIS

UCN polarization can be achieved, as noted at the beginning of this chapter, by transmitting neutrons through a thin magnetized film. For this purpose reflection from a magnetized film may also be utilized, but the reflection is less effective (Egorov *et al.* 1976), mostly. The polarizing power of a film is determined by the relation

$$P = (N_\uparrow - N_\downarrow)/(N_\uparrow + N_\downarrow), \tag{7.3}$$

where $N_{\uparrow,\downarrow}$ is the number of neutrons passing through the film with spins directed upwards or downwards, respectively. To analyse the polarization the same kind of film is used, in front of which a spin flipper (SF) is established, i.e. a device that makes it possible to rotate the neutron spin through 180°. Experimentally the polarization ratio is measured:

$$\varepsilon = (N_- - N_+)/(N_- + N_+), \tag{7.4}$$

where $N_\pm$ is the counting rate of the detector obtained with the SF switched on and switched off, respectively. Let us establish the relationship between P and ε. To this end we shall describe the neutron beam by the two-dimensional vector

$$n = \begin{matrix}\psi_\uparrow \\ \psi_\downarrow\end{matrix}\begin{pmatrix} n_\uparrow \\ n_\downarrow \end{pmatrix}, \qquad n_0 = \begin{pmatrix} 1 \\ 1 \end{pmatrix},$$

where the components $n_{\downarrow,\uparrow}$ are proportional to $N_{\downarrow,\uparrow}$ and n_0 describes the initial non-polarized beam. Such a representation permits simplification of many formulae. For instance, $N_\uparrow + N_\downarrow \sim nn_0$, $N_\uparrow - N_\downarrow \sim n\sigma_z n_0$, where σ_z is the Pauli matrix, and $\varepsilon = n\sigma_z n_0/nn_0$. The vector $n = T_1 n_0$ is made to correspond to the beam passing through the polarizer, where T_1 is the transmission matrix of the polarizer film with the elements $(T_1)_{ij}$ $(i, j = 1, 2)$. T_{ij} are the transition probabilities of a neutron from the state with spin j before passing through the film to the state with spin i after the film. The polarizing power of the film, eqn (7.3), can be represented in the form

$$P_2 = (n_0\sigma_z n)/(n_0 n) = (n_0\sigma_z T_1 n_0)/(n_0 T_1 n_0)$$

If, along the path from the polarizer film to the analyser, the polarization transforms in accordance with the matrix Q, then the beam intensity that passes through the analyser is proportional to

$$T_Q \approx n_0 T_2 Q T_1 n_0,$$

where T_2 is the transmission matrix of the analyser. We shall now denote by $F_\pm$ the matrix corresponding to the SF switched on and switched off, respectively. Then ε (see (7.4)) can be represented in the form

$$\varepsilon = n_0 T_2(F_- - F_+)T_1 n_0/(n_0 T_2(F_- + F_+)T_1 n_0).$$

Hence, we can see that ε depends not only on the polarizing powers of the films P and A, but also on the parameters of the matrix F, i.e. on the properties of the SF.

Let us now clarify which properties the transmission matrix Q must possess in the region between the films. First of all, its matrix elements must, evidently, be non-negative, and transmission from left to right and vice versa can be described, generally speaking, by different matrices $\vec{Q}$ and $\overleftarrow{Q}$, respectively. From the principle of detailed balance, however, it follows that in the absence of absorption the matrix $\overleftarrow{Q}$ is equal to the transposed matrix $\vec{Q}$, i.e. $\overleftarrow{Q} = (\vec{Q})'$. Indeed, if the region under study is immersed in a non-polarized equilibrium neutron gas, then polarization should appear by itself neither to the right nor to the left of this region, i.e. $\vec{Q}_{ij} = \overleftarrow{Q}_{ji}$.

From the conditions of detailed balance it is also possible to obtain more detailed information on the matrix elements. Generally, the matrix elements depend on the wavevectors $\mathbf{k}_0$ and $\mathbf{k}$ of the incident and transmitted neutrons, respectively:

$$Q_{ij} = Q_{ij}(\mathbf{k}_0, \mathbf{k})\mathrm{d}\Omega_k.$$

The principle of detailed balance requires the film inserted in the isotropic non-polarized neutron gas not to create polarization or anisotropy. This will not happen when $k_{0\perp} Q_{ij}(\mathbf{k}_0, \mathbf{k}) = k_\perp Q_{ji}(\mathbf{k}, \mathbf{k}_0)$. Below, however, we shall not consider angular dependences and we shall assume

$$Q_{ij} = \iint Q_{ij}(\mathbf{k}_0, \mathbf{k}) k_{0\perp}\, \mathrm{d}\Omega_k\, \mathrm{d}\Omega_0/\pi k,$$

i.e. the matrix element is integrated over the angles of the outgoing neutron, Ω_k, and averaged over the isotropic angular distribution of the incident flux.

Besides the principle of detailed balance, it is necessary also to take into account the unitarity condition, in accordance with which neutrons must not vanish or be created. If the region under study does not reflect and does not absorb neutrons, then the intensity of the transmitted neutrons of both polarizations must be equal to the incident intensity. Let

$$Q = \begin{pmatrix} a & b \\ c & d \end{pmatrix}.$$

Then the unitarity condition requires

$$a + c = 1, \quad b + d = 1, \quad a + b = 1, \quad c + d = 1,$$

from which it follows that Q has the form

$$Q = \vec{Q} + \overleftarrow{Q} = \begin{pmatrix} a & 1-a \\ 1-a & a \end{pmatrix} = a + (1-a)\sigma_x = p_x + q\bar{p}_x, \qquad (7.5)$$

where $q = 2a - 1$ and p_x, $\bar{p}_x$ are projection operators. In the following we shall also use the operators p_z, $\bar{p}_z$. Projection operators are defined with the help of the relations

$$p_i = (1 + \sigma_i)/2, \quad \bar{p}_i = (1 - \sigma_i)/2, \quad i = x, z,$$

where $\sigma_{x,z}$ are Pauli matrices.

Let us now consider the transmission of a magnetic foil (for definiteness we choose the direction from left to right). The film is assumed to be single domain, i.e. we assume it can be mentally divided into three parts: two near the entrance and exit sides and the third in the middle, the latter being homogeneously magnetized. The transmission of the left side, taken separately, can be represented in the most general form by the expression

$$\text{a)}\ \vec{T}_l = t_l p_z(p_x + \tau_l \bar{p}_x), \quad \text{b)}\ \overleftarrow{T}_l = t_l(p_x + \tau_l \bar{p}_x)p_z, \qquad (7.6)$$

where the matrix (7.6b) is the transposed matrix (7.6a) and the arrow indicates the direction of transmission. The parameter $t_l(1 + \tau_l)/2$ is the probability for a neutron with its spin up to penetrate the film, and the parameter $t_l(1 - \tau_l)/2$ is the probability for a neutron with its spin down to reverse the direction of its spin and also to penetrate the film. The magnetization of the film is supposed to be such that neutrons with spin down cannot penetrate the film.

In a similar way we can write the transmission matrices of the right side of the film:

$$\text{c)}\ \vec{T}_r = t_r(p_x + \tau_r \bar{p}_x)p_z, \quad \text{d)}\ \overleftarrow{T}_r = t_r p_z(p_x + \tau_r \bar{p}_x). \qquad (7.6)$$

The transmission matrices contain two parameters. It is evident that $t_{l,r} \leqslant 1$. If these parameters are not due to losses, then the corresponding side of the film must partly reflect neutrons incident on it from the inside of the film, otherwise the number of particles will not be conserved. The corresponding reflection matrices are

$$\text{a)}\ \overleftarrow{R}_l = r_l p_z, \quad \text{b)}\ r_l = 1 - t_l, \quad \text{c)}\ \vec{R}_r = r_r p_z, \quad \text{d)}\ r_r = 1 - t_r. \qquad (7.7)$$

At $t_l = t_r = 1$ they are all zero.

Through the middle part of the film only neutrons of one polarization are transmitted equally well in both directions. So the transmission matrix can be represented by $\overleftrightarrow{T}_m = t_0 p_z$ with $t_0 < 1$ being totally due to losses in the film.

The transmission matrix of the whole film is obtained with the help of the recursive relation method accounting for multiple reflection inside the film:

$$\vec{T} = \vec{T}_r(1 - \overleftarrow{T}_m \overleftarrow{R}_l \vec{T}_m \vec{R}_r)^{-1} \vec{T}_m \vec{T}_l.$$

Substituting here (7.6a), (7.6b), (7.7a), and t_0 for the matrix T_m we obtain a relation which (see Appendix 7.6) can be represented in the form:

$$\vec{T} = t[p_x + \tau_l\tau_r\bar{p}_x + \sigma_z(\tau_r p_x + \tau_l\bar{p}_x)], \tag{7.8}$$

where the parameter t is a complicated function of three parameters t_l, t_r, and t_0. Upon transmission through a magnetic film a non-polarized beam of neutrons is transformed into

$$n = \vec{T}n_0 = t(1 + \tau_r\sigma_z)n_0,$$

so the parameter τ_l plays the part of the polarizing power p. After transmission of a polarized beam $n = (1 + p_0\sigma_z)n_0$ through the magnetic film, it is transformed into the beam

$$\vec{T}n = t(1 + p_0\tau_l)(1 + \tau_r\sigma_z)n_0,$$

whence it follows that τ_l plays the part of the analysing power A of the film.

If a film is not a single-domain one, i.e. it is composed of different parts exhibiting different directions of magnetisation, then the expression (7.8) must be averaged over different domains, which reduces it to the form (Taran 1986)

$$\vec{T} = t[p_x + S\bar{p}_x + \sigma_z(Pp_x + A\bar{p}_x)],$$

where $P = (s\uparrow - s\downarrow)\tau_r$, $A = (s\uparrow - s\downarrow)\tau_l$, $S = (s\uparrow - s\downarrow)\tau_r\tau_l$, and $s\uparrow$, $s\downarrow$ are relative areas of the film surface with up and down directions of magnetization in the middle. The polarized beam is transformed after transmission through such a film into the beam

$$\vec{T}n = t[1 + P_0A + (P + SP_0)\sigma_z]n_0.$$

Now it is not difficult to write down the expression for the transmission of two independent films with an SF between them. Indeed, upon transmission of an unpolarized beam through the first film it acquires the form $n_1 = t_1(1 + P_1\sigma_z)n_0$. The SF transforms it, with the help of the matrix $F = p_x + \varphi\bar{p}_x$, into $n_{1F} = t_1(1 + P_1\varphi\sigma_z)n_0$. After transmission through the second film (analyser) the beam becomes

$$n_{1F2} = t_1t_2[1 + P_1\varphi A_2 + (P_2 + S_2P_1\varphi)\sigma_z]n_0.$$

The counting rate J_d of the detector is

$$J_d = J_0t_1t_2(1 + P_1A_2\varphi)/2,$$

where J_0 is the intensity of the primary beam. If the SF is perfect ($\varphi = \pm 1$), then measurement of the polarization ratio (7.4) gives $\varepsilon = P_1A_2$. If both films are identical and symmetric, then $P_1A_2 = P_1^2 = \varepsilon$, and the polarizing power (7.3) is obtained as $P_1 = \varepsilon^{1/2}$ (Lobashov and Serebrov 1978). In principle, we can imagine the analyser to be perfect ($A_2 = 1$), so $P_1 = \varepsilon$ (Herdin *et al.* 1978).

The latter expression yields a lower, and thus a safer, estimate of the film polarizing power.

It is interesting to note that the value of $P_1 A_2$ can be measured without any SF, if the transmissions of two films $t_{1,2}$ are first measured separately. In this case the product $P_1 A_2$ is defined simply from the relation

$$P_1 A_2 = 2J_d/J_0 t_1 t_2 - 1.$$

Experimentally such a procedure is not convenient. The measurement with the SF is less laborious, but it introduces some uncertainty into the measurement, since the flipping efficiency $(1 - \varphi)/2$ is not generally known beforehand. Nevertheless it can be determined if, instead of one, two SFs are used in series (Herdin *et al.* 1978; Podgoretskii and Tzulaja 1985; Taran 1986).

Indeed, after measuring the detector counting rates $J_{d0}\alpha_1 + P_1 A_2$ with both SFs switched off, $J_1\alpha_1 + \varphi_1 P_1 A_2$, $J_2\alpha_1 + \varphi_2 P_1 A_2$ with one or the other switched on, and $J_{12}\alpha_1 + \varphi_1\varphi_2 P_1 A_2$ with both SFs switched on it is possible to determine uniquely all the three parameters φ_1, φ_2, and $P_1 A_2$. Let us define the ratios $\alpha_i = J_i/J_{d0}$, where $i = 1, 2, 12$. They are

$$\alpha_i = a + (1 - a)\varphi_i,$$

where $a = 1/(1 + P_1 A_2)$ and $\varphi_{12} = \varphi_1\varphi_2$. With their help it is not difficult to obtain

$$\varphi_1 = (\alpha_1 - \alpha_{12})/(1 - \alpha_2), \qquad \varphi_2 = (\alpha_2 - \alpha_{12})/(1 - \alpha_1),$$

$$p_1 A_2 = (1 - \alpha_1)(1 - \alpha_2)/(\alpha_{12} - \alpha_1\alpha_2).$$

A similar analysis can be applied for the transmission of three, and even more, independent films, but it can be considered to be correct only for a single transmission of neutrons through the film, i.e. in the case of beam or in time-of-flight experiments. In the case of a stationary flow of UCN through a system of several films it is necessary to take into account multiple reflection between the films. To do this we need to know the reflection from a single film.

Now we shall consider the reflection of UCN from a film. The reflection matrix of the left part of the film taken separately can be represented, in general, in the form

$$\vec{R}_1 = p_x + a_1\bar{p}_x - t_1(1 + \tau_1\sigma_z)/2,$$

where the depolarization at the surface (the parameter a_1) and the transmission into the middle part (the second summand) are taken into account. This expression can be reduced to the form

$$\vec{R}_1 = r_1(p_x + b_1\bar{p}_x + s_1\sigma_z),$$

involving three other parameters.

Let us consider the total reflection from the left accounting for multiple reflections inside the film. In accordance with the recursive relations we have

$$\vec{R} = \vec{R}_l + \overleftarrow{T}_l \overleftarrow{T}_m \vec{R}_r (1 - \vec{T}_m \overleftarrow{R}_l \overleftarrow{T}_m \vec{R}_r)^{-1} \vec{T}_m \vec{T}_l,$$

which can be rewritten (see Appendix 7.6) in the form

$$\vec{R} = R_l(p_x + \rho_l \bar{p}_x + \xi_l \sigma_z),$$

involving three other parameters. The reflection matrix from the right can be given in a similar way with the help of three other parameters.

Having found the reflection matrices, it is not difficult to obtain the transmission T_{1F2} of a two-film system taking into account the multiple reflection between them:

$$\vec{T}_{1F2} = \vec{T}_2(1 - \vec{F}\overleftarrow{R}_1\overleftarrow{F}\vec{R}_2)^{-1}\vec{F}\vec{T}_1.$$

The calculation given in Appendix 7.7 shows that the counting rate of the detector, in the case of an ideal SF ($\varphi = \pm 1$), is given by the expression

$$J_d = (1 + \beta_1 \varphi x + \beta_2 x^2)/(\beta_3 + \beta_4 \varphi x) + \beta_5 x^2),$$

where $x = P_1 A_2$ and β_i are constants defined by film parameters. Now the polarizing ratio (7.4) is not equal to $P_1 A_2$, and it is not difficult to show that the expression $P_q = \varepsilon^{1/2}$ used above for identical and symmetric films gives an underestimated value for the film polarizing power P_1, since it does not take into account multiple reflection between the films.

We note that every film is characterized by a minimum of seven parameters: two triplets correspond to two regions near the surfaces, and one to losses in the middle. This set of parameters is not sufficient, in general, if it is necessary to take into account interference phenomena, but it is the most general set from a phenomenological point of view if the interference phenomena are neglected. In this respect the models describing a film with the help of three parameters, considered by Pokotilovskii (1978) and Ignatovich and Taran (1982, 1983, 1984), are convenient from the point of view of the investigation of multiple reflection but are not sufficient physically.

Measurement of ε was performed by Luschikov *et al.* (1978), Egorov *et al.* (1974), and Altarev *et al.* (1982), but only Herdin *et al.* (1978) utilized the time-of-flight technique. The following respective values were obtained in these works: $\varepsilon = 0.36, 0.57, 0.91$, and 0.95. The question arises as to why the value of ε is so small, especially in Luschikov *et al.* and Egorov *et al.* It may be, first of all, that this is due to multiple reflection between films. Indeed, in a simple model of the film (Ignatovich and Taran 1982) it was shown that $P_1 A_2 = [(1 + 3\varepsilon^2)^{1/2} - 1]\varepsilon$, i.e. in Luschikov *et al.* $P_1 A_2 = 0.64$, while in Egorov *et al.* $P_1 A_2 = 0.70$. In more complicated cases rescattering can affect the value of $P_1 A_2$ even more. All its values, however, turn out to be too low to be satisfactory. In Egorov *et al.* iron films were used as the polarizer, with an isotopic content of $^{56}Fe_{0.2}{}^{54}Fe_{0.8}$, in which case $V_{lim} = 0$ for one spin component, as well as films of a natural isotopic mixture for which this is not so. In both cases the value of ε is, within experimental error, the same, although the intensity I_Q in the second case naturally fell by 20% because the transmission was worse. It is worth noting that films vapour-deposited on to titanium foil 1.5 μm thick and on to the cover glass (100 μm), both 2000 Å thick, yielded the same value of ε, although the neutron intensity in the second case was $\sim$1.5 times lower than in the first and reflection should be significant, at least because of the potential of glass. This testifies either in favour of the rescattering in Egorov *et al.* being insignificant, or that the losses in between the polarizer and the analyser were large, leading to a decrease in the rescattering effect.

It is possible that a general depolarization occurs in the gap between the polarizer and the analyser. In Egorov *et al.* (1974) to test this hypothesis a piece of a glass neutron guide 25 cm long was placed between magnetic films, in one case, and a glass neutron guide 110 cm long with its walls vapour-coated with nickel was used in another case. The polarization ratio ε did not change by more than 10%. This means that in either case depolarization in the intermediate region was not significant, or it was independent of the distance. The depolarization could be caused by inhomogeneities of the field or by the fields in the films being non-parallel. In the first case the probability of depolarization ($w = (1 - q)/2$) is determined by the magnitude of the fluctuating field, by the correlation length, and by the time the neutron is to be found in the fluctuating field (Appendix 7.8). The corresponding investigation with UCN was performed by Efimov and Ignatovich (1974) and Taran (1973*a*, *b*). If B is the amplitude of the fluctuating field, l the dimension of the correlation region, and v the neutron velocity, then the probability $w_{\uparrow\downarrow}$ of spin flip during the time t the neutron is in the fluctuating field can be represented by the expression

$$w_{\uparrow\downarrow} = \tfrac{2}{3}\gamma^2 B^2 t t_0 \tilde{K}(2\omega_0 t_0), \tag{7.9}$$

where $t_0 = l/v$ and γ determines the frequency of precession in the field B: $\omega = \gamma B$. $\omega_0 = \gamma B_0$ is the precession frequency of the neutron spin in the homogeneous leading field B_0, and $\tilde{K}(s)$ is the Fourier transform of the correlation function $\tilde{K}(s) = \int_{-\infty}^{\infty} K(t'/t_0)\exp(2i\omega_0 t')\,d(t'/t_0)$, where $s = 2\omega_0 t_0$. The presence of the factor t in (7.9) signifies that $w_{\uparrow\downarrow}$ is the higher, the longer the time the UCN are found in the gap between the polarizer and the analyser, i.e. the longer is this gap itself.

In principle, the quantity $w_{\uparrow\downarrow}$ may not depend on the length of the gap. For instance, if the directions of the fields in P and A are at a certain angle θ to each other, then as Byrn (1979) demonstrated, the polarization ratio for a UCN beam transmitted through the polarizer film is characterized by a polarization vector $\mathbf{P}_1$, and the analyser film characterized by a similar vector

$\mathbf{P}_2$ when the SF operation is perfect, $\varepsilon = \mathbf{P}_1\mathbf{P}_2 = P_1P_2\cos\theta$, i.e. the quantity $\cos\theta$ plays the part of a characteristic parameter w. But such considerations are not very satisfactory, in the case of UCN since, due to the small velocity of a neutron, its spin follows adiabatically the direction of the field (Byrn 1979). Indeed, let the angle between the film fields be $\pi/3$ (which is, naturally, overstated) and the distance between them be 10 cm. In accordance with Vladimirskii (1960) the probability of depolarization can be estimated from the quantity $w_{\uparrow\downarrow} = \exp(-\pi\omega\tau)$, where ω is the precession frequency in the leading field and τ is the characteristic rotation time of the field. In the case under consideration, when the velocity of the UCN is $\sim 5\ \mathrm{m\,s^{-1}}$, $\tau = 0.02$ s. For the frequency $\dot{\omega} = 10^5\ \mathrm{s}^{-1}$, which corresponds to a leading field of about 1 G, we obtain an extremely small $w_{\uparrow\downarrow}$. Thus, both experiment and theory testify against the suggestion that the relatively small value of ε observed by Egorov *et al.* (1974) could be due to depolarization in the intermediate region. Hence it can be concluded that the low values of ε are due to the polarizing power differing from unity.

If we set aside the possibility of the film being non-uniform, when part of film is magnetized in the direction opposite to the main part of it, or because of the presence in the film of holes through which neutrons of both polarizations pass, then two effects may be pointed out as depolarizing factors: the depolarization on inhomogeneities of the magnetic field near the surface (Stepanov and Shelagin 1977*b*; Taran 1976); and the non-collinearity of the fields inside and outside the film (Ignatovich 1978*d*).

In Appendix 7.9 the depolarization is computed for UCN having passed through the magnetic film when on the exit surface a field fluctuation of magnitude B_1 is present with the correlation function $K(r) = v_0\delta(r)$, where $\delta(r)$ is the Dirac δ function and v_0 is the mean volume occupied by the field inhomogeneity. Since the expected depolarization is small, the calculation may be performed by applying perturbation theory. The probability of depolarization can be estimated by the quantity

$$w = cv_0(\mu 2m/\hbar^2)^{3/2}B_1^2/\sqrt{B_0}, \quad \varphi_r = 1 - 2w,$$

where $c = 0.03$. Thus, even if the fluctuation amplitude B_1 exhibits the value B_0 and the dimension of the fluctuation is $v_0 \sim B_0^{-3/2} \approx k_{\mathrm{lim}}^{-3}$, the polarizing power of the film will, nevertheless, not be less than 94%.

In Appendix 7.10 the polarizing power is considered of a film with internal and external fields on the interface that are not collinear. In this case $P = \cos\theta$, where θ is the angle between the internal and external fields B. To explain the low polarizing power of the film in this case the angle θ must be considered to be quite large: $\theta \approx 30°$. It is interesting that in the case of non-collinear internal and external fields the non-polarized neutrons impinging upon the film undergo a triple beam reflection, since, besides the specular reflection of the non-polarized beam, there are two lateral non-specular polarized beams reflected at angles differing from the specular angle ϑ by the quantity

$$d\vartheta = \pm(\mu B/E)\operatorname{tg}\vartheta,$$

where E is the energy of the incident neutrons.

The reflection at non-specular angles is due to the component of the neutron momentum parallel to the reflecting surface being conserved, while the normal component changes when a spin flip occurs in the external field.

7.4 ROTATION OF THE NEUTRON SPIN

To rotate the spin of a neutron two principles may be applied: the resonance method and the adiabatic method.

The resonance method consists of turning the spin by a radio-frequency field B_1 of frequency ω in a constant homogeneous magnetic field B_0. In Appendix 7.10 it is shown that the probability of a spin flip occurring in a time interval t is given by

$$w_{\uparrow\downarrow} = (\omega_1/\sqrt{(\omega-\omega_0)^2+\omega_1^2})^2\sin^2(\sqrt{\omega_1^2+(\omega-\omega_0)^2}\,t), \quad \omega_i = 2\mu B_i/\hbar.$$

This expression was obtained by Rabi (see *Krupchitskii (1985)) and it has the form of a resonance that is narrow when the field B_1 is low. The probability of a spin flip depends on the time t the neutron is in the region occupied by the radio-frequency field, and consequently on its velocity, so the efficiency of an SF is lower, the less monochromatic the neutron spectrum. In the case of UCN one has to deal with quite a broad neutron spectrum and therefore the resonance method of spin flipping does not always turn out to be the most effective one.

The adiabatic method consists of turning the spin by a radio-frequency field B_1 in an inhomogeneous constant field $B_0(x)$. If the radio-frequency field rotates with a frequency ω, then in the reference system rotating with it the spin is submitted to the action of an effective field, $B_{eff} = [(B_0 - B_\omega)^2 + B_1^2]^{1/2}$, where $B_\omega = \hbar\omega/\mu$. If at a certain point in space in the path of the neutron $B_0 = B_\omega$, while at the end points of the path $|B_0 - B_\omega| \gg B_1$, then the effective field in the rotating reference system turns about 180°. The spin of the neutron follows the effective field adiabatically, if the precession frequency in the field B_1 is higher than the velocity with which the effective field rotates at the resonance point $B_0 = B_\omega$, i.e. if the condition of adiabaticity,

$$d\omega_0/dt \ll \omega_1^2, \tag{7.10}$$

is satisfied. The calculations performed in Appendix 7.11 (see also Ezhov *et al.* (1976); Luschikov and Taran (1984)) show that $w_{\uparrow\downarrow} = \sin^2\chi/2$, where χ is the total rotation angle of the effective field in the rotating reference system. This

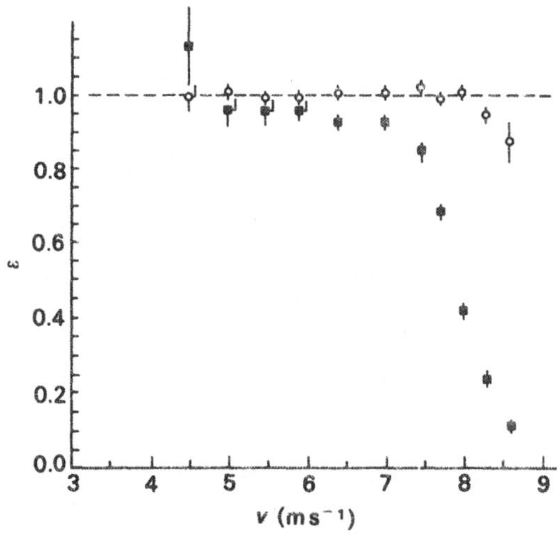

Fig. 7.8 Experimentally measured efficiency of the flipper (○) and polarization ratio (■) as function of the neutron velocity. The rapid fall of the polarization ratio in the vicinity of the limiting velocity represents an apparatus effect (Herdin *et al.* 1978).

angle is determined only by the magnetic fields of the system and is the same for neutrons of arbitrary energies. A restriction is imposed on the neutron energy by the condition of adiabaticity (7.10), which is not rigorous in the case of UCN. If the field B_0 changes linearly, at $\chi \approx 180°$ it is possible to take into account a correction due to non-adiabaticity. In Vladimirskii (1960) the probability of a spin flip was shown in this case to be

$$w_{\uparrow\downarrow} = 1 - \exp(-\pi\omega_0/\omega_1),$$

i.e. the closer to unity, the better satisfied is the condition (7.10).

To conclude we present in Fig. 7.8 the experimental measurements of the efficiency of an SF based on the method of adiabatic rotation. These data were obtained by Herdin *et al.* (1978) by a time-of-flight method.

APPENDICES

Appendix 7.1 Depolarization of a neutron in an inhomogeneous magnetic field without the influence of the field on its velocity

$$\text{a)}\ i\partial\xi/\partial t = \mathbf{B}\sigma\xi, \quad \text{b)}\ \mathbf{B} = \mu[\mathbf{B}]/\hbar, \quad \text{c)}\ \mathbf{B} = (B_x, B_z), \quad \text{d)}\ B_x = B_0. \qquad (1)$$

$$\text{a)}\ B_z = -\dot{B}_0 t, \quad \text{b)}\ t_0 = \dot{B}_0^{-1/2}, \quad \text{c)}\ \lambda = B_0 t_0, \quad \text{d)}\ t/t_0 \to t. \qquad (2)$$

$$\text{a)}\ i\partial\xi/\partial t = (-t\sigma_x + \lambda\sigma_z)\xi, \quad \text{b)}\ \partial^2\xi/\partial t^2 = [i\sigma_z - (t^2 + \lambda^2)]\xi. \qquad (3)$$

$$\text{a) } \xi = \xi_- + \xi_+, \quad \text{b) } \sigma_z \xi_\pm = \pm\, \xi_\pm, \quad \text{c) } (\partial^2/\partial t^2 + t^2 + \lambda^2 \mp \mathrm{i})\xi_\pm = 0. \tag{4}$$

$$\xi_+ = \exp(\mathrm{i}t^2/2)[c_1 F(\alpha, \tfrac{1}{2}, -\mathrm{i}t^2) + c_2 \mathrm{i}\lambda t F(\alpha + \tfrac{1}{2}, \tfrac{3}{2}, -\mathrm{i}t^2)]. \tag{5}$$

$$\xi_- = \exp(\mathrm{i}t^2/2)[c_3 \mathrm{i}\lambda t F(\alpha + 1, \tfrac{3}{2}, -\mathrm{i}t^2) + c_4 F(\alpha + \tfrac{1}{2}, \tfrac{1}{2}, -\mathrm{i}t^2)]. \tag{6}$$

$$\text{a) } \alpha = \lambda^2/4\mathrm{i}, \quad \text{b) } t \to -\infty, \quad \text{c) } |\xi_-(t)|^2 \to 0, \quad \text{d) } |\xi_+(t)|^2 \to 1. \tag{7}$$

$$\text{a) } t \to +\infty, \quad \text{b) } |\xi_-(t)|^2 \to 1 - \exp(-\pi\lambda^2), \quad \text{c) } |\xi_+(t)|^2 \to \exp(-\pi\lambda^2). \tag{8}$$

$$\text{a) } \lambda^2 = B_0^2/\dot{B}_0 = \omega_0/\omega_1, \quad \text{b) } \omega_0 = B_0, \quad \text{c) } \omega_1 = \dot{B}_0/B_0. \tag{9}$$

$$\text{a) } \lambda \gg 1, \quad \text{b) } \exp(-\pi\lambda^2) \simeq 0. \tag{10}$$

$$\text{a) } B_0 \propto r^l, \quad \text{b) } \dot{B}_0 \simeq r^{l-1}\dot{r} \simeq r^{l-1}\sqrt{E}, \quad \text{c) } \lambda^2 \simeq r^{l+1}/\sqrt{E}. \tag{11}$$

In the reference system connected with the neutron the equation for the spinor ξ reduces to the form (1a), where σ are the Pauli matrices and the vector $\mathbf{B}$ is related to the external magnetic field $[\mathbf{B}]$ by the relation (1b). Let us choose zero time to coincide with the neutron passing through the minimum field and the direction of the coordinate axes to be such that the field can be represented in the form (1c, d) and (2a). We introduce the notation (2b, c) and pass to the dimensionless time by substitution of (2d). Then eqn (1a) will be rewritten in the form (3a). Differentiating (3a) with respect to time we obtain (3b). We represent the spinor ξ in the form of the sum (4a), the summands of which correspond to certain projections of the spin on to the z axis, satisfy the condition (4b) and the independent equations (4c) following from (3b) and (4b). The solutions of eqns (4c) are represented in the form of the combinations (5) and (6) of degenerate hypergeometric functions F with the parameter (7a). The constants c_i must be chosen so as to satisfy the initial conditions (7b–d) and the unitarity condition. The probability of a spin flip is then determined by the expressions (8) (Vladimirskii 1960). The parameter (2c) can be represented in the form (9a), where the notations (9b, c) are applied. In the case of large λ, (10a) and (10b) hold and from (8b, c) it follows that the neutron spin follows the field direction adiabatically. In the field of a $(2l + 1)$-pole magnet we have (11a, b) along the trajectory of a neutron, where $\mathbf{r}$ is the radius vector of the trajectory and the parameter determining the probability of depolarization is proportional to (11c).

Appendix 7.2 UCN storage time in the magnetic field of a $2(l + 1)$-pole magnet

$$\text{a) } \mathbf{B} = B_0(r/r_0)^l \mathbf{e}, \quad \text{b) } \mathbf{e} = (\sin l\varphi, \cos l\varphi), \quad \text{c) } \operatorname{rot}\mathbf{B} = \operatorname{div}\mathbf{B} = 0. \tag{1}$$

$$\text{a) } B_0 = (2\mu m/\hbar^2)[B_0], \quad \text{b) } \kappa^2 = B_0 r_0^2, \quad \text{c) } t = (\hbar/2m)[t]\kappa^{2\alpha}/r_0^2. \tag{2}$$

$$\text{a) } r = [r]\kappa^{\alpha}/r_0, \quad \text{b) } \alpha = 2/(l+2). \tag{3}$$

$$\text{a) } [\mathrm{i}\partial_t + \Delta - r^l(\mathbf{e}\boldsymbol{\sigma})]\Psi(\mathbf{r}, t) = 0, \quad \text{b) } \Delta = \partial^2/\partial x^2 + \partial^2/\partial y^2. \tag{4}$$

$$\mathbf{e}\boldsymbol{\sigma} = \exp(\mathrm{i}l\varphi\sigma_z/2)\sigma_y\exp(-\mathrm{i}l\varphi\sigma_z/2). \tag{5}$$

$$\Psi(\mathbf{r}, t) = \exp(\mathrm{i}m\varphi + \mathrm{i}l\varphi\sigma_z/2)\psi_m(r, t)/\sqrt{2\pi}. \tag{6}$$

$$\text{a) } [\mathrm{i}\partial_t + \Delta_r - \mu^2/r^2 - \sigma_y r^l - (lm/r^2)\sigma_z]\psi_m = 0, \quad b) \ \Delta_r = (1/r)\partial_r r\partial_r. \tag{7}$$

$$\text{a) } \mu^2 = m^2 + l^2/4, \quad \text{b) } \delta u = lm\sigma_z/r^2, \quad \text{c) } \sigma_y\zeta_\pm = \pm\zeta_\pm, \quad \text{d) } \sigma_z\zeta_\pm = \zeta_\mp. \tag{8}$$

$$\psi_m(r, t) = \int \mathrm{d}E a_m(Et)g_m(E, r)\zeta_- \exp(-\mathrm{i}Et)$$
$$+ \sum_n a_{mn}(t)g_{mn}(r)\zeta_+ \exp(-\mathrm{i}E_n t). \tag{9}$$

$$\text{a) } \int g_m(E, r)g_m(E', r)r\,\mathrm{d}r = \delta(E - E'), \quad \text{b) } \int g_{mn}(r)g_{mn'}(r)r\,\mathrm{d}r = \delta_{nn'}. \tag{10}$$

$$\text{a) } a_{mn}(0) = \delta_{nn_0}, \quad \text{b) } a_m(E, 0) = 0. \tag{11}$$

$$\mathrm{d}a_m(E, t)/\mathrm{d}t = M(E, E_n)\exp[-\mathrm{i}(E_n - E)t]. \tag{12}$$

$$M(E, E_n) = lm \int g_m^*(E, r)g_{mn}(r)r\,\mathrm{d}r/r^2. \tag{13}$$

$$|a_m(E, t)|^2 = |M(E, E_n)|^2 \frac{\sin^2[(E - E_n)t/2]}{[(E - E_n)^2/4]}. \tag{14}$$

$$\sin^2 \alpha t/\alpha^2 \xrightarrow[t\to\infty]{} \pi t\delta(\alpha). \tag{15}$$

$$w_-(t) = \int |a_m(E, t)|^2\,\mathrm{d}E \xrightarrow[t\to\infty]{} \pi t|M(E, E_n)|^2_{E=E_n} = t/\tau. \tag{16}$$

$$\tau = 1/\pi|M(E, E_n)|^2_{E=E_n}. \tag{17}$$

In a $2(l+1)$-pole magnet, the axis of which is directed along the z direction, the field is written in the form (1a), where $\mathbf{e}$ is the unit vector (1b) and φ is the azimuthal angle. The field (1a) satisfies eqns (1c). For convenience the dimensionality of B_0 is chosen in the form (2a), where $[B_0]$ exhibits the same magnetic field but of the natural dimensionality. We shall introduce the dimensionless parameter (2b) and pass to the dimensionless variables (2c) and (3a), where $[t]$ and $[r]$ possess their usual dimensionalities. The parameter α is indicated in (3b). The Schrödinger equation for the neutron spinor wave-function Ψ reduces to the form (4a), where the Laplacian (4b) is given in the two-dimensional space. Let us take advantage of the transformation (5) and of (6). Substituting (5) and (6) into (4a) leads to eqn (7a, b), where the parameter μ is defined in (8a). We assume the quantity (8b) to be the perturbation. We introduce the spinors $\zeta_\pm$ satisfying the conditions (8c, d). We represent the solution of eqn (7a) in the form (9), where the functions

$g_{mn}(r)$ and $g_m(E, r)$ satisfy the non-perturbed equations. In this case $g_{mn}(r)$ corresponds to the bound state with the spin directed along the non-perturbed field, while the function $g_m(E, r)$ pertains to the continuous spectrum with the spin opposite to the field. These functions are normalized in accordance with (10a, b). We shall assume the neutron to be in a certain bound state n_0 initially, i.e. we assume the coefficients in the expansion (9) to be equal to (11a, b). Substitution of (9) into (7) to the first order of perturbation theory leads to eqn (12) for the coefficients $a_m(E, t)$, determining the relative number of states in the continuous spectrum and consequently the probability of leaving the trap. The matrix element is indicated in (13). Upon solving eqn (12) we find (14). Taking advantage of relation (15) we represent the probability of the UCN leaving the trap during the time t in the form (16), from which the storage time τ, eqn (17), is determined.

Appendix 7.3 Quasi-classical approximation

$$g_m(E, r) = (1/\sqrt{\pi r})c \cos[\theta(r) - \pi/4]/\sqrt{p(r)}. \tag{1}$$

$$\text{a) } p(r) = \sqrt{E - v(r)}, \quad \text{b) } v(r) = \mu^2/r^2 \pm r^l, \quad \text{c) } \mu^2 = m^2 + l^2/4. \tag{2}$$

$$\text{a) } \theta(r) = \int_{r_<}^{r} p(r')\mathrm{d}r', \quad \text{b) } p_n(r) = \sqrt{E_n - \mu^2/r^2 - r^l}, \quad \text{c) } \int_{r_<}^{r_>} p_n(r')\mathrm{d}r' = \pi\nu. \tag{3}$$

$$\text{a) } \nu = n_r + \tfrac{1}{2}, \quad \text{b) } \mu \to \lambda\mu, \quad \text{c) } \nu \to \lambda\nu, \quad \text{d) } E_n \to \lambda^\gamma E_n. \tag{4}$$

$$\text{a) } r' = \lambda^s r, \quad \text{b) } \int p_n(r')\mathrm{d}r' = \int \lambda^s \mathrm{d}r \sqrt{E_n\lambda^\gamma - \lambda^{2(1-s)}\mu^2/r^2 - \lambda^{ls}r^l} = \pi\lambda\nu. \tag{5}$$

$$\text{a) } ls = \gamma = 2(1 - s), \quad \text{b) } s = 2/(l + 2), \quad \text{c) } \gamma = 2l/(l + 2). \tag{6}$$

$$\text{a) } E_n = \mu^\gamma f(\nu/\mu)E_0 = \nu^\gamma \varphi(\mu/\nu)E_0, \quad \text{b) } E_0 = (\hbar^2/2mr_0^2)[2m\mu B_0 r_0^2/\hbar^2]^s. \tag{7}$$

$$\text{a) } \mu \gg \nu, \quad \text{b) } f(\nu/\mu) = a + b\nu/\mu, \quad \text{c) } \int_{r_<}^{r_>} \mathrm{d}r \sqrt{f - 1/r^2 - r^l} = \pi\nu/\mu. \tag{8}$$

$$\text{a) } v(r) = 1/r^2 + r^l = v_1(r_1) + v_1''(r - r_1)^2/2, \quad \text{b) } v'(r_1) = 0. \tag{9}$$

$$\text{a) } r_1 = \alpha^{s/2}, \quad \text{b) } \alpha = 2/l, \quad \text{c) } v_1(r_1) = \alpha^{1-s}/s, \quad \text{d) } v_1'' = 4\alpha^{-2s}/s. \tag{10}$$

$$\int_{r_<}^{r_>} \mathrm{d}r \sqrt{f - v_1 - v_1''(r - r_1)^2/2} = \tfrac{1}{2}(f - v_1)\sqrt{2/v_1''}\,\pi = \pi\nu/\mu. \tag{11}$$

$$f \approx v_1 + \sqrt{2v_1''}\,\nu/\mu = (1/s)\alpha^{1-s}(1 + \sqrt{2l\gamma}\,\nu/\mu). \tag{12}$$

$$\text{a) } \nu \gg \mu, \quad \text{b) } \pi = \int \mathrm{d}r \sqrt{\varphi - (\mu/\nu)^2/r^2 - r^l} \approx \int_0^{r_>} \mathrm{d}r \sqrt{\varphi - r^l}, \quad \text{c) } r_> = \varphi^{1/l}. \tag{13}$$

$$\text{a)}\ \int_0^{r_>} \mathrm{d}r\sqrt{\varphi - r^l} = \frac{\sqrt{\pi}}{4}\frac{\gamma\Gamma(1/l)\varphi^{1/\gamma}}{\Gamma(1/\gamma)}, \quad \text{b)}\ \varphi = \left(\frac{4\sqrt{\pi}}{\gamma}\frac{\Gamma(1/\gamma)}{\Gamma(1/l)}\right)^{\gamma}. \tag{14}$$

$$\text{a)}\ l = 2, \quad \text{b)}\ E_n = 2(\mu + 2\nu) = 2(\mu + 2n + 1). \tag{15}$$

$$g_m(E, r) = c_1 r^{\mu}\exp(-\mathrm{i}r^2/2)F(\mu/2 + \tfrac{1}{2} + \mathrm{i}E/4, \mu + 1, r^2). \tag{16}$$

$$g_{mn}(r) = g_m(E_n, r) = c_2 r^{\mu}\exp(-r^2/2)F(-n, \mu + 1, r^2). \tag{17}$$

The quasi-classical wavefunctions have the form (1), where c is a normalizing constant, $p(r)$ is the quasi-classical momentum, (2a), $v(r)$ is the generalized potential including the centrifugal potential μ^2/r^2, (2b), with the parameter μ^2 indicated in (2c). The angle θ is presented in (3a), where $r_<$ is the turning point closest to the centre. For a bound state the momentum is represented in the form (3b), from which the quantization condition (3c) is obtained, where $r_>$ is the second turning point (apogee). The parameter ν is indicated in (4a) and n_r is the radial quantum number. We shall demonstrate the energy of the bound state, E_n, to be a homogeneous function of μ and ν. To this end we perform the scaling transformation (4b, c) and show E_n to transform in accordance with (4d). Performing in (3c) the change of variables (5a) we obtain (5b). Expression (5b) is identical to (3c), if the equalities (6a) are satisfied, from which (6b, c) follow. Thus, the energy levels are determined by one or the other expressions (7a). The characteristic parameter of the same dimensionality as energy is indicated in (7b). Under the condition (8a) the function f in (7a) can be represented in the form (8b). The coefficients a and b are determined from the integral (8c). We now expand the potential in the vicinity of the minimum point (in the case of large μ the trajectory is nearly circular, i.e. the displacements along r are insignificant). The minimum is determined by the condition (9b). The minimum point is indicated in (10a, b). The coefficients of the expansion (9a) are given in (10c, d). Substitution of (9a) into (8c) leads to (11), from which (12) is found; the parameter s is indicated in (6b). When the condition (13a) is fulfilled we reduce expression (3) to the form (13b, c). Integration in (13b) leads to (14a), from which for φ in (7a) the asymptotic value (14b) is obtained. In the case of a six-pole magnet we have $l = 2$, eqn (15a); for the energy levels we have the exact values (15b) and the exact wavefunctions for the non-perturbed problem, (16) and (17), respectively, where $F(a, b, x)$ are degenerate hypergeometric functions.

Appendix 7.4 Using quasi-classical functions to calculate τ when the UCN are kept in a $2(l + 1)$-pole magnet

$$g_{mn}(r) = \sqrt{1/\pi r}\, c_n \cos[\theta_{mn}(r) - \pi/4]/\sqrt{p_n(r)}. \tag{1}$$

$$g_m(E, r) = \sqrt{1/\pi r}\, c_E \cos[\theta_m(E, r) - \pi/4]/\sqrt{p(E, r)}. \tag{2}$$

$$\text{a) } \theta_m(E, r) = \int_{r_<}^{r} p(E, r')\,\mathrm{d}r', \quad \text{b) } p(E, r) = \sqrt{E - \mu^2/r^2 + r^l}. \tag{3}$$

$$\text{a) } \int g_m(E, r) g_m(E', r) r\,\mathrm{d}r = |c_E|^2 \delta(E - E'), \quad \text{b) } c_E = 1. \tag{4}$$

$$\text{a) } \int g_{mn}^2(r) r\,\mathrm{d}r = |c_n|^2 E^{(2-l)/2l} A, \quad \text{b) } c_n = E^{-(2-l)/4l}. \tag{5}$$

$$\text{a) } A = \int \mathrm{d}x/2\pi\sqrt{1 - \beta^2/x^2 - x^l} \approx 1, \quad \text{b) } \beta^2 = \mu^2 E^{-(2+l)/l} \lesssim 1. \tag{6}$$

$$\text{a) } M(E, E_n) = \int g_m(E, r) g_{mn}(r) r\,\mathrm{d}r/r^2, \quad \text{b) } r = xE^{1/l}. \tag{7}$$

$$\text{a) } E = E_n, \quad \text{b) } p_n(r) = \sqrt{E}p_-(x), \quad \text{c) } p(E, r) = \sqrt{E}p_+(x). \tag{8}$$

$$\text{a) } p_\pm(x) = \sqrt{1 - \beta^2/x^2 \pm x^l}, \quad \text{b) } \theta(r) = E^{1/\gamma}\theta_\pm(x). \tag{9}$$

$$\text{a) } \theta_\pm(x) = \int_{x_<^\pm}^{x} p_\pm(x')\,\mathrm{d}x', \quad \text{b) } \gamma = 2l/(l+2), \quad \text{c) } x_<^+ - x_<^- \approx \beta^{l+1}. \tag{10}$$

$$\text{a) } \mu \gg \nu, \quad \text{b) } E_n \approx \mu^\gamma, \quad \text{c) } \beta \approx 1, \quad \text{d) } M(E, E_n) = M. \tag{11}$$

$$M \approx \int_{x_<^+} \cos(\mu\theta_+(x))(x\,\mathrm{d}x/\mu x^2\sqrt{\pi x p_+(x)})g_{mn}(x). \tag{12}$$

$$M \approx g_{mn}(x_<^-)\int_0^\infty \cos(\mu y^{3/2})\,\mathrm{d}y/\mu y^{1/4} \approx \mathrm{e}^{-\mu}/\mu^{3/2}. \tag{13}$$

$$\text{a) } \nu \gg \mu, \quad \text{b) } \beta \ll 1, \quad \text{c) } x = \beta y, \quad \text{d) } p_\pm(x) = p_\pm(y) = \sqrt{1 - 1/y^2 \pm \beta^l y^l}. \tag{14}$$

$$M \approx (\tfrac{1}{2}\pi\nu\beta)\int_{y_<^+}^\infty \cos\{\nu\beta/[\theta_{m+}(y) - \theta_{m-}(y)]\}y\,\mathrm{d}y/y^3\sqrt{p_+(y)p_-(y)}. \tag{15}$$

$$M \approx (\tfrac{1}{2}\pi\nu\beta)\int_0^\infty \cos\left(\nu\beta^{l+1}\int_0^z (x+1)^{l/2}\,\mathrm{d}x/\sqrt{x}\right)\mathrm{d}z\Big/\sqrt{z}(z+1). \tag{16}$$

$$\text{a) } \nu\beta^{l+1} \ll 1, \quad \text{b) } M \approx \tfrac{1}{2}\nu\beta, \quad \text{c) } \nu\beta^{l+1} \gg 1, \quad \text{d) } M \approx 1/\pi\nu^2\beta^{l+2}. \tag{17}$$

$$\text{a) } \tau = 1/\pi l m M, \quad \text{b) } \tau \sim \sqrt{\mu}\mathrm{e}^\mu, \quad \text{c) } \tau \sim 2/\pi l, \quad \text{d) } \tau = \nu\beta^{l+1}. \tag{18}$$

The radial functions for the discrete and continuous spectra are, respectively, equal to (1) and (2). For the continuous spectrum the notations (3a, b) are utilized and for the discrete spectrum we adopt the notations (3a, b) from Appendix 7.3. From the normalization condition (4a), into which the functions (2) are substituted, follows (4b). Requiring the norm (5a) to be equal to unity we obtain (5b), because, in accordance with (6a), the factor A is close to unity, since the relation (6b) is always valid. Transition between two states is determined by the matrix element (7a). Since the transition is elastic, $E = E_n$ and (7b) can be substituted into the integral (7a). Upon performing such substitution the quantities p transform in accordance with (8b, c) and (9a), while the phases θ transform in accordance with (9b) and (10a, b). The difference between the lower limits in the integrals (10) can be estimated by the quantity (10c). When (11a) holds, equalities (11b, c) hold approximately. We shall adopt the notation (11d). Expression (7a) is written in the form (12), and it can be approximated (this approximation can only yield an upper estimate) by the expression (13). In the case when (14a) is valid, (14b) also holds. Performing the substitution (14c) we obtain (14d), and then the matrix element (7a) reduces to the form (15). It may be approximated by the expression (16). In the case (17a) expression (16) may be estimated by the quantity (17b), while if (17c) is valid (16) can be estimated by the quantity (17d). The storage time without taking account of the self-decay is determined by formula (18a). When $\mu \gg \nu \gg 1$, substituting (13) here we obtain the estimate (18b), while in the case of (11a) for (17a) we obtain (18c) and for (17c) we obtain the estimate (18d).

Appendix 7.5 Bound UCN states in the field of a straight current

$$\text{a) } \mathbf{B} = 0.2J\mathbf{e}/\mathbf{r}, \quad \text{b) } \mathbf{e} = (-\sin\varphi, \cos\varphi), \quad \text{c) } \mathbf{r} = \mathbf{r}(\cos\varphi, \sin\varphi). \tag{1}$$

$$\text{a) } r = [r]\sqrt{E}, \quad \text{b) } (2m/\hbar^2)0.2\mu J/\sqrt{E} = 2n. \tag{2}$$

$$[-\Delta + 2n(\boldsymbol{\sigma}\mathbf{e})/r + 1]\Psi(\mathbf{r}) = 0. \tag{3}$$

$$\text{a) } \boldsymbol{\sigma}\mathbf{e} = \mathrm{i}(\boldsymbol{\sigma}\mathbf{r})\sigma_z/r, \quad \text{b) } \Psi(\mathbf{r}) = \sigma_z(\boldsymbol{\sigma}\mathbf{r})\int \psi(\mathbf{p})\exp(\mathrm{i}\mathbf{p}\mathbf{r})\,\mathrm{d}^2p. \tag{4}$$

$$\text{a) } [(p^2+1)\sigma_z(\boldsymbol{\sigma}\nabla_p) + 2n]\psi(\mathbf{p}) = 0, \quad \text{b) } \sigma_z(\boldsymbol{\sigma}\mathbf{p}) = \mathrm{i}p(\boldsymbol{\sigma}\mathbf{e}_p). \tag{5}$$

$$\text{a) } \mathbf{e}_p = (-\sin\varphi_p, \cos\varphi_p), \quad \text{b) } (\boldsymbol{\sigma}\mathbf{p})(\boldsymbol{\sigma}\nabla_p) = \mathbf{p}\nabla_p + \mathrm{i}\boldsymbol{\sigma}[\mathbf{p}\nabla] = p\partial_p + \mathrm{i}\sigma_z\partial_{\varphi p}. \tag{6}$$

$$\{(p^2+1)[\partial/\partial p + \mathrm{i}(\sigma_z/p)\partial/\partial\varphi_p] + 2\mathrm{i}n(\boldsymbol{\sigma}\mathbf{e}_p)\}\psi(\mathbf{p}) = 0. \tag{7}$$

$$\psi(p) = \exp(-\mathrm{i}\sigma_z\varphi_p/2 + \mathrm{i}m\varphi_p)p^{m-\frac{1}{2}}\begin{pmatrix} f_+(p) \\ pf_-(p)\end{pmatrix}. \tag{8}$$

$$\begin{aligned}&\text{a) } [(1+p^2)\mathrm{d}f_+/\mathrm{d}p^2 + nf_- = 0,\\ &\quad \text{b) } (1+p^2)[(m+\tfrac{1}{2}) + p^2\mathrm{d}/\mathrm{d}p^2]f_- = nf_+.\end{aligned} \tag{9}$$

$$\text{a) } x = p^2/(p^2+1), \quad \text{b) } x(1-x)f''_+ + (m+\tfrac{1}{2}-x)f'_+ - nf_+ = 0. \tag{10}$$

$$\text{a) } f_+(p) = c_{nm}F(n, -n, m+\tfrac{1}{2}, x), \quad \text{b) } E_n = \frac{(0.2\,J\mu/h)^2 m_n}{2n^2}. \tag{11}$$

$$\begin{pmatrix} f_+(p) \\ pf_-(p)\end{pmatrix} = c_{nm}\begin{pmatrix}(m+\tfrac{1}{2})F(n, -n, m+\tfrac{1}{2}, x) \\ [np/(p^2+1)]F(n+1, 1-n, m+\tfrac{3}{2}, x)\end{pmatrix}. \tag{12}$$

The magnetic field of a straight-line wire along which a current J flows is represented in the form (1a), where **e** is a unit vector, eqn (1b). We shall pass to the dimensionless coordinate (2a) and introduce the dimensionless parameter n, eqn (2b). The Schrödinger equation in this case reduces to the form (3). We take advantage of the identity (4a) and represent the wavefunction in the form (4b). Substituting (4) into (3) we obtain eqn (5a). We introduce the vector $\mathbf{e}_p$ by means of the equality (5b); it is indicated in (6a). We multiply eqn (5a) on the left by $\boldsymbol{\sigma}\mathbf{e}_p$, then take advantage of equality (5b) and of (6b), and, as a result, obtain eqn (7). We represent its solution in the form (8). Substituting (8) into (7) we obtain a set of equations for the components $f_{\pm}(p)$, eqn (9). We introduce the variable (10a) and reduce the set of equations to the second-order equation (10b). Its solution is indicated in (11a), where c_{nm} is a normalizing constant and $F(a, b, c, x)$ is a hypergeometric function. The energy levels are indicated in (11b). Having determined $f_{-}(p)$ from (9a) we find, upon substituting (11a), the expression (12) for the spinor in (8) (*Pron'ko and Stroganov 1977).

Appendix 7.6 Phenomenology of polarizing films

a) $\vec{T}_l = t_l p_z(p_x + \tau_l \bar{p}_x)$, b) $\overleftarrow{T}_l = t_l(p_x + \tau_l \bar{p}_x)p_z$, c) $\overleftrightarrow{T}_m = t_0$. (1)

a) $\vec{T}_r = t_r(p_x + \tau_r \bar{p}_x)p_z$, b) $\overleftarrow{T}_r = t_r p_z(p_x + \tau_r \bar{p}_x)$. (2)

a) $p_i = (1 + \sigma_i)/2$, b) $\bar{p}_i = (1 - \sigma_i)/2$, c) $p_i^2 = p_i$, d) $p_i + \bar{p}_i = 1$. (3)

a) $\overleftarrow{R}_l = r_l = 1 - t_l$, b) $\overleftarrow{R}_r = r_r = 1 - t_r$. (4)

a) $\vec{T} = \vec{T}_r(1 - \vec{T}_m \overleftarrow{R}_l \overleftarrow{T}_m \vec{R}_r)^{-1} \vec{T}_m \vec{T}_l$, b) $\vec{T} = 2t(p_x + \tau_r \bar{p}_x)p_z(p_x + \tau_l \bar{p}_x)$. (5)

a) $t = t_0 t_l t_r / 2[1 - t_0^2(1 - t_l)(1 - t_r)]$, b) $p_i \bar{p}_i = 0$, c) $p_x \sigma_z = \sigma_z \bar{p}_x$. (6)

a) $\vec{T} = t[p_x + \tau_l \tau_r \bar{p}_x + \sigma_z(\tau_r p_x + \tau_l \bar{p}_x)]$, b) $t = 1$, c) $n = (1 + P_0 \sigma_z)n_0$. (7)

a) $p_x n_0 = n_0$, b) $\vec{T}n = (1 + \tau_l P_0)(1 + \tau_r \sigma_z)n_0$, c) $\vec{T}n_0 = (1 + \tau_r \sigma_z)n_0$. (8)

a) $\tau_r = \vec{P}$, b) $\tau_l = \overleftarrow{P}$, c) $\vec{T} = [p_x + S\bar{p}_x + \sigma_z(\vec{P}p_x + \overleftarrow{P}\bar{p}_x)]$, d) $S = \vec{P}\overleftarrow{P}$. (9)

a) $\vec{T}n = (1 + \overleftarrow{P}P_0)(1 + \vec{P}\sigma_z)n_0$, b) $n\vec{T} = n_0(1 + \overleftarrow{P}\sigma_z)(1 + \vec{P}P_0)$. (10)

a) $\vec{R}_l = p_x + a_l \bar{p}_x - t_l(1 + \overleftarrow{P}\sigma_z)/2$, b) $\vec{R}_l = r_l(p_x + b_l \bar{p}_x + s_l \sigma_z)$. (11)

a) $r_l = 1 - t_l/2$, b) $b_l = (a_l - t_l/2)/r_l$, c) $s_l = -t_l \overleftarrow{P}/2r_l$. (12)

a) $\vec{R} = \vec{R}_l + \overleftarrow{T}_l \overleftarrow{T}_m \vec{R}_r(1 - \vec{T}_m \overleftarrow{R}_l \overleftarrow{T}_m \vec{R}_r)^{-1} \vec{T}_m \vec{T}_l$,

b) $\vec{R} = R_l(p_x + \rho_l \bar{p}_x + \xi_l \sigma_z)$. (13)

a) $R_l = r_l + t t_0 r_r t_l / t_r$, b) $\rho_l = [r_l b_l + (R_l - r_l)\overleftarrow{P}^2]/R_l$. (14)

a) $\xi_l = \overleftarrow{P} t_l(2 t t_0 r_r / t_r - 1)/2R_l$, b) $\overleftarrow{R} = R_r(p_x + \rho_r \bar{p}_x + \xi_r \sigma_z)$. (15)

Each single-domain magnetic film can be considered to consist of three parts: the left surface part with the left-to-right transmission matrix (1a) and with (1b) for the opposite direction; the middle part which only transmits neutrons of a single polarization and is characterized by the same transmission coefficient (1c) in both directions; and the right surface part with the left-to-right transmission matrix (2a) and the right-to-left matrix (2b). Utilized here are projection operators of the form (3a, b) with the Pauli matrices σ_i ($i = x, z$) possessing the properties (3c, d). Reflection from the surface parts within the film is characterized by the coefficients (4a, b), as follows from (1b), (2a), and from unitarity, and neglecting surface absorption. The total transmission matrix of the film from the left to the right is written with the aid of recursive relations in the form (5a). Substituting here expressions (1), (2), and (4) we obtain (5b), where the transmission coefficient t is (6a). Now we shall take advantage of the properties of the Pauli matrices, from which the relations (6b, c) follow, and reduce (5b) to the form (7a). Since in polarization analysis relative quantities are measured, for simplicity we can assume (7b). A neutron beam of polarization P_0 is described by expression (7c), where n_0 corresponds to a non-polarized beam and satisfies the condition (8a). The action of matrix (7a) on the vector (7c) and n_0 with (7b) taken into account leads to the expressions (8b, c). We introduce the notations (9a, b); then (7a), taking into account (7b), can be rewritten in the form (9c, d). The action of (9c, d) on (7c) from the right and from the left leads to the expressions (10). Using (1a), the reflection matrix from the left surface part of the film is represented in the form (11a), where a_1 is an arbitrary parameter: $|a_1| \leqslant 1$. With the aid of (3d) we reduce (11a) to the form (11b) where the new parameters (12) are introduced. With the aid of recursive relations the total reflection matrix from the left is written in the form (13a). Substituting here (11b), (1), and (4) we reduce (13a) to the form (13b) with the parameters indicated in (14) and (15a). It is possible in a similar way to write the reflection matrix (15b) from the right side of the film. As follows from (7a), (13b), and (15b), the total number of parameters describing transmission of the film and reflection from it from both sides is nine: t, τ_i, R_i, ρ_i, ξ_i ($i = \mathrm{r}, 1$). However, if absorption in the surface regions of the film is neglected, then these nine parameters are expressed through the independent parameters t_0, t_i, τ_i, and a_i ($i = \mathrm{r}, \mathrm{l}$), the number of which is seven.

Appendix 7.7 The transmission of two or three films with and without rescattering

$$\text{a) } \overrightarrow{T}_{21} = \overrightarrow{T}_2(1 - \overrightarrow{T}_\mathrm{f}\overleftarrow{R}_1\overleftarrow{T}_\mathrm{f}\overrightarrow{R}_2)^{-1}\overrightarrow{T}_\mathrm{f}\overrightarrow{T}_1, \quad \text{b) } \overrightarrow{T}_\mathrm{f} = \overleftarrow{T}_\mathrm{f} \equiv F = (p_x + \varphi\bar{p}_x). \tag{1}$$

$$\text{a) } \overrightarrow{R}_2 = R_{21}(p_x + \rho_{21}\bar{p}_x + \xi_{21}\sigma_z), \quad \text{b) } \overleftarrow{R}_1 = R_{1\mathrm{r}}(p_x + \rho_{1\mathrm{r}}\bar{p}_x + \xi_{1\mathrm{r}}\sigma_z). \tag{2}$$

$$\text{a) } \overrightarrow{T}_i = [p_x + \overleftarrow{P}_i\overrightarrow{P}_i\bar{p}_x + \sigma_z(\overrightarrow{P}_i p_x + \overleftarrow{P}\bar{p}_x)], \quad \text{b) } \hat{R}_i = R_i(p_x + \rho_i\bar{p}_x + \xi_i\sigma_z). \tag{3}$$

$$\text{a) } \hat{} = \overleftarrow{}, \overrightarrow{}, \quad \text{b) } (1 - F\overleftarrow{R}_1 F\overrightarrow{R}_2)^{-1} = (1 - \tilde{R}_2\tilde{F}\tilde{R}_1\tilde{F})/N, \quad \text{c) } \tilde{F} = (\bar{p}_x + \varphi p_x). \tag{4}$$

$$\text{a) } \tilde{R}_i = R_i(\bar{p}_x + \rho_i p_x - \xi_i\sigma_z),$$
$$\text{b) } N = (1 - F\overleftarrow{R}_1 F\overrightarrow{R}_2)(1 - \tilde{R}_2\tilde{F}\tilde{R}_1\tilde{F}) \equiv n_0 N n_0/2. \tag{5}$$

$$\text{a) } N = n_0(1 - F\overleftarrow{R}_1 F\overrightarrow{R}_2 - \tilde{R}_2\tilde{F}\tilde{R}_1\tilde{F} + F\overleftarrow{R}_1 F\overrightarrow{R}_2\tilde{R}_2\tilde{F}\tilde{R}_1\tilde{F})n_0/2,$$
$$\text{b) } \hat{R}\tilde{R} = R^2(\rho - \xi^2). \tag{6}$$

$$N = 1 + R_1^2R_2^2(\rho_1 - \xi_1^2)(\rho_2 - \xi_2^2)\varphi^2 - R_1R_2(1 + 2\xi_1\xi_2\varphi + \rho_1\rho_2\varphi^2). \tag{7}$$

$$\text{a) } J = n_0\vec{T}_{12}n_0/2, \quad \text{b) } J = n_0\vec{T}_2(1 - \tilde{R}_2\tilde{F}\tilde{R}_1\tilde{F})F\vec{T}_1 n_0/2N. \tag{8}$$

$$J = \{1 + \overleftarrow{P}_2\vec{P}_1\varphi - R_1R_2\varphi[\varphi(\rho_2 - \overleftarrow{P}_2\xi_2)(\rho_1 - \vec{P}_1\xi_1)$$
$$+ (\overleftarrow{P}_2 - \xi_2)(\vec{P}_1 - \xi_1)]\}/N. \tag{9}$$

$$J_\pm = \{1 \pm \overleftarrow{P}_2\vec{P}_1 - R_1R_2[(\rho_2 - \overleftarrow{P}_2\xi_2)(\rho_1 - \vec{P}_1\xi_1)$$
$$\pm (\overleftarrow{P}_2 - \xi_2)(\vec{P}_1 - \xi_1)]\}/N_\pm. \tag{10}$$

$$N_\pm = 1 + R_1^2R_2^2(\rho_1 - \xi_1^2)(\rho_2 - \xi_2^2) - R_1R_2(1 \pm 2\xi_1\xi_2 + \rho_1\rho_2). \tag{11}$$

$$\text{a) } \xi_1 = \mu_1\vec{P}_1, \quad \text{b) } \xi_2 = \mu_2\overleftarrow{P}_2, \quad \text{c) } \rho_1 = \nu_1\vec{P}_1^2, \quad \text{d) } \rho_2 = \nu_2\overleftarrow{P}_2^2. \tag{12}$$

$$\text{a) } J_\pm = (1 \pm Ax - Bx^2)/(C \mp Dx + Ex^2), \quad \text{b) } A = 1 - R_1R_2(1 - \mu_2)(1 - \mu_1). \tag{13}$$

$$\text{a) } B = R_1R_2(\nu_2 - \mu_2)(\nu_1 - \mu_1), \quad \text{b) } C = 1 - R_1R_2, \quad \text{c) } D = 2R_1R_2\mu_1\mu_2. \tag{14}$$

$$\text{a) } E = R_1^2R_2^2(\nu_1 - \mu_1^2)(\nu_2 - \mu_2^2) - R_1R_2\nu_1\nu_2, \quad \text{b) } x = \vec{P}_1\overleftarrow{P}_2. \tag{15}$$

$$\text{a) } \varepsilon = (J_+ - J_-)/(J_+ + J_-), \quad \text{b) } \varepsilon = x(\alpha + \beta x^2)/(1 + \gamma x^2 - \delta x^4). \tag{16}$$

$$\text{a) } \alpha = A + D/C, \quad \text{b) } \beta = (AE - DB)/C,$$
$$\text{c) } \gamma = (AD + E)/C - B, \quad \text{d) } \delta = BE/C. \tag{17}$$

$$\text{a) } W = \vec{J}_{321} = n_0\vec{T}_3F_2\vec{T}_2F_1\vec{T}_1n_0/2, \quad \text{b) } W = t(1 + \overleftarrow{P}_3\vec{P}_2\varphi_2)(1 + \overleftarrow{P}_2\vec{P}_1\varphi_1). \tag{18}$$

$$\text{a) } \vec{T}_i = [p_x + S_i\bar{p}_x + \sigma_z(\vec{P}_ip_x + \overleftarrow{P}_i\bar{p}_x)], \quad \text{b) } n = (1 + P_0\sigma_z)n_0. \tag{19}$$

$$\text{a) } T_in = [1 + P_0\overleftarrow{P}_i + (\vec{P}_i + S_iP_0)\sigma_z]n_0, \quad \text{b) } x = \overleftarrow{P}_2\vec{P}_1, \quad \text{c) } y = \overleftarrow{P}_3\vec{P}_2. \tag{20}$$

$$\text{a) } W = t(1 + \varphi_1x + \varphi_2y + \varphi_1\varphi_2z), \quad \text{b) } z = \overleftarrow{P}_3S_2\vec{P}_1. \tag{21}$$

$$\text{a) } a = W(++), \quad \text{b) } b = W(-+), \quad \text{c) } c = W(+-), \quad \text{d) } d = W(--). \tag{22}$$

$$\text{a) } t = (a + b + c + d)/4, \quad \text{b) } y = (a + b - c - d)/(a + b + c + d). \tag{23}$$

$$\text{a) } x = (a - b + c - d)/(a + b + c + d), \quad \text{b) } (a - b - c + d)/(a + b + c + d). \tag{24}$$

With the aid of recurrent relations the transmission of two films with a spin flipper in between them is written in the form (1a), where the transmission of the spin flipper is represented in (1b). The reflection matrices of the films are indicated in (2). We shall represent the transmission of the films in the form (3a). For convenience we introduce for (2) the notations (3b) and (4a). The inverse matrix in (1a) is readily verified to be written in the form (4b) where the notations (4c) and (5a) are used. The numerical factor N equals (5b) and it is most conveniently calculated as a matrix element in between n_0 states. Removing the brackets in (5b) we obtain (6a). Substituting the expression (6b), which is readily verified, we reduce (6a) to the form (7). The total transmission of a non-polarized beam through two films equals (8a). Substituting here (1a) and (4b) we obtain (8b) from which (9) is obtained by straightforward calculations. When $\varphi = \pm 1$, the expression (9) reduces to the form (10), (11). We now perform the transformation (12); then (10) will be rewritten in the form (13a), where the notations (13b), (14), and (15) are applied. The physically measured ratio (16a) is reduced, upon substituting (13a), to the form (16b) with the coefficients (17). Now we shall consider direct transmission of a non-polarized beam through three films in between which spin flippers are inserted. The probability of registering a neutron is determined by the expression (18a) which, upon performing the substitutions (1b) and (3a) and elementary calculations, is reduced to the form (18b). In the case of a poly-domain magnetic film the transmission matrix can be of the more general form (19a), and the beam (19b) is transformed by it into (20). Substituting (19b) into (18a) we obtain (21a), with the notations (20b, c) and (21b). Setting $\varphi_i = \pm 1$ and introducing the notation (22), where the first argument is related to φ_1 and the second to φ_2, we obtain (23) and (24).

Appendix 7.8 Depolarization of UCN in a weakly inhomogeneous magnetic field

$$\text{a) } i(\partial/\partial t)\xi(t) = [B_0\sigma_z + \boldsymbol{\sigma}\mathbf{B}(t)]\xi(t), \quad \text{b) } B = \gamma[B], \quad \text{c) } \gamma = \mu/\hbar. \tag{1}$$

$$\text{a) } \xi(t) = \exp(-iB_0\sigma_z t)\eta(t), \quad \text{b) } i(\partial/\partial t)\eta(t) = \boldsymbol{\sigma}\tilde{\mathbf{B}}(t)\eta(t). \tag{2}$$

$$\boldsymbol{\sigma}\tilde{\mathbf{B}}(t) = \exp(iB_0\sigma_z t)\boldsymbol{\sigma}\mathbf{B}(t)\exp(-iB_0\sigma_z t). \tag{3}$$

$$\text{a) } \eta(t) = \eta(0) + \int_0^t \boldsymbol{\sigma}\tilde{\mathbf{B}}(t')\,dt'\eta(0), \quad \text{b) } \eta(0) = \xi_\uparrow, \quad \text{c) } \sigma_z\xi_{\uparrow,\downarrow} = \pm\,\xi_{\uparrow,\downarrow}. \tag{4}$$

$$M_{\uparrow\downarrow} = (\xi_\downarrow\xi(t)) = \exp(iB_0 t)\int B_-(t')\exp(-2iB_0 t')\,dt'. \tag{5}$$

$$\text{a) } B_- = B_x - iB_y, \quad \text{b) } \langle B_i\rangle = 0, \quad \text{c) } \langle B_i(t)B_j(t')\rangle = (\bar{B}^{-2}/3)\delta_{ij}K(t-t'). \tag{6}$$

$$w_{\downarrow\uparrow} = \langle|M_{\uparrow\downarrow}|^2\rangle = \int_0^t dt'\int_0^t dt''\langle B_-^*(t')B_-(t'')\rangle\exp[2iB_0(t'-t'')]. \tag{7}$$

$$\text{a)}\ t \to \infty, \quad \text{b)}\ w_{\downarrow\uparrow} = \tfrac{2}{3}\overline{B^2}t\tilde{K}(2B_0), \quad \text{c)}\ \tilde{K}(x) = \int_{-\infty}^{\infty} K(t)\exp(ixt)\,dt. \tag{8}$$

$$\text{a)}\ K(t) = \exp(-|t|/t_0), \quad \text{b)}\ \tilde{K}(2B_0) = 2t_0/(1 + 4B_0^2 t_0^2). \tag{9}$$

$$\text{a)}\ t_0 = l/v, \quad \text{b)}\ t = L/v, \quad \text{c)}\ w_{\uparrow\downarrow} = \tfrac{4}{3}\overline{B^2}\,Ll/(v^2 + 4B_0^2 l^2). \tag{10}$$

$$\text{a)}\ K(t) = \exp\left(-\frac{t^2}{2t_0^2}\right), \quad \text{b)}\ w_{\downarrow\uparrow} = \frac{\sqrt{2\pi}}{3}\,\overline{B^2}\,\frac{Ll}{v^2}\exp(-2B_0^2 t_0^2). \tag{11}$$

$$\text{a)}\ w_{\downarrow\uparrow} = \lim_{t\to\infty}(w_{\uparrow\downarrow}/t), \quad \text{b)}\ d^3k_\uparrow N_\uparrow w_{\downarrow\uparrow} = d^3k_\downarrow N_\downarrow w_{\uparrow\downarrow}. \tag{12}$$

$$\text{a)}\ d^3k = d\Omega dEk/2, \quad \text{b)}\ dE_\uparrow = dE_\downarrow, \quad \text{c)}\ t_0 = l/k_{\uparrow\downarrow}. \tag{13}$$

$$p = (N_\uparrow - N_\downarrow)/(N_\uparrow + N_\downarrow) = (w_{\uparrow\downarrow}k_\downarrow - w_{\downarrow\uparrow}k_\uparrow)/(w_{\uparrow\downarrow}k_\downarrow + w_{\downarrow\uparrow}k_\uparrow). \tag{14}$$

In a reference system related to the moving neutron the equation of motion for the spin reduces to the form (1a), if the quantity (1b) is assumed as the field B, where $[B]$ exhibits the natural dimensionality of the magnetic field and γ is indicated in (1c). In eqn (1a) B_0 is the homogeneous external field and B is the fluctuating field. Let us perform substitution (2a); then (1a) is reduced to the form (2b), where the field $\tilde{B}$ is determined by the relation (3). In the first approximation in $\tilde{B}$ the solution of (2b) has the form (4a). Consider the spin to be initially polarized along the field, i.e. (4b) and (4c) hold. The matrix element describing the spin flip equals (5), where the quantity B_- is defined in (6a). We shall consider the field B to be random, i.e. to satisfy the conditions (6b, c), where the angle brackets signify averaging over all possible trajectories, $\overline{B^2}$ is the average square fluctuation, and $K(t)$ is the correlation function. The probability of a spin flip occurring is determined in (7). For large times (8a) the expression (7) reduces to the form (8b), involving the Fourier transform (8c) of the correlation function. For a correlation function of the form (9a) we have (9b). The parameter t_0 is related to the dimension l of the field inhomogeneity by the expression (10a), where v is the neutron velocity. The time t is related to the neutron path range L in the inhomogeneous field (10b). Substituting (10a, b) into (9b) and (8b) we obtain (10c). For a correlation function of the form (11a) we obtain (11b), instead of (10b). We can introduce the probability of spin flip per unit time (12a). In equilibrium the relation (12b) should hold, where $N_{\uparrow,\downarrow}$ is the number of neutrons with spins up and down, respectively, and $d^3k_{\uparrow,\downarrow}$ is the elementary volume in momentum space, (13a), for neutrons with spins up and down, respectively. Taking into account (13b) and substituting, instead of t_0, expression (13c) it is possible to find the probability of spin flip both in one and in the other direction. In equilibrium a polarization is established which can be calculated in accordance with (14).

Appendix 7.9 Depolarization of UCN passing through a magnetic film in the presence of magnetic inhomogeneities on the surface

$$\text{a) } [-\Delta - E + (u_0 - B_0\sigma_z)\theta(z<0) - \boldsymbol{\sigma}\mathbf{B}(\mathbf{r})]\Psi(\mathbf{r}) = 0, \quad \text{b) } u_0 = B_0. \tag{1}$$

$$\text{a) } B = (2m\mu/\hbar^2)[B], \quad \text{b) } \hat{G} = \vec{p}_{z\uparrow}G_\uparrow\vec{p}_{z\uparrow} + \vec{p}_{z\downarrow}G_\downarrow\vec{p}_{z\downarrow},$$

$$\text{c) } p_{z\uparrow,\downarrow} = (1 \pm \sigma_z)/2. \tag{2}$$

$$\text{a) } [\Delta + E - (u_0 \mp B_0)\theta(z<0)]G_{\uparrow,\downarrow}(\mathbf{r},\mathbf{r}') = \delta(\mathbf{r}-\mathbf{r}'), \quad \text{b) } \delta u = \boldsymbol{\sigma}\mathbf{B}(\mathbf{r}). \tag{3}$$

$$\text{a) } \sigma_\perp = (\pi/k_{\uparrow\perp})\int \overline{|\delta u|^2_{\uparrow\downarrow}}\,\delta(k_\uparrow^2 - k_\downarrow^2)\mathrm{d}^3k_\downarrow/(2\pi)^3, \quad \text{b) } w_{\downarrow\uparrow} = \sigma_\perp/S_s. \tag{4}$$

$$\text{a) } \Psi_{\uparrow,\downarrow} = \xi_{\uparrow,\downarrow}f_{\uparrow,\downarrow}(k_\perp, z)\exp(\mathrm{i}\mathbf{k}_\parallel\mathbf{r}_\parallel), \quad \text{b) } \sigma_z\xi_{\uparrow,\downarrow} = \pm\,\xi_{\uparrow,\downarrow}, \quad \text{c) } \mathbf{k}_\uparrow = \mathbf{k}_0. \tag{5}$$

$$\text{a) } f_\uparrow(k_\perp, z) = \exp(\mathrm{i}k_{0\perp}z), \quad \text{b) } \mathbf{k}_\downarrow = \mathbf{k}, \quad \text{c) } k'_\perp = \mathrm{i}k''_\perp,$$

$$\text{d) } k''_\perp = \sqrt{u_0 + B_0 - k_\perp^2}. \tag{6}$$

$$f_\downarrow(k_\perp, z) = t_\downarrow\exp(k''_\perp z)\theta(z<0) + [\exp(-\mathrm{i}k_\perp z) + r_\downarrow\exp(\mathrm{i}k_\perp z)]\theta(z>0). \tag{7}$$

$$\text{a) } t_\downarrow = 2k_\perp/(k_\perp + \mathrm{i}k''_\perp), \quad \text{b) } r_\downarrow = (k_\perp - \mathrm{i}k''_\perp)/(k_\perp + \mathrm{i}k''_\perp). \tag{8}$$

$$\text{a) } \delta u_{\downarrow\uparrow} = \int B_-(\mathbf{r})\exp(\mathrm{i}\mathbf{q}_\parallel\mathbf{r}_\parallel)f_\downarrow(k_\perp, z)f_\uparrow(k_{0\perp}, z)\mathrm{d}^3r, \quad \text{b) } q_\parallel = \mathbf{k}_{0\parallel} - \mathbf{k}_\parallel. \tag{9}$$

$$\text{a) } B_- = B_x - \mathrm{i}B_y, \quad \text{b) } \bar{B}_i = 0, \quad \text{c) } \overline{B_i(\mathbf{r})B_j(\mathbf{r}')} = (\bar{B}^2/3)\delta_{ij}K(\mathbf{r}-\mathbf{r}'). \tag{10}$$

$$\text{a) } K(r) = v_0\delta(r), \quad \text{b) } \overline{|\delta u|^2_{\downarrow\uparrow}} = \tfrac{2}{3}B^2 v_0|t_\downarrow|^2 S_s/2k''_\perp. \tag{11}$$

$$w_{\downarrow\uparrow} = \frac{2}{3}\frac{B^2 v_0}{k_{0\perp}(u_0 + B_0)}\int k_\perp^2\delta(k_0^2 - k^2)\frac{\mathrm{d}^3k}{(2\pi)^2 k''_\perp}. \tag{12}$$

$$\overline{w_{\downarrow\uparrow}} = \int\frac{1}{\pi}\cos\theta_0\,\mathrm{d}\Omega_0\int\frac{4k_0^3\mathrm{d}k_0}{(u_0 + B_0)^2}w_{\downarrow\uparrow}(k_0). \tag{13}$$

$$\overline{w_{\downarrow\uparrow}} = \frac{8}{3}\frac{\overline{B^2}v_0}{(2\pi)^2\sqrt{u_0 + B_0}}\int_0^1 k_0^2\,\mathrm{d}k_0\,2\pi\int_0^{k_0}\frac{k_\perp^2}{k''_\perp}\mathrm{d}k_\perp. \tag{14}$$

$$\overline{w_{\downarrow\uparrow}} = \frac{\sqrt{2}}{9\pi}\frac{\overline{B^2}v_0}{\sqrt{B_0}}\int_0^1(1 - x^{3/2})x^{1/2}(1-x)^{-1/2}\,\mathrm{d}x. \tag{15}$$

$$\overline{w_{\downarrow\uparrow}} = \frac{\sqrt{2}}{9\pi}\frac{\overline{B^2}v_0}{\sqrt{B_0}}[B(\tfrac{3}{2},\tfrac{1}{2}) - B(3,\tfrac{1}{2})]. \tag{16}$$

$$\text{a) } \overline{w_{\downarrow\uparrow}} = c\overline{B^2}v_0/\sqrt{B_0}, \quad \text{b) } c = (1/9\sqrt{2\pi})(\pi - 32/15) \approx 0.03. \tag{17}$$

$$\text{a) } \vec{P} = \varphi_r = 1 - 2\bar{w}_{\downarrow\uparrow} \approx 0.94, \quad \text{b) } \bar{B}^2 = B_0^2 = u_0^2, \quad \text{c) } v_0 u_0^{3/2} \approx 1. \tag{18}$$

The Schrödinger equation for the wavefunction can be reduced to the form (1a). Further, for simplicity we shall assume the nuclear and magnetic potentials to be equal to each other, eqn (1b). The dimensionality of the field is determined in accordance with (2a). The depolarization problem can be solved applying perturbation theory, assuming the field B to be the perturbation and taking advantage of the Green formalism, which can be represented in the form (2b), where $p_{z\uparrow,\downarrow}$ represent projection operators (eqn (2c)). The arrows over p_2 indicate the direction of the action of the operator. The scalar functions $G_{\uparrow,\downarrow}$ satisfy eqns (3a). Consecutive application of the Green functions for the perturbation (3b) yields the same result as when the scattering cross-section (4a) is calculated directly, from which the probability of depolarization is determined in accordance with (4b), where S_s is the total surface of the film. The wavefunctions of the initial and final states are written in the form (5a), involving spinors satisfying eqn (5b). The initial state is indicated by an arrow directed upward and by the wavevector $\mathbf{k}_0$. Its function $f(k_\perp, z)$ has the form (6a), if condition (1b) is fulfilled. The wavevector of the final state is $\mathbf{k}_\downarrow$, eqn (6b), and its normal component is $k'_\perp$ defined in (6c, d). The function $f_\downarrow$ for the final state is equal to (7), which involves the amplitudes (8). The matrix element for the spin flip is indicated in (9a). It depends on the vector $\mathbf{q}_\parallel$ (eqn (9b)) and on a combination of the components of the field B_- (eqn (10)). We shall consider the field B to be random, i.e. its components satisfy the conditions (10b, c). As the correlation function $K(r)$ we assume the expression (11a). The parameter v_0 characterizes the fluctuation volume. The averaged square of the matrix element is written in the form (11b). Substituting (11b) into (4) we obtain (12). We average (12) over the angles and the incident neutron spectrum, assuming the angular distribution to be isotropic and the spectrum to be a segment of a perfect Maxwellian spectrum. As a result we obtain expression (13), the calculation procedure of which leads to the expression (17), from which (18) is determined for the worst relations (18b, c).

Appendix 7.10. Transmission of UCN through the interface of two media with non-collinear magnetic flux densities

$$\text{a) } B = B_1\theta(x<0) + B_2\theta(x>0), \quad \text{b) } B_{1\perp} = B_{2\perp}, \text{ c) } H_{1\parallel} = H_{2\parallel}. \tag{1}$$

$$\text{a) } \Psi(\mathbf{r}) = \exp(\mathrm{i}\mathbf{k}_\parallel \mathbf{r}_\parallel)\psi(x), \quad \text{b) } \mathbf{k}_\parallel = (k_z, k_y), \quad \text{c) } \mathbf{r}_\parallel = (z, y). \tag{2}$$

$$[\mathrm{d}^2/\mathrm{dx}^2 + E_\perp + (u_0 - \boldsymbol{\sigma}\mathbf{B}_1)\theta(x<0) - \boldsymbol{\sigma}\mathbf{B}_2\theta(x>0)]\,\psi(x) = 0. \tag{3}$$

$$\text{a) } E_\perp = E - k_\parallel^2 = k_\perp^2 = k_x^2, \quad \text{b) } B = (2\mu m/\hbar^2)[B], \quad \text{c) } \mathbf{e}_i = \mathbf{B}_i/B_i. \tag{4}$$

$$\Psi(x) = [\exp(\mathrm{i}\hat{k}_1^- x)\xi_0 + \exp(-\mathrm{i}\hat{k}_1^- x)\xi_r]\theta(x<0) + \exp(\mathrm{i}\hat{k}_2^- x)\xi_t\theta(x>0). \tag{5}$$

$$\text{a) } \hat{k}_i^\pm = \sqrt{k_{i\perp}^2 \pm \sigma \mathbf{B}_i} = k_i^\pm p_{i\uparrow} + k_i^\mp p_{i\downarrow}, \quad \text{b) } k_i^\pm = \sqrt{k_{i\perp}^2 \pm B_i}. \tag{6}$$

$$\text{a) } k_{1\perp}^2 = k_1^2 + u_0, \quad \text{b) } k_{2\perp} = k_\perp, \quad \text{c) } p_{i\uparrow,\downarrow} = [1 \pm (\sigma \mathbf{e}_i)]/2. \tag{7}$$

$$\text{a) } \xi_0 + \xi_r = \xi_t, \quad \text{b) } \hat{k}_1^-(\xi_0 - \xi_r) = \hat{k}_2^- \xi_t, \quad \text{c) } \xi_r = \hat{r}\xi_0, \quad \text{d) } \xi_t = \hat{t}\xi_0. \tag{8}$$

$$\text{a) } \hat{r} = (\hat{k}_1^- + \hat{k}_2^-)^{-1}(\hat{k}_1^- - \hat{k}_2^-), \quad \text{b) } \hat{t} = (\hat{k}_1^- + \hat{k}_2^-)^{-1} 2\hat{k}_1^-. \tag{9}$$

$$\text{a) } (\hat{k}_1^- + \hat{k}_2^-)^{-1} = (\hat{k}_1^+ + \hat{k}_2^+)/N, \quad \text{b) } N = (\hat{k}_1^+ + \hat{k}_2^+)(\hat{k}_1^- + \hat{k}_2^-). \tag{10}$$

$$\text{a) } N = (\xi p_{1\uparrow} N p_{1\uparrow} \xi)/(\xi p_{1\uparrow} \xi), \quad \text{b) } \hat{k}_i^+ \hat{k}_i^- = k_i^+ k_i^-. \tag{11}$$

$$\text{a) } \hat{k}_i^\pm p_{i\uparrow} = k_i^\pm p_{i\uparrow}, \quad \text{b) } p_{i\uparrow} p_{j\uparrow,\downarrow} p_{i\uparrow} = p_{i\uparrow}(1 \pm \mathbf{e}_1\mathbf{e}_2)/2, \quad \text{c) } \mathbf{e}_1\mathbf{e}_2 = \cos\theta. \tag{12}$$

$$N = k_1^+ k_1^- + k_2^+ k_2^- + (k_1^+ k_2^- + k_1^- k_2^+)\cos^2(\theta/2) + (k_1^+ k_2^+ + k_1^- k_2^-)\sin^2(\theta/2). \tag{13}$$

$$\text{a) } [p_{i\downarrow} p_{j\uparrow,\downarrow}]_- = \mp i\sigma[e_i e_j]/2, \quad \text{b) } p_{i\downarrow} p_{j\uparrow,\downarrow} p_{i\uparrow} = \mp i\sigma[\mathbf{e}_i \mathbf{e}_j] p_{i\uparrow}/2. \tag{14}$$

$$\text{a) } p_{2\uparrow} = \Phi^{-1} p_{1\uparrow} \Phi, \quad \text{b) } \Phi = \exp(i\theta\sigma \mathbf{e}_{12}/2), \quad \text{c) } \mathbf{e}_{12} = [\mathbf{e}_1 \mathbf{e}_2]/|[e_1 e_2]|. \tag{15}$$

$$\text{a) } r_{\uparrow\uparrow} = (\xi p_{1\uparrow} \hat{r} p_{1\uparrow} \xi)/(\xi p_{1\uparrow} \xi) = r_1 + r_2, \quad \text{b) } r_{\downarrow\downarrow} = r_1 - r_2,$$

$$\text{c) } r_1 = (k_1^- k_1^+ - k_2^- k_2^+)/N. \quad \text{d) } \xi \text{ is an arbitrary spinor.} \tag{16}$$

$$r_2 = [(k_1^- k_2^+ - k_2^- k_1^+)\cos^2(\theta/2) - (k_1^+ k_2^+ - k_1^- k_2^-)\sin^2(\theta/2)]/N. \tag{17}$$

$$r_{\downarrow\uparrow} = (\xi p_{1\downarrow} \hat{r} p_{1\uparrow} \xi)/\sqrt{(\xi p_{1\downarrow} \xi)(\xi p_{1\uparrow} \xi)} = k_1^-(k_2^+ - k_2^-)\sin\theta/N. \tag{18}$$

$$t_{\uparrow\uparrow} = (\xi p_{2\uparrow} \hat{t} p_{1\uparrow} \xi)/\sqrt{(\xi p_{2\uparrow} \xi)(\xi p_{1\uparrow} \xi)} = 2k_1^-(k_1^+ + k_2^+)\cos(\theta/2)/N. \tag{19}$$

$$\text{a) } t_{\downarrow\downarrow} = 2k_1^+(k_1^- + k_2^-)\cos(\theta/2)/N, \quad \text{b) } t_{\downarrow\uparrow} = 2k_1^-(k_1^+ + k_2^-)\sin(\theta/2)/N. \tag{20}$$

$$\text{a) } t_{\downarrow\uparrow} = 2k_1^+(k_1^- + k_2^+)\sin(\theta/2)/N, \quad \text{b) } T_{\downarrow\uparrow} = (k_2^+/k_1^-)|t_{\downarrow\uparrow}|^2. \tag{21}$$

$$\text{a) } B_2 \ll B_1, \quad \text{b) } k_2^- \approx k_2^+ \approx k_2, \quad \text{c) } P = (T_{\uparrow\uparrow} - T_{\downarrow\uparrow})/(T_{\uparrow\uparrow} + T_{\downarrow\uparrow}) \approx \cos\theta. \tag{22}$$

Consider a magnetic material occupying the semi-space $x < 0$; then the field B throughout the entire space can be represented in the form (1a). At the interface the normal component of B, (1b), and the tangent component of the magnetic field H, (1c), are continuous. The wavefunction of a neutron propagating from the medium into vacuum can be represented in the form (2a); the vectors $\mathbf{k}_\parallel$ and $\mathbf{r}_\parallel$ parallel to the interface have components (2b) and (2c), respectively. The Schrödinger equation can be reduced to the form (3), if the energy of normal motion (4a) is introduced and the magnetic flux density is expressed in the units (4b). In (4c) the unit vector is defined along the magnetic flux on both sides of the interface. The function $\psi(x)$ entering into (2a) can be represented in the form (5), the exponents of the exponential functions containing the operators of the wavevector (6a) expressed through the normal components of the wavevector (6b) inside, (7a), and outside, (7b), of the medium, as well as the projection operators (7c). Since the wavefunction and its derivative are continuous, eqns (8a, b) can be derived for the spinor amplitudes. We define the reflection operator $\hat{r}$, eqn (8c), and the transmission operator (8d). From (8a, b) it follows that they have the form (9a, b). The validity of (10a) is readily verified, where N is calculated in accordance with (10b) or (11a). Taking advantage of the identities (11b) and (12) and substituting (10b) into (11a) we obtain (13). Further, utilizing the identities (14) and introducing the transformation (15b) relating the operators $p_{1;2\uparrow}$, (15a), and the unit vector (15c) we find the diagonal matrix elements of the operator (16a, b). They are expressed through the quantities (16c) and (17). The non-diagonal matrix elements (18) are found in a similar manner. The matrix elements of the transmission operator are presented in (19), (20), and (21a). The transmission coefficient involving a spin flip is indicated in (21b). If the condition (22a) is fulfilled, then for $k_\perp^2 \gg B$ (22b) holds and the polarization of the transmitted beam, (22c), equals the cosine of the angle between the two magnetic fluxes.

Appendix 7.11 Resonance spin flip

$$\text{a) } \mathrm{i}d\xi(t)/\mathrm{d}t = [B_0\sigma_z + \mathbf{B}\boldsymbol{\sigma}]\xi(t), \quad \text{b) } B = \gamma[B], \quad \text{c) } \gamma = \mu/\hbar. \tag{1}$$

$$\text{a) } \mathbf{B} = B(\cos 2(\omega t + \varphi), \sin 2(\omega t + \varphi), 0), \quad \text{b) } \boldsymbol{\sigma}\mathbf{B} = B\hat{f}_1(t, \varphi)\sigma_x \hat{f}_1^{-1}(t, \varphi). \tag{2}$$

$$\text{a) } \hat{f}_1(t, \varphi) = \exp(-\mathrm{i}\sigma_z(\omega t + \varphi)), \quad \text{b) } \xi(t) = \hat{f}_1(t, \varphi)\eta(t). \tag{3}$$

$$\text{a) } \mathrm{i}d\eta/\mathrm{d}t = (\boldsymbol{\sigma}\boldsymbol{\Omega})\eta, \quad \text{b) } \boldsymbol{\Omega} = (B, 0, B_0 - \omega), \quad \text{c) } \eta(t) = \hat{f}_2(t)\eta(0). \tag{4}$$

$$\text{a) } \hat{f}_2(t) = \exp(-\mathrm{i}\boldsymbol{\Omega}\boldsymbol{\sigma}t), \quad \text{b) } \eta(0) = \hat{f}_1^{-1}(0, \varphi)\xi(0) = \exp(\mathrm{i}\sigma_z\varphi)\xi(0). \tag{5}$$

$$\text{a) } \xi(t) = \hat{F}_1(t, \varphi)\xi(0), \quad \text{b) } \hat{F}_1(t, \varphi) = \hat{f}_1(t, \varphi)\hat{f}_2(t)\hat{f}_1^{-1}(0, \varphi). \tag{6}$$

$$\text{a) } \hat{U}\exp(\mathrm{i}\mathbf{b}\boldsymbol{\sigma})\hat{U}^{-1} = \exp(\mathrm{i}\boldsymbol{\sigma}\mathbf{b}_u), \quad \text{b) } \boldsymbol{\sigma}\mathbf{b}_u = \hat{U}\boldsymbol{\sigma}\mathbf{b}\hat{U}^{-1}. \tag{7}$$

$$\text{a) } \hat{f}_1(t, \varphi)\hat{f}_2(t)\hat{f}_1^{-1}(t, \varphi) = \exp(-\mathrm{i}\boldsymbol{\Omega}_t\boldsymbol{\sigma}t), \quad \text{b) } \boldsymbol{\Omega}_t = \mathbf{B} + \mathbf{B}_0 - \omega\,\mathbf{e}_0. \tag{8}$$

$$\text{a) } \mathbf{e}_0 = \mathbf{B}_0/B_0, \quad \text{b) } \hat{F}_1(t, \varphi) = \exp(-\mathrm{i}\boldsymbol{\Omega}_t\boldsymbol{\sigma}t)\exp(-\mathrm{i}\sigma_z\omega t). \tag{9}$$

$$\text{a) } \xi(0) = \xi_\uparrow, \quad \text{b) } \sigma_z\xi_{\uparrow,\downarrow} = \pm\,\xi_{\uparrow,\downarrow}, \quad \text{c) } F_{\downarrow\uparrow} = (\xi_\downarrow \hat{F}_1(t, \varphi)\xi_\uparrow). \tag{10}$$

$$\text{a) } \exp(\mathrm{i}\mathbf{A}\boldsymbol{\sigma}) = \cos A + \mathrm{i}[(\boldsymbol{\sigma}\mathbf{A})/A]\sin A, \quad \text{b) } A = \sqrt{A_x^2 + A_y^2 + A_z^2}. \tag{11}$$

$$\text{a) } F_{\downarrow\uparrow} = \mathrm{i}(B/\Omega_t)\exp[\mathrm{i}(\omega t + 2\varphi)]\sin(\Omega_t t),$$

$$\text{b) } \Omega_t = \sqrt{B^2 + (B_0 - \omega)^2} = \Omega. \tag{12}$$

$$w_{\downarrow\uparrow} = |F_{\downarrow\uparrow}|^2 = [B/\sqrt{B^2 + (B_0 - \omega)^2}]^2 \sin^2[\sqrt{B^2 + (B_0 - \omega)^2}\,t]. \tag{13}$$

The equation of motion for the spinor can be reduced to the form (1a) by choosing the magnetic field units (1b) with the constant γ, (1c). We consider the field B_0 to be constant and B to be radio-frequency, while its components have the form (2a). We shall take advantage of the transformation (2b) with the operator (3a) and perform the substitution (3b) in (1a), as a result of which we arrive at eqn (4a) with the vector $\mathbf{\Omega}$, eqn (4b). The solution of eqn (4a) is written in the form (4c) with the operator (5a) and the initial spinor value $\eta(0)$ expressed through $\xi(0)$ with the aid of (5b). Substituting (5b) into (4c) and (3b) we obtain (6). We shall now take advantage of the transformation rules (7) with the aid of an arbitrary unitary operator U. From them it follows that (8a) is valid and, in accordance with (2), (8b) also holds, where the unit vector (9a) is introduced. Thus, with the aid of (8a) the operator (6b) is represented in the form (9b). Now let (10a) be valid initially, and let the spinor $\xi_{\uparrow,\downarrow}$ satisfy the condition (10b). Then the amplitude of the state with a spin flip occurring is calculated in accordance with (10c). Making use of the identities (11) and substituting (9b) into (10c) we obtain (12). The probability of a spin flip is determined in accordance with the Rabi formula presented in (13).

Appendix 7.12 Adiabatic spin flip

$$\text{a) } i\dot{\xi} = [\sigma_z B_0(t) + \boldsymbol{\sigma}\mathbf{B}(t)]\xi, \quad \text{b) } \xi = \hat{f}_1(t,\varphi)\eta. \tag{1}$$

$$\text{a) } i\dot{\eta} = (\boldsymbol{\sigma}\mathbf{\Omega}(t))\eta, \quad \text{b) } \mathbf{\Omega}(t) = \Omega(t)(\sin 2\chi, 0, \cos 2\chi). \tag{2}$$

$$\text{a) } \Omega(t) = \sqrt{B^2 + [B_0(t) - \omega]^2}, \quad \text{b) } \sin 2\chi = B/\Omega(t),$$

$$\text{c) } \cos 2\chi = [B_0(t) - \omega]/\Omega(t). \tag{3}$$

$$\text{a) } \boldsymbol{\sigma}\mathbf{\Omega}(t) = \Omega(t)\hat{g}_1(t)\sigma_z\hat{g}_1^{-1}(t), \quad \text{b) } \hat{g}_1(t) = \exp(-i\sigma_y\chi). \tag{4}$$

$$\text{a) } \eta(t) = \hat{g}_1(t)\zeta(t), \quad \text{b) } i\dot{\zeta} = [\Omega(t)\sigma_z - \dot{\chi}\sigma_y]\zeta \approx \Omega(t)\sigma_z\zeta. \tag{5}$$

$$\text{a) } \dot{\chi} = B\dot{B}_0(t)/\Omega^2(t) \approx \dot{B}_0/B \ll \Omega, \quad \text{b) } B_0(t_r) = \omega. \tag{6}$$

$$\text{a) } \zeta(t) = \hat{g}_2(t)\zeta(0), \quad \text{b) } \hat{g}_2(t) = \exp(-i\sigma_z\gamma), \quad \text{c) } \gamma = \int_0^t \Omega(t')\,dt'. \tag{7}$$

$$\text{a) } \xi(t) = \hat{G}(t,\varphi)\xi(0), \quad \text{b) } \hat{G}(t,\varphi) = \hat{f}_1(t,\varphi)\hat{g}_1(t)\hat{g}_2(t)\hat{g}_1^{-1}(0)\hat{f}_1^{-1}(0,\varphi). \tag{8}$$

$$\hat{G}(t,\varphi) = e^{i\alpha_1\sigma_z}e^{i\beta_1\sigma_y}e^{i\alpha_2\sigma_z}e^{i\beta_2\sigma_y}e^{i\alpha_3\sigma_z}. \tag{9}$$

$$\alpha_1 = -(\omega t + \varphi), \quad \beta_1 = -\chi(t), \quad \alpha_2 = -\gamma(t), \quad \beta_2 = \chi(0), \quad \alpha_3 = \varphi. \tag{10}$$

$$\text{a) } \xi(0) = \xi_\uparrow, \quad \text{b) } \sigma_z\xi_{\uparrow,\downarrow} = \pm\,\xi_{\uparrow,\downarrow}, \quad \text{c) } \xi(t) = (G_{\uparrow\uparrow} - i\sigma_y G_{\downarrow\uparrow})\xi_\uparrow. \tag{11}$$

$$(\xi_\downarrow \xi(t)) = G_{\downarrow\uparrow} = -e^{i(\alpha_3-\alpha_1)}(\cos\beta_2 e^{i\alpha_2}\sin\beta_1 + \sin\beta_2 e^{-i\alpha_2}\cos\beta_1). \tag{12}$$

$$w_{\downarrow\uparrow} = |G_{\downarrow\uparrow}|^2 = \cos^2\alpha_2\sin^2(\beta_1+\beta_2) + \sin^2\alpha_2\sin^2(\beta_1-\beta_2). \tag{13}$$

$$\text{a) } \cos^2\alpha_2 \simeq \sin^2\alpha_2 \simeq 1/2, \quad \text{b) } \omega_{\downarrow\uparrow} \simeq (1 - \cos 2\beta, \cos 2\beta_2)/2. \tag{14}$$

$$\text{a) } B_0(t) = \omega - \dot{B}_0 t, \quad \text{b) } i\dot{\eta} = (-t\dot{B}_0\sigma_z + B\sigma_x)\eta. \tag{15}$$

$$\text{a) } \eta(t) = \hat{f}(t, -\infty)\eta(-\infty), \quad \text{b) } \xi(t) = \hat{f}_1(t, \varphi)\hat{f}(t, -\infty)\eta(-\infty). \tag{16}$$

$$\text{a) } \eta(-\infty) = \xi_\uparrow, \quad \text{b) } w_{\uparrow\downarrow} = |f_{\downarrow\uparrow}|^2 = |\xi_\downarrow \xi(t)|^2. \tag{17}$$

$$\text{a) } \lambda = B/\sqrt{\dot{B}_0}, \quad \text{b) } w_{\downarrow\uparrow} \underset{t\to\infty}{=} 1 - \exp(-\pi\lambda^2). \tag{18}$$

Assume the field B_0 in (1a) of Appendix 7.11 to vary slowly with time in magnitude, but not in direction. The equation for the spinor has the form (1a), with the notation adopted in Appendix 7.11. Now we perform the substitution (1b), similar to (3b) of Appendix 7.11, and obtain eqn (2a) with the vector (2b), (3) which, contrary to (4b) of Appendix 7.11, depends on time. We shall make use of the representation (4a) with the operator (4b). We perform the substitution (5a) into eqn (2a) and obtain eqn (5b) in which the term proportional to $\dot{\chi}$ can be neglected under the condition (6a) even at the point $t = t_r$ determined by the condition of resonance (6b). The solution of the approximate equation (5b) has the form (7a) with the operator (7b, c). Taking into account (5a) and (1b) we obtain the final solution of (1a) in the form (8a) with the operator (8b) represented by the product of exponentials (9) with the exponents (10). We now assume the initial conditions in the form (11a) satisfying eqn (11b). Then the solution (8a) can be represented by the expression (11c). The amplitude of the state with spin flip is determined in (12), while the probability for a spin flip to occur relative to the field $B_0(t)$ is given by the expression (13). Under condition (14a) the formula (13) is reduced to the form (14b), taking into account (10). If the field B_0 depends linearly on the time, eqn (15a), then eqn (2a) transforms into (15b). Its solution is sought in the form (16a) and, using (1b), the solution of eqn (1a) is (16b). Assuming (17a) as the initial conditions we find the spin-flip probability in the form (17b) and, as shown in (8c) of Appendix 7.1, it can be calculated without the approximation (5b) and is equal to (18b), i.e. it is expressed through the parameter (18a), which characterizes the degree to which the adiabaticity condition (6a) is satisfied.

8

The application of UCN

UCN can be utilized as tools for investigations of fundamental problems (studies of the properties of the neutron itself) and in applications (studies of substances). We shall first discuss the fundamental problems.

8.1 USING UCN TO STUDY PROPERTIES OF THE NEUTRON ITSELF

8.1.1 Searching for the electric dipole moment (EDM) of the neutron

The EDM of elementary particles essentially implies, instead of the classical EDM, the interaction of the magnetic moment or the spin of the particle with the electric field:

$$\alpha\boldsymbol{\mu}\mathbf{E} = \beta\mathbf{s}\mathbf{E}, \tag{8.1}$$

where the constant α is dimensionless and β has the dimensionality of the EDM, i.e. β can be represented in the form $2eD$, where e is the elementary charge and D is a certain length. In principle, the neutron exhibits internal structure, so a real EDM may exist, such as that of the NaCl molecule, although it is not yet being researched.

The interaction (8.1) violates spatial and time symmetry (i.e. changes sign) when the space or time is inverted. To verify this suppose the magnetic moment to be in the form of a coil with current and let the axis of the coil be directed along the electric field. Now let the space be reflected in the plane of the coil. Then $\boldsymbol{\mu}$ will retain its direction, since the current is not changed, while **E** will point in the opposite direction, so expression (8.1) will change sign. If, on the other hand, we substitute $-t$ for t, then all the particle velocities will change sign, resulting in the current in the coil and in μ changing direction and consequently (8.1), also changing sign. (We note that the field **E** can be created by motionless charges, so it does not change when the time is inverted.)

The issue of measuring the EDM of the neutron was raised by E.M. Purcell and N.F. Ramsey in 1950 (*Purcell and Ramsey 1950). The first experimental results were obtained in 1957 (*Smith *et al.* 1957). In the same year violation of spatial parity in the β-decay of nuclei was observed (*Lee and Yang 1957; *Wu *et al.* 1957). L.D. Landau demonstrated (*Landau 1957) that for the

EDM of elementary particles to exist not only spatial symmetry, but also symmetry with respect to time must be violated. In 1964 the non-conservation of time symmetry was discovered under the assumption that the CPT theorem be satisfied. Thus, searches for the EDM of the neutron turned out to be fully justified. The most precise experiment without utilizing UCN was performed in 1977 (*Dress *et al.* 1977), and from this experiment it was determined that $D < 3 \times 10^{-24}$ cm. With the aid of UCN the upper limit for D could be expected to be significantly reduced. This was especially desirable since theoretical estimations yielded EDM values close to the attainable experimental sensitivity. These estimates changed with time, but they always turned out to be slightly lower than the precision achieved in the experiment. The hopes put on UCN turned out, on the whole, to be justified. With their aid it has already been possible to lower the limit for the EDM of the neutron by nearly an order of magnitude (Lobashov and Serebrov 1984; Altarev *et al.* 1988*a*; Morse 1984), and the achieved sensitivity can be expected to be improved upon.

Before we proceed to discuss the advantages of UCN compared with neutrons of higher energies let us consider the general scheme for measuring the EDM of the neutron (Fig. 8.1). The experiment is based on the fact that in

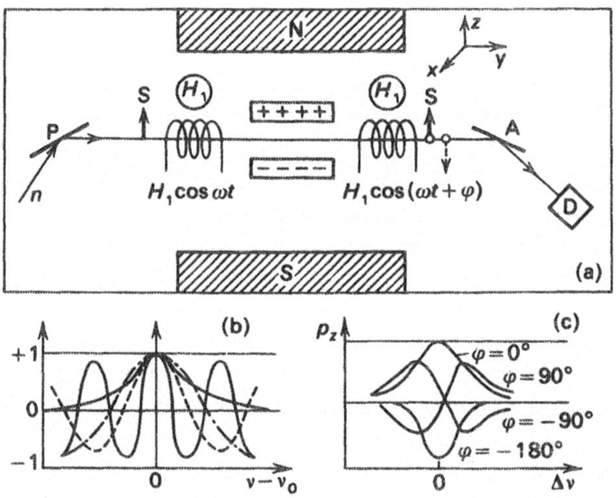

Fig. 8.1 Experiment to search for the EDM. (a) Layout of flow mode experiment with separated radio-frequency fields; (b) several resonance curves corresponding to neutrons with various velocities and a phase shift $\varphi = 0$. The Lorentzian curve is obtained upon averaging the resonance curves over various times during which the neutron is to be found within the intermediate region under the assumption that the time distribution is described by an exponential function; (c) several averaged resonance curves for different phase shifts φ, indicated in the figure, between the two radio-frequency fields (Lobashov and Serebrov 1978).

the presence of interaction (8.1) the spin precession frequency of a neutron in a magnetic field changes by the quantity $\pm\Delta\Omega = eDE/\hbar$ when the electric field is switched on. In practice the change in the precession angle $\Delta\varphi$ is measured when the electric field is made to change direction. If the angle $\Delta\varphi$ is measured with a fixed uncertainty, then the longer the measurement continues (i.e. the longer the neutron is to be found in the electric field) the higher the precision with which the change in the precession frequency is determined.

Measurement of the precession angle $\Delta\varphi$ is performed with the aid of two radio-frequency fields separated in space (or in time) and differing in phase. The first radio-frequency field rotates the spin of the polarized neutrons in the plane perpendicular to the leading homogeneous magnetic field, whereupon the neutrons arrive in the region where the radio-frequency field is absent. Here, during a time T, their spin precesses about the homogeneous field B_0. Then they come once again to the region under the influence of the radio-frequency field. If the phase of the second oscillating field coincides with the phase of the first field, and the frequency equals the spin precession frequency, then the second field rotates the neutron spin through another 90°, as if no intermediate region existed at all between the radio-frequency fields. If the phase of the second radio-frequency field is shifted with respect to the first by 90°, then it turns out to be parallel to the spin and does not force the spin out from the plane perpendicular to the leading field B_0, so the polarization analyser which determines the spin polarization on to the direction of B_0 will indicate a zero value. If the spin precession frequency alters, for example under the influence of an electric field, this results in the generation of a polarization with a magnitude proportional to the change in precession frequency. If we denote by N_+ the number of neutrons registered by the detector and exhibiting a spin parallel to the field B_0, when the magnetic and electric fields in the intermediate region between the oscillating fields are parallel to each other (and by N_- the similar number of neutrons detected when B_0 and E are antiparallel), then the quantity D can be determined from the relation

$$(N_+ - N_-)/(N_+ + N_-) = 2eEDT/\hbar.$$

If the measured difference $N_+ - N_-$ is within the statistical uncertainty $2N^{1/2}$, then an upper limit is imposed on D:

$$D < \hbar/2eET\sqrt{N} = \hbar/2eE\sqrt{J}\,T\sqrt{t}, \qquad (8.2)$$

where J is the detector counting rate and t is the total run time of the experiment ($N = Jt$).

To determine the resonance frequency, i.e. the spin precession frequency in the field B_0, the dependence of the counting rate of the detector registering neutrons with the spin parallel to B_0 on the frequency of the oscillating field,

ω, is measured. The corresponding curves for different phase shifts between the first and second radio-frequency fields separated in space are of the form presented in Fig. 8.1(c). The resonance frequency is determined from the intersection of the two curves (see Fig. 8.1(c)) obtained by shifting the phases by $\pm 90°$. The resonance curve for oscillating fields separated in time is presented in Fig. 8.2.

We note that the sweep of the resonance curves is not 100%, i.e. their minima do not reach zero. This is due to a number of reasons: incomplete polarization, imperfect rotation of the spin through 90°, incorrect phasing of the radio-frequency fields, depolarization of neutrons on inhomogeneities of the magnetic field in the apparatus, and so on. All these imperfections lead to a decrease in the sensitivity of the experiment. The change in the detector counting rate related to a change in the precession frequency in the intermediate region of $\pm \Delta\Omega$ is determined by the expression $N_+ - N_- = 2\Delta\Omega \mathrm{d}N(\omega)/\mathrm{d}\omega$. The derivative $\mathrm{d}N(\omega)/\mathrm{d}\omega$ can be represented in the form $\mathrm{d}N(\omega)/\mathrm{d}\omega = tJ\mathrm{d}w_{\uparrow\uparrow}/\mathrm{d}\omega$, where J is the counting rate of the detector at the operation point, i.e. at the resonance frequency, and $w_{\uparrow\uparrow}$ is the probability for the neutron to exhibit the same spin direction at the moment of detection, as at the beginning. In the ideal case $\mathrm{d}w_{\uparrow\uparrow}/\mathrm{d}\omega = T$, where T is the time required for the neutron to pass through the intermediate region. In reality this derivative is somewhat smaller, and if the difference $N_+ - N_-$ does not exceed the statistical uncertainty, then, instead of expression (8.2), the following expression is utilized for the upper limit of D:

$$D < \frac{h\sqrt{J}}{eE\sqrt{t}(\mathrm{d}J/\mathrm{d}\omega)}, \tag{8.3}$$

where $\mathrm{d}J/\mathrm{d}\omega$ is the derivative of the counting rate with respect to the frequency of the oscillating field at the operation point. From expression (8.3)

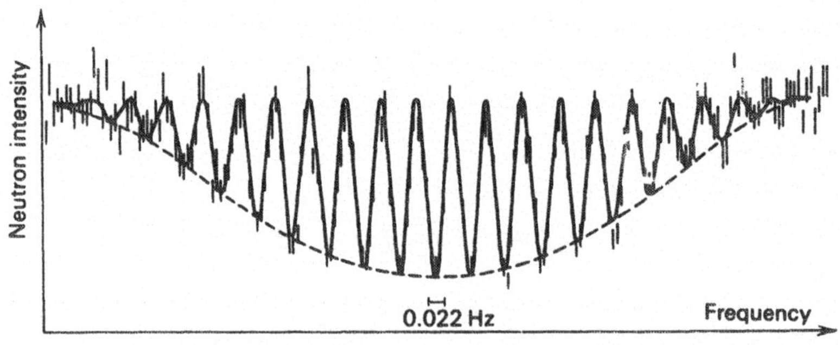

Fig. 8.2 Resonance curve obtained in the accumulation mode experiment (Pendlebury *et al.* 1984), when the radiofrequency fields were separated in time but not in space.

and the form of the curves in Fig. 8.1(c) it clearly follows that the highest sensitivity can be achieved when the phase shift between the radio-frequency fields is $\pm 90°$.

Now we shall discuss the advantages and disadvantages of UCN, which manifest themselves in the search for the EDM of the neutron. One of the most notable disadvantages of UCN is their low intensity. In this connection it is necessary to mention the commonly encountered assertion that the lowest measurable value of D is determined by the uncertainty relation $\Delta E \Delta t \geqslant h$, i.e. when $\Delta E = eDE$ and $\Delta t = T$ the upper limit for D cannot be reduced below the quantity h/eET. This assertion, generally speaking, is incorrect. From expressions (8.2) and (8.3) it follows that the minimum measurable value of D involves the factor $1/N^{1/2} = 1/(Jt)^{1/2}$, which can be infinitesimal (here N is the total number of neutrons registered during the experiment in time t and J is the average counting rate). In relation to the mean counting rate J UCN are worse than faster neutrons. For example, compare the intensity of UCN and that of neutrons with velocities of 150 m s^{-1} which are utilized in the most sensitive beam experiment (*Dress *et al.* 1977). We shall consider the neutron source to be one and the same and the spectrum to be a perfect Maxwellian one. In a beam experiment the intensity is proportional to $v_\parallel \Delta v_\parallel \Delta^2 v_\perp$, where $v_\parallel = 150 \text{ m s}^{-1}$ is the velocity component parallel to the beam direction and $v_\perp$ represents the perpendicular components. The collimation necessary in the beam experiment must be chosen such that the interval $\Delta^2 v_\perp$ is of the order of magnitude of v_{lim}^2, where v_{lim} is a quantity of the order of magnitude of the UCN limiting velocity. Requirements of monochromaticity led the interval $\Delta v_\parallel$ to be limited to values of $\sim 50 \text{ m s}^{-1}$. Comparing the beam intensity with the UCN intensity, proportional to v_{lim}^4, and setting $v_{\text{lim}} \approx 5 \text{ m s}^{-1}$, we find the intensity in the beam experiment to be ~ 300 times higher than in the case of UCN.

UCN, however, do have the advantage that in their case the time T is large. Indeed, when the flight base is 1 m (the length of the intermediate region between the radio-frequency fields) the time T in the beam experiment is ~ 0.01 s. If the UCN were utilized in the form of a beam too, then their time of flight along this base would be 20 times longer, because of their smaller velocity, and this would compensate for their low intensity. Actually, owing to total reflection the UCN can be retained within the intermediate region significantly longer (in principle, up to $\tau_\beta \approx 1000$ s), and so the sensitivity can correspondingly be increased by four orders of magnitude. At present the storage times actually achieved of neutrons with velocities of $\sim 5 \text{ m s}^{-1}$ are significantly smaller, ~ 50 s, and the UCN intensities may also be somewhat lower. Nevertheless the T factor is decisive for UCN and permits a sensitivity of one or two orders of magnitude higher than the sensitivity of the beam experiment to be obtained.

The electric field E in experiments with UCN is somewhat lower than in the beam experiment. The field E is created with the aid of two electrodes to which a high voltage is applied. The distance between the electrodes in the beam experiment (*Dress *et al.* 1977) was about 1 cm. In the case of UCN the distance between the electrodes is several times larger (7–10 cm (Altarev *et al.* 1982)), so the corresponding discharge occurs more readily. In the main the discharge proceeds along the surface of the insulator separating the electrodes, and it occurs easier when the mean free path of the ions and electrons in the vacuum is long. In the beam experiment it was possible to raise the field strength to $E = 100\ \mathrm{kV\,cm^{-1}}$, while in the case of UCN the upper limit of $25\ \mathrm{kV\,cm^{-1}}$ had to be imposed. Thus, relative to the field strength E the sensitivity of UCN is four–five times lower than the sensitivity of the beam experiment. Nevertheless, this circumstance does not entirely dismiss the general gain of sensitivity obtained in utilizing UCN.

The greatest advantage of UCN is their insensitivity to the relativistic effect which limits the precision of a beam experiment. The essence of the relativistic effect is that the electric field E in a reference frame related to the neutron creates a magnetic field $\mathbf{B} = [\mathbf{E}\mathbf{v}]/c$. If $\mathbf{E}$ is not parallel to $\mathbf{B}_0$, then the additional magnetic field has a non-zero component along $\mathbf{B}_0$ which changes sign when E is switched over, thus altering the precession frequency and imitating the EDM. In *Dress *et al.* (1977) for the imitation of $D = 10^{-23}$ cm it was sufficient for the angle between $\mathbf{E}$ and $\mathbf{B}_0$ to be about 2°. This effect is difficult to control or discard and thus it imposes the main limit on the sensitivity of a beam experiment. In the case of UCN the effect is significantly smaller—not only because the velocity v is lower, but also because the magnitude of the relativistic effect, since the motion of UCN is chaotic, is determined not by the velocity v but by the drift velocity through the region occupied by the electric field. This drift velocity turns out to be significantly lower than v. Furthermore, in the case of UCN an experiment that totally excludes the relativistic effect is feasible.

We shall now turn to the description of an actual experiment in search of the EDM performed with the aid of UCN and described by Altarev *et al.* (1982) (Fig. 8.3). Several peculiarities may be noted in this experiment. First, the neutrons were transmitted in a continuous flow through two vessels with electric fields which were antiparallel. The presence of two vessels permitted the elimination of false effects due to the motion of external magnetic objects, especially when it was correlated with the switching of the high voltage. For example, the switch itself, when moving in space, may alter the field B_0. Since a change in the homogeneous magnetic field introduces opposite contributions to D in the upper (D^{up}) and lower (D^{dn}) vessels, the quantity $D^{\mathrm{up}} - D^{\mathrm{dn}}$ depends entirely on the change in the magnetic field. Thus, by using the difference $D^{\mathrm{up}} - D^{\mathrm{dn}}$ it is possible to control the change in the magnetic field occurring when the field E is switched.

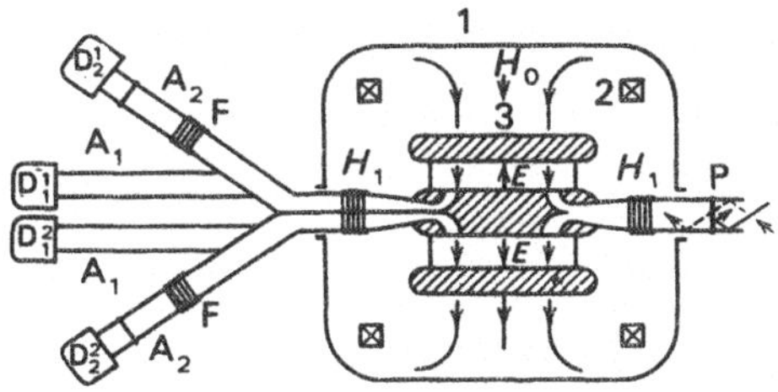

Fig. 8.3 Experimental set-up (Altarev *et al.* 1979) for searching for the EDM of the neutron with the aid of UCN. A liquid hydrogen converter in the vertical channel served as the source of UCN. The UCN density at the entrance was 2.5 cm^{-3}; P, polarizer; $A_{1,2}$, analysers; F, spin-flippers; $D_{1,2}^{1,2}$, UCN detectors; H_1, radio-frequency fields; H_0, uniform magnetic field, E, electric field; 1, magnetic screen; 2, coils producing the magnetic field gradient; 3, two chambers.

Secondly, the neutrons transmitted through each of the two vessels were registered by two detectors, one of which registered UCN polarized along the field B_0 while the other registered oppositely polarized neutrons. Since the neutrons polarized in both directions carry the same information on the polarization arising under the influence of the field E, registration of UCN with both spin projections (Taran 1974, 1977) allows the experimental sensitivity to be enhanced by a factor of $2^{1/2}$.

A set of four detectors made it possible to change in an independent way the four quantities $D_i^{\text{up, dn}}$, $i = 1, 2$, where the subscript indicates the detector number and the superscripts the vessel. Of these four quantities we can make up four combinations:

$$\tfrac{1}{4}(D_1^{\text{up}} + D_2^{\text{up}} + D_1^{\text{dn}} + D_2^{\text{dn}}) = D,$$

$$\tfrac{1}{4}(D_1^{\text{up}} + D_2^{\text{up}} - D_1^{\text{dn}} - D_2^{\text{dn}}) = D_\alpha,$$

$$\tfrac{1}{4}[(D_1^{\text{up}} - D_2^{\text{up}}) - (D_1^{\text{dn}} - D_2^{\text{dn}})] = D_\beta,$$

$$\tfrac{1}{4}[(D_1^{\text{up}} - D_2^{\text{up}}) + (D_1^{\text{dn}} - D_2^{\text{dn}})] = D_\gamma.$$

The first defines the quantity sought, D, the second defines the false effect due to the magnetic field changing when E is switched, the third defines the influence of the electric field on the counting rates of the various detectors, and the fourth defines the degree to which the various effects are compensated.

Polarization and its analysis were achieved by transmitting the neutrons through a 2 μm copper foil with an isotopic content of $^{54}Fe_{0.8}{}^{56}Fe_{0.2}$ exhibiting equal magnetic and nuclear scattering amplitudes. The spin was rotated adiabatically. The probability of a spin flip occurring upon passage through the system with two radio-frequency fields separated in space and with adiabatic flippers is calculated in Appendix 8.1. It can be represented by the expression (Altarev *et al.* 1980*a*)

$$w_{\uparrow\downarrow} = \frac{\omega_1^2}{(\omega_0 - \omega)^2 + \omega_1^2} \cos^2 \{\tfrac{1}{2}[(\omega - \omega_0)T + \psi]\},$$

where $\omega_i = \gamma B_i$, T is the time of flight in the field B_0, and ψ, ω are the phase shift and frequency of tne fields B_1. In the case of resonance flippers we have an additional factor f, depending on time t_1 of exposure to field B_1:

$$f = \sin^2(2\sqrt{(\omega - \omega_0)^2 + \omega_1^2}\, t_1)$$

The adiabatic method represents an advantage for UCN, since it works identically for all the neutrons with velocities $v < v_{\max}$ ($> v_{\lim}$), for which the condition of adiabaticity holds (Appendix 7.12). This condition can be represented in the form

$$\frac{v_{\lim}}{\gamma B_1} \frac{d}{dz} \frac{\omega_0(z) - \omega}{\sqrt{(\omega_0(z) - \omega)^2 + \omega_1^2}} \ll 1;$$

if the left-hand side is set equal to a constant $\alpha \ll 1$ (Ezhov *et al.* 1976), then for the magnetic field $B_0(z)$ inside the flipper we obtain the dependence of its variation on the coordinate along the flipper, z:

$$B_0(z) = B_1 az/\sqrt{1 - a^2 z^2},$$

where a $= \alpha\gamma B/v_{\lim}$. The flipper in the above experiment was 80 cm long, the amplitude of the radio-frequency field in it was $B_1 = 0.0069$ Oe and its frequency was 83 Hz. The field B_0 varied from 6×10^{-2} to 2.8×10^{-2} Oe.

The neutron guides supplying the neutrons to the vessels were made of a non-magnetic material (a 45% Cu + 55% ^{58}Ni alloy) by plasma vapour deposition on to glass in a vacuum. The vessels represented quartz chambers coated with beryllium oxide; the electrodes in them were copper and coated with the 45% Cu + 55% ^{58}Ni alloy. The limiting velocity for the whole system is determined by this alloy and amounts to $v_{\lim} = 7\ \mathrm{m\,s^{-1}}$. The time the neutron is in the chamber is $T \approx 5$ s. It was measured from the filling and emptying curves for the vessel closed by shutters. The half-width of the resonance curves for this storage time was about 0.08 Hz. The homogeneity and the stability of the magnetic field during the time between the switchings of the field (200 s) were, respectively, $(1\text{–}2) \times 10^{-5}$ and 1.5×10^{-7} s.

The source of UCN was a liquid hydrogen converter with the beryllium reflector situated in the vertical channel at the VVR-M reactor (Altarev *et al.* 1980*a*). At the entrance to the spectrometer the UCN counting rate was $5 \times 10^4\ \mathrm{s^{-1}}$, while at the exit from the chambers it was $4.5 \times 10^3\ \mathrm{s^{-1}}$. The detectors were proportional counters filled with ^{3}He. The sensitivity in the measurements of the EDM for D was $\approx 8 \times 10^{-25}$ cm per 24 h. The upper limit $D < 4.0 \times 10^{-25}$ cm was established (Lobashov and Serebrov 1984) at a 90% confidence level. One of the main problems in the further enhancement of the sensitivity was the fight against leakage currents, which change when the field E is switched and imitate the EDM. Studies revealed, however, that they can be dealt with and their effect can be diminished to the level of $D \approx 2 \times 10^{-26}$ cm.

The experiment of Altarev *et al.* (1980*a*) was performed in the continuous flow mode of UCN through the intermediate region between the separated radio-frequency fields. In Grenoble (Pendlebury *et al.* 1984) the limit for $D \approx (0.3\text{–}4.8) \times 10^{-25}$ cm was achieved in the accumulation version (the corresponding resonance curve is shown in Fig. 8.2), when the neutrons are exposed to the stationary field in the closed accumulation vessel ($\tau \approx 60$ s) in between the two radio-frequency fields. A similar experiment was carried out at the Leningrad Institute of Nuclear Physics (LNPI) (Gatchina). The accumulation version was shown by calculations (Taran 1973*b*) to possess a sensitivity three times higher than the continuous flow mode. In Gatchina (Altarev *et al.* 1988*a*) the upper limit $(-1.4 \pm 0.6) \times 10^{-25}$ cm e for the EDM has already been achieved (2.6×10^{-25} cm e at the 95% confidence level). The results obtained in Grenoble seem to testify in favour of the absence of the EDM of the neutron at a level of $(-0.6 \pm 0.6) \times 10^{-25}$ cm e (Grenoble Conference on Fundamental Research in Neutron Physics; private communication 1989). One can hope that in the near future, even if the EDM is not observed, the upper limit for D will be reduced to 10^{-25} cm. A review of the problems relevant to the EDM and a discussion of the prospects for utilizing UCN in this regard can be found in Ramsey (1988) or, for example, in Serebrov (1979).

An entirely different experimental layout for searching for the EDM is discussed in Golub (1984*a*). The author proposes to take advantage of the reaction n + ^{3}He, which is essentially dependent on the mutual polarization of the neutron and the helium nuclei. Therefore, if polarized UCN are stored in liquid ^{4}He with an admixture of ^{3}He nuclei, and if the polarization of the neutrons is changed by an external electric field, then if the EDM exists the UCN spins will be rotated and thus a change in the UCN capture rate will be observed as the external field is changed.

For the readers' convenience we list all the references relevant to the above issue and UCN: Shapiro (1968), Golub and Pendlebury (1972), Porsev and Serebrov (1973*b*), Taran (1973*a*, *b*, 1975*a*, *b*, 1977), Egorov *et al.* (1976), Ezhov *et al.* (1976), Altarev *et al.* (1977, 1979, 1980*a*, 1981, 1982, 1985, 1986*b*, 1988*a*), Serebrov (1979), Summer (1979), Kozlov *et al.* (1980), Golub *et al.* (1981), Nikitenko and Taran (1981), Pokotilovskii (1981), Lobashov and Serebrov (1978, 1984), Golub (1983, 1984*a*, 1986), Morse (1984), Pendlebury *et al.* (1984), Ramsey (1988), Miranda (1985), *Purcell and Ramsey (1950), *Smith *et al.* (1957).

The Berry phase

This is a suitable place to consider new notion, appeared in general physics which touched upon UCN too. The corresponding experiment (Richardson *et al.* 1988) was performed on the EDM equipment which has just been described. To understand the meaning of 'the Berry phase' it is necessary to consider this experiment.

The electric field in the EDM chamber of ILL was switched off, and three coils, with mutually perpendicular axis, were mounted around it. These coils provide a magnetic field inside the chamber, whose vector precesses around a direction with the frequency ω, which is small enough, and neutron spin can adiabatically follow the direction of the field. If the magnetic field $\mathbf{B}$ does not rotate in time the phase of neutron spin precession would be $\phi_0 = \gamma Bt$. But the tip of the vector $\mathbf{B}$ rotates, and this affects the phase ϕ_0. After one revolution of the tip, the phase ϕ acquires an increment $\delta\phi$. Just this increment was measured by Richardson *et al.* (1988). The Berry phase is $\theta = 2\pi + \delta\phi$.

To understand the effect that was measured and its magnitude, see Appendix 7.11, where resonance spin flip is considered. The solution of the problem is straightforward. Indeed, the vector $\mathbf{B}$ can be broken down in two parts: a constant projection $\mathbf{B}_0$ on the axis of precession and a perpendicular component $\mathbf{B}_1$ rotating with angular velocity ω. Vector $\mathbf{B}_1$ can be considered as a radio frequency field. The evolution of the neutron spinor ζ is described by the operator

$$F = \exp(-i\boldsymbol{\Omega}\boldsymbol{\sigma}t)$$

(see the Appendix 7.11), where

$$\boldsymbol{\Omega} = \mathbf{B}_0 + \mathbf{B}_1 - \omega\mathbf{B}_0/B_0,$$

and for simplicity γ has been taken equal to unity. In time t the total spin precession phase becomes

$$\phi = \Omega t = t\sqrt{(1 - \omega/B_0)^2 B_0^2 + B_1^2} = \phi_0 + \delta\phi.$$

Since ω is supposed to be small we can use the series expansion in powers of ω, and in the first order approximation we obtain

$$\delta\phi = -\omega t B_0/B = -\omega t \cos\beta$$

where

$$B = \sqrt{B_0^2 + B_1^2}$$

and β is the angle between the magnetic vector B and axis of its precession.

All these things were well known; however, Berry gave a new interpretation of his angle. The magnitude of $\theta = \omega t(1 - \cos\beta)$ after one revolution of the vector $\mathbf{B}$ is

$$2\pi(1 - \cos\beta),$$

which is equal to a solid angle traced by the tip of the revolving vector **B**.

In the experiment performed by Richardson *et al.* (1988) the variation of spin flip was observed. The probability of the spin flip in the presence of radio frequency field is shown in section 7.4 and is calculated in Appendix 7.11:

$$W = [B_1/\sqrt{(1-\omega/B_0)^2 B_0^2 + B_1^2}]^2 \sin^2(t\sqrt{(1-\omega/B_0)^2 B_0^2 + B_1^2}).$$

From the time variation of the value W the phase ϕ was obtained and angle θ was deduced. The results of the experiment are in good agreement with theoretical considerations. (See also: *Bitter and Dubbers 1987 and Dubbers 1988.)

8.1.2 The gravitational dipole moment (GDM)

If P and T parities are violated, then an interaction is possible between the neutron spin and the gravitational field in the form $\alpha s \nabla \varphi$, where φ is the gravitational potential and the constant α has the same dimensionality as the GDM: $\alpha = mD_g$. Here m is the mass of the neutron and D_g is the effective GDM length. In the field of the Earth $\nabla \varphi = \mathbf{g}$, where $\mathbf{g}$ is the gravitational acceleration vector and the interaction of the spin with the gravitational field is represented in the form $mD_g sg$. A short review on the GDM can be found in Golub (1978*b*). The upper limit of D_g for the proton was established by B.V. Vasil'ev (*Vasil'ev 1969). The proton magnetic resonance was measured in this experiment when the magnetic field was parallel and antiparallel to the Earth's gravitational field. The nuclear resonance frequencies differed in these two cases by 1.4 ± 0.7 Hz; hence $D_g < 10^{-5}$ cm. Measurement of the EDM at a level of $D \approx 10^{-25}$ cm when the field strength is 100 kV cm^{-1} is equivalent to achieving a sensitivity in the energy shift of $\sim 10^{-20}$ eV. If the same sensitivity can be achieved in measuring the GDM, this will permit the limit for D_g to be lowered to 10^{-11} cm. In this case, however, the gravitational field cannot be switched, so the magnetic field must be switched with an accuracy of 10^{-8} G, which is quite a difficult technical task to perform.

8.1.3 The electric charge of the neutron, q_n

In principle, although no rigorous foundations exist for this, the neutron can be assumed to have a charge constituting a small part of the elementary charge. In this case, if the neutrino is considered completely neutral, it follows from charge conservation in β-decay that the charges of the proton and of the electron should differ. If the difference amounts at 10^{-20} e, then the cosmological expansion of the Universe can be attributed to Coulomb repulsion. Indirect data relevant to the degree of neutrality of molecules reveal (*Dylla and King 1973) that $q_n < 4.3 \times 10^{-21}$ e. One of the most direct and precise experiments was performed by *Shull *et al.* (1967). Here the upper limit for q_n was found to be $q_n < 4 \times 10^{-18}\ e$. In Gähler *et al.* (1982) this limit was reduced to $q_n \approx (-1.5 \pm 2.1) \times 10^{-20}\ e$, and in Baumann *et al.* (1988) it was reduced even more.

The layout of the optical system for neutrons of 1.2–3.0 nm wavelength used in this experiment is shown in Fig. 8.4(a). An image of the slits was obtained and a shift of the image under the action of an electric field was measured. The image of a 31 slit object with 30 μm wide slits separated by 30 μm wide absorbing zones is shown in Fig. 8.4(b). The image was scanned with a slit of 10 μm width. The measured resolution of 28.5 μm showed the upper limit of the neutron charge to be $(-0.4 \pm 1.1) \times 10^{-21}\ e$ (Baumann *et al.* 1988).

At LNPI a team of Bulgarian and Soviet specialists performed an experiment in search of q_n utilizing UCN; the experimental layout is shown in Fig. 8.5(a, b). The UCN source is in the form of a vertical slit and is established in between two well-polished perfectly horizontal glass plates. The neutrons fan out from the source and, upon undergoing specular reflection from the horizontal plates, reach a concave vertical cylindrical mirror which reflects them specularly and focuses them in the form of a vertical band representing an image of the slit on to a screen situated in the same plane as the source itself. A high voltage is applied to the lateral walls which are coated with an absorber to avoid reflection from them. The

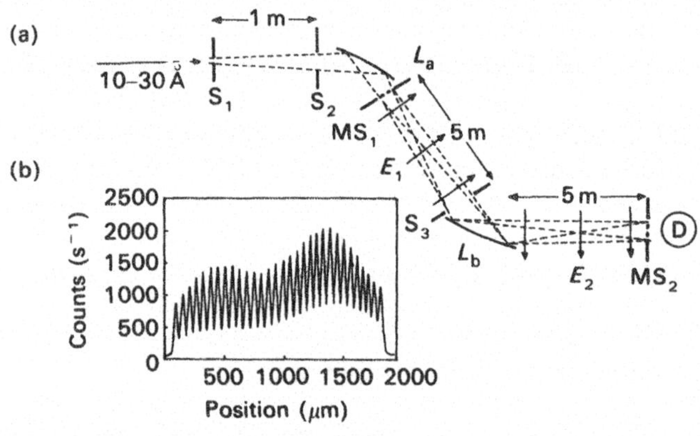

Fig. 8.4 (a) The design of the optical imaging system. MS_1 and MS_2 are the object and the image, respectively. (b) The image of the multislit system scanned with a single 10 μm slit MS_2 (Baumann *et al.* 1988).

---

track detector of Kashukeev and Chikov (1979*a*): photograph of tracks. (d) Plot of track density variation across the slit image. The widths of the slits and the distances are indicated in mm; the channel is 0.225 mm in width.

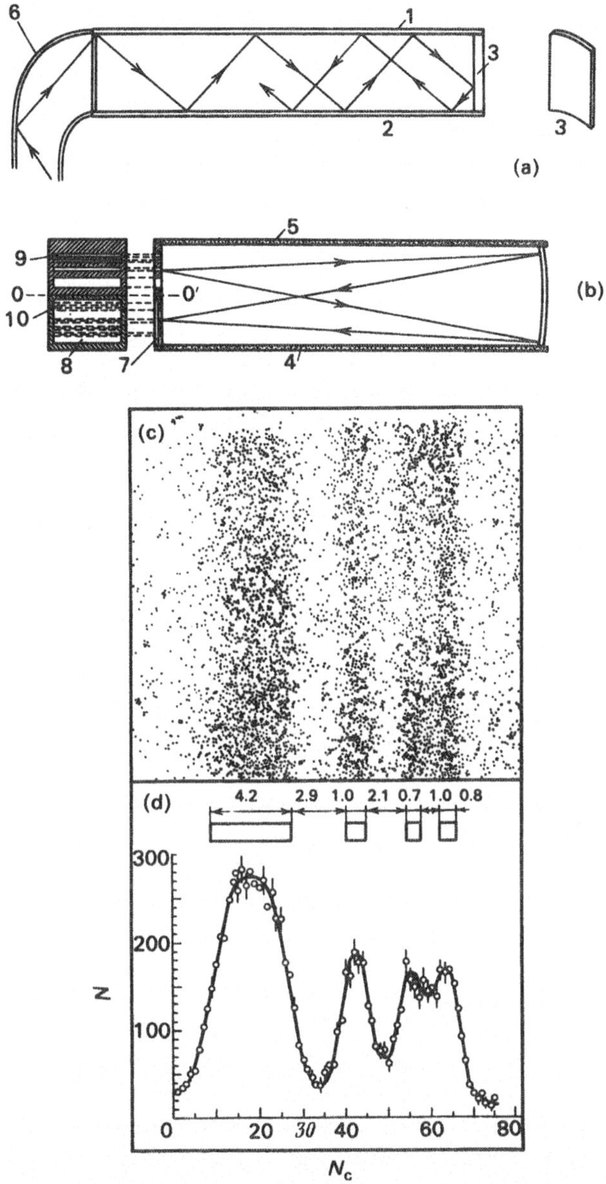

Fig. 8.5 Experimental set-up for searching for the electric charge of the neutron: (a) lateral view; (b) top view: 1 and 2, horizontal plane mirrors; 3, vertical cylindrical mirror; 4 and 5, lateral polyethylene absorbers and high-voltage electrodes; 6, neutron guide; 7, detector shielding; 8, track detector; 9, source in the form of a slit (here four slits of the model set-up are shown); 10, image of the source; 0–0′, symmetry plane of the set-up of Kashukeev and Chikov (1979*a*, *b*). (c) Image of slits on the surface of the

neutrons are deflected in the electric field E in a direction perpendicular to their trajectory by the quantity $\Delta l = q_n E t^2/2m$, where t is the time the neutrons take to travel to the concave mirror and back. The result is a displacement of the slit image. If we take into account that as a result of acceleration along the field E the angle of incidence onto the focusing mirror is altered, then the displacement of the image turns out to be two times smaller. Measuring the displacement Δl we find the charge:

$$q_n = 4m\Delta l/Et^2.$$

First a model experiment (Kashukeev and Chikov 1979*b*) was performed. Four vertical slits served as the source of UCN. The same number of images was obtained. The screen was a track detector representing a photographic glass plate coated with a uranium–titanium radiator. Figure 8.5 demonstrates the image of the slits obtained during a 72.5 h exposure, and the change in track density along the screen is plotted in Fig. 8.5(b). From the plot it follows that the resolution of the method in this case is 1 mm.

So far only a sensitivity to the charge at a level of $(4.3 \pm 7.1) \times 10^{-20}$ e has been achieved in a full-scale experiment (Borisov *et al.* 1987*a*). One of the major difficulties in the experiment was related to the deformation of the system under the influence of the high-voltage electrostatic field.

In Ioffe and Zabiyakin (1988) a method is discussed of lowering the upper limit for q_n to 3.2×10^{-23} e with the aid of experiments on the diffraction of slow neutrons on diffraction gratings. Such a limit can be achieved in 10 days of measurement at the LNPI reactor.

For the readers' convenience we list here all the references on the neutron charge relevant to UCN: Kashukeev (1971, 1981, 1982), Kashukeev and Yaneva (1978), Kashukeev and Chikov (1979*a*, *b*, 1982), Frank (1983), Kalchev *et al.* (1984), Kalus and Gähler (1984), Baumann *et al.* (1988), Borisov *et al.* (1987*a*, *b*, 1988), Ioffe and Zabiyakin (1988), *Shull *et al.* (1967).

8.1.4 The neutron lifetime τ_β

In the three experiments aimed at measuring the half-life of the neutron, $T_{1/2} = \tau_\beta \ln 2$, the charged decay products on a section of the beam trajectory were investigated (*Christensen *et al.* 1972; *Bondarenko *et al.* 1978; *Byrne *et al.* (1980). The results obtained for $T_{1/2}$ of 10.82(0.21), 10.13(0.09), and 10.61(0.16) min differ by nearly 10%, although the statistical accuracy is 1%. The figure 10.13 has been revised recently (*Spivak 1988). The new value, 10.27(0.1), decreases the gap somewhat, but does not discard it. (A short review of the relevant problems may be found, for example, in the book by *Alexandrov (1982).) UCN make the determination of τ_β possible by registration of the neutrons themselves, instead of their decay products. The idea of the experiment is to measure the storage time of the UCN, τ, by varying the loss time τ_l. Since $1/\tau = 1/\tau_\beta + 1/\tau_l$, we obtain $\tau_\beta = \tau$ by extrapolating τ_l to infinity. In doing so it is necessary to understand clearly the problem of storage in order to control τ_l reliably, or to decrease the loss coefficient to neglect the quantity $1/\tau_l$ with the desired precision.

The quantity τ_l can be varied either by changing the flight path of the UCN between two collisions with the walls (Kosvintsev *et al.* 1980*a*), or by altering the loss coefficient (Antonov *et al.* 1982*a*). In the ideal case we can, without taking gravity into account, represent τ_l in the form $\tau_l = 4V/S\eta v f(v)$, where V and S represent the volume and the surface, respectively, of the trap walls, v is the neutron velocity, and $f(v)$ is a known function. Here, the neutron distribution in the trap is assumed to be homogeneous and isotropic, while the reflection from the walls is taken to be perfectly specular. If the latter requirement is dropped, then the function $f(v)$ turns out to be dependent on the reflection indicatrix, but the quantity τ_l always remains proportional to the factor $V/S\eta$. Therefore, if for a given volume V the surface of the walls S is changed and extrapolated to zero, then $\tau \rightarrow \tau_\beta$, independently of the form of function $f(v)$. In this determination of τ_β there remains one difficult point related to the dependence of the number of UCN in the trap on the exposure time not being exponential. The storage curve can be considered exponential only in the case of a sufficiently monochromatic spectrum of the accumulated neutrons and low specular reflection from the walls. In the case of purely specular reflection and a perfectly monochromatic spectrum the neutrons that first leave the trap are those with small angles of incidence, followed by those with large angles, which is precisely what leads to a non-exponential storage curve. For its description, in this case, it is necessary to know the initial UCN angular distribution, which generally speaking is impossible. If the UCN spectrum is not monochromatic, then its form must be known, otherwise methodical uncertainties may arise.

The quantity τ_l is large when the velocity of the accumulated neutrons is low, and therefore it is reasonable to measure τ_β by utilizing low-energy UCN. In the experiment of Kosvintsev *et al.* (1980*a*) neutrons with energies of 5–18 neV were used. At low energies it is necessary to take gravity into account, which leads to a more complex dependence of τ_l not only on the velocity, but also on the form of the trap. The contribution of the bottom of the trap to τ_l differs from that of the lateral walls (see Appendices 6.1 and 6.2). If the times τ_b and τ_w, related to the losses occurring in collisions of the neutrons with the trap bottom and its walls respectively, are introduced, then $1/\tau_l = 1/\tau_b + 1/\tau_w$. For $\tau_{b,w}$ the following equalities hold:

$$\tau_{b,w} = 4V/\eta\, v S_{b,w} f_{b,w}(v),$$

where $S_{b,w}$ are the surfaces of the trap bottom and its walls, and $f_{b,w}(v)$ is the corresponding function of the velocity. These functions are readily computed in the case of an isotropic angular distribution and perfectly specular reflection from the walls.

Shown in Fig. 8.6 is the trap in which UCN were stored for measuring τ_β. It is a vertical cylindrical aluminium vessel into which aluminium ribs were inserted to increase its surface. From the dependence of τ on the number of ribs it is possible, assuming the reflection from the walls to be perfectly specular, to find η and $1/\tau_\beta + 1/\tau_b$, whereupon τ_b can be calculated and then τ_β determined. Naturally, uncertainties related to the reflection being non-specular and to the storage curve not being exponential turn out to be non-removable. In the case considered they limited the accuracy of the experiment to 10%. The measured half-life turned out to be $T_{1/2} = \ln 2 . \tau_\beta = 10.11 \pm 1.10$ min.

By taking advantage of a trap of D_2O ice it is possible to decrease the experimental uncertainty to 1%. The storage time in the D_2O ice trap was $\tau = 950 \pm 60$ s, indicating the insignificant contribution of τ_l to the total storage time τ.

At present an experiment is underway at LNPI (Alfimenkov *et al.* 1988), in which τ_β will be measured by the storage time in a slitless trap (dipper) proposed by P.E. Spivak. The dipper can be represented as an empty sphere

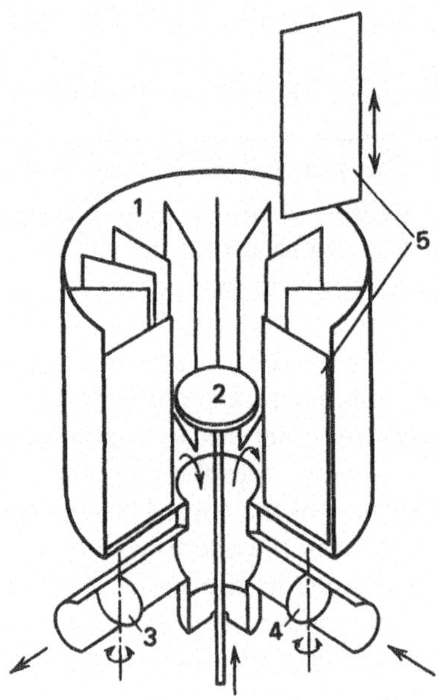

Fig. 8.6 Vessel for storage of UCN used in experiment on measurement of τ_β: 1, cylindrical vessel made of aluminium; 2, platelike shutter; 3, exit shutter; 4, entrance shutter; 5, insertable ridges of aluminium (Kosvintsev *et al.* 1980).

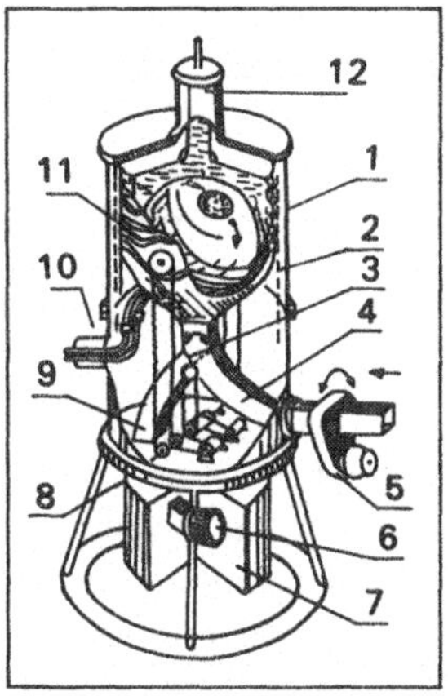

Fig. 8.7 Experimental set-up for measuring τ_β (Alfimenkov *et al.* 1988): 1, scoop; 2, nitrogen screen; 3, distributing valve; 4 and 9, neutron guides for input and output of UCN, respectively; 5, input shutter; 6, detector; 7, detector shielding; 8, drive mechanism for shutter and trap; 10, cryosupplies; 11, cooled volume; 12, sluice of system for performing vapour coating inside the trap.

(Fig. 8.7) with its upper segment cut off and which can rotate about the horizontal axis. The dipper is placed inside an evacuated cavity and, when the cavity is being filled with neutrons, its open segment is turned downwards. If neutrons are let into the cavity, they will also enter the sphere through the open segment, and the UCN density inside and outside the sphere will be the same. When the open segment of the sphere is turned upwards, part of the UCN that do not succeed in reaching the top will be scooped up by the sphere and hence be confined. The storage of these UCN is utilized to measure τ_β. The more energetic neutrons will leave the trap through the opening and will be registered by the detector. From the counting rate of the detector it can be seen whether the outflow of the fast UCN has been completed or not. When the given exposure time expires, the dipper is inclined and the neutrons remaining in it are poured out. The storage time τ is determined from the amount of UCN poured out. To suppress heating of the UCN the dipper will be cooled to 4 K. To reduce absorption of the UCN the walls of the dipper will be coated with either D_2O, O_2, or CO_2. Depending on how close the storage curve is to being exponential, it will be possible to judge how effectively the UCN losses are suppressed. Taking into account the remaining losses τ_β is derived from τ and the uncertainty with which the lifetime of the neutron is measured is determined.

It is noteworthy (this was pointed out by A. Steyerl) that the amount of UCN confined within the sphere is independent of the position of the axis about which the sphere rotates—whether it passes through the centre of the sphere, or is lower or higher than the centre—if only the rotation proceeds slowly enough. Then, if the centre of mass rises or is lowered, in accordance with the equivalence principle there arises a small additional gravitational field in the motion taking place in the reference system connected with the centre of mass of the sphere. This gravitational field only slightly influences the spectrum of the confined neutrons. Thus, all the captured neutrons move, as a whole, together with the sphere.

Bokun (1984) proposed a method to measure τ_β by storing very slow UCN in the funnel of rotating ^{4}He. Such a vessel is good because it does not absorb neutrons. However, questions arise here related to the interaction of UCN with the surface excitations of a rotating liquid, which are interesting from the point of view of the physics of the helium itself.

Measurement of τ_β can also be performed when the UCN are stored in a magnetic trap. An experiment involving a toroidal trap (Kügler *et al.* 1985) revealed that the storage time can be brought close to τ_β.

At the last ILL conference in Grenoble (1989) new results on the measurement of τ_β were reported (V.I. Mampe *et al.* 1989). The value of τ_β in a UCN bottle, its walls being covered with Fomblin oil, was found to be 887.6 ± 3 s. The bottle was a rectangular vessel with one wall replaced by a piston moving in a horizontal direction. Changing the position of the piston changes the volume V and the surface area S of the trap and as a result changes the mean free path $l_f = 4V/S$ of the neutron between collisions with the trap walls. In that respect, the experiment carried out by Mampe *et al.* (1989) is similar to the one of Kosvintsev *et al.* (1980*a*). Now, suppose we have measured the storage time for different volumes of the bottle. It gives us a relation

$$1/\tau = 1/\tau_\beta + 1/\tau_1.$$

If we substitute for $1/\tau_1$ its expression $\mu v/l_f$, then we obtain for $1/\tau$ a linear dependence on $1/l_f$. Extrapolation $1/l_f$ to zero gives the intercept $1/\tau = 1/\tau_\beta$. So it should be. But it is necessary to be very careful. In the gravitational field the neutrons feel the ceiling and the floor of the trap differently. And moreover, the mean free path due to free fall effect is not so simply expressed through the volume. But simple considerations (see appendix 6.2) show that $1/l_f = S/4V + \alpha/h_g$, where h_g is the maximum height reachable by neutrons of given energy E: $h_g = E/mg$, α is of the order h/h_g, and h is the height of the bottle. So, if one extrapolates $S/4V$ to zero one obtains an intercept that contains not only $1/\tau_\beta$, but also $\alpha\mu v/h_g$, where v is the mean velocity of the neutrons. It requires some corrections to get τ_β from the intercept. This correction depends on the energy spectrum of UCN, which changes during the exposure time. Determination of the gravity correction is a difficult problem. Luckily, it is not large; in the experiment by Mampe *et al.* (1989) it is reported to be less than 1 per cent.

The preliminary figure obtained in the experiment at LNPI with the dipper was reported to be $\tau_\beta = 868 \pm 9.2$ s. The result obtained with the toroidal magnetic trap is $\tau_\beta = 876.7 \pm 9.9$ s.

All these results obtained with UCN are useful for comparison with the figure $\tau_\beta = 877 \pm 8$ s given in *Spivak (1988) and with $\tau_\beta = 876 \pm 21$ s reported at the same ILL conference and obtained in PERKEO experiments in which the decay products of neutron bunches are detected.

For the readers' convenience we list here all the references on the measurement of τ_β related to UCN: Antonov *et al.* (1982*a*), Abov *et al.* (1984), Bokun (1984), Ageron *et al.* (1986), Koerting (1986), Kosvintsev *et al.* (1977*b*, 1980*a*, 1981, 1982*a*, 1986, 1988), Yoshiki (1988), Alfimenkov *et al.* (1988), Mampe *et al.* (1989).

8.1.5 Wave properties of the neutron

Since, in accordance with quantum mechanics, the neutron has a de Broglie wave corresponding to it, verification of the wave properties of UCN is equivalent to verification of the laws of quantum mechanics. We shall now consider interference and diffraction of UCN.

Here the interference of UCN will be considered mainly in connection with the transmission through thin films and multilayer systems. The interaction of a neutron with a film can be described by a potential exhibiting the form of a rectangular barrier of height u and width d equal to the width of the film. Now, let $k^2 > u$ (k^2 represents the energy of a neutron travelling in a direction normal to the film). The UCN transmission amplitude of the film is given by

$$T = \exp(ik'd)(1 - r^2)/[1 - r^2 \exp(2ik'd)], \qquad (8.4)$$

where $r = (k - k')/(k + k')$ is the reflection amplitude from a single potential step, and $k' = (k^2 - u)^{1/2}$. If the film is sufficiently thin, absorption in it, i.e. the imaginary part u'', can be neglected. From (8.4) it follows that as $k \to \infty$ the amplitude $|T| \to 1$, but it oscillates between unity and the quantity $(1 - r^2)/(1 + r^2)$. The oscillation amplitude $(k - k')^2/[k^2 + (k')^2]$ tends towards zero as $k^2 \to \infty$. The minima of the transmitted wave amplitude $|T|$ occur at $k^2 = [\pi(2n + 1)/2d]^2 + u$. These minima are caused by the interference of waves undergoing reflections from the two edges of the potential.

Presented in Fig. 8.8(a, b) are the results of an experiment on transmission through a carbon film 450 Å thick and through a gold film 645 Å thick (Steyerl 1972*b*). The experimental points are in good agreement with the theoretical curve, taking the resolution into account.

Interference influences not only the amplitude of the transmitted wave, but also the amplitude of the wave reflected from the potential barrier:

$$R = r[1 - \exp(2ik'd)]/[1 - r^2 \exp(2ik'd)]. \qquad (8.5)$$

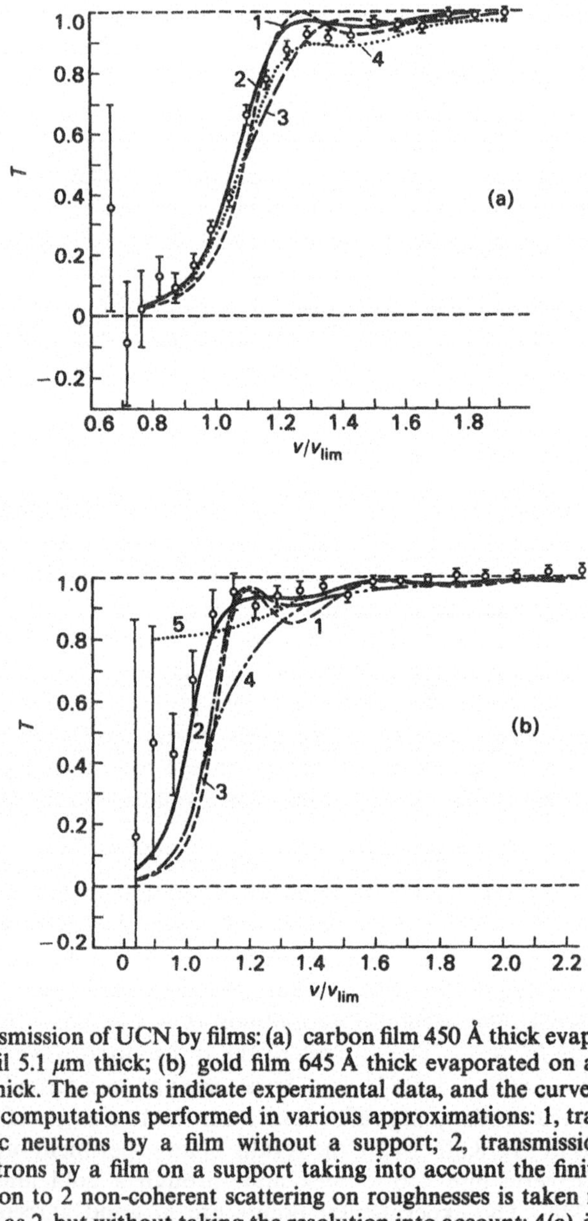

Fig. 8.8 Transmission of UCN by films: (a) carbon film 450 Å thick evaporated on an aluminium foil 5.1 μm thick; (b) gold film 645 Å thick evaporated on a hostaphane foil 4.01 μm thick. The points indicate experimental data, and the curves correspond to theoretical computations performed in various approximations: 1, transmission of monoenergetic neutrons by a film without a support; 2, transmission of monoenergetic neutrons by a film on a support taking into account the finite resolution; 3(a), in addition to 2 non-coherent scattering on roughnesses is taken into account; 3(b), the same as 2, but without taking the resolution into account; 4(a), in addition to 2 coherent scattering is taken into account on roughnesses (the root-mean-square height of the roughnesses is 30 Å, their transverse cross-section is 200 Å); 4(b), the lower tangent to the curve 3(b); 5, in addition to 2 possible distortion is taken into account in the resolution of the time-of-flight spectrometer due to the diffuse reflection of neutrons from the walls in the input neutron guide (Steyerl 1972*b*).

As k increases, R oscillates between zero and $2r(1 + r^2)$. The maxima occur at those k which correspond to the minima of the amplitude $|T|$.

An experiment to observe oscillations was performed using a diffractometer (Scheckenhofer and Steyerl 1977) (Fig. 8.9). The sample (a gold film 2680 Å thick) was substituted for the lower horizontal mirror. As its height varied the energy of the neutrons incident upon the film, as well as the reflection coefficient, also varied (Fig. 8.10).

More interesting is the interference of UCN in a system of three films, which can be represented by a double-humped potential so that a quasi-stationary state may occur in the middle film (Seregin 1977, 1980*a*). The transmission of such a system can be represented in the form

$$T_{123} = \tilde{T}_{12}\exp(\mathrm{i}k_2'd_2)\,\tilde{T}_{23}/[1 - \tilde{R}_{21}\tilde{R}_{23}\exp(2\mathrm{i}k_2'd_2)], \tag{8.6}$$

where, if the films are separated by an infinitesimal gap, the amplitudes $\tilde{T}_{ij}$ and $\tilde{R}_{ij}$ can be represented as follows:

$$\tilde{T}_{12} = T_1 t_2^+/(1 - r_2 R_1), \quad \tilde{R}_{21} = -(r_2 - R_1)/(1 - r_2 R_1), \tag{8.7}$$

$$\tilde{T}_{23} = t_2^- T_3/(1 - r_2 R_3), \quad \tilde{R}_{23} = -(r_2 - R_3)/(1 - r_2 R_3). \tag{8.8}$$

As before, T_i and R_i are still determined by expressions similar to (8.4) and

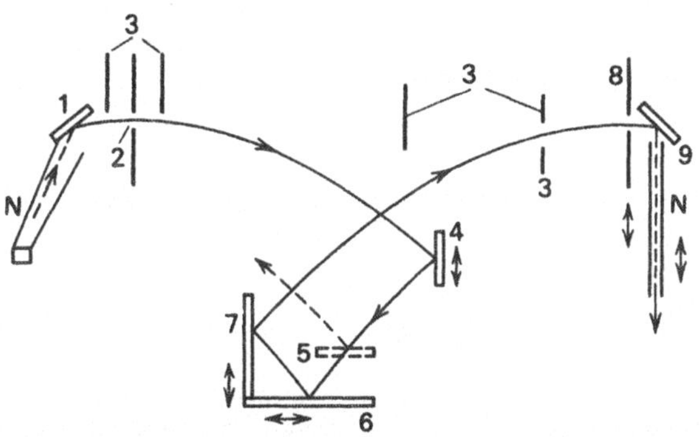

Fig. 8.9 Layout of gravitational spectrometer (Scheckenhofer and Steyerl 1977). The dashed line indicates the neutron's trajectory. The UCN arriving from the turbine impinge upon mirror (1), then pass through the input slit (2) past the beam stoppers (3) and arrive at mirror (4), from which they are reflected to arrive at the sample (5). When the transmission of the sample is investigated, the neutrons leaving it are reflected in succession from mirrors (6) and (7) and, having passed through the exit slit (8), they are reflected from mirror (9) and counted by the detector. When the reflection is being studied the sample is utilized as the mirror (6). If a diffraction grid is substituted for mirror (4), then a neutron reflected from it acquires a vertical velocity component and, correspondingly, the maximum height to which it is raised alters. By using the slit (8) to scan the height, the maximum height reached by a neutron can be measured, i.e. the energy of its vertical motion. The small arrows indicate the directions of possible displacements of various elements of the spectrometer.

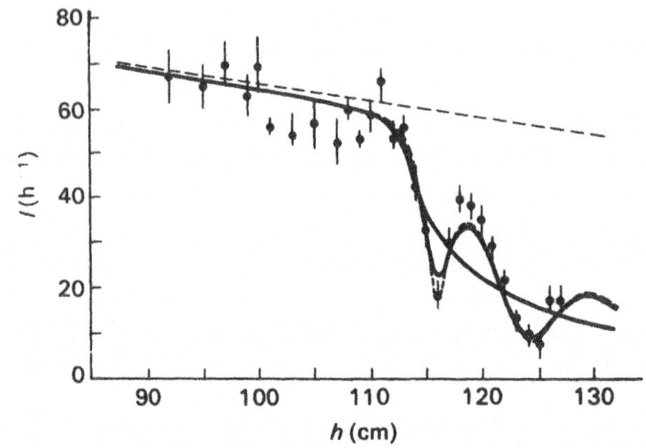

Fig. 8.10 Reflection of UCN from a gold film 2680 Å thick deposited on a glass plate and substituting mirror (6) in the diffractometer in Fig. 8.9. The points are experimental data, and the curves are computed in various approximations: the monotonic curve represents reflection from an infinitely thick layer of gold (no interference effects are present), the oscillating curve represents reflection from a homogeneous gold film of the indicated width deposited on an infinitely thick support (the dashed line is for the case of monoenergetic neutrons, and the solid line takes into account the finite energy resolution). The common slope of the curves on the plot reflects the change in the neutron intensity incident on the mirror (6) occurring when it is displaced in the vertical direction. The limiting energy of gold corresponds to the height $h_{\text{lim}} = 113 \pm 0.2$ cm (Scheckenhofer and Steyerl 1977).

(8.5), in which u_i should be substituted for u_0. r_2 is the reflection amplitude from a potential step of height u_2, t_2^+ is the penetration amplitude into the film, and t_2^- out of the film:

$$t_2^+ = 2k_0/(k_0 + k_2'), \quad t_2^- = 2k_2'/(k_0 + k_2').$$

If we set $\tilde{R}_{ij} = |\tilde{R}_{ij}| \exp(i\psi_{ij})$, then expression (8.6) will exhibit maxima at those k for which $2k_2'd_2 + \psi_{21} + \psi_{23} = 2\pi n$. The formulae for ψ_{ij} are readily obtained from (8.7) and (8.6), and we shall not present them here.

The transmission of a system of three films was investigated experimentally by Steinhauser *et al.* (1980) and Novopoltzev and Pokotilovskii (1981) (Fig. 8.11). The results shown in Fig. 8.11(a) were obtained when the sample was over the lower mirror and those in Fig. 8.11(b) when the sample was substituted for the lower mirror in the diffractometer in Fig. 8.9. It can be seen that the maxima of the transmission curve corresponding to the two resonance states correspond to the same energies that are related to the minima of the reflection curve.

Novopoltzev and Pokotilovskii (1981) studied the transmission of a three-layer system by applying the correlation time-of-flight method.

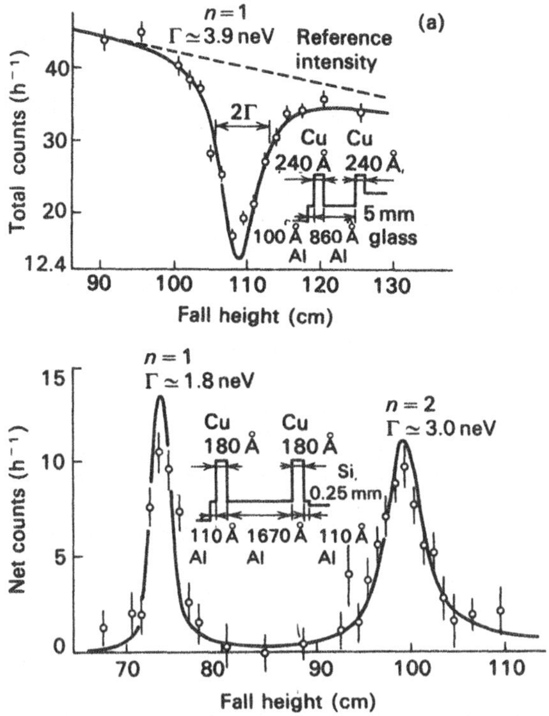

Fig. 8.11 Reflection and transmission of UCN by multilayer systems. (a) Reflection. The $n = 1$ resonance is clearly seen. The next resonance is shallow and therefore cannot be seen. The solid line is obtained by fitting the theoretical curve to the experimental points. The fitted parameters are the thicknesses of the three layers. They turned out to be the following: 225 ± 5 Å for copper and 881 ± 2 Å for the intermediate aluminium layer, in good agreement with the values indicated in the figure. The half-width of the resonance dip is 3.9 neV. The theoretical and experimental depths of the resonance dip differ slightly. The dashed line indicates the initial spectrum of the neutrons incident on the system. (b) Transmission. The two $n = 1$ and $n = 2$ resonances are seen. In fitting the theoretical curve (solid line), computed by taking the energy resolution into account, the layer thicknesses once again served as the fitted parameters. The following values were obtained for copper and aluminium: 1663(3) and 190(5) Å, respectively (Steinhauser *et al.* 1980).

A system consisting of n alternating layers is not actually more complicated. In this case each of the resonance levels is split into $n/2$ sublevels. In the case of transmission by a system of five films, interaction with which can be represented by a three-hump potential, a splitting of several neV was obtained (Fig. 8.12) (Steyerl *et al.* 1981).

If n is large the sublevels ($n/2$) merge into a zone. This circumstance can be utilized in the construction of filters. In Fig. 8.13 the computer-calculated

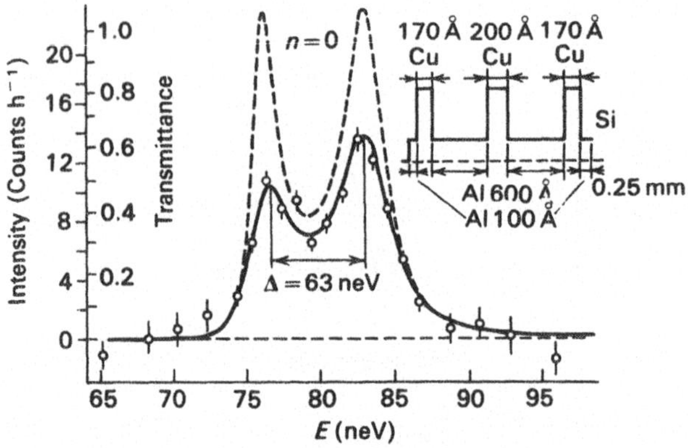

Fig. 8.12 UCN transmission by the multilayer system shown in the inset. The dashed line is computed for monoenergetic neutrons. The solid line is computed by taking the resolution into account. The thickness of the middle layer served as the fitted parameter in fitting the theoretical curve to the experimental data; it turned out to be 208 Å. A split in the resonance level is clearly observed. The decrease in the resonance amplitude is due to the finite resolution and to scattering on roughnesses of the surfaces (Steyerl *et al.* 1981).

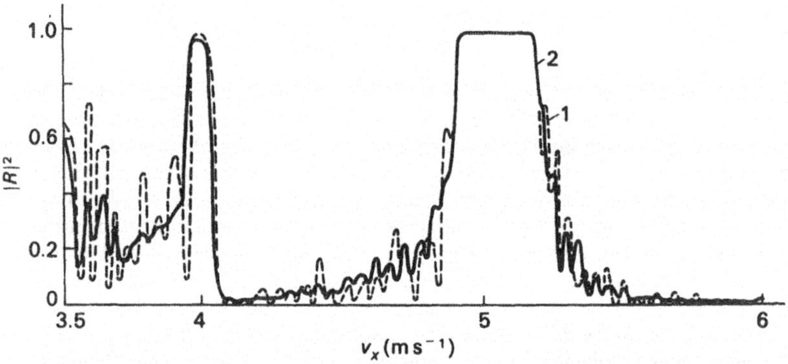

Fig. 8.13 Computer calculation of the dependence of the UCN reflection coefficient on the normal velocity component of the neutron incident upon a multilayer system: 1, random set of thicknesses of alternating layers of Ti and ^{118}Sn for a medium layer thickness equal to 800 Å and a root-mean-square spreading out of 8 Å. The total number of layers is 20; 2, the average for 10 random sets (Antonov *et al.* 1974*a*).

transmission is shown for one such filter involving a small random spreading out of the thicknesses of the individual layers (Antonov *et al.* 1974*a*, *f*, 1988). There filters can be utilized in studies of inelastic UCN scattering with low-energy transfers (Pokotilovskii *et al.* 1980). Filters can also be used as mirrors (*Mezei 1976). If the widths of the alternating layers are gradually varied as the distance from the surface increases, then the region of total reflection can be extended significantly (Antonov *et al.* 1974*a*, *f*; Utsuro *et al.* 1981). Thus, for instance, a system consisting of 360 alternating layers of nickel and titanium, with $u_{Ni} > 0$ and $u_{Ti} < 0$, vapour-deposited on to glass (Ebisawa *et al.* 1979) so that the width of the $(i + 1)$th layer, t_{i+1}, is smaller than the width t_i of the ith layer by

$$\delta = t_i - t_{i+1} = 20\ \text{Å}(200\ \text{Å}/t_i)^{-4.5}, \quad i = 1, 2, \ldots,$$

and so that the first titanium and nickel layers, t_0 and t_1, equal 500 Å and 200 Å, respectively, reflects neutrons nearly totally down to the wavelength λ_m, which is three times smaller than the λ_{lim} for nickel. Such supermirrors are used as turbine blades to extend the range of the UCN obtained with their aid (Utsuro *et al.*, 1983).

If the widths of the layers are the same, then a system of n layers (for instance nickel–titanium layers) may exhibit not only very large (Bragg) reflection (forbidden zones) for $v_\perp > v_{lim}$ of nickel, but also nearly total transmission (conductivity zones) for $v_\perp < v_{lim}$ (Steyerl *et al.* 1985).

One of the objects of interest for investigation is strongly absorbing thin films. As indicated in Chapter 1, reflection of UCN from the surface of a strongly absorbing substance may be stronger, the greater the absorption, but when a neutron does penetrate the absorbing substance it is absorbed extremely effectively. The absorption efficiency is high mainly because of the low velocities of the UCN inside the substance. But this efficiency can be enhanced still further by making use of the interference occurring in the reflection from the two edges of the film. It can readily be demonstrated (Ebisawa *et al.* 1981), for instance, that if a gold film is deposited on nickel, for which $E_{Au} < E_{Ni}$, then for certain energies of the normal motion, $E_{Au} < E < E_{Ni}$, interference within the gold film (see Chapter 5) will lead to a strong decrease in reflection from the nickel.

Interference can also be utilized for the construction of a Michelson-type interferometer (Steyerl *et al.* 1985). The neutron beam in it is decomposed into two components which are subsequently brought together, resulting in the formation of an interference pattern which changes as the path difference for the two components is varied (see also, Ioffe 1988; Ioffe and Zabiyakin (1988)).

8.1.6 Bound neutron states

In the case of interference filters, i.e. multilayer systems, one deals with one-dimensional quasi-stationary states. Now let us consider the motion of UCN within a narrow plane neutron guide with a width that amounts to several hundred ångströms. The motion of a neutron normal to the walls turns out to be quantized, but if the wall is not too thick, then the neutron may tunnel through it and its quantum state will turn out to be quasi-stationary. The width of this state depends on the tunnelling probability.

As the walls of the neutron guide become thicker, the probability of tunnelling decreases and can be neglected when the walls are very thick. The quantum states, however, do not become stationary but remain quasi-stationary, as before, since they disintegrate owing to neutron β-decay, losses in the walls, and scattering on roughnesses. It is worth noting that all the quantum levels, including the lowest one, will have a finite width, even in the absence of β-decay and of losses of neutrons in the walls. This finite width is due to transitions between the levels that occur as a result of scattering on the roughnesses. If, on the other hand, the total energy of a neutron, E, turns out to be higher than the energy of the lowest state, but lower than that of the next quantum state, then transition between the states is impossible and scattering on roughnesses only leads to the neutron momentum turning in the plane parallel to the walls of the neutron guide.

Now let us consider a neutron travelling in a gravitational field along the horizontal plane. Since vertical motion of the neutron is limited from the bottom by the plane and from the top by the gravitational field, motion in the vertical direction is subject to quantization, which means that the energy of vertical motion cannot be less than a minimum energy E_0. This problem is solved in the book by *Flügge (problem N 40) (1971) and in Luschikov and Frank (1978) it was again considered in relation to UCN. It was pointed out that, since UCN possess magnetic moments, their gravitational interaction can be compensated with the aid of a non-homogeneous magnetic field.

Let us estimate E_0 for UCN in the gravitational field. This can be done by using the uncertainty principle, $\Delta z \Delta p_z \geqslant h/2$. Setting $\Delta p_z = mv_z$, $\Delta z = v_z^2/2g$, we obtain $v_z^3 \geqslant gh/m$, from which it follows that $E_0 = mv_z^2/2 = (m^{1/2} gh)^{2/3}/2 \sim 10^{-12}$ eV and, correspondingly, $v_z = 1 \text{ cm s}^{-1}$. In the case of polarized neutrons the gravitational interaction may be compensated with the aid of a non-homogeneous magnetic field. Correspondingly, the lowest energy E_0 will also change.

It is possible to imagine a two-dimensional bound state of the neutron. It is realized in channels of radius $r \approx \lambda_{\text{lim}}$. These channels serve as waveguides for UCN. Thick layers of a substance pierced by a large number of waveguides can serve as monochromatizers and collimators for UCN. If the energy E of a neutron is such that $E_1 < E < E_2$, where $E_{1,2}$ are two adjacent bound states, then in being scattered on roughnesses of the walls of a waveguide the neutron can only change the sign of the wavevector component parallel to the axis. When undergoing scattering on roughnesses, neutrons with high energies may leave the waveguides through the walls and perish inside the substance. The lifetime of UCN in waveguide systems was considered by Kazarnowskij and Stepanov 1976).

Three-dimensional quasi-stationary states are formed in pores of radius $r \approx \lambda_{lim}$. This is discussed in Chapter 6, when the influence of pores on the loss coefficient is analysed. If a pore is deep enough in the material, then the lifetime inside the pore is typical of the UCN storage time in a vessel. The problem of a bound state in a pore was solved by Kagan (1970) in connection with the announcement in the literature (*Grant and Cobble 1969) of a bound state of the neutron being observed in crystalline LiF at liquid helium temperatures. This announcement was subsequently refuted. In Kagan (1970) the probability for UCN to be produced in pores when thermal neutrons are slowed down, and the UCN storage time in a pore, are calculated.

8.1.7 Diffraction

Diffraction of UCN was realized on the diffractometer shown in Fig. 8.9 (Scheckenhofer and Steyerl 1977). The right vertical mirror was replaced by a diffraction grid of dimensions $10 \times 20\ \text{cm}^2$, the number of grooves amounting to 1200 per mm. The profile of the grooves was such that the neutrons were mostly diffracted in the direction of the first order of the diffraction pattern. Diffraction on a grid leads to the incident beam being split into two with differing vertical velocity components. Upon undergoing reflection from the lower horizontal and the left vertical mirrors these two beams are focused at the exit slit into two spots separated in height. By changing the position of the exit slit it is possible to observe the change in the counting rate of the detector. Shown in Fig. 8.14 is the measured dependence of the counting rate of the detector on the position of the exit slit. The width of the lines does not exceed the resolution of the apparatus. This means that the spreading out of the beam occurring in diffraction is not great (i.e. a neutron undergoing diffraction interacts with no less than 100 grooves), from which it follows that the coherence length of the neutron is not smaller than 10^{-2} cm, which exceeds by a factor of 10 the lower limit for the coherence length established by *Shull (1969).

In the works of Ioffe (1983, 1988), Ioffe and Zabiyakin (1988), and also Frank (1983) the utilization of diffraction gratings for the construction of an interferometer for UCN is discussed. In such a device it is necessary to take into account the gravitational deflection of the neutron paths, leading to abberations. With the aid of four diffraction gratings arranged in a special manner these abberations can be eliminated, as a result of which the interferometer becomes an extremely sensitive device. As already mentioned above, it can be applied, for example, in the search for the electric charge of the neutron.

So far we have dealt only with plane straight gratings. A zone mirror represents a somewhat different type of grid. It is dealt with in Schutz *et al.* (1980), Kearney *et al.* (1980), and Klein *et al.* (1981). A zone mirror is utilized

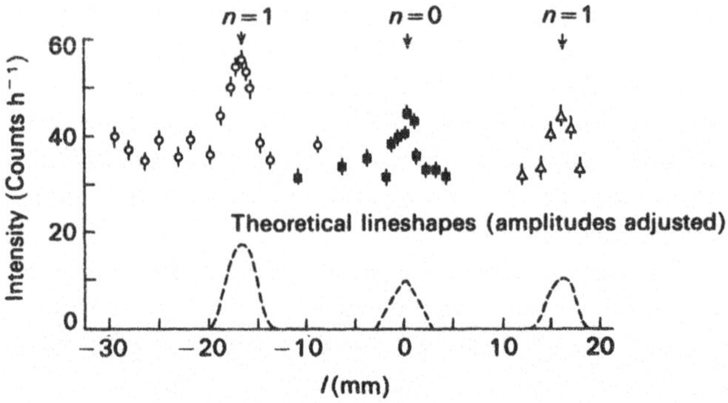

Fig. 8.14 UCN diffraction peaks obtained with the aid of the diffractometer presented in Fig. 8.9, in which the mirror (4) is replaced by a diffraction grid. The profile of the grid grooves is chosen so that only one diffraction peak with $n = 1$ remains. The $n = -1$ peak is obtained by rotation of the grid through 180°. The theoretical curves (the dashed lines) are obtained by taking account of the experimental resolution. The horizontal axis indicates the position of the exit slit. The dimensions of the grid are $10 \times 20\,\text{cm}^2$, the density of grooves on it is $1200\,\text{mm}^{-1}$, and the neutron wavelength is 1500 Å (Scheckenhofer and Steyerl 1977).

in the microscope, which is a truly optical element that has its own focal distance and serves as a lens in the neutron microscope.

Diffraction of UCN on two holes is considered in Surdin (1981). Here also gravity must be taken into account, but the precise influence of gravity on the interference is of the utmost interest in problems of this kind. In principle, the interference of UCN can be used advantageously in studies of gravity, which are of interest (Purica 1979) in connection with the development of the theory of supersymmetry.

For the readers' convenience we list here all the references relevant to the issue of the wave properties of UCN: Steyerl (1972*b*, 1975), Antonov *et al.* (1974*a*, *f*, 1976*b*, 1981*a*, 1986*b*, *c*, *h*, 1987), Ebisawa *et al.* (1979, 1981), Seregin (1977, 1980*a*, 1981), Ignatovich (1977, 1978*a*, *b*, 1985, 1986, 1987*b*, *c*, 1988), Luschikov and Frank (1978), Steyerl *et al.* (1979, 1981, 1985, 1988*b*), Pokotilovskii *et al.* (1980), Steinhauser *et al.* (1980), Surdin (1981), Novopoltzev and Pokotilovskii (1981), Lynn *et al.* (1983), Ioffe (1983, 1988), Ioffe and Zabiyakin (1988), Kaipov and Kadykonor (1989), Novopoltzev *et al.* (1988).

8.1.8 Non-stationary quantum-mechanical phenomena

Quantum-mechanical questions were considered above that were relevant to stationary states of the neutron. In Gerasimov and Kazarnovskij (1976) a proposal is made to study non-stationary processes, for example the speading out in time of a wave packet prepared in advance. Suppose the neutron is in a certain bound state between two potential barriers. If the barriers are removed sufficiently fast, then the state will start to spread out, and this spreading out can be investigated by a study of the distribution of the neutron in space and time. The most interesting thing is the possibility of registering the neutron at a small time interval t at distances $x > vt$, where v is the velocity of the neutron corresponding to the initial stationary state. Within a classical framework this registration is not feasible.

Similar non-stationary processes are considered in Gähler and Golub (1984) and Felber *et al.* (1988), where three-dimensional diffraction is computed of neutrons on a slit with a width which varies in time. From these calculations it is possible to determine the neutron intensity after the slit and the neutron spectrum. The intensity after the slit is shown to exhibit a time structure similar to the spatial structure when diffraction occurs on a stationary slit, and the transmitted neutron spectrum is shown to be modulated. The modulation depends on how the width of the slit varies in time.

In principle, all these peculiarities can be observed in experiments. For example, the diffractive structure can be observed with the aid of a time analyser with a resolution of about 1 μs, using a sufficiently thin scintillation counter. One can even observe a structure similar to the interference pattern from two slits, which arises as a result of a single slit being opened two times in succession. For the observation of the spreading out in time a frame-overlap spectrometer is proposed which makes use of the spin-echo principle.

In this spectrometer there are two choppers. They are situated in such a way that the second chopper turns out to be open when it is reached simultaneously by neutrons from different cycles of the first chopper. This is illustrated in Fig. 8.15. The detector is established in a similar manner with respect to the second chopper. If the neutron spectrum incident upon the spectrometer is sufficiently monochromatic, then the detector registers the

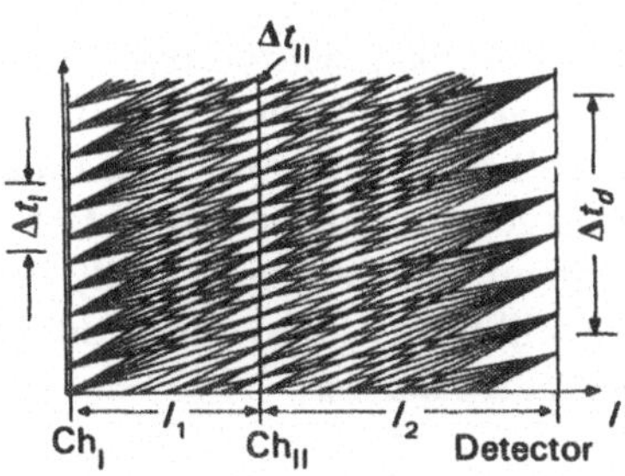

Fig. 8.15 Principle of the frame-overlap time-of-flight spectrometer: velocity diagram for neutrons passing through the two choppers. The slope of the lines is inversely proportional to the velocity of the transmitted neutrons. Time focusing at the detector is achieved for $\Delta t_I/\Delta t_{II} = (l_1 + l_2)/l_2$, $\Delta t_d/\Delta t_{II} = (l_1 + l_2)/l_1$ (Gähler and Golub 1984).

neutrons at discrete moments of time. The slit variable in time (a third fast chopper) established between the detector and the second chopper will lead to a broadening of the spectrum and to registration by the detector of neutrons outside the indicated discrete moments. Estimations reveal that the experiment is technically quite feasible.

In Gerasimov and Kazarnovskij (1976) a proposal is put forward also to observe inelastic scattering of UCN from an oscillating plate. In the case of small oscillation amplitudes of frequency ω the neutron energy changes in inelastic scattering by $\pm \hbar\omega$. When the oscillation amplitudes are large, the velocity of a reflected neutron is determined by the velocity of the wall at the time of reflection. It is interesting to follow the transition from quantum-mechanical to classical reflection. Naturally, the period of oscillations must be significantly greater than the collision time of the neutron with the wall, that is $\sim 10^{-8}$ s.

In dealing with non-stationary processes one must also raise the question of involving a delayed-choice experiment (Miller 1983). This experiment is performed with an interferometer inside which the path of the neutrons is divided by a semi-transparent mirror into two parts, and interference is observed as a result of the phase difference acquired by the wavefunction when the neutron travels along different paths.

According to present-day theory the neutron travels simultaneously along two paths, and if one of the paths is closed, then the wavefunction collapses and the neutron will travel only along the remaining path, whence the interference disappears. In Miller (1983) the following question is raised: what will happen if the experiment is carried out with a pulsed source and the decision whether or not to partition off one of the paths is taken after the neutron, in agreement with classical kinematics, passes the branching point of the paths and starts travelling simultaneously along two paths? It may turn out to be possible, in this way, to study the dynamics of the collapse of the wavefunction.

The answer to the above question is readily obtained by direct computation using the non-stationary Schrödinger equation, as was done, for instance, by Ignatovich and Ostanevich (1985). Therefore the results of the experiment are not difficult to predict, and since quantum mechanics has so far always maintained its good reputation in all tests, it can be foreseen that this time too the experiment, which is most likely best performed with slow neutrons such as UCN, will testify in favour of the prescriptions inherent in the apparatus of quantum mechanics.

8.1.9 Other quantum-mechanical problems

Among the problems of quantum mechanics which can be studied by making use of UCN there is the verification of the non-linearity of quantum mechanics (Gähler *et al.* 1981), the problem of spin precession (Bernstein and Zeilinger 1980), and the investigation of possible hidden states of the neutron (Ignatovich and Luschikov 1981; Kosvintsev *et al.* 1982*b*).

The non-linearity of quantum mechanics is tested by using an interferometer with a partially absorbing plate established in one of its paths. If the phase of the wavefunction depends on its amplitude (e.g. the larger the phase, the greater the amplitude), then the enhancement of the phase along that path, which includes the absorber, will depend on the position of the absorber. In the case under consideration it will be small when the absorber is close to the beginning of the path. At present it has been established (Gähler *et al.* 1981) that if the non-linearity in the Schrödinger equation is described by the expression $-\sigma \ln|\psi|^2$, then the magnitude of σ does not exceed 3.3×10^{-15} eV.

Precession of the spin, usually observed in experiments, is not pure precession (Bernstein and Zeilinger 1980). The point is that for a neutron happening to be in a constant magnetic field the spinor components directed along and opposite the field are characterized by differing momenta k, since their interaction potentials with the field differ, while the total energy is the same. Thus, the phase of $\exp(ikr)$ turns out to be different for the two components, which is precisely what leads to the effective precession.

Precession of the spin, for example for testing the spinor properties of the wavefunction, is observed by using an interferometer in which different magnetic fields are established along different paths. For the observation of pure precession a pulsed source of neutrons is needed, and the magnetic field in one of the paths should be switched on just when the neutron passes the splitting point of the path and starts travelling along different paths. In this case the velocities of different spinor components remain the same and unaltered, while the energies turn out to be different. The most convenient neutrons for this experiment are neutrons with UCN energies.

The question of hidden states, as well as the packet properties of the neutron, were discussed in Chapter 6, and it is partly relevant to the problem of N–$\bar{\text{N}}$ oscillations (the $\bar{\text{N}}$ state can be interpreted as hidden). Therefore, we shall not dwell upon this issue. In addition to these problems we can mention several others which have not yet been touched upon from the point of view of UCN applications, but which can be investigated with UCN in the near future. These problems are: interaction of the neutron with a strong oscillating field and the problem of dressed neutrons (*Fox *et al.* 1987; *Muskat *et al.* 1987); bunching, antibunching, and squeezed states (*Leuchs 1986; *Nieto 1986); non-tunnel transmission (*Leinson *et al.* 1979). Belomytsev *et al.* (1988) proposed to use UCN measurement of correlation coefficients in neutron β-decay.

For the readers' convenience we provide here a list of references on the applications of UCN to fundamental quantum-mechanical problems:

Gerasimov and Kazarnovskij (1976), Ignatovich (1974*b*, 1978*c*), Bernstein and Zeilinger (1980), Gähler *et al.* (1981), Ignatovich and Luschikov (1981), Kosvintsev *et al.* (1982*b*), Miller (1983), Gähler and Golub (1984), Masalovich and Frank (1984), Novopoltsev and Pokotilovskii (1985), Ignatovich and Ostanevich (1985), Felber *et al.* (1988), Richardson *et al.* (1988), *Werner *et al.* (1975).

8.1.10 Neutron–antineutron oscillations

Lately, grand unification models in gauge field theories have been under intense investigation. From these models, for instance, follow the non-conservation of baryon charge and the possibility of transition from the neutron state to the antineutron state. (Note that, if the neutron has electric charge and electric charge is rigorously conserved, then n–n̄ oscillations are forbidden (Gähler *et al.* 1982).) Mathematically, n–n̄ oscillations are described in the same way as the precession of a spin. If the n state is represented by the spinor $\xi_\uparrow$, corresponding to the spin directed along the z axis, and the n̄ state by the spinor $\xi_\downarrow$ corresponding to the spin directed opposite to the z axis, then oscillations can be described by the Hamiltonian $(h/\tau_{n\bar{n}})\sigma_x$, where σ_x is the Pauli matrix and $\tau_{n\bar{n}}$ is the period of precession. Theoretical estimations yield a value of $\sim 10^9$ s for $\tau_{n\bar{n}}$. If the energies of the n and n̄ states (in other words, their masses) are identical, then the probability of observing transition from n to n̄ in time t is $\sin^2(t/\tau_{n\bar{n}}) \approx (t/\tau_{n\bar{n}})^2$, when $t \ll \tau_{n\bar{n}}$. Different interactions which lead to a split between the n and n̄ masses, for example magnetic or nuclear interaction, may be described by an additional term in the Hamiltonian, $\Delta m\sigma_z$. In the case the transition probability from n to n̄ is determined by the expression

$$w_{n\bar{n}}(t) = \sin^2(\sqrt{1+\omega^2\tau_{n\bar{n}}^2}\,t/\tau_{n\bar{n}})/(1+\omega^2\tau_{n\bar{n}}^2) \approx (t/\tau_{n\bar{n}})^2, \tag{8.9}$$

where $\omega = \Delta m/h$ and the equality is obtained under the assumption that $\omega\tau_{n\bar{n}} \gg 1$ and $\omega t \ll 1$. To satisfy the last inequality the experiment must be performed in a magnetic vacuum, which should be deeper, the longer is t. If $\omega t \gg 1$, then the probability of transition to the antineutron state becomes $2(\omega t)^2$ times less than (8.9). The condition imposed on the external magnetic field may be weakened if the field is made variable (Yoshiki and Golub 1988). Indeed, let us apply such a magnetic field that $\omega t \gg 1$. If we change its sign every quarter of the period $t_1 = \pi/2\omega$, then the probability of the neutron transition to the antineutron state during the time $t = Kt_1$, where K is an integer, will be

$$w_{n\bar{n}} = [K\sin(t_1\sqrt{1+\omega^2\tau_{n\bar{n}}^2}/\tau_{n\bar{n}})]^2/(1+\omega^2\tau_{n\bar{n}}^2) = (2/\pi)^2(t/\tau_{n\bar{n}})^2,$$

which is almost the same as (8.9).

Searching for n–n̄ oscillations is equivalent to looking for annihilation processes of antineutrons with nuclei in a neutron beam. In the case of a thermal neutron flux of Φ_T and a flight base L, the expected antineutron flux at the end of the flight base amounts to $\Phi_T w_{n\bar{n}}$, where $w_{n\bar{n}}$ is the probability of n–n̄ transition during the time $t_f = L/v$ and v is the velocity of the neutron.

From the point of view of looking for n–n̄ oscillations UCN possess two advantages compared with thermal neutrons. First, their velocity is small, so there is no need for a long flight base; secondly, during its confinement in the vessel a neutron undergoes numerous collisions with the walls. If the number of collisions is N, then the transition probability of a given neutron into an antineutron during the confinement time in the vessel increases by a factor of N. If the interaction potentials of the neutron and the antineutron with the walls are about the same, then reflection from the wall does not alter the phase relationship between the n and n̄ states. The time t in (8.9) is, correspondingly, N times greater than the free path time t_f between two collisions with the wall, and consequently $w_{n\bar{n}}$ increases by a factor of N^2. In this case experiments utilizing UCN could be more sensitive than those with thermal neutrons, which have a much higher intensity. There may also occur an intermediate version when the reflection phase of the antineutron differs from that of the neutron, although the phase difference is not significant. In this case one can introduce the time needed for correlating the phases, τ_φ, or in other words the correlation number of collisions with the wall n_c during which the phase accumulates. Then the total probability of finding an antineutron in the trap during the storage time Nt_f is $Nn_c w_{n\bar{n}}$, i.e. once again there remains the possibility of gain for the UCN. Unfortunately, existing data on the antineutron scattering amplitude on nuclei are still insufficiently reliable, so a final conclusion concerning the advantage of UCN cannot be made.

For the readers' convenience we list here all the references on UCN in which N–N̄ oscillations are dealt with: Yoshiki (1981), Chetyrkin *et al.* (1981), Kazarnowskij *et al.* (1981), Petschek (1983), Casella (1984), Yoshiki and Golub (1988), *Luschikov *et al.* (1981).

8.2 UTILIZATION OF UCN IN STUDIES OF MATTER

8.2.1 Scattering and absorption cross-sections in homogeneous substances

It is known that the cross-sections of inelastic scattering and capture grow, as the velocity v of the neutron decreases, proportionally to $1/v$. This fact follows from the definition of the cross-section itself as the ratio between the number of events per unit time to the incident flux density. It is interesting to verify the above inverse proportionality in the UCN energy region. The cross-sections are measured by transmitting neutrons through a substance. Only neutrons with energies higher than E_{lim} can pass through the substance, however, so strictly speaking neutrons of higher energies are dealt with in these experiments, instead of UCN; they are sometimes called VCN (very cold neutrons) to distinguish them from UCN. Obviously, transmission in homogeneous substances must differ from transmission in non-homogeneous substances, since in the latter additional elastic scattering on inhomogeneities occurs. The $1/v$ dependence can be tested only with homogeneous substances.

In investigating the transmission by a substance of neutrons with energies close to E_{lim} it is necessary to take into account their reflection from the interface (Steyerl and Vonach 1974) and the difference between the neutron velocity inside the vacuum and the substance. Let J_{in} be the incident and J_{out} the transmitted neutron flux; then the transmission τ is

$$\tau = J_{\text{out}}/J_{\text{in}} = \alpha t^2/(1 - \alpha^2 r^2), \tag{8.10}$$

where

$$\alpha = \exp(-N\sigma_{\text{tot}}L), \quad t = 4k\,\text{Re}\,k'/|k + k'|^2,$$
$$r = |k - k'|^2/|k + k'|^2, \quad k' = \sqrt{k^2 - k_{\text{lim}}^2},$$

k is the wavevector in vacuum, and L is the thickness of the plate of the substance under study. The neutrons are assumed to impinge normally upon the interface. The cross-section sought, σ_{tot}, is related to a single atom. Expression (8.10) is obtained from the exact quantum-mechanical transmission coefficient upon averaging over the spreading out of the width along the sample and over the finite resolution. Denoting $\beta = t^2/2\tau r^2$ and resolving (8.10) with respect to α, we obtain

$$\alpha = \sqrt{\beta^2 + 1/r^2} - \beta,$$

from which $\sigma_{\text{tot}} \sim 1/k'$ is determined for homogeneous substances. We note that for $k/k_{\text{lim}} = 1.5$, taking into account the reflection r yields a correction to τ of about 4%.

Experiments performed with homogeneous substances on the transmission of neutrons with velocities from 5 to 140 m s^{-1} are presented in many publications. In all cases the $1/v$ law is confirmed and the obtained cross-sections extrapolated to the thermal point are in good agreement with known total absorption and inelastic scattering cross-sections (*Hughes and Schwartz 1958). In those cases when the absorption cross-section is small, it was possible to measure both the absorption and inelastic scattering cross-sections individually by determining the transmission at room temperature and at liquid nitrogen temperature, when inelastic scattering can be considered totally suppressed. We note that the elastic scattering cross-section does not vary as the neutron velocity decreases, and at low energies it can be totally neglected.

8.2.2 Investigation of inhomogeneities in non-homogeneous substances

Investigation of the transmission of non-homogeneous substances is of great interest. The total macroscopic scattering cross-section $\Sigma_{tot} = N_0 \sigma_{tot}$ in a non-homogeneous substance involves three summands, $\Sigma_{tot} = \Sigma_{el}^{inc} + A/k + \Sigma_{inh}(k)$. The first term is due to elastic incoherent scattering on individual nuclei and the second includes the absorption and inelastic scattering cross-sections, so the first two terms are precisely the same as in the case of homogeneous substances. The third term is directly related to scattering in inhomogeneities.

Inhomogeneities are characterized by a variation of the scattering density $\tau = \Delta(N_0 b)$ and by a volume V_{inh} or characteristic radius R. The cross-section $\Sigma_{inh}(k)$ can be represented in the form $\sigma_{inh}(0)\nu_{inh}g(k)$, where $\sigma_{inh}(0) = 4\pi\tau^2 V_{inh}^2$, ν_{inh} is the number of inhomogeneities per unit volume, and the function $g(k)$ characterizes the energy dependence of the cross-section. Obviously, at small k, i.e. at large wavelengths, when $\lambda \gg R$, each inhomogeneity serves as an independent pointlike scatterer, so $g(0) = 1$. At very small wavelengths the neutrons do not feel the inhomogeneities and, therefore, $g(k \to \infty) = 0$. In Appendix 8.2 it is shown that in the case of spherical inhomogeneities the behaviour of the function $g(k)$ changes, as k decreases, from $1/k^4$ to $1/k^2$. Since the cross-section is determined by the transmission method, it is necessary to take into account the aperture of the detector θ. When $kR\theta \gg 1$ we have $g(k) = 9/8k^4R^4\theta^2$; when $kR\theta < 1 < kR$, the function $g(k)$ can be approximated by the expression $9/8k^2R^2$; and, finally, when $kR \ll 1$, we have $g(k) \approx 1$.

Usually inhomogeneities are investigated using small-angle scattering of thermal neutrons, but, as shown in Steyerl (1974*a*), they can also be studied by investigating the energy dependence of the total cross-section. For example, if the deviation from the $1/v$ law is approximated by a piecewise smooth function exhibiting a behaviour such as $1/k^4R^4\theta^2$ at high velocities and $1/k^2R^2$ at lower velocities, then the radius of the inhomogeneity can be determined from the matching condition of the two pieces, $1/k^4R^4\theta^2 = 1/k^2R^2$. If the opening angle θ and the intersection point k are known, then the radius can be immediately determined: $R = 1/k\theta$. This method was tested by Lengsfeld and Steyerl (1977) with a colloidal solution representing a suspension of SiO_2 spheres of radius 130 Å in a mixture of H_2O and 1.5%D_2O. The resulting analysis of the neutron experiment revealed that this technique permits the determination of the radii of the SiO_2 spheres with a good precision.

Engelmann *et al.* (1979) carried out a comparison of the VCN transmission technique and the small-angle scattering of thermal neutrons in studies of the AlZn alloy containing 5.5% of zinc. In this alloy phase precipitation takes place at temperatures lower than the temperature of homogenization

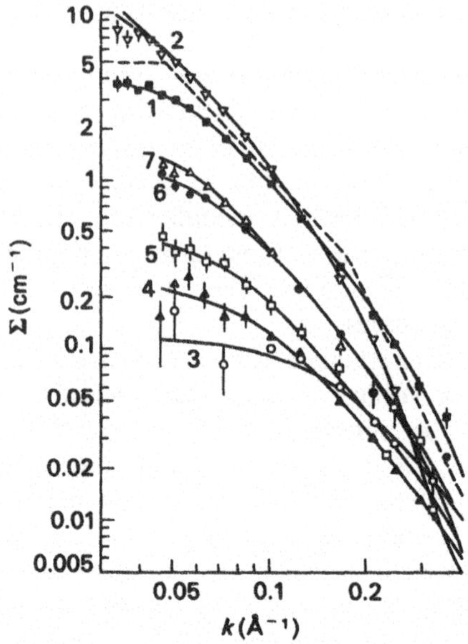

Fig. 8.16 Total macroscopic neutron scattering cross-section in the Al + 5.5 at. % Zn alloy after various thermal treatments. The absorption cross-section and the inelastic and elastic non-coherent scattering cross-sections are subtracted from Σ: 1 and 2, samples cooled off slowly and rapidly, respectively, after long ageing at room temperature; 3–7, samples cooled slowly for different exposure times at a temperature of 100°C: 0, 0.5, 2.5, 5, 10–15 h respectively. The solid lines are computed by the model of spherical zones of the enriched phase. The radii of the zones and the number of zones per unit volume served as parameters for fitting the curves to the experimental data. The chain curve is obtained by taking into account the interference of neutron waves in scattering on various zones. The dashed line indicates the asymptotic behaviour of curve 1. For small k it can be approximated by a constant, at large k its behaviour is proportional to $\sim 1/k^4$, and in the intermediate region it is proportional to $\sim 1/k^2$ (Engelmann *et al.* 1979).

(250–350°C). So-called Guinier–Preston zones are formed which are strongly enriched with zinc (65–69%), while the concentration of zinc outside these zones falls to the level of 2–5%. These zones grow (age) at differing rates at temperatures lower than the critical temperature. Both the above techniques make possible the investigation of the ageing dynamics with about the same reliability. Small-angle scattering, however, yields somewhat more complete information. It exhibits advantages in studies of strongly absorbing substances from a statistical point of view. On the other hand, transmission of VCN is attractive owing to the technical simplicity of the experiment, and for this reason it is, in certain cases, preferable. Shown in Fig. 8.16 is the dependence of the total cross-section Σ_{tot} on the wavevector for various ageing regimes together with the curve obtained by fitting the parameters to the experimental points.

Steyerl (1972c) pointed out that the transmission of a film is also influenced by roughnesses. The transmission of a cold-rolled polycrystalline copper foil 0.12 μm thick was measured. The additional scattering that led to a deviation from the $1/v$ law decreased noticeably when the foil underwent mechanical polishing, from which the dimensions of the roughnesses on the non-polished film could be deduced. Their root-mean-square height was 100–240 Å and the correlation length was 200–500 Å.

8.2.3 Transmission of magnetic films

In Chapter 7 the transmission of magnetic films was considered in the approximation when the neutron beam was represented by a two-component column. However, a density matrix provides the most adequate description of a neutron beam. Indeed, a neutron beam with zero polarization along the z axis may have a non-zero polarization along some other axis. Rotation of such a beam polarization with the help of a spin flipper may create a polarization along the z axis which cannot be described with the two-component formalism.

In general, a density matrix can be represented in the form

$$\rho = I(1 + \boldsymbol{P}\boldsymbol{\sigma}),$$

where $\boldsymbol{P}$ is a polarization vector of length not exceeding unity, I is half of the beam intensity $\mathrm{Tr}\,\rho = 2I$, and all the parameters (I and the components of the vector $\boldsymbol{P}$) are real numbers. The projection of the polarization vector on any direction $\boldsymbol{Q}(|\boldsymbol{Q}| = 1)$ is represented as

$$\mathrm{Tr}(\boldsymbol{Q}\boldsymbol{\sigma}\rho)/2I = (\boldsymbol{Q}\boldsymbol{P}).$$

Let us consider a magnetic film with a single-domain central part of magnetization parallel to the film surface. Subsurface parts of the film cannot be magnetized homogeneously because of surface anisotropy and roughnesses. When a beam enters the central part of the film after transmission through the left subsurface part, its density matrix ρ_0 is transformed to ρ in accordance with the following expression:

$$\rho = I(1 + \boldsymbol{P}\boldsymbol{\sigma}) = \vec{T}_1 \rho_0 \vec{T}_1^{+} = \vec{T}_1 I_0(1 + \boldsymbol{P}_0\boldsymbol{\sigma})\vec{T}_1^{+}, \tag{8.11}$$

where the matrix $\vec{T}_1 = \underset{\leftarrow}{\mu}(1 + \boldsymbol{m}\boldsymbol{\sigma})$ may involve complex parameters. We shall express the matrix $\vec{T}_1$ in the form

$$\vec{T}_1 = \vec{t}_1[(1 + \boldsymbol{e}\boldsymbol{\sigma})/2]\exp(\mathrm{i}\boldsymbol{a}_1\boldsymbol{\sigma}), \tag{8.12}$$

where the vectors $\boldsymbol{e}$ and $\boldsymbol{a}_1$ are real and $|\boldsymbol{e}| = 1$. The last factor in (8.12) corresponds to a turn of the beam polarization inside the subsurface part of the film about the vector $\boldsymbol{a}_1$ through an angle $|\boldsymbol{a}_1|$. The second factor in (8.12) represents a projection operator P_e which extracts the polarization along the axis $\boldsymbol{e}$ corresponding to the direction of magnetization inside the film (it may not coincide with the direction of the external magnetic field if some anisotropy is present). The first factor characterizes the transmission of the subsurface part of the film $|\vec{t}_1|^2$. The phase of the factor t_1 is unimportant, so t_1 can be taken to be a real positive number.

Since $|\vec{T}_1|^2_{ij}$ determines the probability of transition from the state j to the state i, then the transmission matrix in the opposite direction of the left subsurface region, from the central part of the film to the outside and taking account the detailed balance theorem, can be represented in the form

$$\overleftarrow{T}_1 = \overleftarrow{t}_1 \exp(i\boldsymbol{a}_1\boldsymbol{\sigma})[(1 + \boldsymbol{e\sigma})/2], \tag{8.13}$$

where $\overleftarrow{t}_1 = (i - \vec{t}_1^{\,2})^{1/2}$.

The transmission matrices of the right subsurface region of the film can be represented in similar forms:

$$\vec{T}_r = \vec{t}_r \exp(i\boldsymbol{a}_r\boldsymbol{\sigma})[(1 + \boldsymbol{e\sigma})/2], \tag{8.14}$$

$$\overleftarrow{T}_r = \overleftarrow{t}_r[(1 + \boldsymbol{e\sigma})/2]\exp(i\boldsymbol{a}_r\boldsymbol{\sigma}). \tag{8.15}$$

The transmission matrices of the whole film, taking into account rescattering at both ends, are

$$\vec{T} = t\exp(i\boldsymbol{a}_r\boldsymbol{\sigma})[(1 + \boldsymbol{e\sigma})/2]\exp(i\boldsymbol{a}_1\boldsymbol{\sigma}), \tag{8.16}$$

$$\overleftarrow{T} = t\exp(i\boldsymbol{a}_1\boldsymbol{\sigma})[(1 + \boldsymbol{e\sigma})/2]\exp(i\boldsymbol{a}_r\boldsymbol{\sigma}), \tag{8.17}$$

where $t = \vec{t}_1\vec{t}_r/(1 - \overleftarrow{t}_1\overleftarrow{t}_r)$. From (8.16) and (8.17) it follows that the transmission of a single-domain film is described, in general, by nine parameters. To describe the transmission of a poly-domain film it is necessary to introduce a distribution of the vectors $\boldsymbol{e}$, which gives at least one additional parameter. The density matrix ρ_0 of the primary beam after transmission of the neutrons through the film transforms into

$$\rho = \vec{T}\rho_0\vec{T}^+ = I(1 + \boldsymbol{P\sigma}),$$

where, for instance,

$$I = J_0 t^2\{(\boldsymbol{eP}_0)\cos(2a_1) - ([\boldsymbol{e}\hat{\boldsymbol{a}}_1]\boldsymbol{P}_0)\sin(2a_1) + 2\sin^2(a_1)(\boldsymbol{e}\hat{\boldsymbol{a}}_1)(\hat{\boldsymbol{a}}_1\boldsymbol{P}_0)\},$$

$$\hat{\boldsymbol{a}} = \boldsymbol{a}/a, \quad a = |\boldsymbol{a}|.$$

It is not difficult to obtain an expression for $\boldsymbol{P}$ also, but we shall not write it out because of its cumbersome form.

From the above it follows that to investigate magnetic films (i.e. to measure all the phenomenological parameters) with the help of UCN transmission it is necessary to perform a three-dimensional polarization analysis. The appropriate experiments have not yet been conducted.

Inhomogeneities may be of a magnetic origin. The transmission method permits the determination (Steyerl 1974*a*; Steyerl and Lerner 1976) of the dimensions of domains and of the Block walls. In Steyerl and Lerner (1976) the dimensions of domains and domain walls were measured in nickel, iron, and cobalt. Usually the domains are very large and the correction to the cross-section Σ_{tot} due to such dimensions is proportional to $\sim 1/k^4$ within the whole range of velocities, while scattering on the domain walls is, within a certain energy range, proportional to $1/k^2$. In the energy range where scattering on the walls is also proportional to $\sim 1/k^4$, the contribution of the walls is approximately two times higher than that of the domains, but to a certain extent it depends on the angle between the magnetizations in adjacent domains. Under the assumption that this angle equalled 90°, the dimensions of the domains in nickel, iron and cobalt samples turned out to be, respectively, 5, 36, and 3 μm, while the widths of the domain walls were 420, 235, and 225 Å. In certain samples the additional scattering on inhomogeneities disappeared when the sample was magnetized up to saturation, while in others this did not happen. It could be, however, that in the latter case the residual additional scattering was of a non-magnetic origin.

8.2.4 Measurement of the limiting energies for various substances

The VCN transmission technique can also be utilized to determine the limiting energy of a substance and to reveal admixtures on the surface. In Steyerl (1972*b*) the limiting energies were measured for graphite, gold, and copper, and in Steyerl and Trusted (1974) they were measured for aluminium, Al_2O_3, and Cu_2O. Scheckenhoffer and Steyerl (1977) measured the limiting velocity of glass, but here the measurement was performed by making use of the reflection of UCN, instead of their transmission. Presented in Fig. 8.17 are the results of a measurement of the dependence of the reflection coefficient of glass on the energy of the incident neutrons. Also shown is the theoretical curve for reflection of UCN from a potential barrier. The experimental points lie on the smoother curve. The discrepancy is due to the presence of admixtures on the surface (Golub and Pendlebury 1974). It is as if the admixtures smear the interface, and this leads to a change in the reflection coefficient near and above E_{lim}. The experimental points in Fig. 8.17 can be described theoretically, if a hydrogen film is assumed to be present on the glass surface and smearing the interface to a depth of ~ 73 Å (Scheckenhoffer and Steyerl 1977).

It is also interesting to consider the transmission curve of magnetic materials. A material magnetized up to saturation is characterized by two differing potentials for the two opposite polarizations. In Fig. 8.18 the transmission curve is shown for a monocrystalline iron film magnetized up to

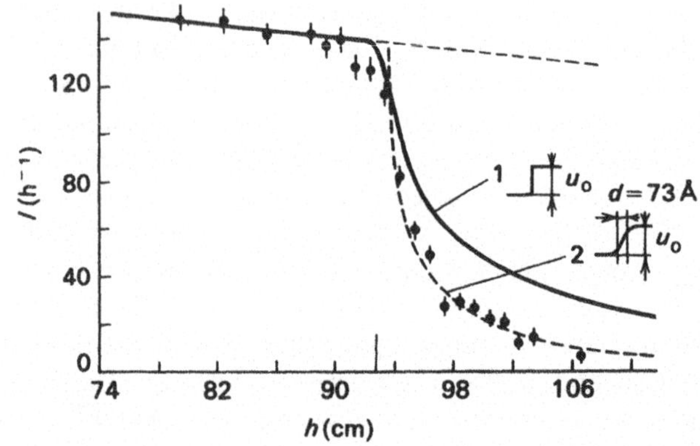

Fig. 8.17 UCN reflection from glass measured with the aid of the diffractometer of Fig. 8.9. The points are experimental. The solid curve represents reflection from an ideal potential step taking into account the experimental resolution. The dashed line represents the calculated reflection of monoenergetic neutrons from a step with a smoothed-out edge. The curve describes the experimental data well, if the depth of the smearing is assumed to be equal to 73(3) Å. The horizontal axis represents the height from which the neutron falls on to the mirror. The short vertical bar indicates the limiting height $h_{\text{lim}} = 93.6$ cm (Scheckenhofer and Steyerl 1977).

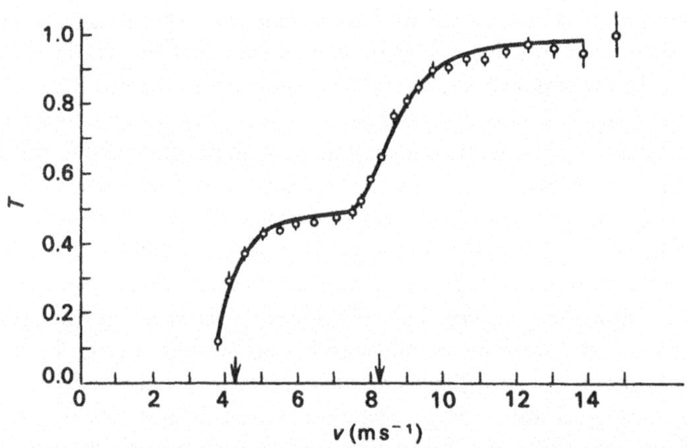

Fig. 8.18 Transmission of UCN by a magnetized monocrystalline iron film 1500–3000 Å in width in the case of normal falling. The solid line is obtained by theoretical fitting to experimental data. The fitted parameters are the two limiting velocities and the broadening of the resolution due to partial diffuse reflection of the UCN from the walls of the supply neutron guide. As a result of the fit the limiting velocities indicated by arrows in the figure turned out to be 8.18(2) and 4.11(4) m s^{-1} respectively (Herdin *et al.* 1978).

saturation. The experiment is presented in Herdin *et al.* (1978). The two steps corresponding to the two limiting energies are clearly seen.

In conclusion we give a complete list of references on studies of the interaction of UCN with matter so as to reflect, also, on those points that were not raised in the text: Gurevich and Nemirovsky (1961), Steyerl (1969, 1972*c*, 1974*a*, *b*), Golub and Pendlebury (1974), Steyerl and Vonach (1974), Steyerl and Lerner (1976), Kashukeev *et al.* (1976, 1977), Scheckenhoffer and Steyerl (1977), Lengsfeld and Steyerl (1977), Akhmetov *et al.* (1977*b*), Gorjachev and Lin'kova (1977), Antonov *et al.* (1978*a*, *c*), Bates (1978, 1981), Engelmann *et al.* (1979), Antonov *et al.* (1980*a*, *b*), Kaipov and Kaspanov (1981), Kaipov *et al.* (1982), Golikov *et al.* (1985*b*, *c*, 1986, 1987, 1988*b*), Steyerl *et al.* (1983), Antonov *et al.* (1984*a*, *c*), Ludecke *et al.* (1985), Antonov *et al.* (1985*a*–*d*), Antonov *et al.* (1986*d*–*g*), Stepanov and Shelagin (1986), Novopoltzev and Pokotilovskii (1987), Morosov *et al.* (1987*a*, 1988*b*), Antonov *et al.* (1988), Barabanov (1988), Isakov *et al.* (1988).

8.2.5 Interaction of UCN with gases

Experiments for studying the interaction of UCN with gases stand somewhat apart from transmission experiments with films. The interaction cross-section of a neutron with atoms or with molecules is not determined by the velocity of the neutron, but by the relative velocity of the molecules and the neutron, which is significantly higher than the velocity of an ultracold neutron. In the case of an atomic gas this is not so essential, since the cross-section is easily computed, but in the case of molecular gases this circumstance becomes essential, since a collision with a molecule may lead to such processes occurring in the molecule that are forbidden by the law of energy conservation, if only the kinetic energy of the incident neutron is taken into account. Kadykenov *et al.* (1982, 1984) computed the interaction of UCN with molecular gases. The transmission of gases has been measured in a number of investigations (Akhmetov *et al.* 1976, 1977*b*, 1981*a*, *b*; and others). In some cases deviations from the predicted interaction cross-sections were observed. These deviations may be due to several reasons. Thus, for example, adsorption of the gas being studied is possible on the walls of the container. Furthermore, the molecules of the gas may form weakly bound clusters. These clusters scatter UCN as a whole and the scattering cross-section by a cluster is n^2 greater than the scattering cross-section on a single molecule, where n is the number of molecules in a cluster. On the whole, however, the interaction of UCN with gases is still insufficiently known, and if the existence of the above-mentioned effects are to be proved, then they will deserve a thorough investigation in their own right.

The interaction with gases can be studied in experiments on UCN storage too and in continuous flow experiments such as those presented in Ignatovich *et al.* (1982). In the latter case it is possible to measure inelastic scattering and absorption of UCN separately, as well as the difference between the scattering cross-section on molecules in a free state and when they are adsorbed on the surface. Most likely, investigations in this field will start to progress as soon as the problem of UCN storage is solved and experimental physicists learn how to treat the walls of vessels for UCN storage, achieving, in doing so, the required degree of perfection.

To conclude, we provide for the readers' convenience a full list of references on the investigation of the interaction of UCN with gases: Kosvintsev *et al.* (1980*d*), Akhmetov *et al.* (1981*a*, *b*, 1984), Kadykenov *et al.* (1982, 1984), Kaipov and Kadykenov (1983), Granada (1985), Kaipov *et al.* (1988).

8.3 INVESTIGATION OF THE INTERACTION OF UCN WITH THE SURFACE OF A MATERIAL

8.3.1 Admixtures

Owing to total reflection a confined neutron undergoes numerous collisions with the walls of the vessel, thus enhancing the efficiency of its interaction with the surface of the material. This fact can be used advantageously to find small amounts of absorbing admixtures present on the surface of a material. Thus, for example, with the aid of UCN some boron was found on the inner surface of the walls of a detector. The boron happened to be there as a result of treating the walls with sandpaper containing boron. The detector was a proportional gas chamber, and it turned out that its counting rate did not drop to zero in the absence of ^{3}He in it. The reason was that the UCN arriving in the inner volume of the detector collided many times with its walls and finally were absorbed by the small number of boron nuclei happening to be on the walls, as a result of the treatment with the boron sandpaper, and were thus registered.

Surface admixtures can also be studied in storage experiments. If a reference sample is placed inside the chamber for UCN storage, and if, subsequently, it is exchanged with a sample containing an admixture on its surface, the UCN storage times in these two cases will differ, and the amount of admixtures can be determined by the difference between the storage times. This technique is, naturally, most sensitive either to strongly absorbing admixtures or to elements exhibiting large inelastic scattering cross-sections. In Morosov (1980) a similar difference method was applied to measure the total coefficient η of a given sample with a known surface area.

8.3.2 Roughnesses

UCN may be utilized for the investigation of surface roughnesses. To this end one must either measure the transmission of the neutron guide for a given treatment of its walls (in this case averaged characteristics are obtained), or measure the reflection indicatrix for a given beam (in this case more detailed information on the roughnesses of a given part of the surface is obtained)

8.3.3 Surface excitations

Since UCN only interact with the surface region of a substance ~100 Å in depth, they can be used to study surface phenomena involving approximately the same region: surface waves, magnetic excitations. Owing to their low energy UCN are sensitive to low-energy transitions. Thus, for example, in storage experiments (Kosvintsev *et al.* 1979) the heating of an ultracold neutron in a single collision with the wall by $\sim 10^{-10}$ eV would be quite registrable by the change in the spectrum of the confined neutrons. Such a high sensitivity is due, here, to multiple collisions of UCN with the walls.

Shown Fig. 8.19 is a spectrometer which is described in Steyerl (1978*a*) and intended for the investigation of inelastic scattering of UCN with a resolution of 3×10^{-8} eV at momentum transfers of ~ 0.03 Å^{-1}. Figure 8.20 demonstrates the results on the observation of quasi-elastic scattering.

8.3.4 The interaction of UCN with powders

Recently the interaction of UCN with powders has been under investigation (Golikov *et al.* 1986, 1987, 1988*a*, *b*). These experiments are interesting because of the technique applied and also from a theoretical point of view (they have been mentioned briefly in Chapter 4). The experiments are performed with a branching neutron guide. At one end of the branching guide a detector is established and at the other end the powder to be studied. Interaction with the powder influences the counting rate of the detector. Changing the temperature, the amount of the powder, and the dimensions of its grains allows the change in the counting rate of the detector to be observed and hence the parameters characterizing the interaction of the neutrons with the surface of the substance to be deduced.

A remarkable feature peculiar to powders is that the UCN diffuse among their grains and interact with them numerous times, as a result of which the effect of the neutron interaction with the surface of the substance is enhanced significantly. Hitherto no theory has existed for the interaction of radiation with finely dispersed media, when the interaction cannot be dealt with by perturbation theory and the grains are large in comparison with the wavelength. Investigations involving UCN have made possible the formulation of a model to solve this problem.

We shall consider the set of grains as a set of straight neutron guides with diffusely reflecting walls. In this case the probability of reflection from the surface of the powder equals the probability of reflection from the entrance

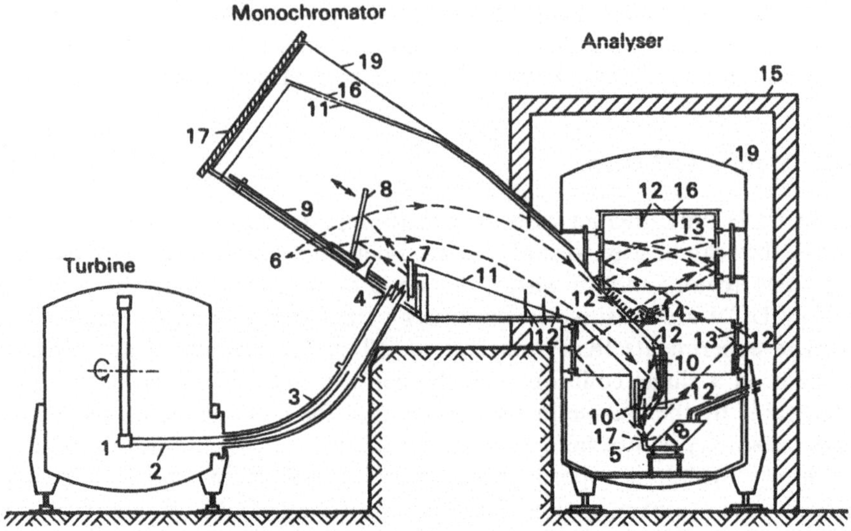

Fig. 8.19 Layout of gravitational spectrometer NESSIE (Neutron-Schwerkraft-Spectrometrie). A neutron turbine serves as the source. Monochromatization and analysis of the spectrum is carried out by utilizing the maximum elevation of the neutron parabolic trajectory. Spatial focusing is performed with the aid of the gravitational field and two curved mirrors: 1, turbine blades; 2 and 3, neutron guides; 4, transverse cross-section of the outgoing beam; 5, sample surrounded by the polyethylene absorber; 6, image of the source; 7, first deflecting mirror; 8, monochromatizing mirror; 9, servomechanism for displacing the monochromatizing mirror; 10, focusing mirrors; 11, glass plates for limiting the beam in the transverse direction; 12, beam stoppers; 13, cylindrical mirrors; 14, double-chamber detector with shielding and collimator; 15, entrance shielding; 16, plastic shielding containing boron; 17, polyethylene shielding; 18, nitrogen cryostat; 19, vacuum chamber (10^{-3} mm Hg) (Steyerl *et al.* 1983).

opening of the neutron guide, which, as shown in Chapter 4, is

$$R = (1 - k\rho\,\mathrm{th}\,kL)/(1 + k\rho\,\mathrm{th}\,kL), \tag{8.18}$$

if the exit opening of the neutron guide is closed with a reflecting wall, and

$$R = (1 - k^2\rho^2)/(1 + k^2\rho^2 + 2k\rho\,\mathrm{cth}\,k\rho), \tag{8.19}$$

if the exit of the neutron guide is completely open. Here $\rho = 2D/v$, where v is the neutron velocity, $D = lv/3$ is the diffusion coefficient, and l is the free flight path between two consecutive collisions with the walls of the imaginary neutron guide, which can be equated to the distance between the grains or to the dimension of a grain itself. L is the length of the neutron guide, which can be considered equal to the width of the layer of powder, $k = 1/L_D$,

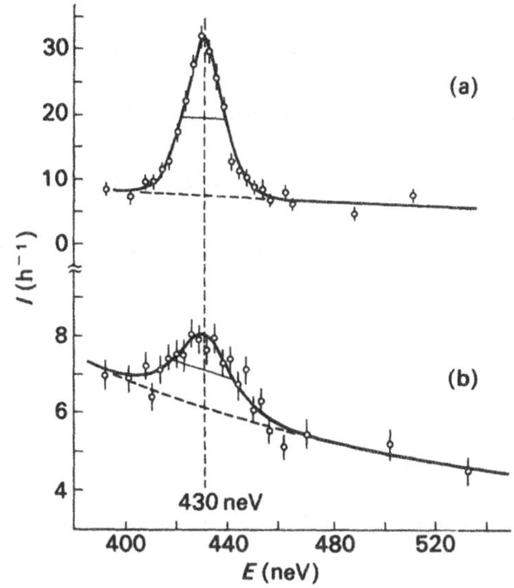

Fig. 8.20 Observation of quasi-elastic scattering with the aid of the NESSIE spectrometer: (a) resolution curve measured by the elastic scattering in graphite; (b) scattering in a solution of polydimethylsiloxane in deuterated benzene at 70°C. The total width of the peak measured at half-height changed from 17.0(6) neV to 22(3) neV (Steyerl *et al.* 1983).

$L_D = l(3/\mu)^{1/2}$ is the diffusion length, and μ is the UCN loss coefficient in a single neutron collision with the surface.

Formulae (8.18) and (8.19) hold when the powder is deposited on both reflecting and absorbing supports. As $L \to \infty$, both formulae tend towards the same limit

$$R = (1 - 2\sqrt{\mu/3})/(1 + 2\sqrt{\mu/3}) \simeq 1 - 4\sqrt{\mu/3}, \qquad (8.20)$$

which is natural, since, as the length of the neutron guide increases, the influence of the exit opening on the reflection from the entrance opening decreases. It must be pointed out that for small μ the reflection coefficient from the surface of the powder differs from unity by a quantity proportional to $\mu^{1/2}$, instead of μ, as in the case of reflection from the surface of a solid-state body. This indicates a significant enhancement of the interaction effect of a neutron with the surface of a substance.

The validity of the diffusion approach to the interaction of UCN with powders is verified experimentally by measurement of the absorption by the powder as the width of the powder is varied. The losses are determined by comparison with the losses in a known sample (polyethylene). By changing the temperature of the sample losses due to heating of the UCN can be determined. (It must be emphasized that in this case only small volumes have to be heated, so temperature studies do not represent a difficult task.)

Straightforward measurements with powders of copper, copper oxide, and graphite showed that upon applying a certain annealing technology the loss coefficient of these substances can be made to approach the theoretical value. Hence the conclusion may seem evident that the problem of UCN storage is already solved. We shall be careful not to make such a conclusion yet, but to set it aside until these results are replicated in other laboratories.

The technique for investigating powders is also applicable to the measurement of the limiting energies of substances and for studying metallic reflection from the surface of strongly absorbing substances (Golikov *et al.* 1985*b*, *c*, 1986).

8.4 THE NEUTRON MICROSCOPE

The refractive index in the case of neutrons, $n = (1 - u_0/k^2)^{1/2}$, differs significantly from unity when $k^2 \approx u_0$, and for such energies, therefore, it is not difficult to construct focusing systems. Focusing can be performed by lenses and mirrors, as well as by magnetic fields. Shapiro (1972*b*) pointed out that UCN can be utilized in neutron microscopy of materials. The advantage of neutrons (UCN) as compared with light is that, since their wavelengths are shorter (100–1000 Å), they can penetrate a non-transparent substance and detect its magnetic structure. Deep penetration of a substance makes the neutron microscope more preferable than the electron microscope, which, although exhibiting a significantly higher resolution, can only be used to investigate very thin layers of material.

Issues of neutron focusing relevant to obtaining a neutron image are dealt with in a number of references indicated below. In constructing an image it is necessary to take into account the influence of the gravitational field on the neutrons. This influence differs for neutrons of different energies, so the optical elements of neutron optics, generally speaking, exhibit chromatic aberration. The aberration can be reduced by arranging the optical elements in a certain way around the vertical axis or by applying a non-homogeneous magnetic field to compensate the gravitational field. Steyerl and Schutz (1978) proposed to use as the focusing element a concave mirror with a zone plate on its surface (Fig. 8.21). The zone plate is a diffraction grating engraved on the surface of the mirror, which has a groove profile such that diffraction occurs mainly in the given order m of diffraction. The chromaticity of the mirror and of the grating in such a system is mutually compensated and the focusing turns out to be achromatic.

With the aid of such a mirror an image of the UCN source enlarged by a factor of 50 was obtained. The layout of the experimental apparatus is shown in Fig. 8.22. It is important to note that the total neutron intensity at the

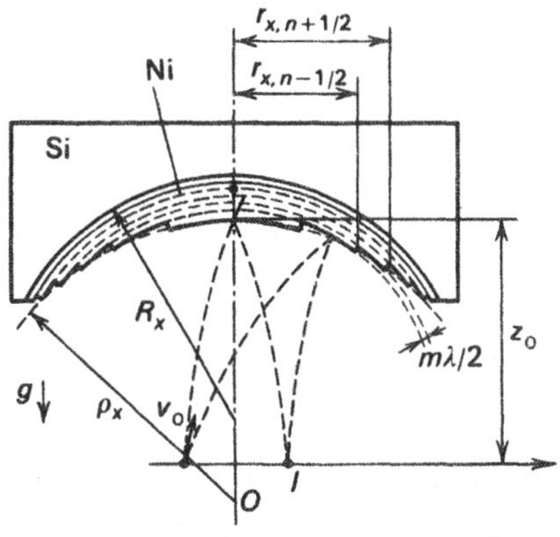

Fig. 8.21 Zone mirror for UCN (Steyerl and Schutz 1978) prepared by depositing nickel layers of decreasing radii on to a concave glass plate. The dashed lines indicate two neutron trajectories for the case when the source and its image are at the same distance from the mirror and the magnification equals unity. *O* indicates the optical axis.

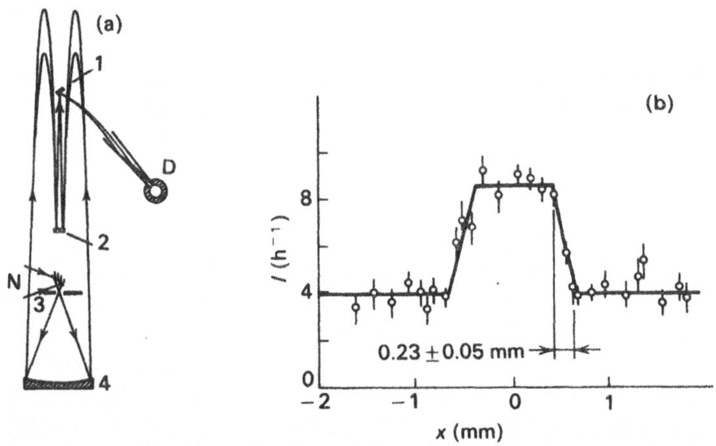

Fig. 8.22 Neutron microscope. (a) Scheme of the microscope: 1, reflecting mirror in the image plane (inclined to this plane at an angle of 15°); 2, the second convex mirror; 3, slit 1.06 mm wide which can be displaced within the range of vision; 4, the first concave mirror. (b) The detector count depending on the position of the slit. The dimensions of the range of vision are characterized by the width of the edge of a trapezoid, i.e. by the distance the edge of the slit is required to be displaced to cover the range of vision. This width can be estimated to be 0.35 mm. Since the dimensions of the mirror 1 are 17 × 17 mm^2, the magnification of the system is 50. The resolution of the system equals 0.1, i.e. it is determined by the uncertainty ± 0.05 mm with which the range of vision is measured (Herrmann *et al.* 1985).

image is the same as the intensity in the source, which means that neutrons of all energies take part in the same way in constructing the image. The creation of a neutron microscope is, naturally, still far from completion, but much has already been achieved. It suffices to say that a magnification of 240 has been obtained for a resolution of 10 μm (Steyerl *et al.* 1988*a*) and it has become clear how chromatic aberrations can be avoided without referring rigorously to the vertical axis (Masalovich and Strepetov 1987; Karaseva *et al.* 1988). We shall not deal with this issue in depth, because it deserves a separate monograph.

An image can also be obtained without any focusing systems. Using the method of neutron radiography Bates (1981) obtained an image of strips of a reflecting substance deposited on to the surface of a scintillation detector provided with a photographic film. In this case the image was obtained simply by exposure of the sample. The images of strongly reflecting or absorbing regions turned out to be lighter than the images of regions more transparent to UCN. Naturally, in this method the magnification was unity.

To conclude we present for the readers' convenience a full list of references on the neutron microscope which are relevant to UCN: Steyerl and Schutz (1978), Frank and Frank (1978), A.I. Frank (1979, 1980, 1981*a*, *b*, 1983, 1987), Schutz *et al.* (1980), Bates (1981), Skachkova and Frank (1981), Klein and Zeilinger (1984), Steyerl (1984), Arzumanov *et al.* (1984*b*), Herrmann *et al.* (1985), Arzumanov *et al.* (1986*a*), Masalovich (1986), Masalovich and Strepetov (1987), Strepetov and Frank (1986*a*, *b*), Terekhov (1977*a*, 1988*b*), Karaseva *et al.* (1988), Steyerl *et al.* (1988*a*), A.I. Frank (1989*a*, *b*), Steyerl (1989*a*).

8.5 CONCLUSION

All the above in no way exhausts the possible applications of UCN. For example, we could add the investigation of the magnetic susceptibility of a substance near the surface by studying the reflection, transmission, and depolarization of the incident polarized neutron beam; and studies of N–N interactions, when, as a target, a vessel is utilized which is filled with a sufficiently dense amount of UCN (Shapiro 1972*a*). In the near future other fields for the application of UCN are certain to be opened. Even now it can be stated that UCN represent a useful tool for investigations in low-energy physics.

APPENDICES

Appendix 8.1 The probability of spin flip in the method of two separate radio-frequency fields

$$i\dot{\xi} = \{[\sigma_z B_0(t) + \boldsymbol{\sigma}\mathbf{B}_1(t)]\theta(t < t_1) + \sigma_z B_0 \theta(t_1 < t < t_1 + T)$$
$$+ [\sigma_z B_0(t) + \boldsymbol{\sigma}\mathbf{B}_2(t)]\theta(t_1 + T < t < t_1 + T + t_2)\}\xi(t). \quad (1)$$

$$\text{a) } \mathbf{B}_i(i) = \mathbf{B}(\cos 2(\omega t + \varphi_i), \sin 2(\omega t + \varphi_i), 0), \quad i = 1, 2, \quad \text{b) } \varphi_2 = \varphi_1 + \psi. \quad (2)$$

$$\text{a) } \xi = \hat{Q}\xi(0), \quad \text{b) } \xi(0) = \xi_\uparrow, \quad \text{c) } \hat{Q} = \hat{G}(t_2, \tilde{\chi}) f_0(T)\hat{G}(t_1, \varphi_1). \quad (3)$$

$$\text{a) } \tilde{\chi} = \omega(t_1 + T) + \varphi_2, \quad \text{b) } \hat{f}_0(t) = \exp(-i\sigma_z B_0 t), \quad \text{c) } B = (\mu/h)[B]. \quad (4)$$

$$\hat{Q} = \prod_{i=1}^{4} [\exp(i\alpha_i\sigma_z)\exp(i\beta_i\sigma_y)]\exp(i\alpha_5\sigma_z). \quad (5)$$

$$\text{a) } \alpha_1 = -\omega t_2 - \chi, \quad \text{b) } \beta_1 = -\chi(t), \quad \text{c) } \alpha_2 = \int_0^{t_2} \Omega(t + t_1 + T)\,dt. \quad (6)$$

$$\text{a) } \beta_2 = \chi(t_1 + T), \quad \text{b) } \alpha_3 = (\omega - B_0)T + \psi, \quad \text{c) } \beta_3 = -\chi(t_1). \quad (7)$$

$$\text{a) } \alpha_4 = \int_0^{t_1} \Omega(t)\,dt, \quad \text{b) } \beta_4 = \chi(0), \quad \text{c) } \alpha_5 = \varphi_1. \quad (8)$$

$$\text{a) } \Omega(t) = \sqrt{[\omega - B_0(t)]^2 + B^2}, \quad \text{b) } \cos 2\chi(t) = [B_0(t) - \omega]/\Omega(t). \quad (9)$$

$$\text{a) } \chi(0) = \chi(t_1 + T + t_2) = 0, \quad \text{b) } \varphi_1 = 0, \quad \text{c) } \xi = (Q_{\uparrow\uparrow} + i\sigma_y Q_{\downarrow\uparrow})\xi_\uparrow. \quad (10)$$

$$\hat{Q} = \exp[i\sigma_z(\alpha_1 + \alpha_2)]\exp(i\beta_2\sigma_y)\exp(i\alpha_3\sigma_z)\exp(i\beta_3\sigma_y)\exp[i(\alpha_4 + \alpha_5)\sigma_z]. \quad (11)$$

$$Q_{\downarrow\uparrow} = \exp i(\alpha_4 + \alpha_5 - \alpha_1 - \alpha_2)[\sin\beta_2 \exp(i\alpha_3)\cos\beta_3$$
$$+ \cos\beta_2 \exp(-i\alpha_3)\sin\beta_3]. \quad (12)$$

$$w_{\downarrow\uparrow} = |Q_{\downarrow\uparrow}|^2 = \sin^2(\beta_2 + \beta_3)\cos^2\alpha_3 + \sin^2(\beta_2 - \beta_3)\sin^2\alpha_3. \quad (13)$$

$$\text{a) } \chi(t_1) = \chi(t_1 + T) \equiv \chi_1, \quad \text{b) } \beta_2 = \beta_3 = \pi/4,$$
$$\text{c) } w_{\downarrow\uparrow} = \sin^2(2\chi_1)\cos^2[(\omega - B_0)T + \psi]. \quad (14)$$

$$\text{a) } B_0(t) \equiv B_0, \quad \text{b) } t_1 = t_2, \quad \text{c) } \Omega(t) = \Omega, \quad \text{d) } \alpha_2 = \alpha_4 = \Omega t_1,$$
$$\text{e) } \beta_1 = -\beta_2 = \beta_3 = -\beta_4 = \chi \simeq \pi/4. \quad (15)$$

$$Q_{\downarrow\uparrow} = \xi_\downarrow e^{-i\beta_2\sigma_y} e^{i\alpha_2\sigma_z} e^{i\beta_2\sigma_y} e^{i\alpha_3\sigma_z} e^{-i\beta_2\sigma_y} e^{i\alpha_2\sigma_z} e^{i\beta_2\sigma_y} \xi_\uparrow e^{i(\alpha_5 - \alpha_1)} \quad (16)$$

$$Q_{\downarrow\uparrow} = \{\sin\beta_2\cos^3\beta_2 \cdot 2i[\sin\alpha_3 - \sin(2\alpha_2 + \alpha_3)] - \sin^3\beta_2\cos\beta_2[\sin\alpha_3$$
$$- \sin(\alpha_3 - 2\alpha_2)]\,2i\}\exp[i(\alpha_5 - \alpha_1)] = i - i\sin 2\beta_2[\sin 2\alpha_2\cos\alpha_3$$
$$- 2\cos 2\beta_2 \sin^2\alpha_2 \sin\alpha_3]\exp[i(\alpha_5 - \alpha_1)]. \quad (17)$$

$$Q_{\downarrow\uparrow} \approx -\mathrm{i}\sin 2\beta_2 \cos\alpha_3 \sin 2\alpha_2 \exp[\mathrm{i}(\alpha_5 - \alpha_1)]. \tag{18}$$

$$w_{\downarrow\uparrow} \approx (B^2/\Omega^2)\sin^2(2t_1\Omega)\cos^2[(\omega - B_0)T + \psi]. \tag{19}$$

In eqn (1) for a spinor the same notation is adopted as in eqn (1) of Appendix 7.11. The rotation of spin in the region where radio-frequency fields are present is assumed to be adiabatic. The vectors of two radio-frequency fields $\mathbf{B}_{1,2}$ are indicated in (2a). The phases of these fields are considered to differ by an angle ψ in (2b). The solution of eqn (1) can be represented in the form (3a). Now let the spin be directed along the field B_0 at $t = 0$, eqn (3b). The operator $\hat{Q}$ in (3a) is equivalent to the product of operators (3c), where the operator $\hat{G}$ is determined by expression (8.9) of Appendix 7.12, the phase $\tilde{\chi}$ is given in (4a), the operator $f_0(T)$ in (4b), and all the fields B are given in the dimension (4c), where $[B]$ has the natural dimension. Substituting eqn (9) of Appendix 7.11 and (4b) into (3c) we obtain expression (5) where the factors are arranged so the number i increases. The parameters α_i and β_i are presented in (6)–(8). The frequency $\Omega(t)$ entering into the integrand in (6c) and (8a) is indicated in (9a), and the phase χ is determined in (9b). Consider (10a, b) to hold. Now represent (3a) in the form (10c). The operator (5) reduces to (11) if (10a, b) are satisfied. Expanding the exponentials $\exp(\mathrm{i}\beta\sigma_y)$ in powers of σ_y and taking advantage of $\sigma_z\xi_{\uparrow\downarrow} = \pm\,\xi_{\uparrow\downarrow}$, we obtain expression (12) for the coefficient $Q_{\uparrow\downarrow}$. The probability of spin flip is determined in (13). When (14a, b) is satisfied, expression (13) transforms into (14c).

When the spin rotates in a resonant manner in the region of radio-frequency fields, the field $\mathbf{B}_0$ is constant everywhere (expression (15a)). Let (15b) hold; then expressions (6)–(8) reduce to the form (15c–e) and expression (12) to the form (16), which can be transformed into expressions (17) and (18), from which expression (19) follows for the probability of spin flip.

Appendix 8.2 Computation of σ for UCN scattering on inhomogeneities taking into account the finite aperture of the detector

$$\text{a) } u = 4\pi N_0 b, \quad \text{b) } \Delta u = 4\pi\tau, \quad \text{c) } \tau = \Delta(N_0 b). \tag{1}$$

$$\text{a) } f(\mathbf{q}) = \tau \int_V \exp(\mathrm{i}\mathbf{qr})\mathrm{d}^3 r, \quad \text{b) } \mathbf{q} = \mathbf{k}_0 - \mathbf{k}. \tag{2}$$

$$f(\mathbf{q}) = \left(\frac{4\pi\tau}{q}\right)\int_0^R \sin(qr)\,r\mathrm{d}r = \left(\frac{\Delta u}{q^3}\right)(\sin qR - qR\cos qR). \tag{3}$$

$$\text{a) } \sigma = \int |f(q)|^2\,\mathrm{d}\Omega_k, \quad \text{b) } \mathrm{d}\Omega_k = \int \mathrm{d}^3 k\,\delta(k^2 - k_0^2)2/k_0. \tag{4}$$

$$\text{a) } \mathrm{d}\Omega_k = (2\pi/k_0^2)\,q\mathrm{d}q, \quad \text{b) } q = 2k_0\sin(\theta/2). \tag{5}$$

$$\sigma = \left(\frac{2\pi}{k_0^2}\right)\int_{k\theta_1}^{2k} |f(\mathbf{q})|^2\,q\mathrm{d}q. \tag{6}$$

$$\sigma = 18\frac{\pi}{k_0^2 R^2}\left(\frac{4\pi R^3}{3}\right)^2 \tau^2 \int_{a\theta_1}^{2a} \frac{1}{x^5}(\sin x - x\cos x)^2\,\mathrm{dx}. \tag{7}$$

$$\text{a) } a = k_0 R, \quad \text{b) } \sigma = (18\pi\tau^2 V^2/k_0^2 R^2)[F(2a) - F(a\theta_1)]. \tag{8}$$

$$F(z) = \int_0^z \frac{1}{x^5}(\sin x - x\cos x)^2\,\mathrm{dx}. \tag{9}$$

$$F(z) = \frac{1}{4}(1 - 1/z^2 + \sin 2z/z^3 - \sin^2 z/z^4). \tag{10}$$

$$F(z) = \begin{cases} \frac{1}{4}(1 - 1/z^2), & z \to \infty, \\ z^2/18, & z \to 0. \end{cases} \tag{11}$$

$$\text{a) } 2k_0 R \ll 1, \quad \text{b) } \sigma = 4\pi\tau^2 V^2. \tag{12}$$

$$\text{a) } k_0 R\theta_1 \ll 1 \ll 2k_0 R, \quad \text{b) } \sigma = (9/2)\pi\tau^2 V^2/k_0^2 R^2. \tag{13}$$

$$\text{a) } k_0 R\theta_1 \gg 1, \quad \text{b) } \sigma = (9/2)\pi\tau^2 V^2/k_0^4 R^4 \theta_1^2. \tag{14}$$

$$\Sigma = \sigma\nu. \tag{15}$$

The average potential in a substance is defined in (1a). Inhomogeneities are characterized by the variation of potential (1b, c). The scattering amplitude on an inhomogeneity is indicated in (2a), where V is the volume occupied by the inhomogeneity and $\mathbf{q}$ is the transferred momentum, (2b). The wavevectors $\mathbf{k}_0$ and $\mathbf{k}$ are related to the incident and scattered neutrons, respectively. For a spherical inhomogeneity of radius R integration in (2a) leads to the amplitude (3). The scattering cross-section is determined in (4a). An elementary solid angle $d\Omega_k$ can be represented in the form (4b) or, taking into account (2b), in the form (5a), the modulus of the vector $\mathbf{q}$ being equal to (5b). Substitution of (5a) into (4a) leads to (6), where the aperture of the detector, θ, is taken into account explicitly. Substitution of (3) into (6) leads to (7), the parameter a of which is defined in (8a). The integral (7) is computed analytically, and expression (7) is represented in the form (8b), where the function $F(z)$ indicated in (9)–(10) is introduced. This function has the asymptotes (11). Substitution of (11) into (8b) reveals that, when conditions (12a, 13a, 14a) are satisfied, the scattering cross-section reduces, respectively, to forms (12b), (13b), and (14b). The macroscopic scattering cross-section on inhomogeneities equals (15), where ν is the number of inhomogeneities per unit volume. The quantity ν is assumed to be sufficiently small to justify neglecting interference terms in the scattering cross-section.

9

Reference data

9.1 GENERAL INFORMATION

We present here, without any detailed comments, certain experimental and theoretical data which should be useful, in the main, to specialists active in the field of UCN.

Table 9.1 contains world constants taken mainly from *Wohl *et al.* (1984). Table 9.2 displays the limiting velocities for a number of substances, calculated by the formula

$$v_{\text{lim}} = (\hbar/m)\left[4\pi\rho N_{\text{A}}\left|\sum_i b_i c_i\right| \Big/ \sum M_i c_i\right]^{1/2}, \tag{9.1}$$

where m is the mass of the neutron, ρ is the density of the substance, N_{A} is the Avogadro number, and b_i, M_i, and c_i are, respectively, the coherent scattering amplitude, the gram atomic mass, and the atomic concentration of the ith element in the substance. In case the amplitude $\Sigma b_i c_i$ is negative, instead of a potential barrier, the substance represents a well for the neutron and the reflection is not complete at any of the neutron velocities v, but the reflection coefficient is the greater, the higher the value of v_{lim} for a given v. Also indicated in Table 9.2 are the coefficients relating various quantities and permitting the determination from v_{lim} of the limiting energy E_{lim} in neV or in kelvin, the height H_{lim} to which the neutrons can rise in the gravitational field exhibiting the energy E_{lim}, the magnetic field B_{lim} which creates a potential barrier of height E_{lim} for neutrons with spin parallel to the field, the reduced wavelength $\lambda\!\!\!^{-}_{\text{lim}} = \lambda_{\text{lim}}/2\pi$ and the wavevector $k_{\text{lim}} = 1/\lambda\!\!\!^{-}_{\text{lim}}$.

In Fig. 9.1 the curves are presented for the probability of a neutron passing through the interface within a substance or through a layer of the substance. Curve 1 corresponds to the function

$$T_1(x) = 4x\sqrt{x^2-1}/(x+\sqrt{x^2-1})^2 = 2\,\text{sh}\,2\varphi\exp(-2\varphi), \tag{9.2}$$

where $x = k_\perp/k_{\text{lim}}$ is the component of the wavevector normal to the interface and φ is determined for all the successive formulae (9.2)–(9.7) by the relation $x = \text{ch}\,\varphi$. Curve 2 is described by the function

$$T_2(x) = (2/x^2)\int_1^x T_1(y)\,y\,dy = (-2 + \text{sh}^3 2\varphi - \text{ch}^3 2\varphi + 3\,\text{ch}\,2\varphi)/3x^2. \tag{9.3}$$

TABLE 9.1 Some physical constants

Physical constants	$\varepsilon\,(10^{-6})$[a]
$m_n = 939.5731(27)$ MeV $= 1.674\,954\,6(97) \times 10^{-24}$ g	2.9; 5.8
$\gamma = \mu_n/\mu_N = -1.913\,042\,11(88)$	0.46
$T_{1/2} = 607.8 \pm 5.4$ s; $T_{1/l} = 876.9(7{,}8)$ s	
$h = 6.582\,173(17) \times 10^{-22}$ MeV s $= 1.054\,588(57) \times 10^{-27}$ erg s	2.6; 5.4
$m_e = 0.511\,003\,4(14)$ MeV $= 9.109\,534(47) \times 10^{-28}$ g	2.8; 5.1
$m_p = 938.2796(27)$ MeV $= 1836.151\,52(70)\,m_e$	2.8; 0.38
$m_1 = 1.007\,276\,470(11)$ amu	0.011
1 MeV $= 1.602\,189\,2(46) \times 10^{-6}$ erg	2.9
1 amu $= 931.501\,6(26)$ MeV $= 1.660\,565\,4(95) \times 10^{-24}$ g	2.8; 5.7
$C = 2.997\,924\,58(1.2) \times 10^{10}$ cm s^{-1}	0.004
$e = 1.602\,189\,2(46) \times 10^{-19}$ C	2.9; 2.9
$\mu_N = eh/2m_pc = 3.152\,451\,5(53) \times 10^{-18}$ MeV G^{-1}	1.7
$\mu_N/h = 4.789\,378(14) \times 10^3$ rad s^{-1} G^{-1}	2.9
$\mu_n/h = 9.162\,281(44) \times 10^3$ rad s^{-1} G^{-1} $= 1.458\,222\,3(70) \times 10^3$ cps G	4.8
$h/m_n = 6.296\,223(35) \times 10^{-4}$ cm^2 s^{-1}	5.5
$h^2/2m_n = 3.319\,963(36) \times 10^{-31}$ erg cm^2.	11
$N_A = 6.022\,045(31) \times 10^{23}$ mol^{-1}	5.1
$k_B = 1.380\,662(44) \times 10^{-16}$ erg K^{-1} $= 8.617\,35(28) \times 10^{-11}$ MeV K^{-1} $= 1$ eV$/11\,604.50(36)$ K	32; 32
$V_0 = 22\,413.83(70)$ cm^3 mol^{-1}	31
$\mu_n = 6.030\,772(13) \times 10^{-18}$ MeV G^{-1} $= 9.662\,438(49) \times 10^{-24}$ erg G^{-1}	2.2; 5.1
$n_L = 2.686\,754(97) \times 10^{19}$	31
$g = 980.62$ cm s^{-2} (45° at sea level)	36

[a] ε is the relative uncertainty.

TABLE 9.2 Limiting velocities for certain substances

Substance	$b_{coh}\,(10^{-12}$ cm)	ρ (g cm^{-3})	v_{lim} (m s^{-1})
D_2 (liquid)	1.30	0.15	3.82
D_2O	1.88	1.1	5.57
^4He (liquid)	0.30	1.114	1.6
^6LiOH	0.383	1.(43)	2.6
$^6LiOH.H_2O$	0.212	2.(3)	1.88
LiF	0.74	2.5	4.71
LiF (natural)	0.346	2.6	3.23
Be	0.774	1.80	6.89
BeO	1.35	3.02	6.99
C	0.665	1.80	5.47

TABLE 9.2 (*Continued*)

Substance	b_{coh} (10^{-12} cm)	ρ ($g\,cm^{-3}$)	v_{lim} ($m\,s^{-1}$)
C (graphite)	0.665	2.25	6.11
C (diamond)	0.665	3.25	7.65
CH	0.291	1.0	2.59
$C_{18}H_{14}$	6.73	1.23	3.25
CO_2 (solid)	1.825	1.53	4.5
Mg	0.53	1.74	3.37
MgO	1.11	3.58	5.47
Al	0.35	2.70	3.24
Al_2O_3	2.42	3.7	5.13
SiO_2 (glass)	1.58	2.3	4.26
SiC	1.085	3.22	5.11
S (rhombic)	0.31	2.07	2.45
^{39}Ar (liquid)	2.43	1.65	5.85
K_2O_3	1.64	4.87	4.0
V_2O_3	2.8	3.36	3.94
Cr	0.35	7.1	3.80
Fe	0.95	7.87	6.34
Stainless steel	0.86	8.03	6.0
Ni	1.03	8.9	6.84
^{58}Ni	1.44	8.9	8.14
Cu	0.76	8.93	5.66
^{65}Cu	1.11	8.93	6.76
CuO	1.34	6.4	5.69
Cu_2O	1.94	6.0	4.85
Zn	0.59	6.92	4.33
ZnS	0.9	4.0	3.3
Ga	0.72	6.09	4.34
Ge	0.84	5.46	4.35
Se	0.80	4.80	3.82
Zr	0.70	6.44	3.85
ZrH	0.322	5.6	2.2
$ZrH_{1.8}$	0.02	5.6	0.60
Sn (white)	0.62	7.29	3.38
Te	0.58	6.25	2.92
Dy	1.69	8.5	5.15
^{164}Dy	4.94	8.5	8.76
Ta	0.70	16.6	4.32
Tl	0.89	11.86	3.94
Pb	0.96	11.34	3.97
Bi	0.26	9.75	3.47

E (neV) = 5.227 080(15), v^2 ($m\,s^{-1}$ = $1.03H$ (cm); H (cm) $5.1v^2$ ($m\,s^{-1}$) = $0.975E$ (neV); H (G) = $866.734v^2$ ($m\,s^{-1}$) = $165.8162E$ (neV) = $170.79H$ (cm); k ($Å^{-1}$) = $1.59 \times 10^{-3}\,v$ ($m\,s^{-1}$); $\lambda\!\!\!^{-}$ (Å) = $6.3 \times 10^2/v$ ($m\,s^{-1}$); λ (Å) = $3.96 \times 10^3/v$ ($m\,s^{-1}$); 1 neV = 1.160 450(36) $\times 10^{-5}$ K.

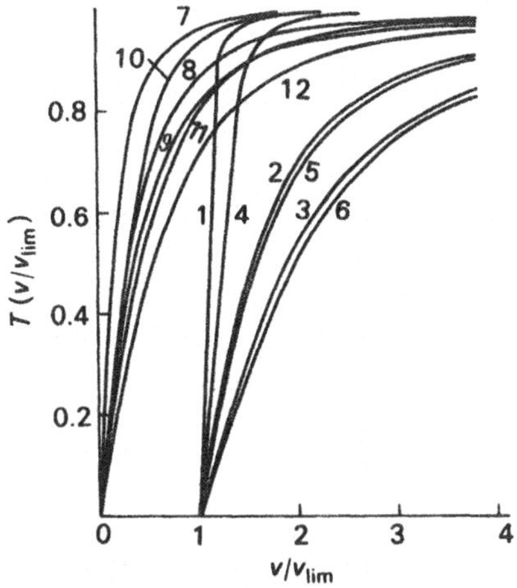

Fig. 9.1 Transmission curves for a potential step barrier (1–3) of a finite-width potential barrier without taking into account losses within the barrier and in the case of averaging over the width (4–6) for a step potential well (7–9) and a finite-width potential well (10–12). Curves 1, 4, 7, 10 are computed for monochromatic neutrons incident normally to the surface of the substance. Curves 2, 5, 8, 11 are computed for monochromatic neutrons with velocity directions distributed isotropically. Curves 3, 6, 9, 12 are obtained by averaging over both the angles and the initial section of the Maxwellian spectrum from 0 to v (see expressions (9.4) and (9.7) in the text).

Here $x = k/k_{\text{lim}}$. Finally, curve 3 is described by the function (9.3) averaged over the section of the Maxwellian spectrum from k_{lim} to k:

$$T_3(x) = \int_1^x \frac{T_2(y)4y^3}{x^4 - 1}\,\mathrm{d}y = \frac{1}{4(x^4-1)}(\varphi - \operatorname{sh} 2\varphi)$$
$$+ \frac{1}{8(x^4-1)}\left[\tfrac{8}{3}\operatorname{ch} 2\varphi(\operatorname{sh}^3 2\varphi - \operatorname{ch}^3 2\varphi + 3\operatorname{ch} 2\varphi - 2) - \tfrac{1}{4}\operatorname{sh} 8\varphi\right]. \quad (9.4)$$

The curves (9.4)–(9.6) are related to the transmission of a layer of substance inhomogeneous in width, when the interference terms vanish as a result of averaging over the width. Absorption, inelastic and incoherent elastic scattering, as well as scattering on various inhomogeneities, are not taken into account. Curve 4 corresponds to neutrons with a fixed $k_\perp = k$:

$$T_4(x) = 2x\sqrt{x^2-1}/(2x^2-1) = \operatorname{th} 2\varphi, \quad x = k_\perp/k_{\text{lim}}. \quad (9.5)$$

Averaging over angles similar to (9.3) leads to curve 5:

$$T_5(x) = [\operatorname{sh} 2\varphi - \operatorname{arctg}(\operatorname{sh} 2\varphi)]/2x^2, \quad x = k/k_{\text{lim}}. \quad (9.6)$$

Finally, curve 6 is described by the expression (9.6) averaged in the same way as (9.4):

$$T_6 = \{4\,\mathrm{ch}\,2\varphi[\mathrm{sh}\,2\varphi - \mathrm{arctg}(\mathrm{sh}\,2\varphi)] + 4\varphi - \mathrm{sh}\,4\varphi\}/8(x^4 - 1). \quad (9.7)$$

All the functions T_1–T_6 tend to zero when $x \leqslant 1$. Curves 7–12 are similar to curves 1–6, but are related to the case when the potential of the substance is negative. The formulae for T_7–T_{12} coincide with the formulae for T_1–T_6 expressed through the hyperbolic angle φ. First, however, φ must be determined from the expression $x = \mathrm{sh}\,\varphi$ and, secondly, for T_9 and T_{12} x^4 must be substituted for $x^4 - 1$ in formulae (9.4) and (9.7), since in this case averaging over the Maxwellian spectrum must be performed from 0 to k. Naturally, the functions T_7–T_{12} do not tend to zero when $x \leqslant 1$.

Table 9.3 contains, for some substances, the coefficients η calculated by formula

$$\eta = 10^{-4} \sum_i c_i \sigma_i / 2.16\, b_{\mathrm{coh}}, \quad (9.8)$$

where $b_{\mathrm{coh}} = \Sigma c_i b_i$, $b_i \times 10^{-12}$, σ_i, barn, are taken from the tables of *Hughes and Schwartz (1958) and referred to the wavelength $\lambda = 1.08$ Å.

9.2 GENERATION OF UCN

In Tables 9.4–9.15 the computed G factors are presented for 15 different substances, the UCN output from polyethylene at $T_c = T_n = 300$ K being adopted as the unit yield in Tables 9.4–9.15. Tables 9.4–9.8 are taken from Golikov *et al.* (1973), Tables 9.9–9.14 from Akhmetov *et al.* (1976), and Table 9.15 from Utsuro and Hetzelt (1977). In Fig. 9.2 the dependence is shown for some substances of $G(T_n, T_c)$ on T_c at $T_n = 300$ K found theoretically, while the experimental data are presented in Figs 9.3 and 9.4. The two theoretical curves in Fig. 9.4 are computed by the Monte Carlo method and indicate the possible corridor of theoretical values of the G factor. The finite

TABLE 9.3 Theoretical values for η at $T = 300$ K

Substance	$\eta\,(10^{-4})$	Substance	$\eta\,(10^{-4})$	Substance	$\eta\,(10^{-4})$	Substance	$\eta\,(10^{-4})$
Cu[a]	1.45	Be	0.05	Teflon	0.05	Zn	0.53
Glass[a]		Ni	1.2	C	0.04	Pb	0.06
boronless	0.05	Fe	0.7	Al	0.21	Bi	0.02
Glass with boron	16						
Quartz	0.1						

[a] The calculation was performed in accordance with the total cross-section measured in Steyerl and Vonach (1974) taking into account later corrections.

TABLE 9.4 Polyethylene

T_n (K)	T_c (K)						
	4	20	40	80	150	220	300
20	6.7	7.2	7.8	5.6	2.8	2.2	1.7
40	13.9	13.9	12.8	7.8	3.9	2.8	2.0
70	20.0	19.4	16.7	9.4	4.4	2.8	2.0
100	22.2	21.7	17.8	9.4	3.9	2.8	1.6
150	21.6	21.1	17.2	8.9	3.3	2.2	1.5
200	20.5	20.0	16.1	7.8	3.3	1.6	1.1
250	18.9	18.3	15.0	7.2	2.8	1.6	1.1
300	17.8	17.2	13.9	6.6	2.8	1.6	1.0

TABLE 9.5 Zirconium hydride, $ZrH_{1.46}$

T_n (K)	T_c (K)					
	4	20	40	80	150	220
20	0.9	1.0	1.3	1.8	2.2	1.7
40	1.4	1.4	1.7	2.0	2.2	1.7
70	1.8	1.8	1.9	2.1	2.1	1.6
100	1.8	1.8	1.9	1.95	1.9	1.4
150	1.8	1.8	1.8	1.8	1.6	1.2
200	1.3	2.3	2.3	2.2	1.8	1.2
250	3.2	3.3	3.3	3.0	2.3	1.4
300	4.3	4.5	4.5	4.1	3.0	1.8
350	–	–	–	–	–	–
400	–	–	–	–	–	–

T_n (K)	T_c (K)				
	300	350	400	500	600
20	1.0	–	–	–	–
40	1.0	–	–	–	–
70	0.9	–	–	–	–
100	0.8	0.6	0.5	0.3	0.2
150	0.6	–	–	–	–
200	0.6	0.5	0.3	0.2	0.15
250	0.7	–	–	–	–
300	0.9	0.6	0.4	0.25	0.17
350	1.0	0.7	0.5	0.3	0.2
400	1.2	0.8	0.5	0.3	0.2

TABLE 9.6 Aluminium

T_n (K)	T_c (K)								
	4	20	40	80	150	220	300	400	500
20	0.02	0.02	0.02	0.03	0.05	0.07	0.08	0.10	0.12
40	0.05	0.05	0.06	0.07	0.09	0.11	0.14	0.16	0.18
70	0.11	0.11	0.11	0.12	0.15	0.18	0.22	0.25	0.28
100	0.14	0.14	0.14	0.15	0.18	0.21	0.25	0.29	0.32
150	0.14	0.14	0.14	0.15	0.18	0.20	0.23	–	–
200	0.12	0.12	0.13	0.13	0.15	0.17	0.20	0.24	0.27
250	0.11	0.11	0.11	0.11	0.13	0.14	0.16	–	–
300	0.09	0.09	0.09	0.09	0.11	0.12	0.14	0.17	0.18

TABLE 9.7 Magnesium

T_n (K)	T_c (K)						
	4	20	40	80	150	220	300
20	0.45	0.45	0.55	0.7	0.8	0.8	0.8
40	1.4	1.4	1.5	1.6	1.6	1.55	1.5
70	1.3	2.3	2.3	2.3	2.2	2.1	2.0
100	2.4	2.4	2.4	2.3	2.2	2.0	1.9
150	2.0	2.0	2.0	1.9	1.7	1.6	1.2
200	1.55	1.55	1.55	1.5	1.3	1.2	1.2
250	1.2	1.2	1.2	1.2	1.0	0.9	0.9
300	1.0	1.0	1.0	0.9	0.80	0.7	0.7

TABLE 9.8 Beryllium

T_n (K)	T_c (K)						
	4	20	40	80	150	220	300
20	0.3	0.3	0.4	0.5	0.28	0.17	0.11
40	0.8	0.8	0.9	0.9	0.4	0.2	0.17
70	2.2	2.2	2.2	2.1	0.9	0.4	0.2
100	5.0	5.0	5.0	4.2	1.6	0.7	0.4
150	10.0	10.0	9.4	8.3	3.0	1.3	0.7
200	13.3	13.3	13.3	11.1	3.9	1.7	0.9
250	15.0	15.0	14.4	12.0	4.4	1.8	1.0
300	15.0	15.0	15.0	12.5	4.4	1.8	1.0

TABLE 9.9 Molecular hydrogen

T_n (K)	T_c (K)				
	20	50	100	200	300
Para-hydrogen					
50	18.3	12.8	6.6	1.2	0.56
100	28.3	19.8	9.9	1.6	0.63
200	33.0	23.3	11.8	1.9	0.77
300	27.4	20.3	10.6	1.8	0.76
600	13.7	12.7	6.6	1.2	0.56
Equilibrium hydrogen					
50	18.9	27.9	17.9	10.4	7.5
100	28.9	14.9	8.3	4.7	3.4
200	33.3	9.3	3.9	2.1	1.5
300	27.8	6.7	2.5	1.3	0.94
600	14.0	3.4	1.1	0.6	0.4
Ordinary hydrogen					
50	60.0	31.0	18.2	10.4	7.4
100	24.0	13.6	8.2	4.8	3.4
200	8.9	5.9	3.7	2.2	1.5
300	5.6	3.5	2.2	1.3	0.94
600	2.2	1.4	1.0	0.6	0.4

TABLE 9.10 Equilibrium gaseous molecular deuterium

T_n (K)	T_c (K)				
	20	50	100	200	300
50	60.0	37.2	22.4	13.5	10.2
100	20.1	13.7	9.1	5.8	4.5
200	6.8	4.9	3.4	2.3	1.8
300	3.5	2.5	1.8	1.2	1.0
600	1.2	0.8	0.6	0.43	0.36

dimensions of the beryllium sample are also taken into account. The G factor was also measured for beryllium in Abov *et al.* (1977) within the range of temperatures from 160 to 300 K. Figure 9.5 shows the dependence of the transmission of three different beryllium samples prepared by differing technologies: crystalline, moulded, and melted. Neutrons with velocities of $\sim 10\ \mathrm{m\,s^{-1}}$ were transmitted (Altarev *et al.* 1982).

TABLE 9.11 Water

T_n (K)	T_c (K)							
	4	10	20	40	80	150	220	300
20	8.2	8.4	9.0	9.6	8.0	5.2	3.7	2.7
40	13.2	13.3	12.3	12.7	9.5	5.7	3.9	2.9
70	14.5	14.5	14.3	12.8	8.8	5.0	3.3	2.4
100	14.2	14.2	13.9	12.0	7.8	4.2	2.8	2.0
150	13.7	13.7	13.3	11.3	6.9	3.5	2.2	1.5
200	13.5	13.0	13.0	10.9	6.4	3.1	1.8	1.3
250	13.3		12.8	10.7	6.1	2.8	1.7	1.1
300	13.3		12.5	10.3	5.8	2.6	1.5	1.0

TABLE 9.12 Heavy water

T_n (K)	T_c (K)				
	4	20	80	200	300
25	178	107	9.29	2.51	1.64
50	386	184	13.55	3.17	2.02
100	486	262	14.1	2.48	1.75
200	523	279	13.39	2.3	1.31
300	463	247	11.53	1.86	1.04

TABLE 9.13 Heavy water with admixture of 1% H_2O

T_n (K)	T_c (K)				
	4	20	80	200	300
25	21.8	22.2	7.9	2.5	1.6
50	47.2	43.2	11.6	3.1	2.0
100	59.3	54.2	12.0	2.8	1.8
200	63.7	57.7	11.5	2.2	1.3
300	56.6	51.2	9.8	1.8	1.0

Experiments with solid-state converters have demonstrated that the ratio of UCN intensities generated in polyethylene at the temperatures 300, 130, and 80 K is $1:2.2 \pm 0.2:4 \pm 0.2$, while theory predicts the ratio 1:3:5.5 (Golikov *et al.* 1973). Zirconium hydride yields the same output of UCN at 300 K as polyethylene, but when cooled to 130 K the output of UCN from it

TABLE 9.14 Monoisopropyl diphenyl, $C_{15}H_{16}$

T_n (K)	T_c (K)						
	4	20	40	80	150	220	300
20	38.5	40.0	35.0	24.4	16.0	12.8	10.5
40	37.8	35.5	28.3	18.3	11.7	8.9	7.2
70	32.8	28.9	21.1	12.8	7.5	5.6	4.7
100	28.3	25.0	17.2	9.4	5.6	4.1	3.3
150	23.9	20.0	13.3	7.2	3.8	2.7	2.2
200	20.5	17.2	11.1	5.6	2.8	2.0	1.5
250	17.8	15.0	9.4	4.7	2.3	1.6	1.2
300	16.0	13.3	8.3	3.9	1.9	1.3	1.0

TABLE 9.15 Deuterium

T_n (K)	T_c (K)		T_n (K)	T_c (K)	
	15	20		15	20
Ortho-deuterium			*Para*-deuterium		
20	247	216	20	158	136
50	134	103	50	106	79.5
75	94.5	72.2	75	79.5	61.9
100	70.1	53.7	100	61.0	47.8

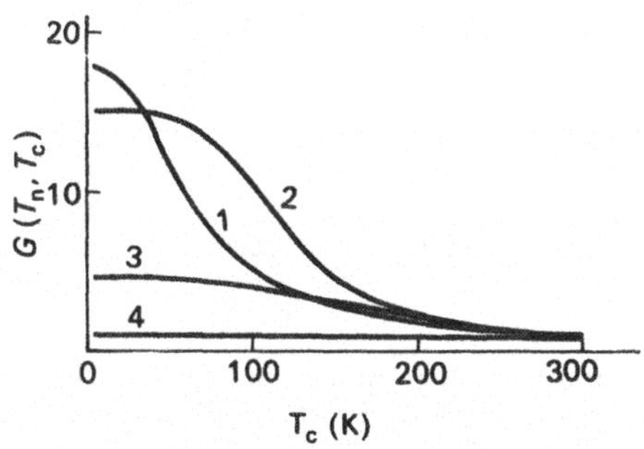

Fig. 9.2 Dependence of the G factor on the temperature T_c at $T_n = 300$ K: 1, polyethylene; 2, beryllium; 3, zirconium hydride; 4, magnesium (Golikov *et al.* 1973).

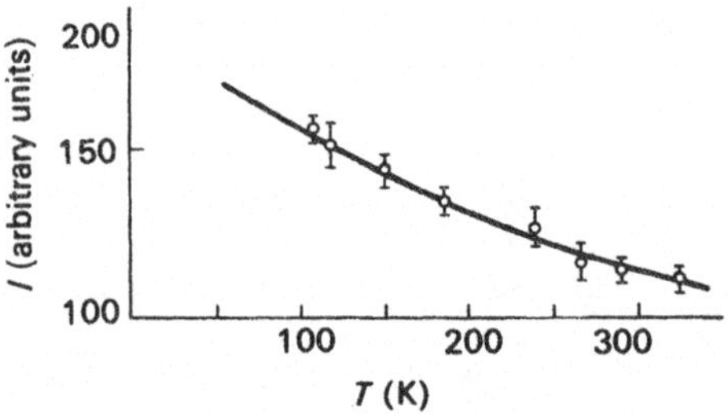

Fig. 9.3 G factor of magnesium at $T_n = 300$ K (Akhmetov *et al.* 1977*a*).

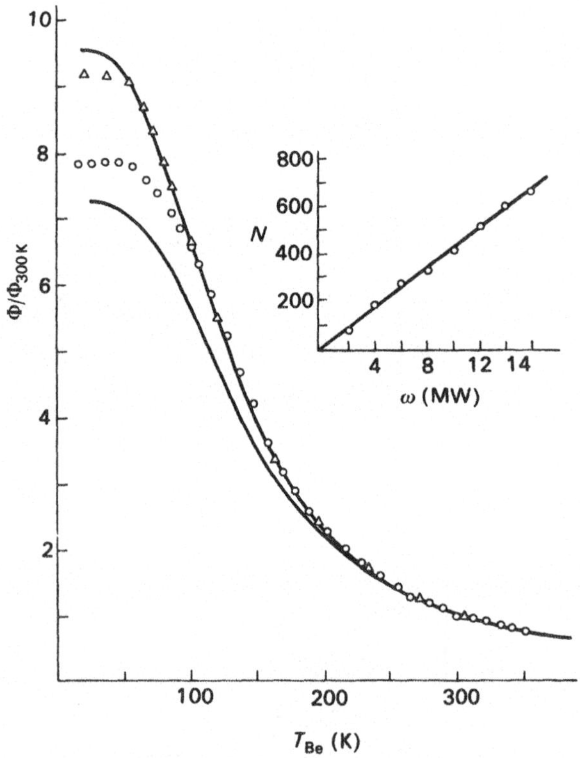

Fig. 9.4 Dependence of the UCN output from beryllium on its temperature (Altarev *et al.* 1976). The solid lines indicate the theoretical limits within which the output from a real sample can be expected to lie. The experimental points correspond to two different beryllium samples. The linear dependence of the UCN output on the power of the reactor, shown in the inset, confirms the efficiency of cooling.

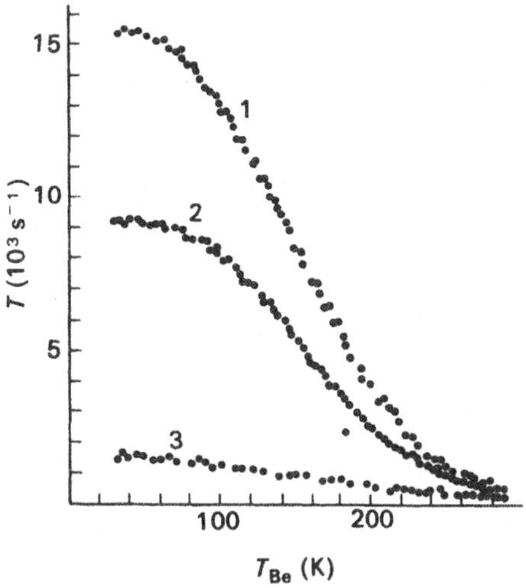

Fig. 9.5 Transmission of various beryllium samples depending on temperature: 1, melted; 2, quasi-monocrystalline; 3, moulded (Altarev *et al.* 1976).

increases by a factor of 2.3 ± 0.2, although theory predicts a 3.4 fold increase. In Groshev *et al.* (1974) the UCN output from $ZrH_{1.9}$, upon cooling from 300 to 100 K, was reported to be enhanced only by 1.6 times (theory gives the value of 3.7). Most likely, this is due to the zirconium hydride being deprived of hydrogen.

The output of UCN from aluminium at 400 K is related to that from magnesium at 300 K and to that from $ZrH_{1.9}$ at 300 K as the ratio 1:2.1:6.5 (Groshev *et al.* 1974), while theory predicts the ratio 1:4.1:8.5.

Steyerl (1972*a*) computed the UCN output from pyrographite at $T_n = 370$ K and $T_c = 510$ K. It was found that cooling the neutron spectrum down to the temperature $T_n = 100$ K leaves the UCN intensity nearly unchanged, while cooling the converter to 0 K increases the UCN output by a factor of 2.5.

Presented in Table 9.9 is the *G* factor calculated for molecular hydrogen involving various concentrations of the para- and ortho-states in accordance with the Boltzmann equilibrium distribution, while at all temperatures ordinary hydrogen contains a distribution of ortho- and para-phases identical to the equilibrium distribution at 300 K. The respective concentrations are indicated in Table 9.16.

Presented in Table 9.10 is the calculated *G* factor for equilibrium gaseous molecular deuterium. The concentrations of the ortho- and para-phases are indicated in Table 9.17.

The UCN output from various gases is given in Fig. 2.3. Shown in Fig. 9.6 is the dependence of the UCN output from hydrogen. From Fig. 9.6 it follows

TABLE 9.16 Percentage of hydrogen molecules at various rotational levels

T_c (K)	Admixture	J				
		0	1	2	3	4
300	Equilibrium concentration	12.89	65.78	11.78	9.07	0.45
	Para-hydrogen	51.40	–	47.0	–	1.59
	Ordinary hydrogen	12.85	65.84	11.75	9.08	0.45
200	Equilibrium concentration	18.65	71.67	7.27	2.38	0.03
	Para-hydrogen	71.82	–	28.05	–	0.13
	Ordinary hydrogen	17.95	72.59	7.01	2.41	0.03
100	Equilibrium concentration	37.36	61.46	1.15	0.02	–
	Para-hydrogen	97.02	–	2.98	–	–
	Ordinary hydrogen	24.25	74.97	0.74	0.03	–
50	Equilibrium concentration	76.88	23.11	0.01	–	–
	Para-hydrogen	99.98	–	0.02	–	–
	Ordinary hydrogen	24.995	75.00	0.005	–	–

TABLE 9.17 Percentage of deuterium molecules at various rotational levels in equilibrium mixture

T_c (K)	J			
	0	1	2	3
300	17.50	20.0	48.5	13.3
200	25.9	25.6	40.8	7.8
100	48.0	31.1	19.8	1.1
50	76.5	20.8	2.7	–
20	97.97	2.03	–	–

that the UCN output from *para*-hydrogen at saturation pressure increases upon cooling from 300 to 80 K only 6 times, instead of the predicted 12-fold enhancement. This, obviously, is explained by a rapid reconversion of the *para*-hydrogen back to the equilibrium state (Akhmetov *et al.* 1977*a*). Reconversion is observed by a change in the UCN output from *para*-hydrogen with time. The reconversion rate also depends on the power of the reactor.

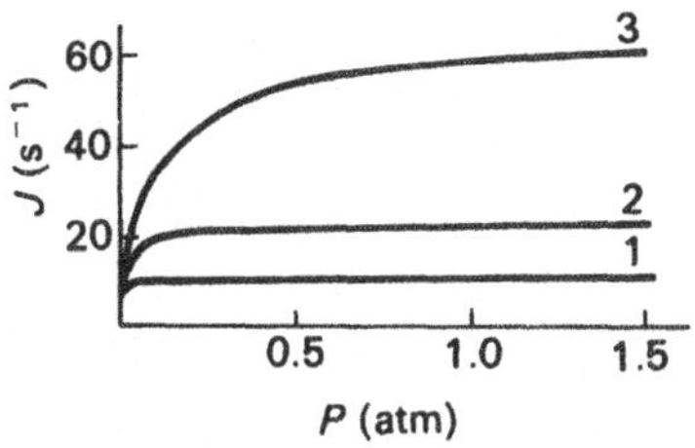

Fig. 9.6 UCN ouput from various modifications of hydrogen at various temperatures and pressures of the gas (Akhmetov *et al.* 1977). *Para*-hydrogen: 1–300 K, 3–80 K. Ordinary hydrogen: 2–80 K; 1–300 K (Akhmetov *et al.* 1977*a*).

The difference between ordinary hydrogen and *para*-hydrogen is well demonstrated by Fig. 9.7.

The experimental dependence of the UCN output from hydrogen and deuterium is also shown in Fig. 2.7.

In Table 9.15 the G factor is indicated for liquid *ortho*- and *para*-deuterium, calculated by taking into account the interference effects in the scattering in the liquid (Utsuro and Hetzelt 1977). The temperature T_c was taken to be 15 and 20 K, respectively, for *ortho*- and *para*-deuterium. The UCN output from water at $T_n = T_c = 300$ K was assumed to be the unit output. The results of the experimental verification are presented in Table 9.18, together with the theoretical values calculated by the gas model (Akhmetov *et al.* 1977*a*) and by the liquid model taking into account interference (Utsuro and Hetzelt 1977).

The UCN output from freeze-deposited converters is shown in Fig. 9.8. The UCN output from water at saturation is 1.3 times higher than from alcohols. Comparison of this output with the UCN output from water

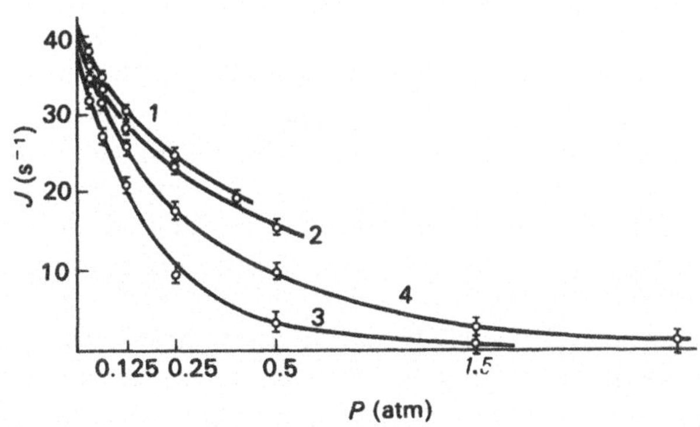

Fig. 9.7 Transmission of UCN by various modifications of hydrogen depending on the gas pressure: *para*-hydrogen: 1, at 80 K; 2, 300 K; ordinary hydrogen: 3, at 80 K; 4, at 300 K (Akhmetov *et al.* 1977*a*).

TABLE 9.18 Factor $G(T_n, T_c)$ in liquid deuterium at $T_c = 20$ K

Model	Substance	T_n (K)		v (m s^{-1})
		300	50	
Gas	*Ortho*-D_2	4.24	87.2	20
	Para-D_2	4.16	81.2	20
Liquid	*Ortho*-D_2	13.6	101	20
	Para-D_2	14.2	89.0	20
Experimental data	Equilibrium D_2	6.88	–	49
		9.63	–	25
		10.2	–	20

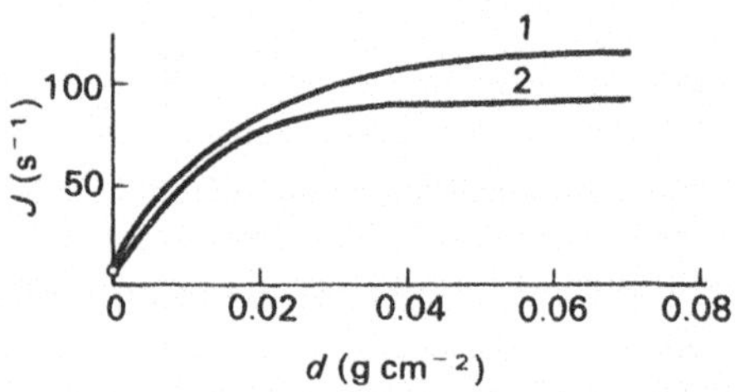

Fig. 9.8 UCN output from a freeze-deposited liquid depending on its mass at a temperature of 80 K: 1, water; 2, ethyl and butyl alcohols (Akhmetov *et al.* 1974*b*).

confined at 300 K in an aluminium ampoule (Fig. 2.2) reveals that cooling the water from 300 to 80 K leads to the intensity generated being increased 3.4 times, instead of the expected 6 times (Akhmetov *et al.* 1977*b*). In Lobashov and Serebrov (1978) it was noted that water frozen on to beryllium also yields an enhancement of the UCN output upon cooling to 20 K by a factor of only 6, instead of the expected 12. In this connection it is interesting to point out the result reported in Akhmetov *et al.* (1976, 1977*a*). Here, water frozen at 80 K on to zirconium hydride led to the UCN output increasing 10 times, which essentially exceeds the predicted value. This, however, may be explained by the zirconium hydride being deprived of hydrogen to a certain extent.

The dependence of the UCN output from heavy water frozen on to zirconium hydride on the amount of freeze-deposited material is shown in Fig. 2.4.

To conclude, we present the full list of references related to UCN converters: Drabkin and Zhitnikov (1960), Kagan (1970), Golikov *et al.* (1973), Namiot (1973), Akhmetov *et al.* (1974*a*, *b*), Akhmetov *et al.* (1976, 1977*a*),

Altarev *et al.* (1976), Taran (1976), Utsuro (1977), Utsuro and Hetzelt (1977), Altarev *et al.* (1980*b*), Golub and Taylor (1983), Altarev *et al.* (1986*a*), Yu *et al.* (1986), Kaipov *et al.* (1988).

It is also appropriate here to indicate the literature: (i) on pulsed shutters: Shapiro (1972*a*), Golikov and Taran (1975), Pokotilovskii *et al.* (1976), Nikitenko and Taran (1978), Novopoltzev and Pokotilovskii (1978, 1979, 1980*a*, 1982), Pokotilovskii (1980), Stoica and Jipa (1984), Novopoltzev *et al.* (1985, 1987); (ii) on UCN generation in liquid helium: Golub (1972, 1979, 1983, 1988), Golub and Pendlebury (1975, 1977), Scott and McClintock (1977), Ageron *et al.* (1978*b*), Jewell *et al.* (1981), Golub and Böning (1981, 1983), Golub *et al.* (1983), Yoshiki (1984), Kilvington *et al.* (1987), Yoshiki *et al.* (1987, 1988); and (iii) on UCN generation with the aid of turbines: Antonov *et al.* (1969a), Kashukeev (1970), Kashukeev *et al.* (1973, 1975*a*, *b*), Steyerl (1975), Dombeck (1977, 1981), Dombeck and Lynn (1979), Steenstrup and Buras (1978), Dombeck *et al.* (1979), Brun *et al.* (1980), Majkrzak and Malik (1980), Utsuro *et al.* (1982, 1983, 1988*b*), Utsuro and Kawabata (1983), Utsuro (1984).

9.3 SOURCES OF UCN

Information is collected in Table 9.19 on some of the existing sources of UCN. To avoid overloading the column containing data on the neutron guides, only the total length of a neutron guide, the diameter of its exit opening, and the material from which it is made are indicated in the vicinity of the exit opening. In many cases the neutron guide is composed of several sections of different diameters and constructed of different materials. The details are to be found in the references presented in the last column. The maximum velocity v_{max} of the neutrons generated in a given source indicates the type of the source. If v_{max} exceeds v_{lim} significantly, this means that the source is mainly intended for work with VCN. The UCN flux density is indicated in column 10, $\Phi_V = J_d/s_d$, where J_d is the counting rate of the detector and s_d is the area of its window. Column 12 contains the flux density of neutrons passing over the shutter, i.e. the ratio of the counting rate of the detector to s_d when the shutter is closed. This information is useful for the estimation of the spectrum of the source. It is clear that in the case of a VCN source this information is superfluous. The efficiency ε of the source (in %) is given in column 13: $\varepsilon = \Phi_U/\Phi_{UM}$, where Φ_{UM} is the UCN flux density in the Maxwellian spectrum at $T_n = 300\,K$. Since cooled converters are utilized in some sources, the G factor is also given in column 13. Thus, the ratio ε/G allows the efficiency of the UCN transportation to be determined. In certain cases the UCN intensity falls with time. The time $T_{1/2}$ during which the intensity is halved may be of interest for constructing UCN channels, so it is

TABLE 9.19 Characteristics of some UCN sources

Institute	Φ_0 $10^{13}\,m^{-2}\,s^{-1}$	Con-verter	T_c (K)	Material of the neutron guide	Size of the neutron guide (cm)	Detector	S_d (cm^2)	v_{max} ($m\,s^{-1}$)	Φ_{Ud} ($cm^{-2}\,s^{-1}$)	v_{lim} ($m\,s^{-1}$)	Φ_{sh} ($cm^{-2}\,s^{-1}$)	ε(%)/G (%)	$T_{1/2}$ (months)	Footnote[a]	Reference
1	2	3	4	5	6	7	8	9	10	11	12	13	14	15	16
FRM,	1	No	–	–	5 × 1300	BF_3	7.5	300	0.8	5.7	–	4/1	6	tof	Steyerl (1969)
Munich, FRG	2	Turbine	–	–	3.4 × 15 × 1100	BF_3	13	20	2.8	5.7	–	4/1	–	1	Steyerl (1975)
Inst. Nuclear Research and Nuclear Energy Sofia, Bulgaria	0.24	Turbine	–	–	4 × 1500	^3He	40	6.2	0.1	6.2	–	0.6/1	–	1,2	Kashoukeev *et al.* (1975*a*)
Argonne National Lab.,	1	Turbine	–	–	10 × 150	^3He	80	20	2.5	6.8	–	4/1	–	1,3 tof	Brun *et al.* (1980)
NRU Chalk River, Canada	0.2	$(CH_2)_n$	330	–	7.3 × 530	Mn	40	5.6	0.7	5.6	–	8/1	–	4,h	Robson and Winfeld (1972)
University Reactor, Risley, UK	0.2	Poly-styrene	300	Glass	5 × 15 × 800	LiF	62	19	0.3	4.3	–	10/1	–	5, tof, h	Roy (1976)
ILL, Grenoble, France	60	H_2O	330	–	7 × 7 × 1500	^3He	35	40	37/360	6.2	260	0.1/1 / 10/1	–	6,i	Ageron *et al.* (1978a)
LNPI, Gatchina, USSR	1	Mg	300	–	5 × 560	LiOH	35	6.2	1.5	6.2	–	2.4/0.5	2	h	Egorov *et al.* (1974)
	3–5	Be	.00	–	7 × 6 × 850	^3He	42	20	36	6.8	140	10/1	12	v	Egorov *et al.* (1974)
	10	Be	.0	–	7 × 6 + 850	^3He	42	20	240	6.8	1500	30/10	12	v	Altarev *et al.* (1982)
	6	H_2	20	–	77 × 6 × 850	^3He	60	–	280	7	280	45/25	–	v	Altarev *et al.* (1980*b*)

Inst. Atomic Physics, Alma-Ata, USSR	0.5	H_2O	80	Cu	17.5 × 600	^3He	30	5.6	4	5.6	–	20/3.4	–	t	Akhmetov *et al.* (1974*a*)
	2	Zr	300	Fe	9 × 400	^3He	30	6.2	46	6.2	–	4.6/1	12	k	Akhmetov *et al.* (1977*a*)
JINR, Dubna, USSR	0.01	ZrH	300	Fe	18.7 × 910	^3He	60	6.2	0.02	6.2	–	4/1	–	h	Golikov *et al.* (1985*a*)
NIRAR, Dimitrovgrad, USSR	20–40	ZrH	300	–	10 × 900	^3He	60	5.6	11	5.6	–	0.9/1	–	h	Kosvintsev *et al.* (1977*c*)
Inst. Atomic Energy, Moscow, USSR	2	H_2O	300	Cu	9.6 × 600	^3He	60	5.6	1.7	5.6	0.17	2/1	8	h	Shapiro (1972*b*)
Inst. Theor. and Exp. Physics, Moscow, USSR	1	Be	160	Cu	10 × 800	^3He	75	5.7	3.0	5.7	0.75	6/3.4	–	v	Abov *et al.* (1981*b*)
Inst. Physics and Engineering, Moscow, USSR	1	Be	300	Fe	5.2 × 800	^3He	9	120	42	–	–	–	–	7v, tof	Antonov *et al.* (1977*a*)

[a] The letters indicate the positioning of the UCN channel: v, vertical; h, horizontal; t, tangential; i, inclined; and r, radial; tof, indicates the presence of a time-of-flight spectrometer. 1, indicated characteristics of the neutron guide leading neutrons to the turbine (if three values are given in column 5, the neutron guide has a rectangular cross-section); 2, whenever the material of the neutron guide is indicated as Fe, the neutron guide is made of electropolished stainless steel; 3, the turbine is a reflecting mica crystal; 4, the detector was an activation Mn–Ni foil; 6, in columns 10 and 13 here two values are indicated (the lower is obtained by conversion based on the estimation of the detector losses and efficiency); 7, the integral flux throughout the spectrum is indicated in the 10th column.

also presented in the table. In addition, Figs 9.9–9.11 display the layouts of the UCN channels established at LNPI, NIRAR, and ILL.

Table 9.19 does not include information about the largest UCN source. It is considered separately below. This source is established on the vertical

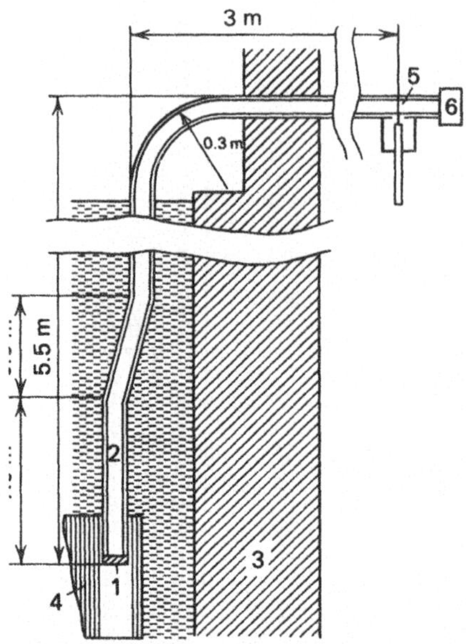

Fig. 9.9 Layout of vertical UCN channel at LNPI (Gatchina) (Egorov *et al.* 1974): 1, converter; 2, neutron guide; 3, biological shielding; 4, reactor zone; 5, shutter; 6, UCN detector.

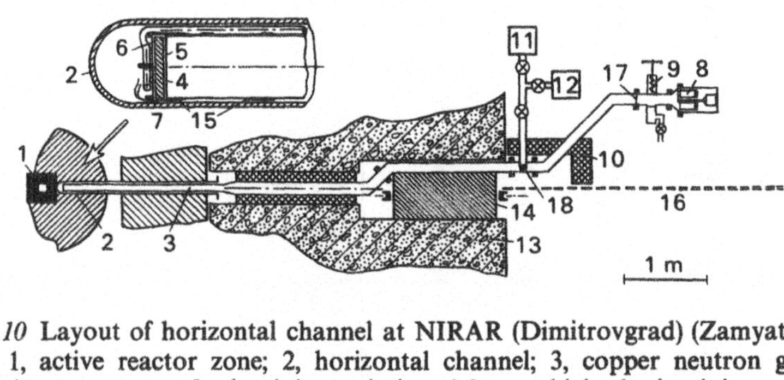

Fig. 9.10 Layout of horizontal channel at NIRAR (Dimitrovgrad) (Zamyatin *et al.* 1974): 1, active reactor zone; 2, horizontal channel; 3, copper neutron guide; 4, magnesium converter; 5, aluminium window 0.2 mm thick, 6, aluminium vessel; 7, welded Al–Cu joint; 8, UCN detector; 9, vacuum valve; 10, additional shielding; 11 and 12, oil-less pumps; 13, reactor shielding; 14, shielding from direct beam; 15, thermocouples; 16, direct beam; 17, foil of aluminium for shielding of vacuum; 18, vertical evacuation pipe.

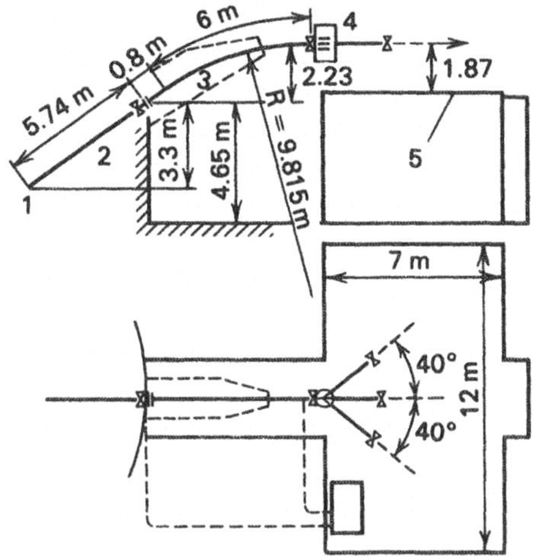

Fig. 9.11 Layout of UCN at ILL (Grenoble) (Ageron *et al.* 1978*a*): 1, converter; 2, section of neutron guide inside the biological shielding; 3, section of neutron guide within the experimental hall; 4, distributor; 5, floor level in the experiments with UCN.

channel of the high-flux reactor at ILL and combines a cold moderator and a Steyerl turbine.

An amount of 25 litres of liquid deuterium at a temperature of 25 K serves as the cold moderator. It is penetrated by thermal neutrons with a flux density of $4.5 \times 10^{14}\ cm^{-2}\ s^{-1}$. The cold moderator and the turbine are connected by a nickel neutron guide with a total length of 18 m. A nickel coating 0.15 mm thick covers the walls of the more solid aluminium neutron guide, the lower hermetically sealed end of which is in the form of a dome and is immersed in the liquid deuterium.

The neutron guide consists of two parts: a lower vertical cryogenic part 5 m long with a round cross-section 7 cm in diameter; and an upper part 13 m long with a square cross-section of $7 \times 7\ cm^2$ and bent with a radius of curvature $R = 13$ m. These parts are partitioned by zirconium films 0.1 mm thick in between which the shiber of the beam is established. For better heat exchange the cryogenic part is filled with helium up to a pressure of ~0.2 mm Hg, while in the upper part a vacuum of $\sim 10^{-5}$ mm Hg is maintained. To disengage the vacuum of the neutron guide and the turbine an aluminium membrane 0.2 mm thick is placed at the end of the neutron guide. The upper part of the neutron guide is divided along all its length into two parts: one, of cross-section $7 \times 3.4\ cm^2$, serves as a supply for the turbine and for the production of UCN; while the other, divided into a series of more narrow sections, serves to bypass the turbine for the extraction of cold neutrons with wavelengths of 20–400 Å.

The transmission of the neutron guide system amounts to 0.24; the reflection coefficient in the subcritical region is 99%. This exceeds the coefficient of 98% for a surface obtained by vapour depositing nickel on to glass.

The stainless steel vessel is filled at the exit opening of the neutron guide up to a UCN density of $\sim 19\ cm^{-3}$ ($v_{lim} = 6.2\ m\,s^{-1}$). Such a density amounts to 2% of the Maxwellian density at the temperature of deuterium. If the aluminium membrane is removed, then the density of accumulated neutrons increases up to $41\ cm^{-3}$.

The neutron flux density at the exit opening measured by the activation of gold is $6.6 \times 10^8\ cm^{-2}\,s^{-1}$. Neutrons with velocities of 40–60 $m\,s^{-1}$ from this flux arrive at the turbine which has 690 blades fixed on a rotor 1.7 m in diameter, the rotation velocity of which (260 rev min^{-1}) amounts to 25 $m\,s^{-1}$ at the periphery. Upon being reflected from the blades these neutrons are transformed into UCN.

The cross-section of the UCN flux from the turbine is $16 \times 8\ cm^2$. Its intensity ($0 < v < 6.2\ m\,s^{-1}$) amounts to $2.6 \times 10^4\ cm^{-2}\,s^{-1}$, corresponding to a density of $87\ cm^{-3}$ and a total flux of $3.3 \times 10^6\ s^{-1}$. In an actual experiment UCN were accumulated in a stainless steel vessel with a density equal to $36\ cm^{-3}$. The accumulation efficiency was, thus, 41%.

The UCN extraction efficiency in the given source can be estimated by the coefficient 10; however, if it is taken into account that the moderator is 10 times cooler and thus equivalent to a G factor ~ 100, then $\varepsilon/G \approx 10\%$.

In conclusion we present the full list of references on UCN sources: Groshev *et al.* (1971), Steyerl, (1972*a*), Kashukeev and Stanev (1973), Akhmetov *et al.* (1974*a*) Zamyatin *et al.* (1974), Egorov *et al.* (1974), Kashukeev *et al.* (1975*b*), Robson (1976*a*), Roy (1976), Akhmetov *et al.* (1977*a*), Antonov *et al.* (1977*a*), Kosvintsev *et al.* (1977*c*), Mampe and Ageron (1978), Ageron *et al.* (1978*a*), Dombeck and Lynn (1979), Abov *et al.* (1981*b*), Ageron and Mampe (1984), Arzumanov *et al.* (1984*a*), Golikov *et al.* (1985*a*), Steyerl *et al.* (1986), Morosov *et al.* (1987*b*), Utsuro *et al.* (1988*b*), Altarev *et al.* (1988*b*).

9.4 SOME DATA ON THE TRANSMISSION OF NEUTRON GUIDES

In Table 9.20 the results are given for the calculation of the transmission of straight cylindrical neutron guides with the following ratios of lengths to radius: $l = 1, 2, 4, 8, 16, 32$, and 64 for the reflection indicatrix (4.8) with $A(\Omega) = 1$ and $A(\Omega) = \cos\theta$ for various parameters of non-specularity, $g = 1$; 0.5; 0.25; and 0.125. The transmission probability W_{tr}^{MC} is computed by the Monte Carlo method, ΔW is the standard root-mean-square deviation, W_{tr}^{D} is

TABLE 9.20 Monte Carlo calculations of the transmission of neutron guides

$A(\Omega)$	l	$G = 1.0$			$G = 0.5$			$G = 0.25$			$G = 0.125$		
		W_{tr}^{MC}	ΔW	W_{tr}^{D}	W_{tr}^{MC}	ΔW	W_{tr}^{D}	W_{tr}^{MC}	ΔW	W_{tr}^{D}	W_{tr}^{MC}	ΔW	W_{tr}^{D}
1	1	0.671	0.007	0.700	0.812	0.005	0.875	0.898	0.004	0.942	0.947	0.003	0.972
	2	0.514	0.007	0.538	0.699	0.007	0.778	0.821	0.005	0.891	0.900	0.004	0.946
	4	0.350	0.007	0.368	0.559	0.007	0.636	0.724	0.006	0.803	0.832	0.005	0.897
	8	0.220	0.006	0.226	0.415	0.007	0.467	0.597	0.007	0.671	0.724	0.006	0.814
	16	0.130	0.005	0.127	0.282	0.006	0.304	0.446	0.007	0.505	0.602	0.007	0.686
	32	0.072	0.004	0.068	0.177	0.005	0.179	0.306	0.007	0.338	0.460	0.007	0.522
	64	0.035	0.003	0.035	0.103	0.004	0.099	0.195	0.006	0.203	0.322	0.007	0.354
$\cos\theta$	1	0.777	0.006	0.704	0.869	0.005	0.803	0.926	0.004	0.865	0.962	0.003	0.907
	2	0.664	0.007	0.614	0.783	0.006	0.732	0.874	0.005	0.814	0.931	0.004	0.870
	4	0.536	0.006	0.511	0.680	0.007	0.645	0.795	0.006	0.746	0.886	0.004	0.819
	8	0.412	0.006	0.403	0.564	0.007	0.542	0.696	0.007	0.660	0.611	0.005	0.752
	16	0.299	0.006	0.300	0.443	0.007	0.431	0.578	0.007	0.557	0.712	0.005	0.752
	32	0.208	0.006	0.210	0.324	0.007	0.323	0.449	0.007	0.445	0.594	0.007	0.565
	64	0.141	0.005	0.139	0.225	0.006	0.227	0.328	0.007	0.334	0.470	0.007	0.452

the probability computed by diffusion theory (Berceanu and Ignatovich 1973) without taking absorption into account: $W_{tr}^{D} = 1/(1 + R)$,

$$R = \tfrac{3}{7} lg/(2 - g), \qquad A(\Omega) = 1$$

$$R = \tfrac{2}{15} lg/\ln[1 + \sqrt{2lg}\exp(-g/3 - 1)], \qquad A(\Omega) = \cos\theta. \tag{9.9}$$

Curves are presented in Fig. 9.12 that correspond to the W_{tr}^{MC} of Table 9.20. The dashed lines correspond to the analytic formulae (9.9). To avoid overloading the figure only the curves for $g = 1$ are shown.

Presented in Fig. 9.13 is the UCN angular distribution at the exit from the neutron guide calculated by Brown *et al.* (1975) using the Monte Carlo method for the reflection indicatrix (4.5)–(4.7) corresponding to a weakly rough surface.

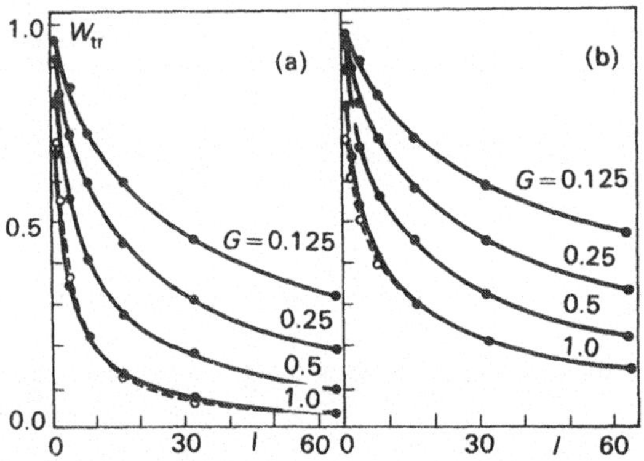

Fig. 9.12 Transmission of straight cylindrical neutron guides depending on the reduced length and on the reflection indicatrix of the neutrons from the walls. Monte Carlo computation (Berceanu and Ignatovich 1973).

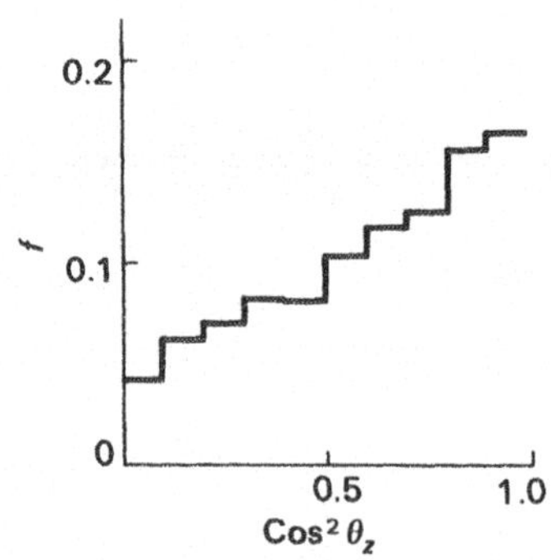

Fig. 9.13 Angular distribution at the exit from the neutron guide for a reflection indicatrix corresponding to small roughnesses (Brown *et al.* 1975).

For the convenience of readers we give here the full list of references on the kinetics of UCN in a gas: Berceanu and Ignatovich (1973), Brown *et al.* (1975), Winfeld and Robson (1975), Frank (1976), Mysenholder (1976), Robson (1976*b*), Kosvintsev *et al.* (1976*b*, 1977*c*, 1984*a*, *b*, 1985), Ignatovich and Terekhov (1978), Kügler *et al.* (1980), Morosov *et al.* (1986), Ignatovich and Nikitenko (1987*a*, *b*), Golikov *et al.* (1988*a*), Morosov *et al.* (1988*a*).

9.5 HYDROGEN ON A SURFACE

Figures 9.14–9.19 demonstrate how the concentration of hydrogen in surface layer of ~200 Å changes in various substances when the surface is cleaned in some way or another. In Table 9.21 it is shown how the hydrogen content changes in certain substances as a result of ion bombardment (La Marche *et al.* 1981) and in Table 9.22 as a result of deuteration.

For the convenience of readers we present the list of references on surface hydrogen relevant to UCN: Lanford and Golub (1977), Bugeat and Mampe (1979), Yates (1979), Mampe and Bugeat (1980), La Marche *et al.* (1981), Mampe *et al.* (1981), Yoshiki and Ueda (1983), Hayashi *et al.* (1986), *Ligeon *et al.* (1980).

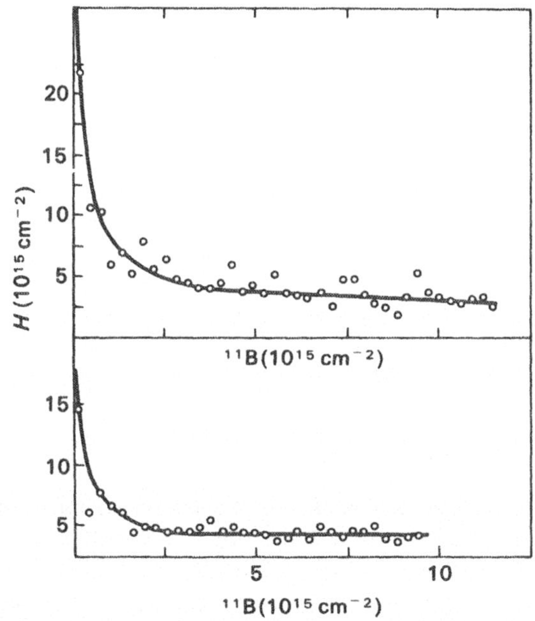

Fig. 9.14 Dependence of the residual hydrogen content in a 200 Å thick surface layer in copper (upper curve) and in stainless steel (lower curve) on the irradiation dose of ^{11}B ions (Bugeat and Mampe 1979).

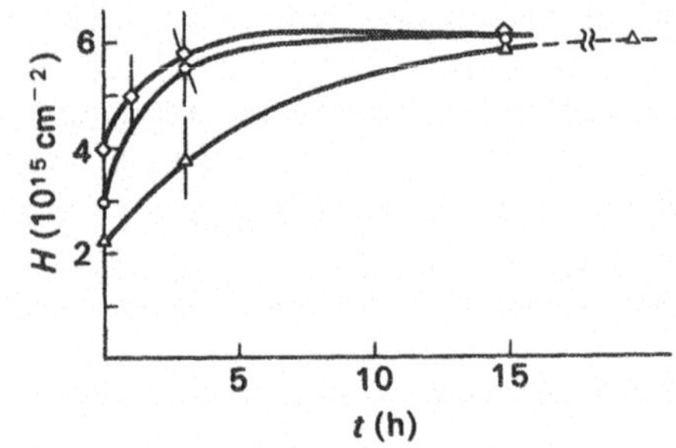

Fig. 9.15 Enhancement of the hydrogen content in a copper surface layer with time under various vacuum conditions, after the surface has been subjected to ion bombardment for cleaning (Bugeat and Mampe 1979): △, 10^{-8} mm Hg; ○, 5×10^{-7} mm Hg; ◇, 5×10^{-6} mm Hg. The last point corresponds to measurements carried out 65 h after the cleaning.

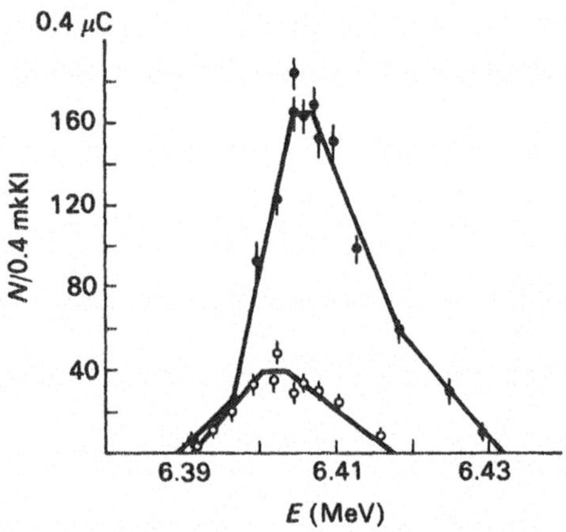

Fig. 9.16 Distribution of hydrogen in a nickel foil 0.127 mm thick: ●, before annealing, total hydrogen concentration 1.6×10^{16} cm^{-2}; ○, after annealing at 1000°C, hydrogen density 3.6×10^{15} cm^{-2} (La Marche *et al.* 1981).

9.6 THE LOSS COEFFICIENT

Table 9.23 presents the loss coefficients for various substances which are derived from experiments on UCN storage. The coefficient η is determined from the storage time by averaging the loss coefficient over the isotropic angular distribution and over a segment of the perfect Maxwellian spectrum with limits indicated in the third column. Also presented is the minimum loss coefficient calculated from the storage time under the assumption that the spectrum represents a monoline corresponding to the hard edge of the

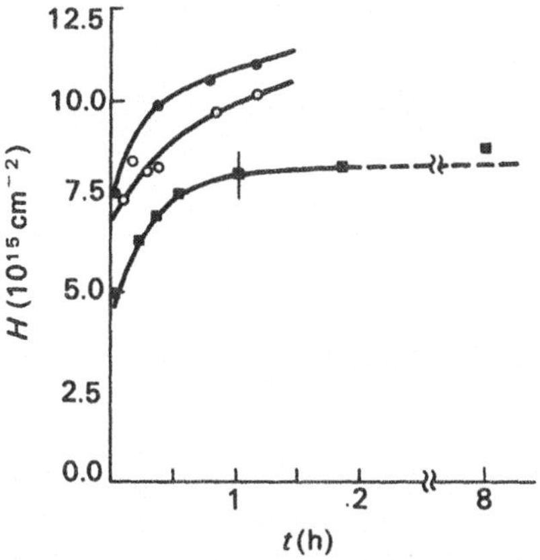

Fig. 9.17 The rate of contamination with hydrogen of the surface of lead at a pressure of (1–3) × 10^{-6} mm Hg when the lead is vapour-deposited at the following rates: ●, 5 monolayer/s; ○, 1 monolayer/s; and ■, 7 monolayers/s (La Marche *et al.* 1981).

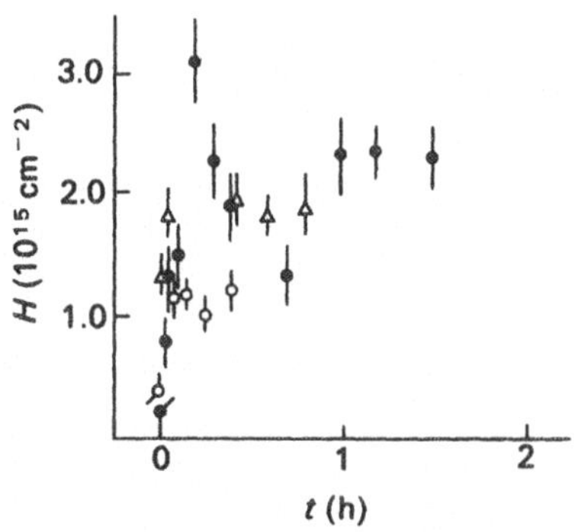

Fig. 9.18 The same as in the previous figure, but for aluminium: ●, 5 monolayers/s; ○, 3 monolayers/s; △, 8 monolayers/s (La Marche *et al.* 1981).

spectrum. Given in the sixth column is the loss cross-section which corresponds to the coefficient η. The cross-section is indicated for $\lambda = 1.08$ Å, i.e. for a neutron velocity of 3700 m s^{-1}.

In addition to this Table, Figs 9.20 and 9.21 demonstrate the storage curves in beryllium and quartz vessels.

In Fig. 9.22 the storage curves in liquid helium at various temperatures are presented.

To conclude we list: (i) the references relevant to the measurement of the loss coefficient: Trüsted *et al.* (1970), Anikolenko *et al.* (1973*b*), Steyerl and Trüsted (1974), Groshev *et al.* (1974, 1976), Antonov *et al.* (1976*a*), Kosvintsev

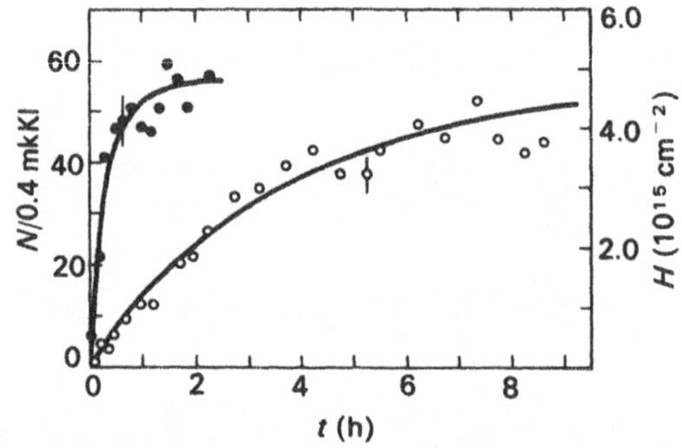

Fig. 9.19 The rate of contamination with hydrogen of a nickel surface (0.127 mm thick and 25 × 38 mm^2 in area) after heating in two differing modes: ○, up to the temperature when the metal starts evaporating; ●, up to 1000°C (La Marche *et al.* 1981).

TABLE 9.21 Hydrogen content n_H (10^{15} cm^{-2}) of the surface of a material before (a) and after (b) ionic spray deposition for 10 min in an atmosphere of O_2 at a pressure of 25 Pa under a current of 50 mA and voltage of 2000 V

Substance	(a)	(b)	Substance	(a)	(b)
Cu	6.5[a]	3.0	Pb	12.	1.7
Ni	8.1[b]	4.0	Cu	58[c]	3.4

[a] Exposed in air after cleaning.
[b] As [a], but with annealing at high temperature.
[c] Initially the sample was not treated in any way.

TABLE 9.22 Deuteration of a substance

Substance	Method of deuteration	Hydrogen content, n_H (10^{15} cm^{-2})	Substance	Method of deuteration	Hydrogen content, n_H (10^{15} cm^{-2})
Cu	Boiling in D_2O	35	C	Etching in DF (15 min)	22
Cu	Etching in DF (15 min)	52	Glass[a]	Etching in DF (15 min)	17–34
C	Boiling in D_2O	31			

[a] Glass has the following composition: 63.5%SiO_2 + 15.5%Al_2O_3 + 4%MgO + 13%CaO + 2% Na_2O + 2%F.

TABLE 9.23 Experimental data

Substance	Sizes $\phi \times l$ (cm^2)	Δv (m s^{-1})	τ (s)	η	η_{min}	σ_1 (barn)	Comments	References
Cu	14 × 174	3.2–5.7	33	5.0	2.2	8.4	1	(a)
Cu	14 × 174	3.2–5.7	14	12	5.3	20	2a	(a)
Cu	8.5 × 250	3.2–5.7	16 ± 1	6.6	2.9	11	3	(b)
Cu	8.5 × 250	3.2–5.7	29 ± 2.2	3.6	1.6	6.0	4	(b)
Cu	8.5 × 250	3.2–5.7	30 ± 2.2	3.4	1.5	5.7	4	(b)
Cu	8.5 × 250	4.5–5.7	27 ± 1.5	3.0	1.7	5.0	4	(b)
Cu	8.5 × 250	0.0–4.6	68 ± 6	3.2	1.7	5.3	4	(b)
Cu	8.5 × 250	1.5–2.9	105	5.0	3.3	8.3	4	(b)
Cu	8.5 × 20	3.2–5.7	30	2.9	1.3	4.8	4	(c)
Cu	8.5 × 200	1.0–2.4	200 ± 17	3.6	2.3	6.0	4	(c)
Cu	8.5 × 200	3.1–3.8	70.4 ± 5.5	3.4	2.7	5.6	4	(c)
Cu	8.5 × 200	3.5–4.1	52 ± 7.3	3.7	3.1	6.2	4	(c)
Cu	8.5 × 200	4.3–4.8	32 ± 2.3	3.9	3.4	6.5	4	(c)
Cu	8.5 × 200	5.3–5.7	20.5 ± 5	3.1	2.3	5.2	4	(c)
Cu	8.5 × 200	5.5–5.7	13.5 ± 4	4.3	3.4	7.2	4	(c)
Cu	9 × 70	3.2–5.7	23 ± 0.3	4.6	2.0	7.6	4	(d)
Cu	8.2 × 96	3.2–5.7	9.4 ± 0.3	11	4.6	18	2b	(d)
Cu	40 × 40	3.2–5.7	135 ± 15	2.1	0.9	3.6	5	(e)
Cu	40 × 40	3.2–5.7	56 ± 2	5.7	2.5	9.5	6	(e)
Cu	17 × 30	3.2–5.7	22 ± 2	7.5	3.3	13	7	(f)
Cu	19 × 224	2.5–5.7	15.5	16	6.5	26	8	(g)
Cu	50 × 24	3.4–4.9	28.7 ± 1.8	15	10	22	9a	(h)
Cu	50 × 24	3.4–4.9	38.2 ± 5.6	11	7.5	18	9b	(b)
Cu	50 × 24	3.4–4.9	52.8 ± 3.2	7.8	5.3	13	9c	(b)
Cu	50 × 24	3.4–4.9	115 ± 8.4	3.3	2.3	5.5	10a	(b)
Cu	50 × 24	3.4–4.9	112 ± 8.6	3.4	2.3	5.7	10b	(h)
Cu	50 × 24	3.4–4.9	91.5 ± 5.8	4.3	2.9	7.2	10c	(b)
Cu	50 × 24	3.4–4.9	94.8 ± 4.4	4.1	2.8	6.9	10d	(h)
Cu	50 × 24	3.43–4.95	88 ± 4.4	4.5	3.1	7.5	10e	(h)
Cu	50 × 24	3.43–4.95	86.2 ± 7.3	4.6	3.1	7.7	10f	(h)
Cu	50 × 24	3.43–4.95	74 ± 12	5.5	3.7	9.1	10g	(h)
Cu	57 × 32	0.0–1.9	338 ± 9	7.5	6.8	13	11[d]	(i)
Cu	61 × 20	0.0–1.9	300 ± 40	11	9.9	18	4[d]	(j)
Glass	10 × 224	2.5–4.2	9	18	8.3	21	12	(g)
Glass	8 × 250	3.2–4.2	25	4.3	2.3	5.0	13	(b)
Glass	8 × 250	3.2–4.2	52 ± 6	2.0	1.1	2.3	14	(b)
Glass	8 × 250	3.3–4.2	104.0 ± 7	0.94	0.51	1.1	15	(b)
Boric glass	8 × 250	1.7–4.3	10	15	5.8	17	16	(b)
Glass	8 × 250	3.2–4.2	75	1.4	0.73	1.6	17a	(k)
Glass	8 × 250	3.2–4.2	145	0.64	0.34	0.73	17b	(k)
Glass	8 × 250	3.2–4.2	140	0.67	0.36	0.77	17c	(k)
Glass	8 × 250	3.2–4.2	160	0.57	0.30	0.65	17d	(k)
Glass	8 × 250	3.2–4.2	170 ± 10	0.53	0.28	0.61	17e	(k)

TABLE 9.23 (*Continued*)

Substance	Sizes $\phi \times l$ (cm^2)	Δv (m s^{-1})	τ (s)	η	η_{min}	σ_l (barn)	Comments	References
Glass	8 × 75	3.2–4.2	40	2.6	1.4	2.9	–	(k)
Glass	8 × 20	3.2–4.2	35	2.6	1.4	3.0	–	(k)
Glass	7 × 120	3.2–4.2	25	3.7	2.0	4.3	–	(l)
Quartz	8 × 75	3.2–4.2	40	2.6	1.4	2.9	–	(k)
Be	8 × 75	3.2–5.77	40	3.3	2.1	5.6	11	(k)
Be	50 × 24	3.43–4.95	63 ± 5	8.5	6.1	14	11	(h)
Be	50 × 24	3.43–4.95	110 ± 21	4.6	3.3	7.7	18a	(h)
Be	50 × 24	3.43–4.95	131 ± 32	3.8	2.7	6.2	18b	(h)
Be	50 × 24	3.43–4.95	121 ± 8.2	4.1	3.0	6.9	18c	(h)
Be	50 × 24	3.4–4.95	113.6 ± 21	4.4	3.2	7.4	18d	(h)
Be	50 × 24	3.43–4.95	115.8 ± 16.3	4.3	3.1	7.2	18e	(h)
Be	50 × 24	3.43–4.95	116.7 ± 10.7	4.3	3.1	7.1	18f	(h)
Be	50 × 24	3.43–4.95	105.0 ± 16.7	4.9	3.5	8.1	18g	(h)
Be	50 × 24	3.43–4.95	102.5 ± 13.8	5.0	3.6	8.3	18h	(h)
Be	50 × 24	3.43–4.95	101.8 ± 15.1	5.0	3.6	8.4	18i	(h)
Be	50 × 24	3.43–4.95	110.3 ± 19.7	4.6	3.3	7.6	18j	(h)
Be	57 × 32	1.4–1.9	645 ± 25	2.0	1.8	3.4	11[d]	(m)
Ni	7 × 120	3.2–6.8	15 ± 1	5.1	2.1	11	19	(l)
Fe	64 × 105	3.2–6.0	47 ± 3	19	10	36	20[d]	
Fe	64 × 105	0.0–5.0	53	28	19	51	20[d]	
Fe	64 × 105	0.0–4.0	85	24	19	45	20[d]	
Fe	64 × 105	0.0–3.4	104	24	19	44	20[d]	
Fe	6.7 × 224	0.0–6.0	52	1.7	0.63	3.1	21	
Teflon	8 × 250	3.2–4.8	10 ± 1.4	11	5.2	13	22	
C	17 × 30	3.2–6.1	35	4.5	1.9	6.4	23	
C	17 × 30	4.2–6.1	32	4.2	2.1	6.1	23	
C	17 × 30	4.4–6.1	30	4.4	2.2	6.3	23	
C	17 × 30	5.4–6.1	3.6	29	19	42	23	
C	14 × 174	3.2–6.1	11	15	6.3	21	–	
Al	57 × 32	1.4–1.9	230 ± 14	7.1	6.2	5.3	24[d]a	
Al	57 × 32	1.4–1.9	170 ± 15	10	9.1	7.9	24[d]b	(m)
Al	57 × 28	0.0–1.9	260 ± 15	6.0	5.2	4.5	25[d]a	(m)
Al	57 × 28	0.0–1.9	410 ± 20	2.9	2.5	2.2	25[d]b	
Al	57 × 28	0.0–1.9	370 ± 20	3.4	3.0	2.6	25[d]c	
Al	57 × 28	0.0–1.9	340 ± 20	4.0	3.5	3.0	25[d]e	
Al	57 × 28	0.0–1.9	335 ± 15	4.1	3.6	3.1	25[d]f	
Al	57 × 28	0.0–1.9	300 ± 20	4.8	4.2	3.6	25[d]g	
Zn	57 × 32	1.4–1.9	298 ± 17	6.8	6.1	8.7	11[d]	(m)
Pb	57 × 32	1.4–1.9	435 ± 23	3.3	2.9	6.8	11[d]	(m)
Bi	57 × 32	1.4–1.9	439 ± 17	2.8	2.4	1.6	11[d]	(m)
CO_2	50 × 24	3.43–4.95	122 ± 12	2.3	1.3	3.2	26[d]	(m)
D_2O	50 × 24	3.43–4.95	132 ± 19	2.8	1.8	3.6	26[d]	(m)
D_2O	50 × 24	3.43–4.95	92 ± 10	4.0	2.7	5.2	26[d]	(m)

References: (a) Groshev *et al.* (1971); (b) Groshev *et al.* (1974); (c) Groshev *et al.* (1976); (d) Strelkov and Hetzelt (1978); (e) Antonov *et al.* (1976*a*); (f) Steyerl and Trüsted (1974); (g) Steyerl (1977); (h) Kosvintsev *et al.* (1980*c*); (i) Kosvintsev *et al.* (1978*a*); (j) Kosvintsev *et al.* (1977*a*); (k) Luschikov *et al.* (1978); (l) Egorov *et al.* (1974); (m) Kosvintsev *et al.* (1978*b*).

In those cases when the comments number is assigned a superscript *d*, η (η_{min}) was calculated by applying the formulae of Appendix 6.3, i.e. in computing η gravity was taken into account. If no superscript *d* is present in the comments number, the calculation was performed by using the formulae of Appendix 6.1. From the comments number one can determine the nature of the treatment, to which the surface of the walls was submitted, and the additional conditions.

1. Chemical polishing. Additional heating at 250°C did not alter the storage time.

2. a, The trap is made of untreated foil; b, electropolished foil cleaned in NaOH and then washed with distilled water.

3. Electropolishing in a $H_3PO_4 + Cr_2O_3$ solution.

4. Electropolishing in H_3PO_4.

5. Electropolishing. The time τ is determined from the slope of the storage curve after exposure for more than 100 s.

6. The time τ is found from the storage curve within the range of exposure times from 50 to 100 s.

7. Electrodeposited copper on to the walls of a pyrographite trap.

8. Washing with isopropylic alcohol. Measurements were also made with a copper cube of side 31.75 cm. After cleaning with isopropylic alcohol the following were obtained, $\tau = 11.2$ s, $\eta = 2.5 \times 10^{-3}$; after cleaning in dilute nitric acid and rinsing in distilled water and isopropylic alcohol, $\tau = 25.5$ s, $\eta = 11.0 \times 10^{-4}$; after chemical polishing in a solution of 50% nitric acid, 25% vinegar, and 25% phosphoric acid and subsequent rinsing in water and alcohol, $\tau = 33$ s, $\eta = 8.6 \times 10^{-4}$; if the range of velocities of the confined neutrons is reduced to 3.93–5.7 m s^{-1}, then the storage time is $\tau = 19$ s, and, accordingly, $\eta = 12.0 \times 10^{-4}$.

9. Electropolishing in dilute H_3PO_4 with subsequent washing in tap water. Storage was carried out at the following temperatures: a, 300 K; b, 170 K; c, 80 K.

10. After treatment 9 degassing was performed at a temperature of 700 K for 6 h. Measurements were performed at the following trap temperatures: a, 80 K; b, 170 K; c, 220 K; d, 300 K; e, 460 K; f, 550 K; g, 800 K.

11. Vapour deposition of copper in a vacuum of 10^{-4} Torr.

12. Rinsing with isopropylic alcohol. The pipe had a rectangular cross-section of 15.24 $\times$ 5.33 cm^2. The calculation presented in the table corresponds to a cylindrical pipe with a diameter chosen so that the area of the cross-section remained unaltered. Measurements were also made with a glass cube of side 31.75 cm. After rinsing with isopropylic alcohol, $\tau = 95$ s and $\eta = 3.4 \times 10^{-4}$; after etching in a 3% solution of HF and subsequent washing in distilled water and isopropylic alcohol, $\tau = 130$ s and $\eta = 2.4 \times 10^{-4}$ ($\eta_{min} = 1.1 \times 10^{-4}$, $\sigma_l = 2.7$ barn).

13. Rinsing in alcohol. The glass composition is as follows: $(SiO_2)_{63.5}(Al_2O_3)_{15.5}(MgO)_4(CaO)_{13}(Na_2O)_2F_2$.

14. Etching in a 3% solution of HF.

15. Multiple additional etchings in a 3% solution of HF.

16. The glass composition is $(SiO_2)_{75}(Al_2O_3)_6(B_2O_3)_7(Na_2O)_6(BaO)_{4.3}(CaO)_{1.7}$. Rinsing in the distillate. The value of η obtained by fitting the experimental storage curve under the assumption of a Maxwellian spectrum is 1.6×10^{-3}.

17. a, The glass is washed and rinsed with OP-7 soap and distilled water; b, thermal treatment was performed under a change of temperature from 400 to 800 K; c–e, measurements were carried out at 80, 300, and 800 K, respectively.

18. Deposition of a 0.35 μm layer of beryllium. Measurements were performed after annealing for 6 h at a temperature of 820 K. Measurements of the trap walls were performed at the following temperatures: a, 20 K; b, 40 K; c, 80 K; d, 150 K; e, 210 K; f, 300 K; g, 400 K; h, 500 K; i, 600 K; j, 700 K.

19. Galvanic coating of copper pipe with nickel. After lengthy thermal treatment at 150°C measurements were performed at room temperature and at 200°C. No change in the storage time was observed.

20. Electropolishing in a 90% solution of H_3PO_4. Calculation of η was performed for a vertical cylinder. In the experiment the cylinder occupied a horizontal position. Taking this circumstance into account should lead to a certain decrease in the value of η.

21. Polished pipe.

22. Rinsing in a hot solution of NaOH.

23. Pyrolytic graphite. The trap underwent annealing for a week at 700 K, but was subsequently in contact with the atmosphere for 6 months.

24. a, Newly spray-deposited aluminium; b, aluminium vessel etched in NaOH and rinsed in distilled water before and after spray deposition and contact with atmosphere for 2 weeks.

25. a, The vessel underwent etching in NaOH and was rinsed in distilled water; b, annealing for 2 h at 250 K. Measurement at room temperature. Further repeated heating to 770 K did not lead to any improvements. The trap temperatures during successive measurements were: c, 400 K; e, 520 K; f, 620 K, g, 800 K.

26. Vapour deposition of coating at 79 K.

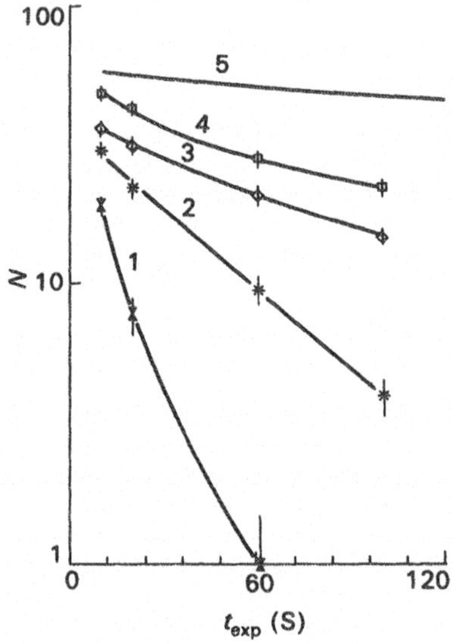

Fig. 9.20 Storage curves in a vertical cylindrical vessel of stainless steel with beryllium spray-deposited on to the walls (6.7 cm in diameter and 1.9 m long): 1, initial curve before deposition of beryllium, $\tau = 20$ s; 2, alternating current discharge in argon for 30 min, $\tau = 50$ s; 3, sputtering of Be by argon ions in a direct current discharge with subsequent cleaning by alternating current in a D_2 atmosphere, $\tau = 140$ s; 4, evaporation of beryllium with subsequent cleaning by D_2 ions in an alternating current discharge, $\tau = 165$ s; 5, computed curve for a perfect surface, $\tau = 420$ s (Mampe *et al.* 1981).

et al. (1976*b*, 1977*a*, 1978*b*, 1979, 1980*a*, *c*, 1983, 1986, 1988), Antonov *et al.* (1978*b*), Luschikov *et al.* (1978), Strelkov and Hetzelt (1978), I.M. Frank (1979), Morosov and Terekhov (1980), Alfimenkov *et al.* (1980), Bates (1982), Akunetz *et al.* (1982), Ignatovich *et al.* (1982), Ageron *et al.* (1985, 1986); and

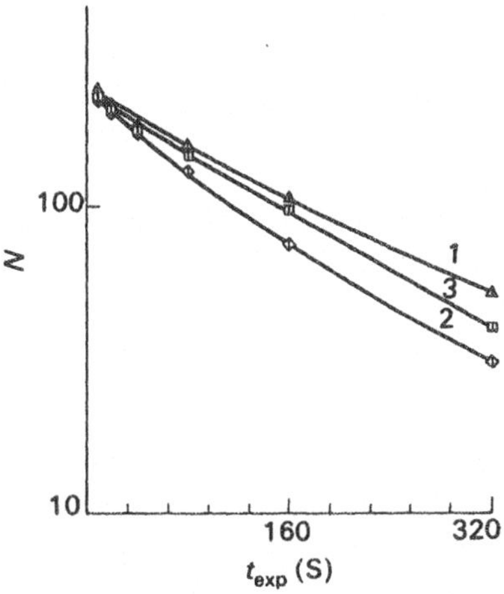

Fig. 9.21 Storage curves for quartz vessel (6.7 cm in diameter and 1 m long) with two stainless steel valves on exposure to an atmosphere of hydrogen: 1, cleaning by a discharge in D_2; 2, the vessel was filled with hydrogen for 1 h at atmospheric pressure; 3, cleaning by a discharge in H_2 (Mampe *et al.* 1981).

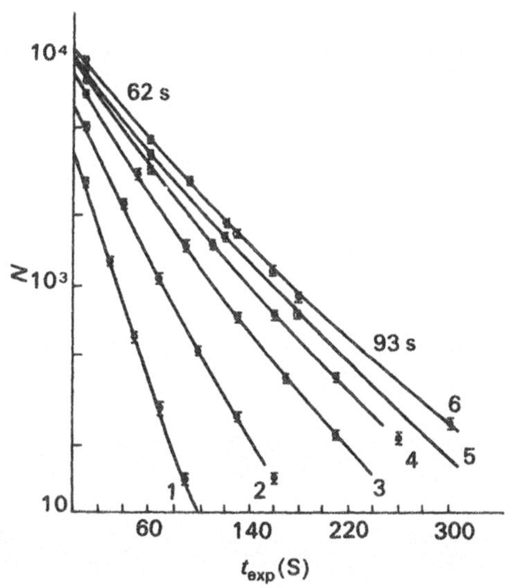

Fig. 9.22 UCN storage curves in liquid helium. The storage times indicated on the upper curve correspond to the slopes of the curve at the indicated points (Golub *et al.* 1983). The curves are obtained at different helium temperatures: 1, 1.16 K; 2, 1.05 K; 3, 0.95 K; 4, 0.85 K; 5, 0.75 K; 6, 0.49 K.

(ii) the references in which attempts were made to explain the nature of the loss coefficient: Zel'dovich (1959), Hennig *et al.* (1970), Ignatovich, (1972*a*, *b*, 1973, 1974*a*, 1975*a*, *c*, 1980*b*), Gerasimov *et al.* (1973), Ignatovich and Stepanov (1974), Stepanov and Shelagin (1974, 1977*a*), Frank (1974*a*, *b*, 1975), Ignatovich and Luschikov (1975), Stepanov (1975, 1976*a*, *b*), Galitskii and Satarov (1976), Ignatovich and Terekhov (1976*a*), Satarov (1976*a*, *b*), Zubarev (1976), Blokhintsev and Plakida (1976, 1977), Holas (1977), Terekhov (1977*b*), Ignatovich and Satarov (1977), Stanev (1978), Seregin (1980*b*), Plakida (1981, 1983).

9.7 ABSORPTION AND INELASTIC SCATTERING CROSS-SECTIONS

In Table 9.24 some absorption and inelastic scattering cross-sections are presented for various substances measured by the transmission of UCN.

TABLE 9.24 Cross-sections measured with VCN

Substance	σ_a (barn)	σ_{ie} (barn)	T (K)	v (m s^{-1})[a]	References[c]
Al	0.124 ± 0.093	0.023 ± 0.004	300	10	(a)
Au	54.1 ± 0.3	–	300	10	(a)
Cu	2.2 ± 0.06	0.12 ± 0.09	300	10	(a)
Cu	2.2 ± 0.08	0.15 ± 0.14	300	10	(b)
Cu	2.01 ± 0.10	0.081 ± 0.076	300	140	(c)
Glass[b]	0.16 ± 0.03	–	300	10	(d)
Ti	3.16 ± 0.4	–	430	10	(b)
Mo	1.5 ± 0.4	0.03 ± 0.014	470	10	(b)
W	11 ± 0.5	–	–	10	(b)
V	2.7 ± 0.4	–	–	10	(b)
Ta	11.6 ± 0.6	–	–	10	(b)
In	104 ± 5.4	–	–	140	(c)
Ag	34.6 ± 2.8	–	–	140	(c)
Ni	3.38 ± 0.16	–	–	140	(c)
Fe	2.1 ± 0.16	–	–	140	(c)
Mo	4.7 ± 0.3	–	–	140	(c)
Al	0.12 ± 0.004	0.023 ± 0.003	300	200	(e)
Ni	2.7 ± 0.1	0.15 ± 0.05	300	200	(e)
Cu	2.08 ± 0.7	0.05 ± 0.04	300	200	(e)
Zn	0.58 ± 0.03	0.07 ± 0.02	300	200	(e)
Ge	1.26 ± 0.02	–	300	200	(e)
Sn	0.30 ± 0.02	–	300	200	(e)
Pb	0.1 ± 0.02	0.06 ± 0.1	300	200	(e)

[a] The velocities at which the measurement was made are indicated.
[b] The glass composition is 71% SiO_2 + 14% Na_2O + 11% CaO, the density is 2.5 g cm^{-3}, and $v_{lim} = 4.26$ m s^{-1}.
[c] References: (a) Steyerl and Vonach (1974); (b) Antonov *et al.* (1978*c*); (c) Kashukeev *et al.* (1976); (d) *Ligeon *et al.* (1980); (e) Zhitarev *et al.* (1981).

TABLE 9.25 The constants $(p\tau)$ at room temperature

Gas	$(p\tau)$ (GPa s)	$\pm\Delta$ (GPa s)	Gas	$(p\tau)$ (GPa s)	$\pm\Delta$ (GPa s)
He	467	33	H_2O	0.36	0.05
Ar	260	6.7	CO_2	20	2.9
H_2	4.90	1.6	D_2O	1.65	0.25
N_2	11.85	1.6	Air	24.7	3.5
O_2	52.6	5.3			

Δ is the experimental uncertainty in the measured $(p\tau)$ value.

These cross-sections are extrapolated to the neutron wavelength $\lambda = 1.08$ Å.

Table 9.25 displays measured values of the constants

$$(p\tau) = k_B T / v\sigma(v)$$

for certain gases. The results are obtained in experiments on UCN storage in traps filled with gases (Kosvintsev *et al.* 1980*d*).

Bibliography on UCN

Abov, Yu.G. *et al.* (1976). Institute of Theoretical and Experimental Physics, Moscow, Preprint 44 [in Russian]. (INIS At. 9: 365712) On a magnet configuration to trap UCN.

Abov. Yu.G. *et al.* (1977). Proc. All-Union Conf. on Neutron Physics, part 1, p. 197 [in Russian]. (A magnetic bottle with estimated storage time $\tau = 100$ s and with three nodal lines is described. The addition of an axis current of 1000 A is proposed to avoid the nodal lines.)

Abov, Yu.G. *et al.* (1981*a*). Institute of Theoretical and Experimental Physics, Moscow, Preprint 16 [in Russian]. (The storage time in the magnetic bottle was estimated in the first experiment to be 141 ± 60 s.)

Abov. Yu.G. *et al.* (1981*b*). Institute of Theoretical and Experimental Physics, Moscow, Preprint 17 [in Russian]. (INIS At. 13: 670803) (The vertical UCN channel is described. The Be converter is cooled with He to 160 K. The neutron guide is of electropolished copper, 10 cm in diameter. $J_{UCN} = 200\ s^{-1}$. The admixture of over-limiting neutrons is 25%.)

Abov. Yu.G. *et al.* (1981*c*). Institute of Theoretical and Experimental Physics, Moscow, Preprint 21 [in Russian]. (The vacuum and cryogenic systems of the vertical channel are described. A vacuum of 10^{-5} Torr is achieved. The results of beryllium *G* factor measurements are given.)

Abov. Yu.G. *et al.* (1982). Institute of Theoretical and Experimental Physics, Moscow, Preprint 21 [in Russian]. (INIS At. 14: 774461) (First experiment on UCN magnetic storage with $\tau = 303 \pm 37$ s, $N(0) = 2.7 \pm 2$ neutrons).

Abov. Yu.G. *et al.* (1983*a*). *Yad. Fiz.*, **38**, 122 (*Sov. J. Nucl. Phys.*, **38**, 70). ($\tau = 303 + 37$ s in the magnetic bottle is reported.)

Abov. Yu.G. *et al.* (1983*b*). Institute of Theoretical and Experimental Physics, Moscow, Preprint 37 [in Russian]. (The results of calculations are given: (1) of UCN depolarization due to scattering on the remnant gas molecules and due to passage near the nodal lines; and (2) of UCN heating due to magnetic field fluctuations and to trembling of the magnetic walls.).

Abov. Yu.G. *et al.* (1984) Proc. All-Union Conf. on Neutron Physics, vol. 3, p. 447 [in Russian]. (INIS At. 17: 001248) Magnetic trap for UCN and neutron lifetime.

Abov. Yu.G. *et al.* (1986*a*). Institute of Theoretical and Experimental Physics, Moscow, Preprint 86–126 [in Russian]. (INIS At. 18: 091609) UCN in magnetic bottle: experiment, estimations, possibilities ($\tau > 500$ s).

The abstracts of papers which have been translated from Russian into English occur under the same volume number of Atom Index as the original paper. INIS At 14: 774461 means an abstract published in INIS Atom Index volume 14 Index 774461.

Abov. Yu.G. *et al.* (1986*b*). *Pis'ma Zh. Eksp. Teor. Fiz.*, **44**, 369 (*JETP Lett.*). Long time storage of neutrons in inhomogeneous magnetic field. (It is shown that a part $N(0) = 0.2$ of the initial UCN has $\tau > 700$ s.)

Abov. Yu.G. *et al.* (1988). *Proc. Int. Conf. on Neutron Physics*, vol. 1, p. 193 [in Russian]. (From the long time tail of the storage curve in a gravimagnetic bottle $\tau = 700$ s, $N(0) = 0.2$ are obtained for pure magnetic storage.)

Ageron, P. and Mampe, W. (1984). *J. Phys. Coll.*, **45**, C3, 279. ILL facilities for fundamental physics experiments.

Ageron, P. *et al.* (1978*a*). In *Neutron inelastic scattering.* IAEA-SM-219/58, p. 53, Vienna. (An inclined UCN channel is described. The converter is water. $\Phi_{UCN} = 300\ cm^{-2}s^{-1}$.)

Ageron, P. *et al.* (1978*b*). *Phys. Lett.*, **A66**, 469. (First experiment on UCN generation in liquid He is described. $\Phi_0 = 3.5 \times 10^8\ cm^{-2}s^{-1}$, $V = 207\ cm^3$, $N_d = 4.8\ s^{-1}$, $T_{He} = 1.15$ K.)

Ageron, P. *et al.* (1978*c*). ILL Technical Report 78AG5. Out of pile UCN source for storage experiment at ILL.

Ageron, P. *et al.* (1985). *Z. Phys.*, **B59**, 261. (The temperature dependence ($6.5\ K < T < 300$ K) of μ in a beryllium bottle is measured; a hydrogen density of $1.5 \times 10^{16}\ cm^{-2}$ on the bottle walls is obtained. The authors claim to have resolved the problem of the excess UCN losses.)

Ageron, P. *et al.* (1986). *Nucl. Instrum. Methods*, **A249**, 261. UCN storage in fluid walled bottles: a method for a neutron lifetime measurement. ($\tau_\beta = 850 \pm 60$ s is obtained in a trap with Fomblin-coated glass walls.)

Aibergenov, T.A. *et al.* (1979). *Kratk. Soobshch. Fiz.*, (7), 30. ($\mu(v/v_{lim}, \eta)$ is calculated. The possibility is discussed of measuring the potential curve measurement for the wall substance with the aid of elastic and inelastic UCN scattering.)

Akhmetov, E.Z. *et al.* (1974*a*). *At. Ehnerg.*, **37**, 35 (*Sov. J. At. Energy*). (A horizontal UCN channel with $ZrH_{1.9}H_2O$, and Mg converters is described. The Mg *G* factor and the He curve are measured.)

Akhmetov, E.Z. *et al.* (1974*b*). Joint Institute for Nuclear Research, Dubna, Preprint P3-8470 [in Russian]. (The theory and experiment on gas and ice converters are reported.)

Akhmetov, E.Z. *et al.* (1976). Proc. All-Union Conf. on Neutron Physics, part 2, pp. 150, 155 [in Russian]. (INIS At. 9: 354000, 354001) (Integral and differential spectra at the horizontal channel and the resistance of neutron guide elbows are reported. The theory and experiment on gas and ice converters are reported.)

Akhmetov, E.Z. *et al.* (1977*a*). Proc. All-Union Conf. on Neutron Physics, part 1, pp. 178, 182 [in Russian]. (INIS At. 9: 413376, 413377) (The radial horizontal UCN channel is described. The neutron guide is a cylindrical electropolished stainless steel tube of length 4 m and internal diameter 9 cm. The converter is $ZrH_{1.9}$. $\Phi_0 = 2 \times 10^{13}$ neutrons/cm^2 s, $J_{UCN} = 139$ neutrons/s. For 6 months J diminished by 20%. UCN spectra and the transmission through hydrogen at 80 and 300 K are reported.)

Akhmetov, E.Z. *et al.* (1977*b*). *Izv. Akad. Nauk Kaz. SSR*, (2), 52. (UCN transmission through Al samples measured with a 30% precision is reported.)

Akhmetov, E.Z. *et al.* (1981*a*). *Izv. Akad. Nauk. Kaz. SSR*, (4), 29. (The v dependence of σ_t for H_2, ^4He, and Ar at different temperatures is reported.)

Akhmetov, E.Z. *et al.* (1981*b*). *Izv. Akad. Nauk. Kaz. SSR*, (6), 14. (INIS At. 14: 797987) (The σ_t for N_2, Ne at 80 K and 300 K, and Xe at 300 K are reported.)

Akhmetov, E.Z. *et al.* (1984). *Izv. Akad. Nauk. Kaz. SSR*, (6), 1. (INIS At. 16: 056472) (The calculation and measurement of D_2 transmission at different pressures and temperatures are reported.)

Akunetz, A.A. *et al.* (1982). *Kratk. Soobshch. Fiz.*, (1), 25. (An experiment on UCN storage in a copper bottle with heavy ice deposited on the walls is reported. $\tau_{D_2O}/\tau_{Cu} = 1.1 \pm 0.9$.)

Alfimenkov, V.P. *et al.* (1980). Joint Institute for Nuclear Research, Dubna, Preprint P3-80-761 [in Russian]. (INIS At. 13: 655384) (The $\tau(T)$ dependence is reported for outgassed and not outgassed Cu and Be bottles in the temperature range 20–800 K. Unsuccessful attempts to get rid of overlosses by D_2O and CO_2 deposition on the bottle walls or by deuteration of the walls with the aid of D_2O, NaOD, DNO_3, and D_2SO_4 are communicated.)

Alfimenkov, V.P. *et al.* (1988). Leningrad Institute for Nuclear Physics, Preprint 1392 [in Russian]. On neutron lifetime measurement with UCN. (A project.)

Altarev, I.S. *et al.* (1976). Leningrad Institute for Nuclear Physics, Preprint 246 [in Russian]. (Investigation of different Be samples for a Be converter optimization is performed.)

Altarev, I.S. *et al.* (1977). *Proc. Workshop on Fundamental Physics with Reactor Neutrons and Neutrinos, ILL, Grenoble*, p. 179.

Altarev, I.S. *et al.* (1979). *Pis'ma Zh. Eksp. Teor. Fiz.*, **29**, 794 (*JETP Lett.*). (The EDM limit of 1.6×10^{-24} e cm is achieved with 90% confidence in a flow experiment with a two chamber spectrometer and double polarization analysis. The parameters of the installation are: $\tau = 5$ s, $E = 25\ \mathrm{kV\,cm^{-1}}$, $H_0 = 0.028$ Oe, $\Delta H(r) = 2 \times 10^{-5}$ Oe, $\Delta H(t) = 10^{-6}$ Oe. The neutron guide is an electropolished stainless steel tube. The Be converter is cooled with He to 30 K. $\Phi_0 = 2 \times 10^{14}$ neutrons/cm^{2}s, $J_{UCN} = 10^4$. During the year J_{UCN} diminished 2–3 times, hypothetically due to ^{6}Li accumulation in the converter.)

Altarev, I.S. *et al.* (1980*a*). *Nucl. Phys.*, **A341**, 269. (The upper EDM limit of 1.6×10^{-24} e cm is achieved.)

Altarev, I.S. *et al.* (1980*b*). *Phys. Lett.*, **A80**, 413. (The T_c dependence of the UCN yield from H_2 and D_2 converters is reported.)

Altarev, I.S. *et al.* (1981). *Phys. Lett.*, **B102**, 13. (The upper EDM limit is lowered to 6×10^{-25}e cm.)

Altarev, I.S. *et al.* (1982). *Neutron-capture gamma-ray spectroscopy, Inst Phys. Conf. Ser. No.* 62, p. 633. (The upper EDM limit is lowered to 4.5×10^{-25} e cm.)

Altarev, I.S. *et al.* (1985). Leningrad Institute for Nuclear Physics, Preprint 1117 [in Russian]. Stabilisation method of the magnetic-resonance spectrometer.

Altarev, I.S. *et al.* (1986*a*). *Pis'ma Zh. Eksp. Teor. Fiz.*, **44**, 269 (*JETP Lett.*, **44**, 344). Universal liquid-hydrogen source of polarized cold and ultracold neutrons on the LNPI reactor VVR-M.

Altarev, I.S. *et al.* (1986*b*). *Pis'ma Zh. Eksp. Teor. Fiz.*, **44**, 360 (*JETP Lett.*) Search for neutron EDM ($d \leqslant 2.6 \times 10^{-25}$ at 95% confidence level).

Altarev, I.S. *et al.* (1988*a*). *Proc. Int. Conf. on Neutron Physics*, vol. 1, p. 50 [in Russian]. (An upper EDM limit of 2.6×10^{-25}e cm at a 95% confidence level is obtained in a storage type UCN experiment with exposition time $t_{exp} = 50$ s.)

Altarev, I.S. *et al.* (1988*b*). *Proc. Int. Conf. an Neutron Physics*, vol. 1, p. 180 [in Russian]. Universal liquid hydrogen source of polarized cold neutrons and UCN at VVR-M reactor ($\Phi_{UCN} = 6 \times 10^3$ neutrons/cm^{2}s, $J_{UCN} = 5 \times 10^5$ neutrons/s).

Anikolenko, V.A. *et al.* (1973*a*). *Kratk. Soobshch. Fiz.*, (9), 13. (A proportional UCN counter with a ^3He radiator and semiconductor detector is described. Its efficiency is estimated to be $\theta \simeq 13\%$.)

Anikolenko, V.A. *et al.* (1973*b*). *Kratk. Soobshch. Fiz.*, (11), 40. (An experiment on UCN storage in a Be bottle at a stationary reactor is reported. $\tau = 20$–50 s.)

Anikolenko, V.A. *et al.* (1974). *Prib. Tekh. Eksp.*, (5), 117 (*Instrum. Exp. Tech.*). (Neutron detection in the presence of the intensive β–γ background.)

Anton, F. *et al.* (1989). *Nucl. Intrum. Methods,* **A284** (1), 101. Measurement of neutron lifetime by magnetic storage of free neutrons.

Antonov, A.V. *et al.* (1969*a*). *Pis'ma Zh. Eksp. Teor. Fiz.*, **9**, 307 (*JETP Lett.*, **9**, 180). (Theory of UCN generation by the reflection of more fast neutrons from runaway mirror is presented.)

Antonov, A.V. *et al.* (1969*b*). *Pis'ma Zh. Eksp. Teor. Fiz.*, **10**, 380 (*JETP Lett.*, **10**, 241). (Theory of UCN accumulation near a stationary reactor is presented.)

Antonov, A.V. *et al.* (1970) *Kratk. Soobshch. Fiz.*, (1), 3. (Theory of UCN accumulation near a stationary reactor is presented.)

Antonov, A.V. *et al.* (1972). Trudy Lebedev Institute in Moscow, vol. 57, p. 270 [in Russian]. (Summary of preceeding works.)

Antonov, A.V. *et al.* (1973). Lebedev Institute in Moscow, Preprint 86 [in Russian]. (Review of UCN counters.)

Antonov, A.V. *et al.* (1974*a*). *Pis'ma Zh. Eksp. Teor. Fiz.*, **20**, 632 (*JETP Lett.*, **20**, 289). (The calculation of multilayer UCN filter.)

Antonov, A.V. *et al.* (1974*b*). *Kratk. Soobshch. Fiz.*, (10), 14. (UCN track detector with U–Ti radiator. $\varepsilon \simeq 60\%$.)

Antonov, A.V. *et al.* (1974*c*). *Kratk. Soobshch. Fiz.*, (11), 11. (A technology of U–Ti radiator preparation.)

Antonov, A.V. *et al.* (1974*d*). *Kratk. Soobshch. Fiz.*, (11), 17. (A UCN detector with a U–Ti radiator and Xe scintillator is described. $\varepsilon \simeq 40\%$.)

Antonov, A.V. *et al.* (1974*e*). *Kratk. Soobshch. Fiz.*, (12), 30. (A UCN detector with a U–Ti radiator and ZnS(Ag) scintillator is described. $\varepsilon \simeq 10$–15%.)

Antonov, A.V. *et al.* (1974*f*). Lebedev Institute in Moscow, Preprint 43 [in Russian]. (Calculation of multilayer systems is presented.)

Antonov, A.V. *et al.* (1976*a*). *Pis'ma Zh. Eksp. Teor. Fiz.*, **24**, 387 (*JETP Lett.*, **24**, 352). (Experiments on UCN accumulation in Cu bottle near aperiodic pulsed IIN reactor are described. $\tau \simeq 45$–50 s.)

Antonov, A.V. *et al.* (1976*b*). *Kratk. Soobshch. Fiz.*, (10), 14. (Numerical calculation of UCN transmission through a vibrating plate is presented.)

Antonov, A.V. *et al.* (1976*c*). Proc. All-Union Conf. on Neutron Physics, part 2, p. 161 [in Russian]. (INIS At. 9: 355493) (Summary of the works of the group for the preceding period.)

Antonov, A.V. *et al.* (1977*a*). *Kratk. Soobshch. Fiz.*, (10), 10. (The time-of-flight spectrometer at the vertical neutron channel of the MIFI reactor is described. Energy range $E = (10^{-8}$–$10^{-4})$ eV, $\Delta E/E = 16\%$, $J = 20$ neutrons/s.)

Antonov, A.V. *et al.* (1977*b*). Trudy Lebedev Institute in Moscow, vol. 94, p. 73 [in Russian]. (Summary of the works of the group during 1972–1976.)

Antonov. A.V. *et al.* (1977*c*). Proc. All-Union Conf. on Neutron Physics, part 1, p. 187 [in Russian]. (INIS At. 9: 413379) (Review of the works of the group during 1976–1977.)

Antonov, A.V. *et al.* (1978*a*). *Kratk. Soobshch. Fiz.*, (9), 43. (The results of measurements of σ_t for Ti, Mo, W, V are presented. The results extrapolated to the thermal point are in agreement with known data. In an unannealed V sample σ_t grows as $v \to 0$ proportionally to $v^{-2.8}$.)

Antonov, A.V. *et al.* (1978*b*). *Kratk. Soobshch. Fiz.*, (10), 3. (The loss coefficient $\eta = 8 \times 10^{-4}$ is extracted from experiments (Antonov *et al.* 1976*a*) taking account of gravity.)

Antonov, A.V. *et al.* (1978*c*). *Kratk. Soobshch. Fiz.*, (11), 13. (The experiments on measurements of σ_a and σ_{ie} for Ta and Cu are reported. The results are not in contradiction with known data.)

Antonov, A.V. *et al.* (1979*a*). *Kratk. Soobshch. Fiz.*, (2), 33. (Numerical Monte Carlo calculations show that overlimiting neutrons live in a bottle approximately 2 s.)

Antonov, A.V. *et al.* (1979*b*). *Kratk. Soobshch. Fiz.*, (6), 41. (The relation $\mu(v) \propto v$ is proved for thin absorbing film on a wall surface.)

Antonov, A.V. *et al.* (1979*c*). *Kratk. Soobshch. Fiz.*, (10), 3. (Some softening of the UCN spectrum in the bottle of the experiment (Antonov *et al.* 1976*a*) is reported. This conclusion is determined from the character of the variation of the ratio of two detector counts with V foil in front of one of them.)

Antonov, A.V. *et al.* (1980*a*). *Kratk. Soobshch. Fiz.*, (8), 36. (Investigation of cross-sections and inhomogenieties in ^{63}Cu and V are reported.)

Antonov, A.V. *et al.* (1980*b*). Proc. All-Union Conf. on Neutron Physics, part 1, pp. 18, 125 [in Russian.] (INIS At. 13: 643772) (σ_t and σ_{ie} are reported for Cu, ^{65}Cu, W, Mo, Ti, V, Be, Ta, Nb, Pd.)

Antonov, A.V. *et al.* (1980*c*). Trudy Lebedev Institute in Moscow, vol. 127, p. 6 [in Russian]. (Summary of the works of the group during 1977–1979.)

Antonov, A.V. *et al.* (1981*a*). *Kratk. Soobshch. Fiz.*, (11), 51. (Numerical calculation of a supermirror.)

Antonov, A.V. *et al.* (1981*b*). Lebedev Institute in Moscow, Preprint 260 [in Russian]. (Monte Carlo simulation of the UCN isotropisation in a bottle in presence of the gravity field.)

Antonov, A.V. *et al.* (1982*a*). *Kratk. Soobshch. Fiz.*, (1), 41. (A method to measure τ_β of the neutron is proposed. It consists of the measurement of UCN storage times, τ, in bottles at different temperatures and of the extrapolation $\mu \to 0$.)

Antonov, A.V. *et al.* (1982*b*). *Kratk. Soobshch. Fiz.*, (8), 41. (Monte Carlo simulation of the UCN emptying curve shows it to become non-exponential with the enlarging of the exit opening.)

Antonov, A.V. *et al.* (1983*a*). *Kratk. Soobshch. Fiz.*, (1), 27. (Numerical simulation of the UCN exit from a bottle taking account of the gravity field is presented.)

Antonov, A.V. *et al.* (1983*b*). *Kratk. Soobshch. Fiz.*, (5), 40. (The spin–orbit part of the UCN potential is estimated to be 10^{-13} eV.)

Antonov, A.V. *et al.* (1984*a*). *Fiz. Tverd. Tela*, **26**, 1585. (From the energy dependence of the UCN transmission through V and Be samples the dimensions and distribution of inhomogeneities are extracted.)

Antonov, A.V. *et al.* (1984*b*). *Kratk. Soobshch. Fiz.*, (11), 28. (INIS At. 17: 076013) (Monte Carlo simulation of the angular distribution of UCN in a bottle showing a 1% anisotropy due to losses in the walls is presented.)

Antonov, A.V. *et al.* (1984*c*). Proc. All-Union Conf. on Neutron Physics, vol 4, pp. 22, 40, 48 [in Russian]. (INIS At. 16: 079630) (VCN transmission through polymers $(-CH_2-)_n$ and $(-CD_2-)_n$ at 293 and 106 K is reported. The σ_a extrapolated to the thermal point is shown to be significantly larger than the known values. UCN reflection from ^{58}Ni^{62}Ni supermirror is measured. Monte Carlo results for the UCN exit from bottles are reported.)

Antonov, A.V. *et al.* (1985*a*). *Pis'ma Zh. Eksp. Teor. Fiz.*, **41**, 210 (*JETP Lett.*, **41**, 256). (UCN transmission through polymer samples for different energies and at different temperatures is reported.)

Antonov, A.V. *et al.* (1985*b*). *Kratk. Soobshch. Fiz.*, (3), 46. (INIS At. 18: 007961) (UCN transmission through ^{58}Ni samples placed in the magnetic field is measured. The σ_t extrapolated to the thermal point is shown to coincide with the known value.)

Antonov, A.V. *et al.* (1985*c*). *Kratk. Soobshch. Fiz.*, (3), 50. (INIS At. 18: 007773) (UCN transmission through quartz glass and Pb samples is measured. The σ_t extrapolated to the thermal point are shown to be larger than the known values.)

Antonov, A.V. *et al.* (1985*d*). *Kratk. Soobshch. Fiz.*, (9), 18.

Antonov, A.V. *et al.* (1985*e*). *Kratk. Soobshch. Fiz.*, (10), 48. (Theory of VCN scattering on inhomogeneities having the shapes of a sphere, cylinder, and disc.)

Antonov, A.V. *et al.* (1986*a*). *Pis'ma Zh. Eksp. Teor. Fiz.*, **44**, 347 (*JETP Lett.*). Cold neutron scattering on the rough interface.

Antonov, A.V. *et al.* (1986*b*). *Zh. Tekh. Fiz.*, **56**, 1241 (see *Sov. Phys.–Tech. Phys.*) (Multilayer systems of Be–Ti bilayers.)

Antonov, A.V. *et al.* (1986*c*). *Zh. Tekh. Fiz.*, **56**, 1588 (*Sov. Phys.–Tech. Phys.*). Neutron energy bands in multilayer interference systems.

Antonov, A.V. *et al.* (1986*d*). *At. Ehnerg.*, **61**, 206 (*Sov. J. At. Energy*, **61**, 738). Investigation of UCN interaction with natural uranium and its U-235 isotope.

Antonov, A.V. *et al.* (1986*e*). *Kratk. Soobshch. Fiz.*, (1), 6. VCN interaction with SiB_x and Nb.

Antonov, A.V. *et al.* (1986*f*). *Kratk. Soobshch. Fiz.*, (2), 15. Total cross-sections of VCN interaction with Pd and PdH_x.

Antonov, A.V. *et al.* (1986*g*). *Kratk. Soobshch. Fiz.*, (4), 15. Method of determination of inhomogeneities' dimensions distribution with the help of VCN.

Antonov, A.V. *et al.* (1986*h*). *Kratk. Soobshch. Fiz.*, (11), 44. Experimental investigation of multilayer interference neutron monochromators.

Antonov, A.V. *et al.* (1986*i*). Lebedev Institute in Moscow, Preprint 273 [in Russian].

Antonov, A.V. *et al.* (1987). *Pis'ma Zh. Eksp. Teor. Fiz.*, **46**, 241 (*JETP Lett.*). Multilayer interference structures of strongly absorbing substances.

Antonov, A.V. *et al.* (1988). *Proc. Int. Conf. on Neutron Physics*, vol. 1, pp. 235, 280, 285, 291 [in Russian]. (The v dependence of total cross-sections of VCN interaction with U, ^{235}U, Pb, SiO_2, SiB_x, PdH_x is reported. Deviations from the $1/v$ law are used to estimate the parameters of inhomogeneities. The excess over the $1/v$ cross-section is interpreted to be due to inelastic scattering. The theory of extraction of size form and distribution of inhomogeneities from experimental data on VCN transmission is presented. Experiments on VCN scattering from magnetic and non-magnetic surface roughnesses and inhomogeneities are reported. The transmission of multilayer systems is studied.)

Arzumanov, S.S. *et al.* (1984*a*). Moscow Kurchatov Institute for Atomic Energy, Preprint 4216/14 [in Russian]; Proc. All-Union Conf. on Neutron Physics, vol. 4, p. 135 [in Russian]. Bent double-channel for VCN on reactor IR-8.

Arzumanov, S.S. *et al.* (1984*b*). Proc. All-Union Conf. on Neutron Physics, vol. 4, p. 140 [in Russian]. (INIS At. 16: 079631); *Pis'ma Zh. Eksp. Teor. Fiz.*, **39**, 486 (JETP Lett., **39**, 590). (The neutron image is obtained using two concave and two plane mirrors. The magnification is 1.375 and 0.725.)

Arzumanov, S.S. *et al.* (1986*a*). *Pis'ma Zh. Eksp. Teor. Fiz.*, **44**, 213 (*JETP Lett.*) Image registration in the neutron microscope (magnification 1.4, resolution 20.2 μm).

Arzumanov, S.S. *et al.* (1986*b*). *Vopr. Nauki Tekh.: Obshch. Yad. Fiz.* **3**, 70. Position sensitive detector for UCN.

Averyanov, P.G. *et al.* (1977). Proc. All-Union Conf. on Neutron Physics, part 1, p. 167 [in Russian]. (INIS At. 9: 413380) (UCN storage in a small magnetic bottle is reported. $\tau = 35 \pm 10$ s, $N(0) = 1$.)

Barabanov, A.L. (1988). *Proc. Int. Conf. on Neutron Physics*, vol. 1, p. 112 [in Russian]. About possible time parity non-conservation in UCN interaction with solids. (UCN transmission through a three-layer system and scattering on a spherical nut-shell potential are considered. No significant effect is found.)

Bates, J.C. (1978). *Nucl. Instrum. Methods*, **150**, 261. Measurement of absorption probabilities for UCN incident on strong absorbers. (It is shown that UCN can diffuse in an inhomogeneous radiator and be counted even if their energy is less than E_{lim} of the radiator.)

Bates, J.C. (1981). *Phys. Lett.*, **A83**, 29. Neutron radiography with UCN.

Bates, J.C. (1982). *Phys. Lett.*, **A88**, 427. UCN storage in a fluid-walled bottle ($\mu = 10^{-4}$), covered with oil $(F_3CCF_2OCF_2CF_5)_n$.

Bates, J.C. (1983). *Nucl. Instrum. Methods*, **216**, 535. (The technology of oil spreading over the bottle walls is described.)

Baumann, J. *et al.* (1988). *Physica*, B&C, **151**, 130. A neutron optical imaging system for the determination of an upper limit of the neutron charge. ($q = (0.4 \pm 1.1) \times 10^{-21}e$.)

Belomytsev, O.B. *et al.* (1988). Leningrad Institute for Nuclear Physics, Preprint 1391 [in Russian]. On measurement of the correlation coefficients in neutron β-decay with UCN.

Belonogov, A.V. *et al.* (1977). *Proc. 7th Int. Vacuum Congr. and the 3rd Int. Conf. on Solid Surfaces, Sept.* 1977, vol. 3, p. 2667. Österreich Studiengeselschaft für Atomenergie, Vienna. Cryosorption pump for UCN source vacuum system oilless pumping.

Berceanu, I. and Ignatovich, V.K. (1973). *Vacuum*, **23**, 441. (Monte Carlo calculation of the transmission of straight tubes without taking account of UCN absorption in the walls is presented.).

Bernstein, H.J. and Zeilinger, A. (1980). *Phys. Lett.*, **A75**, 169. (An experiment is proposed for observing the phase shift due to pure spin precession. In such an experiment one needs to switch on the magnetic field for a short time in one of the paths of the UCN interferometer.)

Binder, K. (1970). PTHM-FRM-110. Wirkungsgerschnitte für die Streung von 'ultrakalten' neutronen.

Blokhintsev, D.I. and Plakida, N.M. (1976). Joint Institute for Nuclear Research, Dubna, Preprint P4-9631 [in Russian]. (INIS At. 8: 329620) (The theory of quasi-elastic UCN scattering from walls without account of the total reflection phenomen is presented.)

Blokhintsev, D.I. and Plakida, N.M. (1977). *Phys. Status Solidi*, (b) **82**, 627. (The theory of inelastic UCN scattering from walls without account of the total reflection phenomen is presented; the effect of hydrogen is mentioned.)

Bokun, R.Ch. (1984). *Yad. Fiz.*, **40**, 287 (*Sov. J. Nucl. Phys.*, **40**, 180). (An experiment on UCN storage in the vortex of rotated ^{4}He for measuring τ_β is discussed.)

Borisov, S.V. *et al.* (1985). *Zh. Tekh. Fiz.*, **55**, 409 (*Sov. Phys.–Tech. Phys.*) Neutron scattering by bubble vibrations.

Borisov, Yu.V. *et al.* (1987*a*) Leningrad Institute for Nuclear Physics, Preprint 1249 [in Russian]. Investigation of UCN application for neutron charge measurement.

Borisov, Yu.V. *et al.* (1987*b*). Joint Institute for Nuclear Research, Dubna, Rapid Communications No. 1[21]-87, p. 40 [in Russian]. Neutron electric charge measurement with UCN. (The value $(4.3 \pm 7.1) \times 10^{-20}e$ is reported.)

Borisov, Yu.V. *et al.* (1988). *Proc. Int. Conf. on Neutron Physics*, vol. 1, p. 244 [in Russian]. (A report on the work (Borisov *et al.* 1987*b*) is given.)

Brizhik, L.S. (1984). *Usp. Fiz. Zh.*, **29**, 492 (*Ukr. Phys. J.*) (Theory of UCN scattering on Davydov solitons.)

Brown, M., Golub, R., and Pendlebury, J. (1975). *Vacuum*, **25**, 61. (A Monte Carlo calculation of UCN transmission through straight tubes with the reflection indicatrix, corresponding to small roughnesses, is reported.)

Brun, T.O. *et al.* (1980). Dept of Physics and Astronomy, Maryland University, Preprint ORO-2504-309; *Phys. Lett.*, **A75**, 223. (Experiments on UCN generation with the runaway crystal are described.)

Bugeat, J.P. and Ligeon, E. (1979). *Nucl. Instrum. Methods*, **159**, 117. Influence of ion bombardment in hydrogen surface layer analysis.

Bugeat, J.P. and Mampe, W. (1979). *Z. Phys.*, **B35**, 273. Reduction of surface hydrogen on UCN container materials by ion bombardment. (Cleaning of Cu, Ni, and stainless steel with the aid of ^{11}B ions and measurement of the hydrogen content on the surfaces and its variation in time are reported.)

Byrne, J. (1979). *Nucl. Instrum. Methods*, **167**, 355. Determination of the efficiency of a neutron polarizer–Analyser combination by the spin-flip method.

Casella, R.C. (1984). *Phys. Rev. Lett.*, **53**, 1033. (It is shown that experiments on $N\bar{N}$ oscillations with phase shift measurement in a magnetic field are less sensitive than experiments involving direct detection of an antineutron.)

Chetyrkin, K.G. *et al.* (1981). *Phys. Lett.*, **B99**, 358. (It is shown that UCN cannot compete with thermal neutrons in $N\bar{N}$ oscillations measurements.)

Deslattes, R.D. (1986). In *The investigation of fundamental interactions with cold neutrons*. NBS Spec. Publ. 711, p. 118. (The application of neutrons for general relativity and quantum gravity investigations is discussed.)

Dombeck, T.W. (1977). Dept. of Physics and Astronomy, Maryland University, Preprint ORO-2504-267. (INIS At. 9:379489) (A UCN bottle for the Argonne pulsed neutron source is proposed.)

Dombeck, T.W. (1981). *Proc. ICANS-IV, KEK, Tsukuba*, KENS Report 11, (ed. Ishikawa, Y. *et al.*), p. 723. National Laboratory for High Energy Physics. (UCN time-of-flight spectrometer with moving crystal is described. The results of experiments of UCN transmission through magnetic and non-magnetic foils and three-layer systems are reported.)

Dombeck, T.W. and Lynn, J.W. (1979). Dept. of Physics and Astronomy, Maryland University, Preprint ORO-2504-291. (INIS At. 10: 489997) (Calculations of UCN production with moving crystal are presented.)

Dombeck, T.W. *et al.* (1979). *Nucl. Instrum. Methods*, **165**, 139. (Calculation of the intensity of the UCN source with a runaway crystal is presented.)

Drabkin, G.M. and Zhitnikov, P.A. (1960). *Zh. Eksp. Teor. Fiz.*, **38**, 1013 (Sov. Phys.–JETP, **38**). (A proposal to slow neutrons with the help of two fields: one periodic in space, the other in time, is proposed. The possibility of UCN production with the help of a travelling wave is mentioned.)

Dvoretzkii, V.N. (1976*a*). Moscow Kurchatov Institute for Atomic Energy, Preprint 2715 [in Russian]. (A differential gravity spectrometer and UCN spectra measured with it are discussed. A hypothesis on the spin-$\frac{3}{2}$ neutron state is advanced.)

Dvoretskii, V.N. (1976*b*). Moscow Kurchatov Institute for Atomic Energy, Preprint

3128 [in Russian]. (A UCN monochromatization method with the help of a 'shuffling' foil is described.)

Dzyublik, A.Ya. and Grigorchuk, N.I. (1983). *Zh. Teck. Fiz.*, **53**, 1167 (*Sov. Phys.–Tech. Phys.*). (Theory of UCN diffraction on a superlattice of pores.)

Ebisawa, T. *et al.* (1979). *J. Nucl. Sci. Technol.*, **16**, 647. Reflectivities of Ni–Mn and Ni–Ti multilayers for monochromators and supermirrors.

Ebisawa, T. *et al.* (1981). *Annu. Rep. Res. Reactor Inst., Kyoto Univ.*, **14**, 10. Neutron–optical properties of multilayer systems with absorption.

Efimov, V.N. and Ignatovich, V.K. (1974). Joint Institute for Nuclear Research, Dubna, Preprint P4-8253 [in Russian]. (INIS At. 9:35403) (The theory of UCN depolarization due to magnetic field inhomogeneities is presented.)

Egorov, A.I. *et al.* (1974). *Yad. Fiz.*, **19**, 300 (*Sov. J. Nucl. Phys.*, **19**, 147). (The horizontal and vertical channels at the VVR-M reactor are described. Al, Mg converters are studied. At the exit of the vertical channel with a Be converter at 40 K and an Ni neutron guide the UCN flux $J_{UCN} = 150$ neutrons/s is achieved. The storage times τ for glass and Ni bottles are measured. No temperature dependence of τ in the case of Ni is found. A UCN polarization of 78% is achieved after UCN transmission through a magnetized foil.)

Egorov, A.I. *et al.* (1975). *Yad. Fiz.*, **21**, 292 (*Sov. J. Nucl. Phys.*, **21**, 153). (Adiabatic spin rotation and the resonance curve with the width 0.08 Hz are described.)

Egorov, A.I. *et al.* (1976). Proc. All-Union Conf. on Neutron Physics, part 2, p. 134 [in Russian]. (INIS At. 9:355525). (First result on EDM: $d = (0.8 \pm 2.6) \times 10^{-23}$ e cm is reported.)

Engelmann, G. *et al.* (1979). *Z. Phys.*, **A35**, 345. (Comparison of small-angle scattering and long-wave neutron transmission methods for investigation of the AlZn sample macroscopic structure is performed.)

Ezhov, V.F. *et al.* (1976). *Pis'ma Zh. Eksp. Teor. Fiz.*, **24**, 39 (*JETP Lett.*). (Adiabatic UCN spin rotation and resonance curve in EDM experiment are reported.)

Felber, J. *et al.* (1988). *Physica*, **B&C 151**, 135. Test of the time-dependent Schrödinger equation with very slow neutrons.

Foldy, L.L. (1966). In *Preludes in theoretical physics*, (ed. A. de Shalit *et al.*), p. 205. North-Holland, Amsterdam. (The possibility of UCN storage in a bottle with He-II walls at 10^{-4} K is discussed.)

Frank, A.I. (1979). Moscow Kurchatov Institute for Atomic Energy, Preprint 3203 [in Russian]. (INIS At. 12:574808). Hamilton analogy and UCN optics.

Frank, A.I. (1980). Proc. All-Union Conf. on Neutron Physics, part 1, p. 150 [in Russian]. (INIS At. 13:643776) (Review on neutron microscope.)

Frank, A.I. (1981*a*). Moscow Kurchatov Institute for Atomic Energy, Preprint 3409/14 [in Russian]. (INIS At. 13:688331). (Optical image production with UCN is discussed.)

Frank, A.I. (1981*b*). *Priroda*, (1), 30. (INIS At. 14:7979890) (Review on UCN.)

Frank, A.I. (1982). *Usp. Fiz. Nauk*, **137**, 5 (*Sov. Phys.–Usp.*). Fundamental properties of the neutron.

Frank, A.I. (1983). *Zh. Tekh. Fiz.*, **53**, 935 (*Sov. Phys.–Tech. Phys.*). About a neutron interferometer placed in a force field. (The sensitivity of neutron-charge experiment with UCN is estimated to be $q_n \simeq 3 \times 10^{-21}$ *e*/day.)

Frank, A.I. (1986). *V. Int. School on Neutron Physics, Alushta*, Joint Institute for Nuclear Research, Dubna, Preprint D3, 4, 17-86-747, p. 192 [in Russian]. (INIS At. 19:010937) (Review on neutron microscope.)

Frank, A.I. (1987). *Usp. Fiz. Nauk*, **151**, 229 (*Sov. Phys.–Usp.*) UCN optics and the neutron microscope problems.

Frank, A.I. (1989*a*). *At. Energ.*, **66**, 93 (*Sov J. At. Energy*). Neutron microscopy with UCN.

Frank, A.I. (1989*b*). Neutron microscopy with UCN [in Russian]. In: *Some Problems of Modern Nuclear Physics* (In honour of I.M. Frank) Nauka, M., p. 213.

Frank, A.I. (1989*c*). *Nucl. Instrum. Methods* **A284** (1), 161. Optics of very slow neutrons and neutron microscopy.

Frank, I.M. (1972). *Priroda*, (9), 24. (Neutron optics.)

Frank, I.M. (1974*a*). Joint Institute for Nuclear Research, Dubna, Preprint P3-7809 [in Russian]; Brookhaven National Laboratory, tr-582 (trans. S.J. Amoretti); in *II Int. School on Neutron Physics*, Joint Institute for Nuclear Research, Dubna, Preprint D3-7991, p. 19 [in Russian]. (INIS At. 7:277192) (UCN optics in terms of the refraction coefficient is presented. A hypothesis of an effective field is advanced.)

Frank, I.M. (1974*b*). Joint Institute for Nuclear Research, Dubna, Preprint P3-7810 [in Russian]; Brookhaven National Laboratory, tr-567 (trans. S.J. Amoretti). (UCN absorption in terms of the imaginary part of a refraction coefficient is presented. Hypothesis of quasi-elastic scattering and anomalous absorption characteristic only for UCN is advanced.)

Frank, I.M. (1975). Joint Institute for Nuclear Research, Dubna, Preprint P4-8851 [in Russian]; Oak Ridge National Laboratory, tr-4129 (trans. M. Gerrard). (UCN scattering on the Rayleigh waves is discussed.)

Frank, I.M. (1976). Joint Institute for Nuclear Research, Dubna, Preprint P3-9846 [in Russian]. (The theory of UCN diffusion along neutron guides is presented.)

Frank, I.M. (1978). In *Nuclear particles interaction with the solid state*, p. 3, MIFI, Moscow [in Russian]. (UCN review.)

Frank, I.M. (1979). Joint Institute for Nuclear Research, Dubna, Preprint P3-12829 [in Russian]. (The flow experiment performed in Grenoble in search of the temperature dependence of μ is reported.)

Frank, I.M. and Frank, A.I. (1978). *Pis'ma Zh. Eksp. Teor. Fiz.*, **28**, 559 (*JETP Lett.*), in *III Int. School on Neutron Physics*, Joint Institute for Nuclear Research, Dubna, Preprint D3-11-787, p. 246 [in Russian]. (Fermi principle in UCN optics.)

Gähler, R. and Golub, R. (1984). *Z. Phys.* **B56**, 5; *J. Phys. Coll.* **45**, C3, 229. Time-dependent neutron optics—quantum mechanical effects on beam chopping and a new type of high resolution neutron spectrometer (FOTOF).

Gähler, R. *et al.* (1981). *Phys. Rev.*, **A23**, 1611. Neutron optical tests of non-linear wave mechanics. ($-b\ln|\psi|^2$, $b < 3.3 \times 10^{-15}$ eV, a lateral shift of the geometrical shadow in the Fresnel diffraction pattern is observed, $\lambda = 20$ Å.)

Gähler, R. *et al.* (1982). *Phys. Rev.*, **D25**, 2887. Experimental limit for the charge of the free neutron. ($q_n < (-1.5\text{–}2.1) \times 10^{-20}\, e$, $\lambda = 20$ Å.)

Galitskii, V.M. and Satarov, L.M. (1976). Moscow Kurchatov Institute for Atomic Energy, Preprint 2657 [in Russian]. (INIS At. 8:533396) (A theory of UCN reflection from a surface with large roughnesses is presented.)

Gerasimov, A.S. and Kazarnovskij, M.V. (1976). *Zh. Eksp. Teor. Fiz.*, **71**, 1700 (*Sov. Phys.–JETP*, **44**, 892). (Some peculiarities of UCN observation and spectrometry in the case of sudden appearance and disappearance of potential walls and upon reflection from a vibrating wall are discussed.)

Gerasimov, A.S. *et al.* (1973). *Kratk. Soobshch. Fiz.*, (8), 56. On the heating of UCN by sound vibrations of the trap walls.

Golikov, V.V. and Taran, Yu.V. (1975). *Prib. Tekh. Eksp.*, (1), 41 (*Instrum. Exp. Tech.*, **18**, 36). (A UCN chopper with a high $E_{\lim}$ foil placed in a high magnetic field is discussed.)

Golikov, V.V. *et al.* (1973). *Zh. Eksp. Teor. Fiz.*, **64**, 73 (*Sov. Phys.–JETP*, **37**, 41). (*G* factor calculations for polyethylene, Al, Mg, ZrH_2, Be is presented. The temperature dependence of the polyethylene *G* factor is reported.)

Golikov, V.V. *et al.* (1977). In *Neutron inelastic scattering*, p. 39. IAEA, Vienna. (Review on UCN generation and storage experiments.)

Golikov, V.V. *et al.* (1985*a*). Joint Institute for Nuclear Research, Dubna, Preprint P3-85-285 [in Russian]. (The horizontal UCN channel at the IBR-2 reactor is described. The neutron guide is an electropolished stainless steel of length 9.6 m and diameter 187 mm. The converter is $(-CH_2-)_n$, $\Phi_0 = 10^{11}$ neutrons/cm^2 s, $J_{UCN} = 1.5$ neutrons/s.)

Golikov, V.V. *et al.* (1985*b*). Joint Institute for Nuclear Research, Dubna, Preprint P3-85-286 [in Russian]. (INIS At. 17:005460) (Methods for measuring integral UCN reflection coefficients are described.)

Golikov, V.V. *et al.* (1985*c*). Joint Institute for Nuclear Research, Dubna, Rapid Communications, N. 9–85, p. 26 [in Russian]. (INIS At. 17:020735) (The results of measurements of integral UCN reflection coefficient from Cd and ^{113}Cd are reported.)

Golikov, V.V. *et al.* (1986). Joint Institute for Nuclear Research, Dubna, Preprint P3-86-767 [in Russian]. (INIS At. 18:092628) UCN reflection from substances with high absorption cross-section.

Golikov, V.V. *et al.* (1987). Joint Institute for Nuclear Research, Dubna, Preprint P3-87-20 [in Russian]. (INIS At. 19:039027) UCN reflection from powder samples.

Golikov, V.V. *et al.* (1988*a*). Joint Institute for Nuclear Research, Dubna, Communication P3-88-48 [in Russian]. UCN diffusion in branched guides.

Golikov, V.V. *et al.* (1989*b*). *Proc. Int. Conf. on Neutron Physics*, vol. 1, p. 274 [in Russian]. (The work of Golikov *et al.* (1987) is reported. σ_1 for Cu, CuO, and graphite powders after annealing was found to be in agreement with theory.)

Golikov, V.V. *et al.* (1989). Joint Institute for Nuclear Research, Dubna, Communication. P3-89-42 [in Russian]. On the measurement of the UCN loss factor in beryllium. (The results of the experiments with beryllium powder outgassed at temperatures higher than 400 K are reported to be in agreement with theory.)

Golub, R. (1972). *Phys. Lett.*, **A38**, 177. (Hydrogenous UCN converters in Debye approximation are calculated.)

Golub, R. (1978*a*). ILL Preprint 78G0114S. (UCN depolarization and deformation of UCN orbits in toroidal trap with account of centrifugal and gravity forces are discussed.)

Golub, R. (1978*b*). In *Fundamental physics with reactor neutrons and neutrinos, Inst. Phys. Conf Ser. No.* 42, p. 104. (A possibility for searching for the neutron gravity dipole moment is discussed.)

Golub, R. (1979*a*). *Phys. Lett.*, **A72**, 387. On the storage of neutrons in superfluid ^{4}He. (τ is calculated for two-phonon and roton scattering.)

Golub, R. (1979*b*). ILL Technical Report 79G025. Lifetime of a neutron at rest in pure superfluid ^{4}He.

Golub, R. (1983). *J. Phys. Lett.*, **4**, L-321. (The application of a superthermal UCN source for EDM experiments is discussed.)

Golub, R. (1984*a*). *J. Phys. Coll.*, **45**, C3, 265. (A possibility of nuclear reaction involving polarized UCN and utilization of polarized ^{3}He nuclei for neutron EDM experiments is discussed.)

Golub, R. (1984*b*). *Nucl. Instrum. Methods*, **A226**, 558. (A UCN review and a discussion of the perspectives of UCN utilization at the TRIGA reactor are presented.)

Golub, R. (1986). *Zh. Tekh. Fiz.*, **56**, 1592 (*Sov. Phys.–Tech. Phys.*). Model of high-voltage system for search of the neutron EDM with UCN.

Golub, R. (1988). *Proc. Int. Conf. on Neutron Physics*, vol. **1**, p. 190 [in Russian]. ICNS' *88 July 1988, Grenoble*, Programme and Abstracts, p. 2–90. Upscattering of UCN by superfluid helium-4. (The UCN production rate of 1/cm^3 s is measured.)

Golub, R. and Böning, K. (1981). *Proc. ICANS-V*, (ed. G.S. Bauer and D. Filges), p. 674. Kernforschungsanlage, Julich. (Review of mechanical and superthermal UCN sources.)

Golub, R. and Böning, K. (1983). *Z. Phys.*, **B51**, 95. (The theory of sub-limiting UCN generation in the walls of a container is presented.)

Golub, R. and Carter, P. (1971). *Nucl. Instrum. Methods*, **91**, 205. Magnetic neutron spectrometers for neutrons with wavelength greater than $\lambda = 40$ Å.

Golub, R. and Pendlebury, J.M. (1972). *Contemp. Phys.*, **13**, 519. EDM of neutrons.

Golub, R. and Pendlebury, J.M. (1974). *Phys. Lett.*, **A50**, 177. Method of studying surface properties relevant to UCN. (A possibility of studying the impurities on surfaces with the aid of the total reflection of over-limiting neutrons is discussed.)

Golub, R. and Pendlebury, J.M. (1975). *Phys. Lett.*, **A53**, 133. Super-thermal sources of UCN. (A proposal for a liquid ^{4}He converter is advanced.)

Golub, R. and Pendlebury, J.M. (1977). *Phys. Lett.*, **A62**, 337. (A UCN density of 4×10^7 neutrons/cm^3 in liquid He is estimated to be attainable for $\Phi_0 = 10^{14}$ neutrons/cm^3 s, $T_n = 20$ K, $T_{He} = 1$ K.)

Golub, R. and Pendlebury, J.M. (1979). *Rep. Prog. Phys.*, **42**, 439. (UCN review).

Golub, R. and Taylor, A.D. (1983). *Nuci. Instrum. Methods*, **212**, 349. (The theory of neutron moderation in a hydrogenous medium with a point monochromatic source is presented.)

Golub, R. *et al.* (1979). *Sci. Am.*, **240**, 106. (Popular UCN review.)

Golub, R. *et al.* (1981). Rutherford Appleton Laboratory Report RL-81-001. (The utilization of UCN for EDM experiments in Grenoble is proposed.)

Golub, R. *et al.* (1983). *Z. Phys.*, **B51**, 187. (The third expriment on UCN generation in liquid He is reported.)

Golub, R., Yoshiki, H., and Gahler, R. (1989). *Nucl. Instrum. Methods*, **A284** (1), 16. From UCN to spin-echo – the bootstrap offers significant gains in sensitivity.

Gorjachev, B.I. and Lin'kova, N.V. (1976). *Kratk. Soobshch. Fiz.*, (7), 32. (The theory of neutron inelastic scattering on a plate taking into account refraction on two boundaries is presented.)

Gorjachev, B.I. and Lin'kova, N.V. (1977). *Kratk. Soobshch. Fiz.*, (9), 8. (A modification of the $1/v$ law taking into account refraction on boundaries is discussed.)

Granada, O.R. (1985). *Phys. Rev.*, **B31**, 4167. (The theory of slow neutron scattering in molecular gases is presented.)

Groshev, L.V. *et al.* (1971). Joint Institute for Nuclear Research, Dubna, Preprint P3-5392 [in Russian]. *Phys. Lett.*, **B34**, 293. (The horizontal UCN channel at the IRT-M reactor is described. $J_{UCN} = 5$ neutrons/s. The experiments on UCN storage in Cu and pyrographite bottles ($\tau \simeq 10$–30 s), as well as results of the He-curve and spectra measurements with the help of gravity spectrometer, are reported.)

Groshev, L.V. *et al.* (1974). Proc. All-Union Conf. on Neutron Physics, part 4, p. 264 [in Russian]. (Experiments on UCN storage in Cu and glass bottles, on UCN generation in Al, Mg, H_2O, and ZrH_2 converters at different T_c, and the construction of the UCN detector DUNya are described.)

Groshev, L.V. *et al.* (1976). Proc. All-Union Conf. on Neutron Physics, part 2, p. 166 [in Russian]. (INIS At. 9:354007); Joint Institute for Nuclear Research, Dubna, Preprint P3-9534 [in Russian]. (INIS At. 9:349137) (The v dependence of $\mu(v)$ for Cu is measured.)

Gurevich, I.I. and Nemirovsky, P.E. (1961). *Zh. Eksp. Teor. Fiz.*, **41**, 1175 (*Sov. Phys.–JETP*, **14**, 838). (The theory of metallic neutron reflection is presented.)

Gurevich, I.I. and Tarasov, L.V. (1965). The physics of low-energy neutrons, p. 287. Nauka, Moscow [in Russian]; see trans. (1968), North-Holland, Amsterdam. (Summary of UCN works.)

Hayashi, S. *et al.* (1986). *Nucl. Instrum. Methods*, **B16**, 377. High sensitivity analysis of deuterium in solids by ^{15}N-induced reactions.

Hennig, K. (1982). *Kernenergie*, **25**, 127. Ultrakalte Neutronen. (Review.)

Hennig, K. *et al.* (1970). Joint Institute for Nuclear Research, Dubna, Preprint P4-5189 [in Russian]. (The theory of UCN interaction with surface excitations without taking into account total reflection phenomena is presented.)

Herdin, R. *et al.* (1977). In *Neutron inelastic scattering*, vol. 1, p. 135. IAEA, Vienna. Polarization and diffraction of UCN.

Herdin, R. *et al.* (1978). *Nucl. Instrum. Methods*, **148**, 353. (A UCN polarization of $P = 98\%$ is achieved by transmission through monocrystal foils.)

Herrmann, P. *et al.* (1985). *Phys. Rev. Lett.*, **54**, 1969. (A neutron microscope with two mirrors and a magnification of 50 is described.)

Holas, A. (1977). *Phys. Status Solidi*, (b), **81**, K-107. (Numerical calculation of the coherent inelastic UCN scattering on a Cu monocrystal is presented.)

Ignatovich, V.K. (1972*a*). Joint Institute for Nuclear Research, Dubna, Preprint P4-6553 [in Russian]. (The theory of UCN specular and non-specular reflection from amorphic media is presented.)

Ignatovich, V.K. (1972*b*). Joint Institute for Nuclear Research, Dubna, Preprint P4-6681 [in Russian]. (Coherent inelastic UCN scattering on the trap walls taking into account the total reflection phenomena is described.)

Ignatovich, V.K. (1973). Joint Institute for Nuclear Research, Dubna, Preprint P4-7055 [in Russian]. (The increase of μ due to surface roughnesses is calculated.)

Ignatovich, V.K. (1974*a*). Joint Institute for Nuclear Research, Dubna, Preprint P4-7831 [in Russian]. Pores effect on UCN absorption in trap walls.

Ignatovich, V.K. (1974*b*). Joint Institute for Nuclear Research, Dubna, Preprint E4-8039. (A hypothesis for the Shrödinger equation modification to explain the anomalous μ is advanced.)

Ignatovich, V.K. (1974*c*). Joint Institute for Nuclear Research, Dubna, Preprint E4-8404. (The quantum theory of UCN storage in a sextupole magnet is presented.)

Ignatovich, V.K. (1975*a*). Joint Institute for Nuclear Research, Dubna, Preprint P4-9007 [in Russian]. (INIS At. 8:329600) (The effect of non-specular reflection on μ is discussed.)

Ignatovich, V.K. (1975*b*). *Izv. Vuzov. Radiofiz.*, **18**, 1552. (The thermodynamic restriction on the form of the reflection indicatrix is considered.)

Ignatovich, V.K. (1975*c*). *Phys. Status Solidi*, (b), **71**, 477. (The theory of UCN inelastic scattering under total reflection is presented.)

Ignatovich, V.K. (1977). Joint Institute for Nuclear Research, Dubna, Preprint P4-10778 [in Russian]. (The theory of reflection from an arbitrary symmetric periodic potential is presented.)

Ignatovich, V.K. (1978*a*). Joint Institute for Nuclear Research, Dubna, Preprint P4-11135 [in Russian]. (Generalization of the periodic potential theory to a non-symmetric period is presented.)

Ignatovich, V.K. (1978*b*). Joint Institute for Nuclear Research, Dubna, Preprint E4-11937. (Some consequences of the periodic potential theory for Laue diffraction are considered.)

Ignatovich, V.K. (1978*c*). *Found. Phys.*, **8**, 565. (Non-spreading wave packets in relativistic and non-relativistic quantum mechanics are discussed.)

Ignatovich, V.K. (1978*d*). *Pis'ma Zh. Eksp. Teor. Fiz.*, **28**, 311 (*JETP Lett.*). (UCN depolarization in transmission through a magnetized foil due to the non-colinearity of internal and external fields is discussed.)

Ignatovich, V.K. (1979). *Proc. Winter School, LNPI, Noorus, March* 1979, p. 200 [in Russian]. (Review on UCN storage problems.)

Ignatovich, V.K. (1980*a*). *Soviet Union patent document* 746 773 OIPOTZ [in Russian]. (Differential gravity spectrometer with an accumulation chamber is described.)

Ignatovich, V.K. (1980*b*). Joint Institute for Nuclear Research, Dubna, Preprint P4-80-261 [in Russian]. (INIS At. 12: 580068) Effect of subsurface clusters on UCN storage time.

Ignatovich, V.K. (1985). Joint Institute for Nuclear Research, Dubna, Preprint P4-85-123 [in Russian]. (New recursive approach to the periodic potential theory is presented.)

Ignatovich, V.K. (1986*a*) *Fiz. Ultrakholodnykh Nejronov*. Nauka, Moscow.

Ignatovich, V.K. (1986*b*). *Usp. Fiz. Nauk*, **150**, 145 (*Sov. Phys.–Usp.*, **150**, 880). Etude on a periodic potential.

Ignatovich, V.K. (1987*a*). Joint Institute for Nuclear Research, Dubna, Preprint P4-87-402, [in Russian]. (INIS At. 19: 030264) Diffusion of UCN in neutron guides in the presence of the gravitational field.

Ignatovich, V.K. (1987*b*). Joint Institute for Nuclear Research, Dubna, Preprint P4-87-878 [in Russian]. New approach to the solution of a one-dimensional schrödinger equation.

Ignatovich, V.K. (1987*c*). Joint Institute for Nuclear Research, Dubna, Preprint E4-87-326. The remarkable capabilities of recursive relations.

Ignatovich, V.K. (1988). Joint Institute for Nuclear Research, Dubna, Preprint P4-88-693 [in Russian]. New approach to dynamical diffraction theory.

Ignatovich, V.K. and Luschikov, V.I. (1975). Joint Institute for Nuclear Research, Dubna, Preprint P3-8785 [in Russian]. (INIS At. 8: 591960) Effect of substance discreteness and of the surface roughness on UCN reflection coefficient.

Ignatovich, V.K. and Luschikov, V.I. (1981). Joint Institute for Nuclear Research, Dubna, Preprint P4-81-77 [in Russian]. (INIS At. 13: 655386) (The hypothesis of secret neutron states is advanced.)

Ignatovich, V.K. and Luschikov, V.I. (1984). *Elem. Chastits At. Jadra*, **15**, 331. (*Sov. J. Part. Nucl.*, **15**, 149). (UCN review.)

Ignatovich, V.K. and Nikitenko, Yu.V. (1987*a*). Joint Institute for Nuclear Research, Dubna, Preprint P3-87-326 [in Russian]. (INIS At. 19: 0300263). UCN diffusion in magnetic neutron guides and powders.

Ignatovich, V.K. and Nikitenko, Yu.V. (1987*b*). Joint Institute for Nuclear Research, Dubna, Preprint P3-87-832 [in Russian]. Physical peculiarities of the gravity spectrometer action.

Ignatovich, V.K. and Nikitenko, Yu.V. (1988). *Yad. Fiz.*, **48**, 1708 (*Sov. J. Nucl. Phys.*). Physical peculiarities of the gravity spectrometer action.

Ignatovich, V.K. and Ostanevich, Yu.M. (1985). Joint Institute for Nuclear Research, Dubna, Rapid Communications No. 7–85 [in Russian]. Quantum mechanics transient processes in an interferometer.

Ignatovich, V.K. and Pokotilovskii, Yu.N. (1976) Joint Institute for Nuclear Research, Dubna, Preprint P4-10145 [in Russian]. UCN magnetic storage in the field of a linear current.

Ignatovich, V.K. and Satarov, L.M. (1977). Moscow Kurchatov Institute for Atomic Energy, Preprint 2820 [in Russian]. (INIS At. 10: 435368) (The theory of UCN inelastic scattering on a two-dimensional hydrogenous gas with the result $\mu(T) \propto \sqrt{T^{1/2}}$ is presented.)

Ignatovich, V.K. and Stepanov, A.V. (1974). Joint Institute for Nuclear Research, Dubna, Preprint P4-7832 [in Russian]. (The theory of μ modification due to films on the UCN reflecting surface is presented.)

Ignatovich, V.K. and Taran, Yu.V. (1982). Joint Institute for Nuclear Research, Dubna, Preprint P3-82-440 [in Russian]. (A possibility to measure the UCN transmission matrix through magnetized foils is considered.)

Ignatovich, V.K. and Taran, Yu.V. (1983). Joint Institute for Nuclear Research, Dubna, Preprint P3-83-180 [in Russian]. (INIS At. 15: 016243) Investigation of the thin film transmission.

Ignatovich, V.K. and Taran, Yu.V. (1984). Proc. All-Union Conf. on Neutron Physics, vol. 4, p. 17 [in Russian]. (INIS At. 16: 076628) (The theory of UCN transmission through a system of two and three foils taking account of multiple scattering between them is presented.)

Ignatovich, V.K. and Terekhov, G.I. (1976*a*). Joint Institute for Nuclear Research, Dubna, Preprint P4-9567 [in Russian]. (The theory of UCN storage taking gravity into account is presented.)

Ignatovich, V.K. and Terekhov, G.I. (1976*b*). Joint Institute for Nuclear Research, Dubna, Preprint P4-10102 [in Russian]. UCN storage in a flat gravimagnetic trap.

Ignatovich, V.K. and Terekhov, G.I. (1978). *Nucl. Instrum. Methods*, **148**, 585. (The theory of UCN stationary and non-stationary diffusion in neutron guides is presented.)

Ignatovich, V.K. *et al.* (1982). Joint Institute for Nuclear Research, Dubna, Preprint P3-82-811 [in Russian]. (The measurement of partial loss coefficients in a flow mode experiment is reported.)

Ioffe, A.I. (1983) *Nucl. Instrum. Methods*, **204**, 565. Diffraction grating interferometer for VCN.

Ioffe, A.I. (1988). *Nucl. Instrum. Methods*, **268**, 169. Aberration-free diffraction grating neutron interferometer.

Ioffe, A.I. and Zabiyakin, V.S. (1988). *Proc. Int. Conf. on Neutron Physics*, vol. 1, p. 163 [in Russian]; *Physica*, B&C **151**, 50. Diffraction-grating neutron interferometers. (It is argued that a UCN interferometer involving four gratings will permit an upper limit of $3.2 \times 10^{-23} e$ of neutron charge to be achieved.)

Isakov, A.I. *et al.* (1988). *Proc. Int. Conf. on Neutron Physics*, vol. 1, p. 294 [in

Russian]. (VCN transmission through H_2O, D_2O, and liotropic liquid crystal $C_{12}H_{25}O(CH_2CH_2O)_nH$ is studied.)
Jewell, C. *et al.* (1981). *Physica*, **B&C**, **107**, 587. (The second experiment on UCN storage in liquid He is reported. $T_{He} = 1.1$–1.2 K, $\tau = 8$–9 s.)
Kadykenov, M.M. *et al.* (1982). *Izv. Akad. Nauk Kaz. SSR*, (6), 64. (UCN transmission through H_2, D_2, N_2 gases is measured.)
Kadykenov, M.M. *et al.* (1984). Proc. All-Union Conf. on Neutron Physics, vol. 4, p. 13 [in Russian]. (INIS At. 17: 000255) (UCN transmission through H, H_2, D_2, N_2, O_2 gases and air is reported.)
Kagan, Yu.M. (1970). *Pis'ma Zh. Eksp. Teor. Fiz.*, **11**, 235 (*JETP Lett.*, **11**, 147). (The theory of sub-limiting UCN generation and UCN traping in pores is presented.)
Kaipov, D.K. (1985). *UCN* Nauka Alma-Ata, [in Russian].
Kaipov, D.K. and Kadykenov, M.M. (1983). *Izv. Akad. Nauk Kaz. SSR*, (6), 36. (The $1/v$ law for neutron interaction with gases is discussed.)
Kaipov, D.K. and Kadykenov, M.M. (1989). *Izv. Akad. Nauk. Kaz. SSR*, (2), 48 [in Russian]. Reflection and transmission of UCN through two step potential barrier, with account of multiple scattering.
Kaipov, D.K. *et al.* (1982). *Izv. Akad. Nauk Kaz. SSR., Ser. Fiz–Mat.*, (2), 62. (INIS At. 14: 766846) Determination of σ_t of UCN interaction with metal atoms. (Transmission through Al, Mo, Nb, Mg is reported.)
Kaipov, D.K. *et al.* (1988). *Proc. Int. Conf. on Neutron Physics*, vol. 1, p. 221 [in Russian]. UCN generation in gas converters. (Gases: H_2, D_2, t_2, N_2, O_2.)
Kaipov, M.D. and Kaspanov, N.K. (1981). *Izv. Akad. Nauk Kaz. SSR*, (4), 25. (INIS At. 14: 757368) (UCN transmission through Al, Mg, Pb foils is reported.)
Kaipov, M.D. *et al.* (1979). Joint Institute for Nuclear Research, Dubna, Preprint P3-12273 [in Russian]. (The characteristics of a rotating UCN counter are reported.)
Kalchev, S.D. and Strelkov, A.V. (1989). Private communication.
Kalchev, S.D. *et al.* (1984). Joint Institute for Nuclear Research, Dubna, Rapid Communications No. 3–84, p. 33 [in Russian]. (INIS At. 16: 072326) (A spectrometer for the search of the neutron charge with UCN is described.)
Kalus, J. and Gähler, R. (1984). *J. Phys. Coll.*, **45**, C3, 17. (Some modifications of the apparatus are discussed to reach a sensitivity of $10^{-21}\,e$ in the experiments in search of the neutron charge.)
Karaseva, I.A. *et al.* (1988). *Proc. Int. Conf. on Neutron Physics*, vol. 1, p. 240 [in Russian]. (The theory of VCN interaction with a variable in space, in particular multilayer magnetic field, is discussed.)
Karaseva, I.A. *et al.* (1988). *Proc. Int. Conf. on Neutron Physics*, vol. 1, 216 [in Russian]. (A possible neutron gravitationally achromatic UCN microscope with a horizontal optical axis is considered. The possibility to achieve $\sim 1\,\mu$m resolution with magnification 50–60 is claimed.)
Kashukeev, N.T. (1970). *Dokl. Bolg. Akad. Nauk*, **23**, 1473. (UCN turbine project.)
Kashukeev, N.T. (1971). *Dokl. Bolg. Akad. Nauk*, **24**, 717. (The idea of the search for the neutron charge using UCN is advanced.)
Kashukeev, N.T. (1981). *Yad. Ehnerg.*, **13**, 3 (INIS At. 13: 658154) (The effect of roughnesses on the silt image in a neutron charge spectrometer is estimated.)
Kashukeev, N.T. (1982). *Dokl. Bolg. Akad. Nauk*, **35**, 325. (With the help of uncertainty relations concerning the entire Universe the minimal neutron charge is estimated to be 5.8×10^{-21}. The idea of charge distribution is advanced.)

Kashukeev, N.T. and Chikov, N.F. (1979*a*). Joint Institute for Nuclear Research, Dubna, Preprint P3-1266 [in Russian]. (INIS At. 11: 526844); *Pis'ma Zh. Eksp. Teor. Fiz.*, **30**, 306 (*JETP Lett.*). (With a model spectrometer the image of the slits is obtained using UCN.)

Kashukeev, N.T. and Chikov, N.F. (1979*b*). *Bolg. Fiz. Zh.*, **6**, 529. (Modifications of the neutron charge spectrometer are discussed.)

Kashukeev, N.T. and Chikov, N.F. (1982). *Dokl. Bolg. Akad. Nauk*, **35**, 45; Joint Institute for Nuclear Research, Dubna, Preprint P3-82-145 [in Russian]. (From the result of the experiment by Kashukeev and Chikov (1979*a*) the specular reflection coefficient, 0.986, of glass is obtained.)

Kashukeev, N.T. and Stanev, G.A. (1973). *Dokl. Bolg. Akad. Nauk*, **26**, 1445. (First measurements at UCN source. $J_{UCN} = 4$ neutrons/s.)

Kashukeev, N.T. and Yaneva, N.B. (1978). *Bolg. Fiz. Zh.*, **5**, 243. (An experiment on neutron charge with UCN is discussed.)

Kashukeev, N.T. *et al.* (1973). *Dokl. Bolg. Akad. Nauk*, **26**, 51, 1449. (Optimization of the rotation frequency of the UCN turbine is discussed.)

Kashukeev, N.T. *et al* (1975*a*). *Yad. Ehnerg.* **2**, 28. (The UCN channel with turbine at the reactor IRT-2000 in Sophia is described.)

Kashukeev, N.T. *et al.* (1975*b*). *Nucl. Instrum. Methods*, **126**, 43. Experiments with UCN turbine.

Kashukeev, N.T. *et al.* (1976). Proc. All-Union Conf. on Neutron Physics, part 2, p. 205 [in Russian]. (INIS At. 9: 353448) (Transmission of neutrons with velocity 140 m s^{-1} through In, Ag, Cu, Ni, Fe, Mo is reported. The $1/v$ law is confirmed.)

Kashukeev, N.T. *et al.* (1977). *At. Ehnerg.*, **42**, 373 (*Sov. J. At. Energy*) (The same as Kashukeev *et al.* (1976).)

Kawabata, Y., Utsuro, M., and Okamoto, S. (1986). *Nucl. Instrum. Methods*, **A245**, 106. Fall-focusing gravity spectrometer with ellipsoidal mirror elements.

Kawabata, Y. *et al.* (1988). *Nucl. Instrum. Methods*, **B30**, 557. Hydrogen concentration on mirror surfaces of neutron bottles.

Kawai, T. *et al.* (1987). *Int. Conf. on Neutron Phys. Sept.* 1987. The instrumental developments for researches on neutron physics at the KUR (5 MW).

Kazarnowskij, M.V. and Stepanov, A.V. (1976). *Kratk. Soobshch. Fiz.*, (7), 38. (The method of the μ determination in terms of the decay of UCN quasi-stationary states in a bottle is described.)

Kazarnowskij, M.V. *et al.* (1981). *Pis'ma Zh. Eksp. Teor. Fiz.*, **34**, 49 (*JETP Lett.*). Possibility of observation of $N\bar{N}$ oscillations by means of UCN.

Kearney, A.D. *et al.* (1980). *Nature*, **287**, 313. (UCN zone plate calculation is presented.)

Kharitonov, A.G. *et al.* (1989). *Nucl. Instrum. Methods*, **A284** (1), 98. Preliminary results of neutron lifetime measurements with gravitational UCN trap.

Kilvington, A.I. *et al.* (1987). *Phys. Lett.*, **A125**, 416. Scattering of UCN by superfluid helium at temperatures around 1 K.

Klein, A.G. and Zeilinger, A. (1984). *J. Phys. Coll.*, **45**, C3, 239. Wave optics with cold neutrons.

Klein, A.G. *et al.* (1981). *Phys. Lett.*, **A83**, 71. Focusing of slow neutrons with cylindrical zone plates.

Koerting, K. (1986). *Nucl. Eur.*, **6**, 41. ILL bottles neutrons to investigate their lifetimes.

Kosvintsev, Yu.Yu. *et al.* (1976*a*). *Pis'ma Zh. Eksp. Teor. Fiz.*, **23**, 135 (*JETP Lett.*, **23**, 118). Storage of UCN in a vessel with a magnetic 'wall'.

Kosvintsev, Yu.Yu. *et al.* (1976*b*). Proc. All-Union Conf. on Neutron Physics, part 2, pp. 188, 193 [in Russian]. (INIS At. 9: 354011, 354012) (UCN storage in a big stainless steel bottle with variable area of slit. Transmission of neutron guides and elbows as well as angular distribution modification with guide length are reported.)

Kosvintsev, Yu.Yu. *et al.* (1976*c*). Dimitrovgrad Institute for Research in Atomic Reactor Science, Preprint 15(281) [in Russian]; *Pis'ma Zh. Tekh. Fiz.*, **2**, 293. (UCN reflection from a magnetic barrier is described.)

Kosvintsev, Yu.Yu. *et al.* (1977*a*). Proc. All-Union Conf. on Neutron Physics, part 1, pp. 152, 156, 173 [in Russian]. (INIS At. 9: 413385, 413386, 413387) (Production of mononenergetic neutrons by UCN acceleration in the gravitational field is described. UCN integral spectra measured with gravity spectrometer in accumulation and flow modes of operation are presented. UCN storage on a horizontal Cu plane is reported. $\tau = 300 \pm 40$ s.)

Kosvintsev, Yu.Yu. *et al.* (1977*b*). *Prib. Tekh. Eks.*, (1), 42 (*Instrum. Exp. Tech.*). (An experiment on UCN τ_β measurement is described.)

Kosvintsev, Yu.Yu. *et al.* (1977*c*). *Nucl. Instrum. Methods*, **143**, 33. (UCN horizontal channel at CM-2 reactor is described. The converter is ZrH_2. $J_{UCN} = 640$ neutrons/s. UCN spectra, transmission through neutron guides and elbows, angular distributions at the exit of the channel, and storage time τ in Cu bottle are reported.)

Kosvintsev, Yu.Yu. *et al.* (1978*a*). *Pis'ma Zh. Eksp. Teor. Fiz.*, **27**, 70 (*JETP Lett.*). (The experiment of Averyanov *et al.* (1977) is reported.)

Kosvintsev, Yu.Yu. *et al.* (1978*b*). *Pis'ma Zh. Eksp. Teor. Fiz.*, **28**, 164 (*JETP Lett.*). (The experiment on UCN storage in bottles with deposition of various materials on the walls in described.)

Kosvintsev, Yu.Yu. *et al.* (1979). *Zh. Eksp. Teor. Fiz.*, **77**, 1277 (*Sov. Phys.–JETP*). (The temperature dependence of μ is discovered.)

Kosvintsev, Yu.Yu. *et al.* (1980*a*). *Pis'ma Zh. Eksp. Teor. Fiz.*, **31**, 257 (*JETP Lett.*, **31**, 236). ($\tau_\beta = 875 \pm 95$ in an Al bottle is achieved.)

Kosvintsev, Yu.Yu. *et al.* (1980*b*). *At. Ehnerg.*, **48**, 166 (*Sov. J. Nucl. Phys.*). (A magnetic integral spectrometer and the spectra measured with its aid are reported.)

Kosvintsev, Yu.Yu. *et al.* (1980*c*). Joint Institute for Nuclear Research, Dubna, Preprint P3-80-91 [in Russian]. (The weak temperature dependence of μ for annealed Cu and Be bottles is reported.)

Kosvintsev, Yu.Yu. *et al.* (1980*d*). Proc. All-Union Conf. on Neutron Physics, part 1, pp. 116, 130 [in Russian]. (INIS At. 13: 645748, 643773) (Review on UCN spectrometers. A spectrum with anomalous step is presented. Interaction of neutrons accelerated in the gravitational field UCN with H_2 gas and the parameters ($p\tau$) for H_2, D_2O, H_2O, He, and others are reported.)

Kosvintsev, Yu.Yu. *et al.* (1981). Dimitrovgrad Institute for Research in Atomic Reactor Science, Preprint 49(502) [in Russian]. Method for τ_β measurement by means of UCN storage in a vessel with variable surface area. ($\tau_\beta = 875 \pm 95$ s.)

Kosvintsev, Yu.Yu. *et al.* (1982*b*). *Pis'ma Zh. Eksp. Teor. Fiz.*, **36**, 346 (*JETP Lett.*, **36**, 424). ($\tau = 950 \pm 60$ s in Al bottle with D_2O deposited at 80 K on the walls is achieved.)

Kosvintsev, Yu.Yu. *et al.* (1982*b*). *At. Ehnerg.*, **53**, 191 (*Sov. J. At. Energy*). (The hypothesis of secret neutron states is checked. No dependence of τ on magnetic field is observed.)

Kosvintsev, Yu.Yu. *et al.* (1983). Dimitrovgrad Institute for Research in Atomic Reactor Science, Preprint 27(542) [in Russian]; *At. Ehnerg.*, **55**, 288 (*Sov. J. At. Energy*). (The experiments with stacks of Cu and Be foils, showing that superfluous μ is due to inelastic scattering, are reported. The energy of final neutrons is estimated to be $E_{Be} = 25$ meV, $E_{Cu} = 50$ meV.)

Kosvintsev, Yu.Yu. *et al.* (1984*a*). Proc. All-Union Conf. on Neutron Physics, vol. 3, p. 328 [in Russian]. (INIS At. 17: 001247) (With the help of a variable-area absorber at the end of the neutron guide the exponential dependence of the UCN transmission on length is observed. For stainless steel and Cu guides $G = 0.1$–0.2, $\mu = 10^{-3}$ are found. The best transmission of electropolished stainless steel neutron guides is established. The diffusion model is checked with a precision of 3–5%.)

Kosvintsev, Yu. Yu. *et al.* (1984*b*). Proc. All-Union Conf. on Neutron Physics, vol. 4, pp. 3, 44 [in Russian]. (INIS At. 17: 001242, 001243) (Transmission of elbows and diaphragms is investigated. The resistance of a diaphragm does not depend on its location inside the neutron guide. The resistance of elbows diminishes when the diffusive part of the reflection coefficient grows.)

Kosvintsev, Yu.Yu. *et al.* (1985). Dimitrovgrad Institute for Research in Atomic Reactor Science, Preprint 14(660) [in Russian]. Neutron gas propagation along neutron guides.

Kosvintsev, Yu.Yu. *et al.* (1986). *Pis'ma Zh. Eksp. Teor. Fiz.*, **44**, 444 (*JETP Lett.*, 44, 571). Neutron decay time measurement in UCN storage experiment.

Kosvintsev, Yu.Yu. *et al.* (1988). *Proc. Int. Conf. on Neutron Physics*, vol. 1, p. 231 [in Russian]. ($\tau = 893 \pm 20$ s in a D_2O ice bottle is achieved.)

Kozlov, A.N., Nikitenko, Yu.V., and Taran, Yu.V. (1980). Joint Institute for Nuclear Research, Dubna, Preprint P13-80-709 [in Russian]. (INIS At. 12: 614235) Two-cascade magnetic field stabilizer of the installation for measuring neutron EDM.

Kügler, K.J. *et al.* (1978). *Phys. Lett.*, **B72**, 422. (Experiment on neutron storage in a magnetic torroidal trap is described.)

Kügler, K.J. *et al.* (1980). *Z. Phys.*, **B39**, 361. (Investigation of slow neutron transmission by curved neutron guides.)

Kügler, K.J. *et al.* (1985). *Nucl. Instrum. Methods*, **A228**, 240. (Detailed description of experiments with toroidal magnetic trap.)

La Marche, P.H. (1981). Ph.D. thesis. Yale University; (INIS At. 14: 790835).

La Marche, P.H. *et al.* (1981). *Nucl. Instrum. Methods*, **189**, 533. (Investigation of efficiency of different surface cleaning methods of hydrogen: deuteration, annealing, deposition of new substance, sputtering by ion bombardment in glow discharge.)

Lanford, W.A. and Golub, R. (1977). *Phys. Rev. Lett.*, **39**, 1509. Hydrogen surface contamination and the storage of UCN. (A profile of the hydrogen distribution near the surface of some solids is measured with the help of the $p + {}^{15}N = {}^{16}O + \gamma$ resonance reaction. The surfaces of Cu, pyrographite, and glass were investigated after different cleaning procedures.)

Lengsfeld, M. and Steyerl, A. (1977). *Z. Phys.*, **27**, 117. (The dimensions of SiO_2 particles are determined from the energy dependence of long-wavelength neutron transmission by silicate glue.)

Lindley, D. (1989). *Nature*, **340**, 503. Neutrons aged in the bottle. (Popular review of Mampe *et al.* 1989.)

Lobashov, V.M. and Serebrov, A.P. (1978). *III Int. School on Neutron Physics*, Joint Institute for Nuclear Research, Dubna, Preprint D3-11-787, p. 225 [in Russian]. (The experiment by Altarev *et al.* (1979) is described.)

Lobashov, V.M. and Serebrov, A. P. (1984). *J. Phys. Coll.*, **45**, C3, 11. (The EDM limit of 4×10^{-25} e cm is reported and the prospects of reaching the value 2×10^{-26} are discussed.)

Ludecke, F. *et al.* (1985). *Physica*, **B132**, 1. (The possibility of observing the liquid-vapour phase transition with the aid of neutron reflection from a thin liquid layer is discussed.)

Luschikov, V.I. (1974). *II Int. School on Neutron Physics*, Joint Institute for Nuclear Research, Dubna, Preprint D3-7991, p. 7 [in Russian]. (INIS At. 7: 277198). Storage of UCN.

Luschikov, V.I. (1977). *Phys. Today*, **30**, (6), 42. (UCN review.)

Luschikov, V.I. (1978). *III Int. School on Neutron Physics*, Joint Institute for Nuclear Research, Dubna, Preprint D3-11-787, p. 183 [in Russian]. (INIS At. 10: 468230) (UCN review.)

Luschikov, V.I. (1986). *Nauchnoe sotrudnichestvo stran v yadernoj fizikie*, (ed. N.N. Bogolubov), p. 210. Ehnergoatomizdat, Moscow. (UCN review.)

Luschikov, V.I. and Frank, A.I. (1978). *Pis'ma Zh. Eksp. Teor. Fiz.*, **28**, 607, (*JETP Lett.*). (The least UCN vertical motion energy 10^{-12} eV in the Earth's gravitational field is pointed out. A possibility to change it with the help of an inhomogeneous magnetic field is noted.)

Luschikov, V.I. and Taran, Yu. V. (1984). Joint Institute for Nuclear Research, Dubna, Preprint D3-84-196 [in Russian]; *Nucl. Instrum. Methods*, **228**, 159. On the calculation of the neutron adiabatic spin-flipper.

Luschikov, V.I. and Taran, Yu. V. (1985). Joint Institute for Nuclear Research, Dubna, Preprint P3-85-429 [in Russian]. (INIS At. 17: 057496) On the propagation of UCN bunches in specular neutron guides.

Luschikov, V.I. *et al.* (1969). *Pis'ma Zh. Eksp. Teor. Fiz.*, **9**, 40 (*JETP Lett.*, **9**, 23). (First UCN experiment. UCN observation and He curve measurement.)

Luschikov, V.I. *et al.* (1978). Moscow Kurchatov Institute for Atomic Energy, Preprint 3066 [in Russian]. (INIS At. 10: 486163) (Unsuccessful attempts to observe temperature dependence of μ in glass bottles. UCN polarization experiment.)

Lynn, J.W. *et al.* (1983). *Physica*, B&C, **120**, 114. (At moving crystal UCN source and time-of-flight spectrometer the UCN transmission through a two-foil system is measured. The intensity of transmitted neutrons when the second foil is non-magnetized is two times less than when it is magnetized. The interference in thin-foil transmission is observed. The spectrometer resolution is estimated to be 40 neV.)

Majkrzak, C.P. and Malik, S.S. (1980). *J. Appl. Cryst.*, **13**, 265. (Numerical calculation of UCN yield from moving crystal.)

Mampe, W. (1982). *Neutron-capture and gamma-ray spectroscopy, Inst. Phys. Conf. Sec. No.* 62, p. 635. (UCN review.)

Mampe, W. and Ageron, P. (1978). *Fundamental physics with reactor neutrons and neutrinos, Inst. Phys. Conf. Ser. No.* 42, p. 148. (The UCN cold source at ILL is described.)

Mampe, W. and Bugeat, J.P. (1980). *Phys. Lett.*, **A78**, 293. (The hydrogen content on a stainless steel surface after cleaning with glow discharge in an Ar atmosphere is reported.)

Mampe, W., Ageron, P., and Gähler, R. (1981). *Z. Phys.*, **B45**, 1. (The storage curves for Al, Be, quartz, and stainless steel bottles after different cleaning procedures are measured.)

Mampe, W. *et al.* (1989*a*). *Phys. Rev. Lett.*, **63**, 593. Neutron lifetime measured with stored ultracold neutrons.

Mampe, W. *et al.* (1989*b*). *Nucl. Instrum. Methods*, **A284** (1), 111. Neutron lifetime from a liquid walled bottle.

Masalovich, S.V. (1986). *Vopr. At. Nauki Tekh. Vyp.*, **3**, (36), 69.

Masalovich, S.V. and Frank, A.I. (1984). Proc. All-Union Conf. on Neutron Physics, **4**, 36 [in Russian]. (In an external field the angle of neutron reflection differs a little from specular. This difference for a vertical mirror in the Earth's gravitational field is equal to 10^{-4} radians.)

Masalovich, S.V. and Strepetov, A.N. (1987). Moscow Kurchatov Institute for Atomic Energy, Preprint 4536/14 [in Russian]. Calculation of UCN optical systems with non-vertical axis in gravitational field.

Matora, I.M. (1969). *At. Ehnerg.*, **27**, 71. (*Sov. J. At. Energy.*). (The problems of neutron focusing in magnetic and gravitational fields are discussed.)

Matora, I.M. (1970). Joint Institute for Nuclear Research, Dubna, Preprint P3-5537 [in Russian]. (Theory of UCN storage in a sextupole magnetic trap taking account of gravity.)

Matora, I.M. (1972*a*). *SU Patent document*, No. 341091 [in Russian]. (A proposal for a toroidal neutron magnetic trap without internal poles.)

Matora, I.M. (1972*b*). *Yad. Fiz.*, **16**, 624 (*Sov. J. Nucl. Phys.*, **16**, 349). (Theory of UCN depolarization in a mgnetic sextupole trap.)

Matora, I.M. (1974). Joint Institute for Nuclear Research, Dubna, Preprint P3-7824 [in Russian]. (A statement that depolarization in a multipole magnetic trap is smaller, if the number of pole pairs is odd, is advanced.)

Matora, I.M. and Strelina, O.A. (1971). Joint Institute for Nuclear Research, Dubna, Preprint P3-5902 [in Russian]. (Numerical calculation of UCN depolarization in magnetic traps and neutron guides.)

Mikerov, V.I. *et al.* (1988). Lebedev Institute in Moscow, Preprint 191 [in Russian]. Methods of reflection coefficient calculation for many-layer interference structures.

Miller, W.A. (1983). *Proc. Int. Symp. on Quantum Mechanics*, (ed. S. Kamefuchi *et al.*), p. 153. Physical Society of Japan, Tokyo. (An experiment on delayed action at UCN interferometer to check quantum mechanics is discussed.)

Miranda, P.C. (1985). *Flavour mixing and violation. Proc. 5th Moriond Workshop, La Peagae (France), Jan.* 1985, (ed. Tran Than Van), p. 233. AFIAP, Paris. Current research for the EDM of the neutron.

Molitor, E.J. (1979). Thesis. Bonn University. (Experiments with magnetic bottles. Sextupole top is described.)

Morosov, V.I. (1978). *At Ehnerg.*, **45**, 442 (*Sov. J. At. Energy*). Review on UCN generation and storage experiments.)

Morosov, V.I. (1980). *Experimental investigations with UCN*. NIRAR, Dimitrovgrad [in Russian]. (Review.)

Morosov, V.I. (1982). *Storage of UCN in closed vessels*. NIRAR, Dimitrovgrad [in Russian]. (Review.)

Morosov, V.I. (1989). *Nucl. Instrum. Methods*, **A284** (1), 108. Neutron lifetime determination by UCN storage.

Morosov, V.I. and Terekhov, G.I. (1980). *Prib. Tekh. Eksp.*, (4), 29 (*Instrum. Exp. Tech.*). (Methods of μ extraction from τ measurements are discussed.)

Morosov, V.I. *et al.* (1986). *Pis'ma Zh. Tekh. Fiz.*, **12**, 12 (*Sov. Tech. Phys. Lett.*). Cryogenic UCN neutron guide with regenerating D_2O surface.

Morosov, V.I. *et al.* (1987*a*). *Pis'ma Zh. Eksp. Teor. Fiz.*, **46**, 301 (*JETP Lett.*). Measurement of UCN reflection from substances with large absorption Cross-section.

Morosov, V.I. *et al.* (1987*b*). Moscow Kurchatov Institute for Atomic Energy, Preprint 4410/2 [in Russian]. Facility to generate UCN at the IR-8 reactor and its efficiency analysis. (Converter—the mixture of light and heavy water. Horizontal channel. $\Phi_{UCN} = 6.7$ neutrons/cm^2 s $\varepsilon = 3$–6%.)

Morosov, V.I. *et al.* (1988*a*). *Proc. Int. Conf. on Neutron Physics*, vol. 1, p. 198 [in Russian]. UCN gas kinetics. (The UCN density distribution along horizontal and vertical neutron guides is measured.)

Morosov, V.I. *et al.* (1988*b*). *Proc. Int. Conf. on Neutron Physics*, vol. 1, p. 273 [in Russian]. Study of thin ice layers with the help of UCN.

Morse, J. (1984). *J. Phys. Coll.* **45**, C3, 13. (The first results on UCN storage, EDM experiment, and achieved limit $(0.3 \pm 4.8) \times 10^{-25}$ e cm are reported.)

Mysenholder, W. (1976). *Nucl. Instrum. Methods*, **13**, 291. (Monte Carlo calculation of UCN transmission by sharp neutron guides and elbows in specular reflection approximation.)

Namiot, V.A. (1973). *Dokl. Akad. Nauk*, **211**, 76 (*Sov. Phys.–Dokl.*, **18**, 481). (A possibility of UCN generation by polarized liquid deuterium in a strong magnetic field is discussed.)

Nikitenko, Yu.V. and Taran, Yu.V. (1978). *Prib. Tekh. Eksp.*, (4), 79 (*Instrum. Exp. Tech.*). (Thermomagnetic shutter for UCN is proposed.)

Nikitenko, Yu.V. and Taran, Yu.V. (1981). Joint Institute for Nuclear Research, Dubna, Preprint P3-81-211 [in Russian]. (The smearing of the UCN Larmor precession phase by magnetic field inhomogeneities is estimated.)

Novopoltsev, M.I. and Pokotilovskii, Yu.N. (1978). Joint Institute for Nuclear Research, Dubna, Preprint P3-11986 [in Russian]. (The experiments on UCN flux chopping with a system of current foils.)

Novopoltsev, M.I. and Pokotilovskii, Yu.N. (1979). *Pis'ma Zh. Tekh. Fiz.*, **5**, 1181 (*Sov. Phys.–Tech. Phys. Lett.*). (The experiment on UCN flux chopping with a system of magnetic foils with different coercive forces.)

Novopoltsev, M.I. and Pokotilovskii, Yu.N. (1980*a*). Joint Institute for Nuclear Research, Dubna, Preprint P3-80-398 [in Russian]. (INIS At. 12: 603865) (The gain of UCN density is calculated for utilization of UCN chopper at pulsed neutron sources.)

Novopoltsev, M.I. and Pokotilovskii, Yu.N. (1980*b*). *Nucl. Instrum. Methods*, **171**, 497. (Optimization of rotating scintillation UCN counter. Measurement of spectral characteristics of different UCN detectors.)

Novopoltsev, M.I. and Pokotilovskii, Yu.N. (1981). Joint Institute for Nuclear Research, Dubna, Preprint P3-81-828 [in Russian]. (The UCN transmission by a three-foil system is measured with a correlation time-of-flight spectrometer.)

Novopoltsev, M.I. and Pokotilovskii, Yu.N. (1982). *Zh. Tekh. Fiz.*, **52**, 1243 (*Sov. J. Tech. Phys.*). (Ferromagnetic UCN chopper performance investigation.)

Novopoltsev, M.I. and Pokotilovskii, Yu.N. (1985). Joint Institute for Nuclear Research, Dubna, Preprint P3-85-843 [in Russian]. Activation UCN experiments. Measurement of the absorption at sub-limiting reflection from Cu.

Novopoltsev, M.I. and Pokotilovskii, Yu.N. (1987). Joint Institute for Nuclear Research, Dubna, Preprint P3-87-408 [in Russian]. (INIS At. 19: 070900). Measurement of UCN reflection from substances with large absorption cross-section.

Novopoltsev, M.I. *et al.* (1985). *Prib. Tekh. Eksp.*, (4), 31 (*Instrum. Exp. Tech.*). (The modulation depths of 4.6 for two-foil and 7 for four-foil chopper are reported.)

Novopoltsev, M.I. *et al.* (1987). Joint Institute for Nuclear Research, Dubna, Preprint P3-87-578 [in Russian]; *Nucl. Instrum. Methods*, **264**, 518. Time-of-flight UCN spectrometry with a thin film ferromagnetic chopper.

Novopoltsev, M.I. *et al.* (1988). *Z. Phys.*, **B70**, 199. Measurement of the absorption coefficient for UCN at a upper surface.

Okolotina, V. (1984). *Rev. Politech.*, (10), 1275. Des neutrons ultrafroids.

Okumura, K. *et al.* (1980). *Annu. Rep. Res. Reactor Inst., Kyoto Univ.*, **12**, 12. (Cold UCN source is described.)

Paul, W. and Trinks, U. (1978). *Fundamental physics with reactor neutrons and neutrinos, Inst. Phys. Conf. Ser. No.* 42, p. 18. A magnetic storage ring.

Pendlebury, J.M. *et al.* (1984). *Phys. Lett.*, **B136**, 327. (Detailed description of the experiment by Morse (1984).)

Petschek, A.G. (1983). *Los Alamos National Laboratory Conf. on Low Energy Tests of Conservation Laws in Particle Physics, Blacksburg, VA, Sept.* 1983, LA-UR-83-2648. (Shows that UCN are competing in $N\bar{N}$ oscillations experiment.)

Plakida, N.M. (1981). Joint Institute for Nuclear Research, Dubna, Preprint P7-81-91 [in Russian]. (The model of tunnel transitions of hydrogen atoms is considered to obtain a weak temperature dependence of μ.)

Plakida, N.M. (1983). *Proc. Semin. Devoted to the 75th Anniversary of D.I. Blokhintsev, Dubna, Jan.* 1983, p. 177 [in Russian]. (INIS At. 19: 046872). UCN heating theory and its experimental testing.

Podgoretskii, M.I. and Tzulaja, M.I. (1985). Private communication.

Pokotilovskii, Yu.N. (1978). Joint Institute for Nuclear Research, Dubna, Preprint P3-11823 [in Russian]. (INIS At. 10: 464924) (Relation between polarization effectiveness and polarization ratio of the foils is established.)

Pokotilovskii, Yu.N. (1980). *Pis'ma Zh. Tekh. Fiz.*, **6**, 1300 (*Sov. Tech. Phys. Lett.*). (A method of pulse UCN generation by shaking them out of a converter with high limiting energy is considered.)

Pokotilovskii, Yu.N. (1981). Joint Institute for Nuclear Research, Dubna, Preprint P3-81-3 [in Russian]. (INIS At. 13: 648375) (The EDM experiment with the help of the spin-echo effect is discussed.)

Pokotilovskii, Yu.N. (1984). Joint Institute for Nuclear Research, Dubna, Preprint P3-84-810 [in Russian]. (INIS At. 16: 056461) (Possible experiments with VCN at a pulse source are discussed.)

Pokotilovskii, Yu.N. *et al.* (1976). *Prib. Tekh. Eksp.*, (3), 32 (*Instrum. Exp. Tech.*). (Three types of ferromagnetic UCN shutters are discussed.)

Pokotilovskii, Yu.N. *et al.* (1980). *Prib. Tekh. Eksp.*, (1), 62 (*Instrum. Exp. Tech.*). (Methods of UCN spectrometry with the use of resonance filters are discussed.)

Porsev, G.D. and Serebrov, A.P. (1973*a*). Leningrad Institute for Nuclear Physics, Preprint 38 [in Russian]. (Calculation of the Be converter in Debye approximation.)

Porsev, G.D. and Serebrov, A.P. (1973*b*). *Proc. IX Winter School, LNPI, Leningrad*, vol. 3, p. 271 [in Russian]. (Review on neutron EDM experiments.)

Purica, I.I. (1979). *Rev. Roum. Phys.*, **24**, 243. (A possibility to measure supergravity effects with the help of UCN is discussed.)

Ramsey, N.F. (1988). *Phys. Scr.*, **22**, 140. The electric dipole moment of the neutron.

Richardson, D.J. and Lanoreaux, S.K. (1989). *Nucl. Instrum. Methods*, **A284** (1), 192. Demonstration of Berry's phase using polarized neutrons.

Richardson, D.J. *et al.* (1988). *Phys. Rev. Lett.*, **61**, 2030. Demonstration of the Berry phase using stored UCN.

Robson, J.M. (1976*a*). *Can. J. Phys.*, **54**, 1277. (UCN generation in polyethylene converter at different temperatures and UCN exit time 4 s out of the total neutron guide system are reported. $J_{UCN}(T_c = 200\text{ K})/J_{UCN}(T_c = 313\text{ K}) = 3$ He curve shows that the UCN reflection from neutron guide walls is perfectly specular. The N_2 curve shows that σ_t ($v = 2200\text{ m s}^{-1}$) = 38.6 barn.)

Robson, J.M. (1976*b*). *Phys. Lett.*, **B65**, 225. (Investigation of UCN angular distribution variation along the neutron guide.)

Robson, J.M. and Winfeld, D. (1972). *Phys. Lett.*, **B40**, 537. (The first experiment on UCN observation in Canada.)

Roy, S. (1976). Ph.D. thesis. Victoria University of Manchester. (Experiments on UCN at the research reactor. Horizontal UCN channel, time-of-flight spectrometer.)

Satarov, L.M. (1976*a*). Moscow Kurchatov Institute for Atomic Energy. Preprint 2680 [in Russian]. (Elastic UCN reflection from a crystal with impurities.)

Satarov, L.M. (1976*b*). Moscow Kurchatov Institute for Atomic Energy. Preprint 2681 [in Russian]. (Theory of UCN heating on phonons.)

Scheckenhofer, H. and Steyerl, A. (1977). *Phys. Rev. Lett.*, **39**, 1310. (Gravity diffractometer description. Observation of UCN diffraction on a diffraction grating. Transverse coherence length is established to be larger than 10^6 Å. Interference in Au thin film transmission and 73 Å smearing of the glass potential near the surface are observed.)

Scheckenhofer, H. and Steyerl, A. (1981). *Nucl. Instrum. Methods*, **179**, 393. (Parameters of the gravity diffractometer are described.)

Schlossbauer, G. (1985). Diploma thesis. Technische Universität Munchen. Untersuchung von Modelmembranen mit sehr langsamen Neutronen.

Schulz, P. (1980). *Umsch.-Wiss. Tech.*, **80**, 437. Die kaltesten Neutronen der Welt.

Schutz, G., Steyerl, A., and Mampe, W. (1980). *Phys. Rev. Lett.*, **44**, 1400. (With the zone mirror the image of the slit is obtained with a magnitification of 6.)

Scott, R. J. and McClintock, P.V.E. (1977). *Phys. Lett.*, **A64**, 205. (A technology for ^{4}He purification to the level $^4\text{He}/^3\text{He} = 2 \times 10^{15}$ is developed.

Serebrov, A.P. (1979). *Proc. Winter School of LIYaF, Leningrad*, p. 3 [in Russian]. Neutron EDM and UCN.

Seregin, A.A. (1977). *Zh. Eksp. Teor. Fiz.*, **73**, 1634 (*Sov. Phys.–JETP*, **46**, 859). (UCN transmission through three-layer system is considered.)

Seregin, A.A. (1980*a*). Proc. All-Union Conf. on Neutron Physics, part. 1, pp. 142, 145 [in Russian]. (INIS At. 13: 643774, 643775) (Surface displacement of reflecting UCN. Review on UCN transmission through multilayer systems experiments.)

Seregin, A.A. (1980*b*). Fiziko-Ehnergeticheskij Institute. Preprint FEI-1116 [in Russian]. (INIS At. 13: 655377) (Hypothesis of UCN bound states at reflecting surface.)

Seregin, A.A. (1981). *Yad. Fiz.*, **33**, 1173 (*Sov. J. Nucl. Phys.*, **33**, 622). Surface displacement of a neutron on reflection.

Shapiro, F.L. (1968). *Usp Fiz. Nauk*, **95**, 145 (*Sov. Phys.–Usp.*, **11**, 345). (UCN EDM experiment in flow mode is discussed.)

Shapiro, F.L. (1972*a*). *Elem. Chastits At. Jadra*, **2**, 973 (*Sov. J. Part. Nucl.*). (A possibility to increase UCN density at pulse reactor is considered.)

Shapiro, F.L. (1972*b*). Joint Institute for Nuclear Research, Dubna, Preprint P3-7135 [in Russian]; In *Int. Conf. on Nuclear Study with Neutrons, Budapest*, 1972, (ed. J. Ero and J. Szucks), p. 259. Plenum, New York. (UCN review.)

Skachkova, O.S. and Frank, A.I. (1981). *Pis'ma Zh. Eksp. Teor. Fiz.*, **33**, 214 (*JETP Lett.*). (The compensation of gravity aberration in a neutron microscope with the help of an inhomogeneous magnetic field is considered.)

Smith, K. (1978). *Fundamental physics with reactor neutrons and neutrinos, Inst. Phys. Conf. Ser. No.* 42, p. 157. (Review of UCN application to fundamental physics problems.)

Smith, K. (1980). *Contemp. Phys.*, **21**, 631. (UCN review.)

Stanev, G.A. (1978). *Yad. Ehnerg.*, **6**, 49. (Calculation of $\mu(v)$ for different forms of potential smearing near the surface.)

Steenstrup, S. and Buras, B. (1978). *Nucl. Instrum. Methods*, **154**, 549. A Monte Carlo simulation of UCN production by Bragg reflection from moving single crystal.

Steinhauser, K.A. *et al.* (1980). *Phys. Rev. Lett.*, **44**, 1306. (Observation of resonances in UCN transmission by three-layer systems.)

Steinhauser, K.A. *et al.* (1985). In *Neutron scattering in the nineties*, p.99. IAEA, Vienna. (Review on UCN generation and applications.)

Stepanov, A.V. (1975). *Teŏr. Mat. Fiz.*, **22**, 425 (*Theor. Math. Phys.*). (The model of roughnesses that lead to μ decreasing is discussed.)

Stepanov, A.V. (1976*a*). *Elem. Chastits At. Jodra*, **7**, 989 (*Sov. J. Part. Nucl.*, **7**, 396; *Sov. J. Plasma Phys.*, **2**, 396). (Theory of optical potential for UCN taking into account surface roughness and inelastic scattering.)

Stepanov, A.V. (1976*b*). *Kratk. Soobsch. Fiz.*, (8), 3. (Theory of UCN elastic scattering on film inhomogeneities and inelastic scattering on phonons.)

Stepanov, A.V. and Shelagin, A.V. (1974). *Kratk. Soobshch. Fiz.*, (1), 12. (The effect of surface roughness on the μ value is calculated. The correction to Ignatovich (1973) is introduced.)

Stepanov, A.V. and Shelagin, A.V. (1977*a*). *Izv. VUZ, Radiofiz.*, **20**, 1575. (The optical potential for UCN taking account of substance inhomogeneities is deduced.)

Stepanov, A.V. and Shelagin, A.V. (1977*b*). *Kratk. Soobshch. Fiz.*, (9), 37. (Theory of UCN depolarization at transmission through inhomogeneously magnetized foil.)

Stepanov, A.V. and Shelagin, A.V. (1984). Proc. All-Union Conf. on Neutron Physics, vol. 4, p. 32 [in Russian]. (INIS At. 16: 082409) (A report of the work by Stepanov and Shelagin (1977*b*).)

Stepanov, A.V. and Shelagin, A.V. (1986). *Kratk. Soobshch. Fiz.*, (3), 23. (INIS At. 18: 034410) Large inhomogeneities' analysis in substances with the help of integral VCN elastic scattering cross-section.

Steyerl, A. (1969). *Phys. Lett.*, **B29**, 33. (The cross-sections of long-wavelength neutrons scattering on ^{79}Au and ^{13}Al are measured on the vertical neutron channel and time-of-flight spectrometer. The law $1/v$ is confirmed.)

Steyerl, A. (1971). Dissertation. Technische Universität, München. (Neutron scattering on surface roughnesses at total reflection and their transmission through a set of substances is considered.)

Steyerl, A. (1972*a*). *Nucl. Instrum. Methods*, **101**, 295. (The time-of-flight spectrometer with graphite converter on the vertical neutron channel is described. The reflection coefficient of electropolished Cu is found to be 0.97.)

Steyerl, A. (1972*b*). *Z. Phys.*, **252**, 371. (The interference effects in the transmission of thin foils are observed.)

Steyerl, A. (1972*c*). *Z. Phys.*, **254**, 169. (Theory of neutron scattering on a rough surface. The deviations of v dependence of transmission through Cu foil from the law $1/v$ are used to deduce the parameters of roughnesses: the height 100–240 Å, width 200–500 Å.)

Steyerl, A. (1974*a*). Preprint PDTUM-FRM-114, München. (Theory of VCN scattering on magnetic domains and Block walls.)

Steyerl, A. (1974*b*). Joint Institute for Nuclear Research, Dubna, Preprint D3-7991. (Review of own works.)

Steyerl, A. (1975). *Nucl. Instrum. Methods*, **125**, 461. (Description of the turbine and of the experiment on UCN transmission through thin magnetized Ni foil.)

Steyerl, A. (1977). *Springer Tracts in Modern Physics*, vol. 80, p. 57. Springer, Berlin. (UCN review.)

Steyerl, A. (1978*a*). *Z. Phys.*, **B30**, 231. A High-resolution neutron gravity spectrometer. (Project NESSIE with resolution 10^{-8} eV.)

Steyerl, A. (1978*b*). Joint Institute for Nuclear Research, Dubna, Preprint D3-11787, p. 209. (Review of own works.)

Steyerl, A. (1982) *The neutron and its applications, the 50th anniversary of the discovery of the neutron, Inst. Phys. Conf. Ser.No.* 64, p. 177. (UCN review.)

Steyerl, A. (1984). *J. Phys. Coll.*, **45**, C3, 255. (Review of neutron microscope works.)

Steyerl, A. (1989*a*). Neutron microscopy. In: *Some problems of modern nuclear physics.* Nauka, M. p. 4.

Steyerl, A. (1989*b*). *Physica B*, **156 & 157**, 528. UCN production and experiment (review).

Steyerl, A. and Lerner R. (1976). *Phys. Status Solidi*, (a), **33**, 531. (Theory and experiment on magnetic domains and Block walls' measurement. Transmission through Fe, Ni, Co is observed.)

Steyerl, A. and Malik, S.S. (1986). *Physica*, B & C, **137**, 270. New development in cold and UCN research.

Steyerl, A. and Malik, S.S. (1989). *Nucl. Instrum. Methods*, **A284** (1), 200. Sources of UCN.

Steyerl, A. and Schutz, G. (1978). *Appl. Phys.*, **17**, 45. (Calculation of concave zone mirror.)

Steyerl, A. and Trüsted, W.D. (1974). *Z. Phys.*, **267**, 379. Experiment with a neutron bottle. (UCN storage in Cu and pyrographite bottles. Measurement of $\mu(v)$ with the help of foils with different E_{lim}. A high albedo of 0.96 is observed from pyrographite.).

Steyerl, A. and Vonach, H. (1974). *Z. Phys.*, **250**, 116. (No deviation from $1/v$ law is found for neutron total cross-section on Au, Al, Cu, mica, glass, air. The Debye temperature 389 ± 50 K for Al is determined.)

Steyerl, A. *et al.* (1979). *Z. Phys.*, **B36**, 109. (The Michelson interferometer is considered for UCN.)

Steyerl, A. *et al.* (1981). *Z. Phys.*, **B41**, 283. (The resonances and their splitting are observed in neutron transmission through a five-layer system.)

Steyerl, A. *et al.* (1983). *Z. Phys.*, **B50**, 281. (Description of the spectrometer NESSIE and experiments performed with it. Resolution 17 neV. The width broadening or quasi-elastic scattering is observed in UCN transmission through polymer solution. The broadening was 6.6 ± 3.2 neV at the moment transfer 0.027 Å^{-1}.)

Steyerl, A. *et al.* (1985). *J. Phys. D: Appl. Phys.*, **18**, 9. Observation of anomalous transmission bands for neutron waves in a multibilayer film.

Steyerl, A. *et al.* (1986). *Phys. Lett.*, **A116**, 347. A new source of cold and ultracold neutrons.

Steyerl, A. *et al.* (1988*a*). *Rev. Phys. Appl.*, **23**, 171. Neutron microscopy.

Steyerl, A. *et al.* (1988*b*). *Physica*, **B151**, 36. Neutron resonators and interferometers for very low energy neutrons. (Review of multilayer systems, EDM resonance curve, diffraction and interference of UCN.)

Stoica, A.D. and Jipa, A. (1984). *Nucl. Instrum. Methods*, **219**, 447. (Theory of UCN generation by shaking out a converter with high E_{lim}. The rate of sub-limiting generation is estimated.)

Stoica, A.D. and Strelkov, A.V. (1978*a*). *Nucl. Instrum. Methods*, **157**, 477. (Measurement of UCN transmission by moving goffered foils.)

Stoica, A.D. and Strelkov, A.V. (1978*b*). Joint Institute for Nuclear Research, Dubna, Preprint P3-11593 [in Russian] (INIS 10: 428625). Energy spectra measurements for upscattered UCN ($E = 10$ meV, $\Phi = 5 \times 10^{-4}$ neutrons/cm^2 s.)

Stoica, A.D., Strelkov, A.V., and Hetzelt, M. (1978). *Z. Phys.*, **B29**, 349. (The energy of a heated neutron after UCN collision with a wall is estimated to be near 10 meV.)

Strelkov, A.V. and Hetzelt, M. (1978). *Zh. Eksp. Teor. Fiz.*, **74**, 23 (*Sov. Phys.–JETP*). (Observation of UCN heating to thermal energies at collision with the wall.)

Strepetov, A.N. and Frank, A.I. (1986*a*). *Pis'ma Zh. Tekh. Fiz.*, **12**, 71 (*Sov. Tech. Phys. Lett.*). Mirror microscope for UCN with aplanatic objective.

Strepetov, A.N. and Frank, A.I. (1986*b*) *Zh. Tekh. Fiz.*, **56**, 1775 (*Sov. Phys.–Tech. Phys.*). Mirror neutron microscope with magnetic correction of aberrations.

Summer, T.J. (1979). Ph.D. thesis. University of Sunsex. (INIS At. 13: 660895) Progress towards a new experiment to search for the EDM of the neutron using UCN.

Surdin, M. (1981). *Lett. Nuovo Cimento*, **31**, 337. (A proposal to measure the UCN interference fringe from two horizontal pinholes for horizontal and vertical neutron rays.)

Taran, Yu.V. (1971). Joint Institute for Nuclear Research, Dubna, Preprint P3-8112 [in Russian]. (Proposal of double polarization analysis for EDM experiment with UCN in storage made with the help of movable analysing foil.)

Taran, Yu.V. (1973*a*). Joint Institute for Nuclear Research, Dubna, Preprint P3-7147 [in Russian]. (Calculation of the EDM experiment with UCN in flow mode.)

Taran, Yu.V. (1973*b*). Joint Institute for Nuclear Research, Dubna, Preprint P3-7149 [in Russian]. (Calculation of the EDM experiment with UCN in storage mode.)

Taran, Yu.V. (1975*a*). Joint Institute for Nuclear Research, Dubna, Preprint P3-8577 [in Russian]. Fast adiabatic spin flip.

Taran, Yu.V. (1975*b*). Joint Institute for Nuclear Research, Dubna, Preprint P3-9307 [in Russian]. (INIS At. 8: 317563) (Discussion of relation between polarization ratio and effectiveness. A proposal of polarization analysis in storage mode experiment with one chamber.)

Taran, Yu.V. (1976). Proc. All-Union Conf. on Neutron Physics, part 6, p. 211 [in Russian]. (Calculation of adiabatic spin rotation and proposal of modification of UCN generation method (Drabkin and Zhitnikov 1960).)

Taran, Yu.V. (1977). Proc. All-Union Conf. on Neutron Physics, part 1, p. 200 [in Russian]. (INIS At. 9: 413391) (The double polarization analysis in EDM experiment with UCN is proposed.)

Taran, Yu.V. (1985*a*). Joint Institute for Nuclear Research, Dubna, Rapid Communications No. 12–85, p. 48 [in Russian]. A new method for the measurement of the polarization characteristics of ferromagnetic films on UCN.

Taran, Yu.V. (1985*b*). ***Prib. Tekh. Eksp.***, (5), 170 (*Instrum. Exp. Tech.*). Multilayer cylindrical shield.

Taran, Yu.V. (1985*c*). SU Patent Document 1293679/A/. Int. cl. G 01 T 1/32, 3/00. 16 July 1985. (INIS At. 19: 046872). Installation for detection of polarized characteristics of ferromagnetic films by means of UCN.

Taran, Yu.V. (1986). Joint Institute for Nuclear Research, Dubna, Preprint P3-86-70 [in Russian]. Polarization analysis with slow neutrons.

Taran, Yu.V. (1987). Joint Institute for Nuclear Research, Dubna, Preprint P3-87-220 [in Russian]. (INIS At. 19: 057392) New possibilities of magnetic films investigation with UCN.

Taran, Yu.V. (1988). *Proc. Int. Conf. on Neutron Physics*, vol. 1, p. 207 [in Russian]. (Report on own work (Taran 1987).)

Terekhov, G.I. (1977*a*). *Pis'ma Zh. Tekh. Fiz.*, **3**, 1275 (*Sov. Tech. Phys. Lett.*). (A magnetic UCN lens is calculated.)

Terekhov, G.I. (1977*b*). Proc. All-Union Conf. on Neutron Physics, part 1, p. 160 [in Russian]. (INIS At. 9: 413392) (The effect of gravity on τ is discussed.)

Terekhov, G.I. (1988*a*). *Proc. Int. Conf. on Neutron Physics*, vol. 1, p. 204 [in Russian]. (The UCN spectrum measured after 3.5 years' exploitation of the neutron channel at CM-2 reactor has no anomalous step.)

Terekhov, G.I. (1988*b*). *Proc. Int. Conf. on Neutron Physics*, vol. 1, p. 204 [in Russian]. Axially symmetric magnetic lens for UCN.

Tervisidis, F., Tsagas, N., and Batas, J.C. (1989). *Nucl. Instrum. and Methods*, **A284** (3), 431. Development of an isotropic velocity distribution in stored UCN and its effect on the measurement of neutron lifetime.

Thompson, D. (1989). *Nucl. Instrum. Methods*, **A284** (1), 40. The search for the EDM of the neutron in ILL.

Trüsted, W.D. *et al.* (1970). Jahrber. Physik Dept, Technische Universität München, BMBW-FB K 72–07, p. 108. (UCN storage in a graphite bottle filled at the reactor and lifted up is reported.)

Utsuro, M. (1977). *Z. Phys.*, **B27**, 111. Slow neutrons scattering and intermolecular excitations in liquid hydrogen.

Utsuro, M. (1982). Kyoto University Research Reactor Institute, Preprint KURRI-TR-225 [in Japanese]. (INIS At. 15: 051584). UCN experiments.

Utsuro, M. (1983). *Nucl. Instrum. Methods*, **213**, 557. A fall-focusing gravity spectrometer for very slow neutrons.

Utsuro, M. (1984). *J. Phys. Coll.*, **45**, C3, 269. (Magnetic slowing and gravity focusing are discussed.)

Utsuro, M. and Hetzelt, M. (1977). In *Neutron inelastic scattering*, p. 67. IAEA, Vienna. (Theory and experiment on UCN generation in liquid D_2.)

Utsuro, M. and Kawabata, Y. (1983). *Physica*, B&C, **120**, 118. (A turbine with supermirror blades and gravity spectrometer with gravity focusing and resolution 20 neV is discussed.)

Utsuro, M. and Yoshiki, H. (1984). *Annu. Rep. Res. Reactor Inst., Kyoto Univ.*, **17**, 94. (INIS At. 17: 029392) (UCN production and application. Review.)

Utsuro, M. *et al.* (1980). *Annu. Rep. Res. Reactor Inst., Kyoto Univ.*, **13**, 161.

Utsuro, M. *et al.* (1981). *Proc. ICANS-IV, Tsukuba, Oct.* 1981, KENS Report 11, p. 743. National Laboratory for High Energy Physics. (Measurement of UCN reflection from a supermirror.)

Utsuro, M. *et al.* (1982). *Proc. ICANS-V, Julich, June* 1982, JÜ1-Conf-45, p. 525. (The proposal of the UCN source at the proton accelerator is advanced. The results of the experiment on neutron reflection from a supermirror are reported.)

Utsuro, M. *et al.* (1983). (INIS At. 16: 003229) UCN facility at KUR with supermirror turbine.

Utsuro, M. *et al.* (1988*a*). *ICNS'88, July* 1988, *Grenoble*, Programme and Abstracts, p. 2–91. VCN facility in horizontal arrangement with deuterium cold source.

Utsuro, M. *et al.* (1988*b*). *Nucl. Instrum. Methods Phys. Res.*, **A270**, 456. A supermirror turbine for an UCN source. ($\Phi_{UCN} = 0.61$ neutrons/cm^2 s, $S = 7 \times 3$ cm^2, $\bar{v} = 10$–20 m s^{-1}.)

Utsuro, M. *et al.* (1988*c*). *Proc. Int. Conf. on Neutron Physics*, p. 185. UCN sources with a supermirror neutron turbine.

Vasil'ev, V.V. (1981). Institute of Theoretical and Experimental Physics, Moscow, Preprint 58 [in Russian]. (Optimization of a measurement cycle in the experiment on UCN storage in a magnetic bottle.)

Vasil'ev, V.V. and Rojnin I.B. (1984). Proc. All-Union Conf. on Neutron Physics, vol. 4, p. 8 [in Russian]. (INIS At. 16: 079627) (It is shown that the correction of the magnetic bottle field with the additional coil current transforms three node lines into nine node points and makes depolarization time reach 30 000 s, which is not sufficient for τ_β measurements.)

Vladimirskii, V.V. (1960). *Zh. Eksp. Teor. Fiz.*, **39**, 1062 (*Sov. Phys.–JETP*, **12**, 740). (Theory of UCN storage in magnetic and gravimagnetic bottles, of depolarization, of heating of UCN due to the magnetic and non-magnetic walls' vibration, of magnetic focusing, of magnetic neutron guides, of UCN generation in horizontal and vertical channels, of sub-limiting UCN generation, and of UCN generation in liquid hydrogen.)

Winfeld, D. and Robson, J.M. (1975). *Can. J. Phys.*, **53**, 667. (Monte Carlo calculation of UCN diffusion along a neutron guide. Main losses are estimated to be due to non-specularity of mechanically polished neutron guide. Air and He curves show large μ and σ_t. Cross-section is 2.5 times higher for air and 5.4 higher for He.)

Yamada, S. *et al.* (1978). *Annu. Rep. Res. Reactor Inst., Kyoto Univ.*, **11**, 8.

Yates, D.J.C. (1979). *Phys. Lett.*, **A72**, 479. (A proposal to deuterate the bottle walls with deuterated methyl spirit.)

Yoshiki, H. (1981). Proc. ICANS-IV, Report II, (eds. Ishikawa, Y. *et al.*), p. 715. National Laboratory for High Energy Physics. (The ^{4}He liquid converter for UCN generation is discussed. For a 30 cm quartz cube, the storage time τ is estimated to be 200 s, if hydrogen density on the walls is 10^{15} cm^{-2}. The feasibility of experiments on N–$\bar{\text{N}}$ oscillations are discussed.)

Yoshiki, H. (1984). KEK Preprint 84-2, Ibaraki [in Japanese]. (INIS At. 17: 82974). Operation of a superthermal UCN converter.

Yoshiki, H. (1988). *ICNS'88, July* 1988, *Grenoble*, Programme and Abstracts, p. 2–88. Monte Carlo calculations on the Serebrov sphere.

Yoshiki, H. and Golub, R. (1988). *Int. Conf. on Neutron Physics, Kiev, Sept.* 1987. Ultracold antineutrons.

Yoshiki, H. and Ueda, H. (1983). KENS Report-IV, KEK-Internal 83–4, Sept. I/E/B Japan, p. 209. (Preliminary results of the effect of D_2 and D_2O on the bottle walls and the value of τ.)

Yoshiki, H. *et al.* (1986). KEK Preprint 86–91, Ibaraki. Feasibility study of a strong UCN source.

Yoshiki, H. *et al.* (1987). *Z. Phys.*, **B67**, 161. Feasibility study of a strong UCN source.

Yoshiki, H. *et al.* (1988). *ICNS'88, July* 1988, *Grenoble*, Programme and Abstracts, p. 2–89. A 3 m long superfluid helium bottle for UCN production.

Yu, Z-ch, Malik, S.S., and Golub, R. (1986). *Z. Phys.*, **B62**, 137. A thin film source of UCN.

Zamyatin, Yu.S. *et al.* (1974). Joint Institute for Nuclear Research, Dubna, Preprint P3-7946 [in Russian]. (UCN horizontal channel with Mg converter at CM-2 reactor is described. $J_{UCN} = 80$ neutrons/s).

Zel'dovich, Ya.B. (1959). *Zh. Eksp. Teor. Fiz.*, **36**, 1952 (*Sov. Phys.–JETP*, **9**, 1389). (The first paper on UCN. A possibility of neutron storage in a closed bottle is proved. UCN yield from liquid He converter and storage time τ are estimated. The heating of UCN by reflection from moving mirrors is mentioned.)

Zhitarev, V.E. *et al.* (1981). *At. Ehnerg.* **50**, 350 (*Sov. J. At. Energy*). (Measurements of σ_t for neutrons with $\lambda = 13$–21 Å and substances: Al, Ni, Cu, Zn, Ge, Sn, Pb.)

Zubarev, A.L. (1976). *Yad. Fiz.*, **26**, (*Sov J. Nucl. Phys.*, **26**, 232). (The solution of a UCN reflection problem when the height of the reflecting potential periodically varies with time.)

Further reading

*Abramov, A.I. (1972). *Measurement of the unmeasurable.* Atomizdat, Moscow [in Russian].

*Albeverio, S., Fenstad, J.E., and Hoegh-Krohn, R. (1988). *Solvable models in quantum mechanics.* Springer, New York.

*Alexandrov, Yu.A. (1982). *Fundamental properties of the neutron.* Energoatomizdat, Moscow [in Russian].

*Avron, J., Grossmann, A., and Hoegh-Krohn, R. (1983). The reflection from a semi-infinite crystal of point scatterers. *Phys. Lett.*, **A94**, 42.

*Barut, A.O. and Božic, M. (1988). Two- and three-dimensional tunneling of neutral spin $\frac{1}{2}$ particles through magnetic fields. *Physica*, **B151**, 180.

*Baryshevskii, V.G., L'uboshits, B.L., and Podgoretskij, M.I. (1965*a*). About the s-wave scattering on a system of several scatterers. Joint Institute for Nuclear Research, Dubna, Preprint P-2111 [in Russian].

*Baryshevskii, V.G., L'uboshits, B.L., and Podgoretskij, M.I. (1965*b*). About the transmission of particles with and without spin through matter. Joint Institute for Nuclear Research, Dubna, Preprint P-2230 [in Russian].

*Bel'aev, V.B. (1978). On the possibility of optical neutrino accumulation in closed vessels. Joint Institute for Nuclear Research, Dubna, Preprint P-1014 [in Russian].

*Berezin, F.A. (1965). *The second quantization method.* Nauka, Moscow [in Russian].

*Berry, M.V. (1984). *Proc., R. Soc.*, **A392**, 45.

*Bitter, T. and Dubbers, D. (1987). *Phys. Rev. Lett.*, **59**, 251. First observation of Berry phase with cold neutrons.

*Bondarenko, L.N. *et al.* (1978). Measurement of the half-life of the neutron. *Pis'ma Zh. Eksp. Teor. Fiz.*, **28**, 329 (*JETP. Lett.*).

*Born, M. and Wolf, E. (1973). *The principles of optics*, p. 57. Nauka, Moscow [in Russian].

*de Broglie, L. (1960). *Nonlinear wave mechanics.* Elsevier, Amsterdam.

*Byrne, J. *et al.* (1980). A new measurement of the neutron lifetime. *Phys. Lett.*, **B92**, 274.

*Christensen, C.J., Nielsen, A., and Bahsen, A. (1972). Free-neutron-beta-decay half life. *Phys. Rev.*, **D5**, 1628.

*Clausing, P. (1971). The flow of highly rarefied gases through tubes of arbitrary length. *J. Vac. Sci. Technol.* **8**, 636.

*Coutinho, F.A.B., Nogami, Y., and Toyami, F.M. (1988). General aspects of the bound-state solutions of one-dimensional Dirac equation. *Am. J. Phys.*, **56**, 904.

*Cowley, R.A. and Woods, A.D.B. (1971). Inelastic scattering of thermal neutrons from liquid helium. *Can. J. Phys.*, **49**, 177.

*Davis, D.H. (1960). Monte Carlo calculation of molecular flow rates through cylindrical elbow and pipes of other shapes. *J. Appl. Phys.*, **31**, 1169.

*Delseme, A.N. and Wegener, A. (1970). Heavy water ice. *Science*, **167**, 44.

*Demkov, Y.N. and Ostrovskii, V.N. (1975). The use of zero-range potentials in atomic physics. Nauka, Moscow [in Russian].
*Dress, W.B., Miller, P.D., and Pendlebury, J.M. (1977). Search for the EDM of the neutron. *Phys. Rev.*, **D15**, 9.
*Dubbers, D. (1988). Measurement of the Berry phase with polarised neutrons. *Physica*, B&C, **151**, 93.
*Dylla, H.F. and King, J.G. (1973). Neutrality of molecules by a new method. *Phys. Rev.*, **A7**, 1224.
*Ezawa, H. (1971). Phonons in a half space. *Ann. Phys.*, **67**, 438.
*Fermi, E. (1972). *Scientific works*, vol. 2, p. 226. Nauka, Moscow [in Russian].
*Fermi, E. and Zinn, W. (1947). *Phys. Soc. Cambridge Conf. Rep.* 92, Chicago.
*Flügge, S. (1971). *Practical quantum mechanics I, Die Grundlehren der Mathematischen Wissenschaften.* Springer, Berlin.
*Foldy, L.L. (1945). The multiple scattering of waves. 1. General theory. *Phys. Rev.*, **67**, 107.
*Fox, R.F. *et al.* (1987). Systematic corrections to the rotating wave approximation and quantum chaos. *Phys. Rev.*, **A36**, 4321.
*Gel'fer, Ya.M., L'uboshits, V.L., and Podgoretskij, M.I. (1975). *The Gibbs paradox*, p. 225. Nauka, Moscow [in Russian].
*Glaser, M.L. (1973). Evaluation of lattice sums. 1. Analytic procedures. *J. Math. Phys.*, **14**, 409.
*Grant, T.J. and Cobble, W. (1969). Evidence for discovery of chemically bound neutrons. *Phys. Rev. Lett.*, **23**, 741.
*Greiner, W. *et al.* (1985). Quantum electrodynamics of strong fields (With an introduction into modern relativistic quantum mechanics). Springer-Verlag, Berlin.
*Grossmann, A., Hoegh-Krohn, R., and Mebkhout, M. (1980*a*). The one particle theory of periodic point interactions. *Commun. Math. Phys.*, **77**, 87.
*Grossmann, A., Hoegh-Krohn, R., and Mebkhout, M. (1980*b*). A class of explicitly soluble, local, many-center Hamiltonians for one-particle quantum mechanics in two and three dimensions I. *J. Math. Phys.*, **21**, 2376.
*Hughes, D. and Schwartz, R. (1958). Neutron cross sections. Brookhaven National Laboratory Report BNL-325.
*Ignatovich, V.K. (1980). Localised solutions of the electrodynamics equations in media with anomaly dispersion. Joint Institute for Nuclear Research, Dubna, Preprint P4-80-103 [in Russian].
*Kagan, Yu.M. and Afanas'ev, A.M. (1966). On nuclear resonance parameters' modifications at the scattering on regular systems. *Zh. Eksp. Teor. Fiz.*, **50**, 271 (*Sov. Phys.–JETP*).
*Kay, G. and Laby, T. (1962). *Tables of physical and chemical constants*, p. 114. Fismatgis, Moscow [in Russian].
*Koelman, J.M.V.A. *et al.* (1988). *Phys. Dev.*, **B38**, 9319. Lifetime of magnetically trapped ultracold atomic deuterium gas.
*Kronig, R. and Penney, W.G. (1931). Quantum mechanics of electrons in crystal lattices. *Proc. R. Soc.*, **A130**, 499.
*Krupchitskii, P.A. (1985). Fundamental investigations with polarized slow neutrons. Energoatomizdat, Moscow [in Russian].
*Landau, L.D. (1957). On the conservation laws for weak interactions. *Nucl. Phys.*, **3**, 127.
*Landau, L.D. and Lifshits, E.M. (1963). *Quantum mechanics*, p. 104. Nauka, Moscow [in Russian].

*Landau, L.D. and Lifshits, E.M. (1965). *Theory of elasticity*, p. 139. Nauka Moscow [in Russian].

*Lax, M. (1951). Multiple scattering of waves. *Rev. Mod. Phys.*, **23**, 287.

*Lax, M. (1952). Multiple scattering of waves. II. The effective field in dense systems. *Phys. Rev.*, **85**, 621.

*Lee, T.D. and Yang, C.N. (1957). Parity nonconservation and a two-component theory of the neutrino. *Phys. Rev.*, **105**, 1671.

*Leinson, L.B. *et al.* (1979). On the possibility of non-tunnel transmission of particles through potential barriers in quantum mechanics. *Dokl. Akad. Nauk URSS*, **248**, 599.

*Leuchs, G. (1986). Photon statistics, antibunching and squeezed states. In *Frontiers of nonequilibrium statistical physics*, (ed. G.T. Moore and M.O. Scully), NATO ASI Series, Series B: Physics, vol. 135, pp. 329–60. Plenum, New York.

*Levenson, L.L., Milleron, M., and Davis, D.H. (1960). Optimisation of molecular flow conductance. *7th Natl Symp. on Vacuum Technology transaction*, (ed. C.R. Meissner), p. 372. Pergamon, Oxford.

*Ligeon, E. *et al.* (1980). *J. Radioanal. Chem.*, **55**, 367. (The fluence of ^{11}B ions and vacuum dependence of hydrogen content on Si surface.)

*Litzman, O. (1986). Dynamical theory of diffraction on a periodic system of point scatterers. *Acta Crystallogr.*, **A42**, 552.

*Luschikov, V.I. *et al.* (1981). An experimental limit for neutron–antineutron oscillations. Joint Institute for Nuclear Research, Dubna, Preprint P3-81-313 [in Russian].

*Masnhara Naoto *et al.* (1988). *Phys. Rev. Lett.*, **61**, 935. Evaporative cooling of spin polarized atomic hydrogen.

*Mayer-Leibnitz, H. and Springer, T. (1963). The use of neutron optical device in beam-hole experiments. *J. Nucl. Energy*, **A/B1**, 217.

*Mezei, F. (1976). Novel polarised neutron devices: supermirror and spin-component amplifier. *Commun. Phys.* **1**, 81.

*Morrow, M. *et al.* (1981). Hyperfine zero-field resonance of atomic hydrogen. *Phys. Rev. Lett.*, **46**, 195.

*Muskat, E., Dubbers, D., and Scharpf, O. (1987). Dressed neutrons. *Phys. Rev. Lett.*, **58**, 2047.

*Nieto, M.M. (1986). What are squeezed states really like? In *Frontiers of nonequilibrium statistical physics*, (ed. G.T. Moore and M.O. Scully), NATO ASI Series, Series B: Physics, vol. 135, pp. 287–307. Plenum, New York.

*Parrat, L.L. (1954). Surface studies of solids by total reflection of X-rays. *Phys. Rev.*, **95**, 359.

*Poyry, H. *et al.* (1975). Principles of reverse neutron time-of-flight spectrometry with Fourier Chapper Applications. *Nucl. Inst. and Meth.*, **126**, 121.

*Pron'ko, G.P. and Stroganov, Yu.G. (1977). A new example of a quantum mechanical problem with a hidden symmetry. *Zh. Eksp. Teor. Fiz.*, **72**, 2048 (*Sov. Phys.–JETP*).

*Purcell, E.M. and Ramsey, N.F. (1950). On the possibility of electric dipole moments for elementary particles and nuclei. *Phys. Rev.*, **78**, 807.

*R'asanov, M.I. (1984). *Electrodynamics of condensed state of matter*, p. 20. Nauka, Moscow [in Russian].

*Rauch, H. (1987). *V School on Neutron Physics, Alusta, October* 1986, Joint Institute for Nuclear Research, Dubna, Preprint D3, 4, 17-86-747, p. 175.

*Rauch, H., Summhammer, J., and Weinfurter, H. (1985). In *Neutron scattering in the nineties*, p. 53. IAEA, Vienna.

*Rutherford, E. (1920). Bakerian lecture: nuclear constitution of atoms. *Proc. R. Soc.*, **A97**, 396.

*Shikin, V.B. (1979). On the cyclotron frequency shift of electrons localized on a liquid helium surface. *Zh. Eksp. Teor. Fiz.*, **77**, 717 (*Sov. Phys.–JETP*).

*Shull, C.G. (1969). Single-slit diffraction of neutrons. *Phys. Rev.*, **179**, 752.

*Shull, C.G., Billman, K.W., and Wedgwood, F.A. (1967). Experimental limit for the neutron charge. *Phys. Rev.*, **153**, 1415.

*Silvera, I.F. and Walzaren, J.B.M. (1986). Spin-polarized atomic hydrogen. In *Progress in low temperature physics*, (ed. D.F. Brewer), Elsevier Science Publisher P.V., North Holland, V. X, p. 139.

*Smith, J.H., Purcell, E.M., and Ramsey, N.F. (1957). Experimental limit for the EDM of the neutron. *Phys. Rev.*, **108**, 120.

*Spivak, P.E. (1988). τ_β from experiment. *Zh. Eksp. Teor. Fiz.*, **94**, 1 (*Sov. Phys.–JETP*).

*Summhammer, J., Niel, L., and Rauch, H. (1986). Focusing of pulsed neutrons by travelling magnetic potentials. *Z. Phys.*, **B62**, 269.

*Thomas, L.E. (1979). Birman–Schwinger bounds for the Laplacian with point interactions. *J. Math. Phys.*, **20**, 1848.

*Turchin, V.F. (1963). *Slow neutrons*, p. 370. Gosatomizdat, Moscow [in Russian] (See Israel Program for Scientific Translations, Jerusalem, 1965.)

*Vasil'ev, B.V. (1969). On the proton gravity moment. *Pis'ma Zh. Eksp. Teor. Fiz.*, **9**, 299 *JETP Lett.*).

*Vlasov, A.D. and Lupandin, O.S. (1979). *From the Ptolemaic epicycles to magic nuclei and plankons. Models in physics*, p. 127. Atomizdat, Moscow [in Russian].

*Walther, H. (1986). The one-atom maser—a test system for simple quantum electrodynamic effects. In *Frontiers of nonequilibrium statistical physics*, (ed. G.T. Moore and M.O. Scully), NATO ASI Series, Series B: Physics, vol. 135, pp. 271–85. Plenum, New York.

*Werner, A. *et al.* (1975). Observation of the neutron phase-shift due to spin precession in a magnetic field. *Phys. Rev. Lett.*, **35**, 1053.

*Whittaker, E.T. and Watson, G.N. (1927). A course of modern analysis. The University Press, Cambridge.

*Wilhelmi, G. (1970). Binary sequences and error analysis for pseudostatistical neutron modulator with different duty cycles. *Nucl. Instrum. Methods*, **81**, 36.

*Wohl, C.G. *et al.* (1984). Review of particle properties. *Rev. Mod. Phys.*, **56**, S1–S304.

*Wu, C.S., Amblar, E., and Hayward, R.W. (1957). Experimental test of parity conservation in beta decay. *Phys. Rev.*, **105**, 1413.

*Young, J.A. and Koppel, J.U. (1964). Slow neutron scattering by molecular hydrogen and deuterium. *Phys. Rev.*, **135**, A603.

*Zahed, I. and Brown, G. (1987). Skyrm model. *Phys. Rep.*, **142**, (1).

Index

The manufacturer's authorised representative in the EU for product safety is Oxford University Press España S.A. of El Parque Empresarial San Fernando de Henares, Avenida de Castilla, 2 - 28830 Madrid (www.oup.es/en or product.safety@oup.com). OUP España S.A. also acts as importer into Spain of products made by the manufacturer.
Printed and bound by CPI Group (UK) Ltd, Croydon, CR0 4YY
14/09/2026
02216290-0008